Mathematical Concepts in Engineering

Mathematical Concepts in Engineering

Editor: Alan Fraser

MURPHY & MOORE
www.murphy-moorepublishing.com

by Murphy & Moore Publishing
ller Plaza,
City, NY 10020, USA
rphy-moorepublishing.com

matical Concepts in Engineering
by Alan Fraser

International Standard Book Number: 978-1-63987-355-5 (Hardback)

Cataloging-in-Publication Data

Mathematical concepts in engineering / edited by Alan Fraser.
p. cm.
Includes bibliographical references and index.
ISBN 978-1-63987-355-5
1. Engineering mathematics. 2. Mathematical analysis.
3. Engineering--Mathematical models. I. Fraser, Alan.
TA330 .M38 2022
620.001 51--dc23

Contents

Preface

Over the recent decade, advancements and applications have progressed exponentially. This has led to the increased interest in this field and projects are being conducted to enhance knowledge. The main objective of this book is to present some of the critical challenges and provide insights into possible solutions. This book will answer the varied questions that arise in the field and also provide an increased scope for furthering studies.

Mathematical concepts in engineering refer to the mathematical techniques and methods which are applied in industry and engineering. The most commonly used mathematical concept in engineering is applied analysis. This involves the use of real and complex analysis such as vector and tensor analysis, approximation theory, differential equations and fourier analysis. Linear algebra and probability are also two areas of mathematics which are widely used in engineering. Engineering optimization refers to the utilization of optimization techniques to achieve the designing goals in engineering. Engineering statistics combines the concepts and techniques of statistics and engineering to analyze data. Such data is primarily related to manufacturing processes such as fabrication process control, component dimensions, type of materials, etc. This book is compiled in such a manner, that it will provide an in-depth knowledge about the theory and practice of mathematical concepts in engineering. From theories to research to practical applications, case studies related to all contemporary topics of relevance to this field have been included herein. The extensive content of this book provides the readers with a thorough understanding of the subject.

I hope that this book, with its visionary approach, will be a valuable addition and will promote interest among readers. Each of the authors has provided their extraordinary competence in their specific fields by providing different perspectives as they come from diverse nations and regions. I thank them for their contributions.

Editor

Probabilistic Analysis of Tunnel Face Stability below River Using Bayesian Framework

Weiping Liu,[1] Xiaoyan Luo,[2] Jinsong Huang,[3] Lina Hu,[1] and Mingfu Fu[1,4]

[1]*School of Civil Engineering and Architecture, Nanchang University, Nanchang 330031, China*
[2]*Jiangxi Science and Technology Normal University, Nanchang 330013, China*
[3]*ARC Centre of Excellence for Geotechnical Science and Engineering, University of Newcastle, Callaghan, NSW 2308, Australia*
[4]*School of Civil Engineering and Architecture, Nanchang Institute of Technology, Nanchang 330099, China*

Correspondence should be addressed to Lina Hu; hulina@ncu.edu.cn

Academic Editor: Eric Feulvarch

A key issue in assessment on tunnel face stability is a reliable evaluation of required support pressure on the tunnel face and its variations during tunnel excavation. In this paper, a Bayesian framework involving Markov Chain Monte Carlo (MCMC) simulation is implemented to estimate the uncertainties of limit support pressure. The probabilistic analysis for the three-dimensional face stability of tunnel below river is presented. The friction angle and cohesion are considered as random variables. The uncertainties of friction angle and cohesion and their effects on tunnel face stability prediction are evaluated using the Bayesian method. The three-dimensional model of tunnel face stability below river is based on the limit equilibrium theory and is adopted for the probabilistic analysis. The results show that the posterior uncertainty bounds of friction angle and cohesion are much narrower than the prior ones, implying that the reduction of uncertainty in cohesion and friction significantly reduces the uncertainty of limit support pressure. The uncertainty encompassed in strength parameters are greatly reduced by the MCMC simulation. By conducting uncertainty analysis, MCMC simulation exhibits powerful capability for improving the reliability and accuracy of computational time and calculations.

1. Introduction

Valid estimation of the tunnel face stability under excavation requires a reliable evaluation of limit support pressure which prevents soil collapse. This issue has been extensively studied with limit equilibrium method [1], numerical methods [2, 3], and experimental methods [4]. The limit equilibrium method is commonly used for predicting limit support pressure, being relatively simple compared with finite element analysis. Investigations showed that soil parameters, such as cohesion and friction angle, are the most important soil properties for influencing the limit support pressure [5]. Therefore, accurate estimation of soil strength parameters is necessary for assessment of the tunnel face stability under tunnel excavation.

Normally the mean values of soil parameters are used in deterministic analysis. Although the method can deliver accurate analytical results of support pressure, it requires the input parameters for every calculation point in situ and cannot control the spatial variability of the input parameters. In fact, geotechnical materials are natural materials, and their properties are affected by various spatially variable factors during their formation processes. The inherent spatial variability has been considered as one of the major sources of uncertainties [6, 7] and can be modeled using stochastic analysis [8].

A major difficulty in estimating accurate limit support pressure arises from the uncertainties incorporated in the input parameters of the computer model. The input strength parameters, such as cohesion and friction angle, are usually determined by direct measurements in laboratory. These samples will be disturbed in the process of testing. Meanwhile, the test conditions in laboratory cannot be exactly the same as in situ. This also leads to a significant uncertainty in predicting the limit support pressure for keeping tunnel face stability. This uncertainty poses challenges for obtaining reliable design of tunnel excavation. The stochastic approach can improve the traditional deterministic methods for taking the uncertainties of the parameters into account.

In practice, tunnel face collapse prediction can be formulated as a classification. Many studies have been performed to analyze the stability of tunnel face [9]. Mollon et al. [10, 11] use response surface method to perform probabilistic analysis of the face stability of tunnel. Li et al. [12] use the first-order reliability method and Monte Carlo simulation to carry out reliability analysis of a circular tunnel subjected to a hydrostatic stress field. Oreste [13] presents a probabilistic design approach for the tunnel taking into account the uncertainties of the rock mass quality index. Celestino et al. [14] evaluate the probability of failure according to load and resistance factor design principles. Langford et al. [15] use a modified point estimate method to design the tunnel. Lu et al. [16] evaluate failure probability of each failure mode of a rock tunnel by the first-order reliability method and the response surface method via an iterative procedure. Eshraghi et al. [17] study the face stability of TBD-driven tunnel in heterogeneous soil using probabilistic approach.

Bayesian approach can update the current state of knowledge about the model parameters based on the measurement data [18]. Many successful applications of Bayesian approach have been reported, e.g., estimation of the parameters of hydrological model [19], confidence interval of SWCC [20], and braced excavation [21]. Although previous studies have been done on the probability analysis of tunnel, a systematic Bayesian framework and Markov Chain Monte Carlo (MCMC) simulation for the probabilistic analysis has not yet been developed. The effect of uncertainties of soil parameters on prediction uncertainty of tunnel face stability below river using a Bayesian framework has not yet been investigated.

In this paper, the probability is associated with the different parameters which are governing the tunnel face stability and furthermore detailed in the following. The soil strength parameters, such as friction angle and cohesion, are assumed as random variables. The tunnel face below river is in the homogeneous soils. The three-dimensional model of tunnel face stability below river is based on the limit equilibrium theory and is adopted for the probabilistic analysis. The probabilities analysis and parameters uncertainty estimation are performed using the Markov Chain Monte Carlo (MCMC) simulation method which is good efficiency for highly nonlinear problem [22], with a delayed rejection adaptive metropolis (DRAM) [23] algorithm. The effects of uncertainty of strength parameters on the limit support pressure are discussed using the proposed Bayesian framework.

2. Limit Support Pressure

Wedge analysis [24] based on the limit equilibrium theory is adopted the limit support pressure for the face stability of a tunnel below river. A simple failure mechanism which idealizes the three-dimensional mode of tunnel face is shown in Figure 1. The collapsing soil in front of the tunnel is schematized as a triangular wedge; the wedge is assumed as a rigid body. B, H, L are the width, length, and height of the prism in Figure 1, respectively. The length L of prism is $L = B/\tan\alpha$, and the inclination of wedge is $\alpha = 45^\circ + \varphi/2$. This three-dimensional model is first proposed by Horn [25], and a right-angled prism extends from the tunnel crown to the surface. The Mohr-Coulomb failure condition with friction angle φ and cohesion c is assumed. Terzaghi's formula [26] has generally been adopted as vertical earth pressure acting the tunnel.

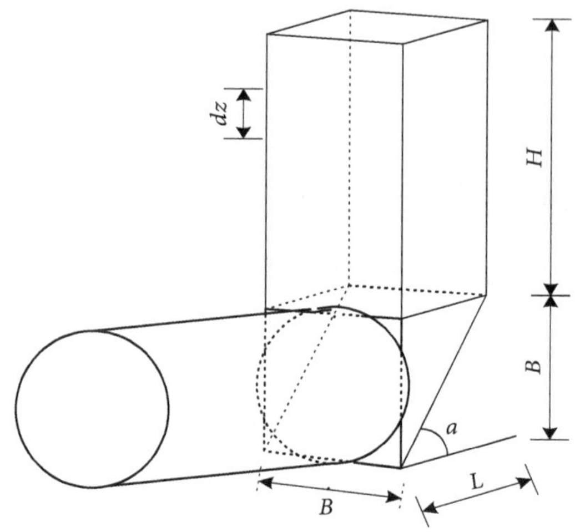

FIGURE 1: Wedge stability model.

The circular cross-section of the tunnel is approximated by a square whose sides are as long as the circular tunnel diameter D and has the same area.

$$B^2 = \frac{1}{4}\pi D^2 \tag{1}$$

Considering an element has dimension dz within prism of height H as shown in Figure 1, the upper and lower vertical force of element are, respectively,

$$P_1 = A\sigma_v, \qquad P_2 = A(\sigma_v + d\sigma_v) \tag{2}$$

where σ_v is the effective earth pressure from the overlying prism and A is the cross-sectional area of prism.

Vertical friction applied to lateral of the element is

$$f = k_0 U \sigma_v \tan\varphi dz + Ucdz \tag{3}$$

where U is the girth of prism; k_0 is coefficient of earth pressure at rest, $k_0 = 1 - \sin\varphi$.

The vertical equilibrium equation of dz reads as follows:

$$k_0 U \sigma_v \tan\varphi dz + Ucdz + A(\sigma_v + d\sigma_v) - A\sigma_v = \gamma A dz \tag{4}$$

where γ is the volumetric weight of the soil.

From the equilibrium equation (4) of the forces acting on the wedge, with boundary conditions $z = 0$, $\sigma_v = p_0$, the effective earth pressure acting upon the wedge can be obtained as

$$\sigma_v = \frac{A\gamma - Uc}{Uk_0 \tan\varphi}\left(1 - e^{-zUk_0\tan\varphi/A}\right) + p_0 e^{-zUk_0\tan\varphi/A} \tag{5}$$

where p_0 is the surcharge on the surface.

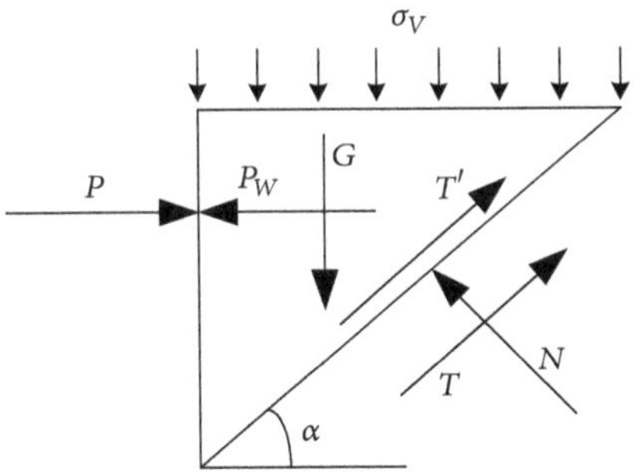

FIGURE 2: Forces acting upon the wedge.

For the sake of simplicity, the soil of wedge is considered to be homogeneous. Assuming the failure criterion holds along the failure face, a static equilibrium equation can be set up. When a shield tunnel is located below the river, the stability analysis of the shield tunnel face needs to consider the influence of pore water pressure. The pore water pressure usually is considered an external force [27] and the forces acting upon the wedge at the tunnel face illustrates in Figure 2. A hydrostatic distribution of pore water pressures along the slip surface is assumed. There are effective earth stress σ_v at the wedge-prism-interface, the self-weight of wedge $G = \gamma B^3/2\tan\alpha$, the support force P at the tunnel face, the normal force N on the inclined sliding surface, the shear forces T on the inclined as well as on the sliding surface, the symmetric normal force N', and the shear force T' on the two lateral surfaces of the wedge.

The vertical mean stress σ_z' on the inclined sliding surface is

$$\sigma_z' = \frac{2\sigma_v}{3} + \frac{B\gamma}{3} \tag{6}$$

The shear forces T on the inclined sliding surface is

$$T = \frac{cB^2}{\sin\alpha} + N\tan\varphi \tag{7}$$

The shear force T' acting on the two sides of the wedge can be written as follows:

$$T' = \frac{B^2}{2\tan\alpha}\left(c + k_0\sigma_z'\tan\varphi\right) \tag{8}$$

Not taking the infiltration in excavation face into account, the overburden strata are assumed to be permeable with high permeability, such as sand and gravel. So a complete hydraulic connection exists between the river water and groundwater. In this sense, the pore water pressure generated by river water can be expressed as

$$P_w = B^2\left(\sum \gamma_w z_i + \gamma_w h\right) \tag{9}$$

where γ_w is the unit weight of water and z_i is the thickness of stratum no. i, which is located above the centre point of tunnel face.

By considering the water pressure and equating force in the vertical and horizontal direction, the limit support force on the tunnel face is

$$P = \varepsilon\left(\sigma_v BL + G\right) - \left(\frac{cB^2}{\sin\alpha} + 2T'\right)\left(\varepsilon\sin\alpha + \cos\alpha\right) + P_w \tag{10}$$

where $\varepsilon = (\sin\alpha - \tan\varphi\cos\alpha)/(\cos\alpha + \tan\varphi\sin\alpha)$ and σ_v is calculated in terms of effective stresses.

The support pressure is simplified and considered uniform, and the minimum support pressure termed as limit support pressure which keeps the tunnel face stable is expressed as

$$\sigma_T = \frac{1}{B^2}\left[\varepsilon\left(\sigma_v BL + G\right) - \left(\frac{cB^2}{\sin\alpha} + 2T'\right)\left(\varepsilon\sin\alpha + \cos\alpha\right) + P_w\right] \tag{11}$$

This would provide a simple design method for limit support pressure on the tunnel face. Effect of a multilayered overburden also can be taken into account. In this paper, this mechanism will be used for the probabilistic analysis.

3. Bayesian Framework

Within a Bayesian framework, inferences are made about the parameters of interest by a probability distribution given the data [28]. Bayesian framework is applied to estimate the values of unknown parameters of a model about which some prior information may be available. According to the Bayes' theorem, the posterior distribution of θ is proportional to the product of the likelihood function and the prior distribution function [29]. Bayes' formula can be written as follows:

$$P(\theta \mid y) = \frac{L(y \mid \theta)P(\theta)}{P(y)} \tag{12}$$

where $P(\theta)$ is the prior probability distribution of the parameters; $L(y \mid \theta)$ is the likelihood function and incorporates the statistical as well as the mechanistic relationships among the predictors and variables; $P(\theta \mid y)$ is the posterior distribution of the parameters; $P(y)$ is the normalizing constant to make the probability density function valid. θ is vector of uncertain input parameters. The prior distribution of θ and measured data can be integrated using a systematic way.

The error or difference between the actual performance and the model prediction is defined as the model correction factor ε which can be used to characterize model uncertainty.

$$y = g(\theta) + \varepsilon \tag{13}$$

where y is the actual limit support pressure; $g(\theta)$ is predicting limit support pressure using the given tunnel face stability model; θ is vector of uncertain input parameters.

The likelihood function gives a measure of the agreement between the available data and the corresponding model output. Assuming that error ε is to be normally distributed, the likelihood function is the conditional probability density function of θ and can be written as

$$L(y \mid \theta) = \phi\left[\frac{y - g(\theta) - \mu_\varepsilon}{\sigma_\varepsilon}\right] \tag{14}$$

where ϕ is the probability density function of standard normal variable; μ_ε and σ_ε are the mean and standard deviation of errors, respectively.

4. Markov Chain Monte Carlo Method

Bayesian updating can be achieved using Bayesian framework, when conjugate priors are given. The posterior distribution function of input parameters and model predicted response cannot be derived through analytical means. An alternative approach is to use a MCMC method to obtain the numerical summarization of the posterior distribution. Therefore, random sampling methods are needed to generate samples from the posterior distribution function. These posterior distributions are obtained using the MCMC simulation, which is an effective random sampling method. The MCMC simulation can maintain adequate sampling density as the number of parameter increases and compute efficiently which has gained popularity in recent years to sample the posterior probability density function [30]. It can handle efficiently problems with a large number of random variables and is very flexible to any type of prior distribution. The idea of the MCMC simulation is to simulate a random walk across the parameter space by drawing samples from the approximate posterior distribution and improving those draws until the distribution converges to the target posterior distribution. Delayed rejection adaptive metropolis algorithm (DRAM) [23] is the widely used algorithms in the MCMC method. A multivariate normal distribution with a mean vector defined by the current value and the covariance matrix C_i is considered as

$$C_i = \begin{cases} C_0 & i \le i_0 \\ s_d \operatorname{cov}(\zeta_0, \dots, \zeta_{i-1}) + s_d \delta I_d & i > i_0 \end{cases} \tag{15}$$

where C_0 is the initial covariance to start the adaptation procedure; s_d is a parameter that depends only the dimension d of the state space, $s_d = 2.4^2/d$; $\operatorname{cov}(\zeta_0, \dots, \zeta_{i-1})$ is the covariance of previous samples; I_d denotes the d-dimensional identity matrix; δ is a small constant.

The definition of the empirical covariance matrix determined

$$\operatorname{cov}(\zeta_0, \dots, \zeta_k) = \frac{1}{k}\left(\sum_{i=1}^{k} \zeta_i \zeta_i^T - (k+1)\bar{\zeta}_k \bar{\zeta}_k^T\right) \tag{16}$$

where $\bar{\zeta}_k = (1/(k+1)) \sum_{i=0}^{k} \zeta_i$.

TABLE 1: Prior distributions.

Parameter	Mean	Std. dev.	COV
φ	30	4.5	0.15
c	5	1	0.20

For $i > i_0$, substituting (16) into (15), the covariance satisfies the recursive equation as

$$C_{n+1} = \frac{n-1}{n} C_n + \frac{s_d}{n}\left(n\bar{\zeta}_{n-1}\bar{\zeta}_{n-1}^T - (n+1)\bar{\zeta}_n \bar{\zeta}_n^T + \zeta_n \zeta_n^T + \varepsilon I_d\right) \tag{17}$$

In the delayed rejection algorithm, the proposed sample $\zeta_i^{(1)}$ is accepted with the probability.

$$\alpha_1\left(\zeta_{i-1}, \zeta_i^{(1)}\right) = \min\left\{1, \frac{p\left(\zeta_i^{(1)} \mid D\right) q_1\left(\zeta_i^{(1)}, \zeta_{i-1}\right)}{p\left(\zeta_{i-1} \mid D\right) q_1\left(\zeta_{i-1}, \zeta_i^{(1)}\right)}\right\} \tag{18}$$

where q_1 is the first proposal distribution.

The second stage proposal is accepted with probability if the $\zeta_i^{(1)}$ is rejected.

$$\alpha_2\left(\zeta_{i-1}, \zeta_i^{(2)}\right) = \min\left\{1, \frac{p\left(\zeta_i^{(2)} \mid D\right) q_1\left(\zeta_i^{(2)}, \zeta_i^{(1)}\right) q_2\left(\zeta_i^{(2)}, \zeta_i^{(1)}, \zeta_{i-1}\right)}{p\left(\zeta_{i-1} \mid D\right) q_1\left(\zeta_{i-1}, \zeta_i^{(1)}\right) q_2\left(\zeta_{i-1}, \zeta_i^{(1)}, \zeta_i^{(2)}\right)} \cdot \frac{\left[1 - \alpha_1\left(\zeta_i^{(2)}, \zeta_i^{(1)}\right)\right]}{\left[1 - \alpha_1\left(\zeta_{i-1}, \zeta_i^{(1)}\right)\right]}\right\} \tag{19}$$

where q_2 is the second proposal distribution.

This process of delaying rejection can be iterated.

5. Probabilistic Analysis

Numerous test investigations and numerical studies of tunnel face stability have illustrated that the friction angle φ and cohesion c have significantly influence the accuracy of the predicting limit support pressure. In this study, only the friction angle and cohesion are considered as random variables. The updating process starts with assumed with the prior distributions for the soil strength parameter of friction angle and cohesion, which are considered important for limit support pressure on tunnel face.

The friction angel and cohesion are considered as random input parameters with normal variables. In this study, the respective means of friction angle and cohesion are denoted as $\mu_\varphi = 30^\circ$, $\mu_c = 5$ kPa, respectively. The prior coefficient of variation (COV) values of the friction angle and cohesion are assumed to be $\mathrm{COV}_\varphi = 15\%$, $\mathrm{COV}_c = 20\%$ [31], respectively. The prior distributions and basic statistics of the soil parameters are summarized in Table 1.

For probabilistic analysis of support stress on tunnel face, variation of friction angle and cohesion are considered. Other

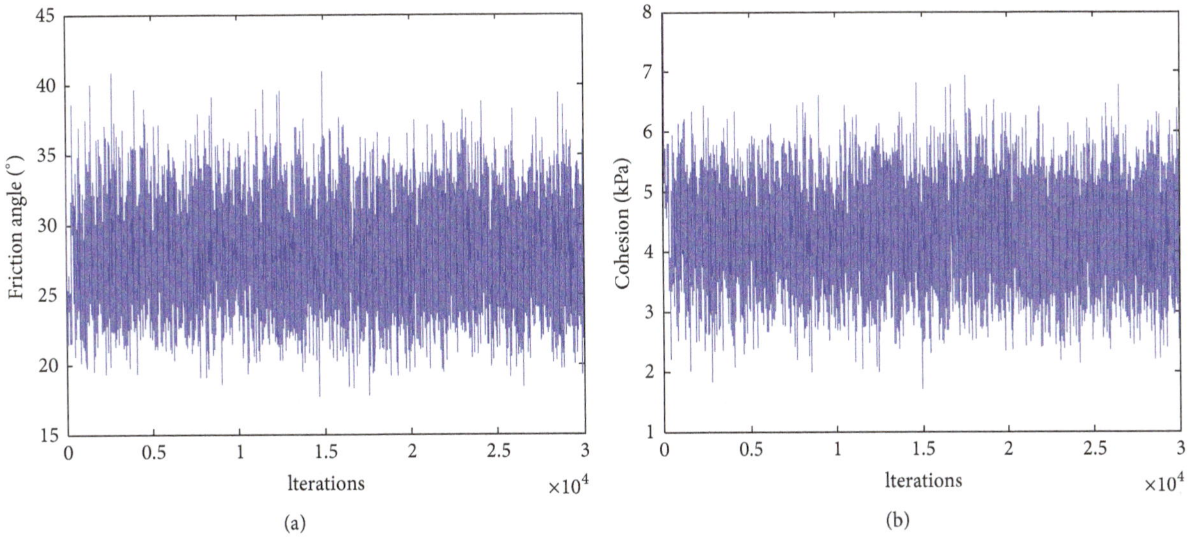

FIGURE 3: Plot of samples with MCMC simulation steps.

parameters come from the crossing Ganjiang River tunnel and are used as follows: saturated unit weight $\gamma_{sat} = 20$ KN/m^3, the tunnel diameter D = 6m, the tunnel cover depth H = 12m, and the water depth in river h = 10m. We consider a measurement outcome of limit support stress $\sigma_T = 254.68$ kPa which corresponds to the measured mean values of friction angle 28° and cohesion 4 kPa, respectively, according to the laboratory test results of soil in site. The adopted model functions are limit support stress which is provided in (11). Based on the Bayesian framework described in the previous section, the material properties of the soils are updated using the assumed the limit support pressure 254.68kPa.

The chains with total number of evaluations equal to 30,000 are used in MCMC simulation with a DEAM algorithm, and the posterior samples generated are shown in Figure 3. From the Figure 3, it shows that the chains can quickly converge to a stable posterior distribution. The Markov chain is effective in collecting representative samples. After verifying that the distribution of the values converged to the posterior distribution, the distribution of the parameters can be used to produce optimum simulations and generate an estimation of the model uncertainty. The initial 5,000 samples are discarded because the Markov chain has not stabilized. The remaining 25,000 samples are used as the posterior distribution samples.

6. Results and Discusses

6.1. Posterior Distribution of Input Parameters. Since the samples are generated by MCMC method, the posterior distributions of the parameters are computed using the chains of the two MCMC simulations. Figures 4 and 5 illustrate the prior and posterior distributions of the friction angle and cohesion, respectively. When additional knowledge is available, the prior distributions of the input parameters can be reformulate. From Figures 4 and 5, it is obvious that the parameters of friction angle and cohesion are concentrated

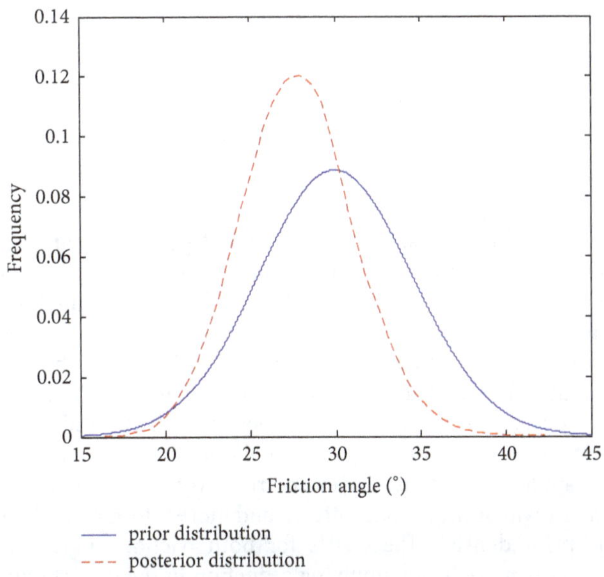

FIGURE 4: Prior and posterior distribution of friction angle.

within a small region and show normal shape. The posterior mean, standard deviation, and COV value are summarized in Table 2 using the remaining 25,000 samples. The "best" parameter set is typically created by computing the mean for each input parameters [32, 33]. The mean of posterior distribution for the friction angle and cohesion is also smaller than that of prior distribution, the updated mean values of the friction angle, and cohesion decrease from 30° to 27.816° and from 5 kPa to 4.268 kPa, respectively. The means of values are getting increasingly closer to the measured parameters (friction angel 28° and cohesion 4 kPa) which express the ability of the MCMC method to converge to the measured parameters.

TABLE 2: Statistic of posterior distribution parameters.

Parameter	Mean	Std. dev.	COV
φ	27.816	3.181	0.114
c	4.268	0.697	0.163

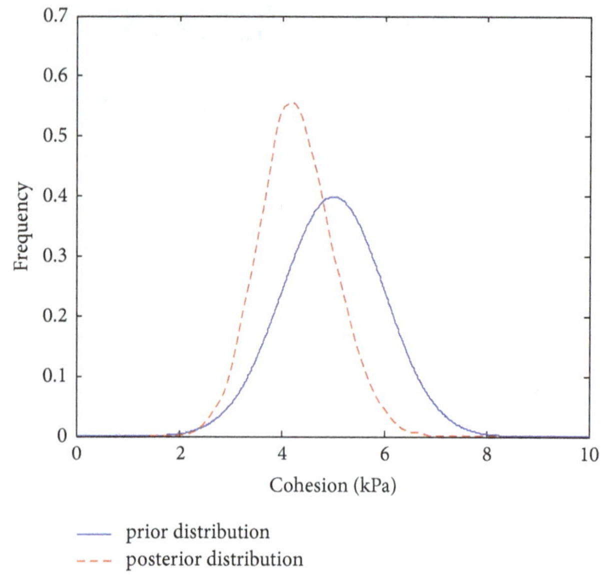

FIGURE 5: Prior and posterior distribution of cohesion.

According to Table 2, the prior and posterior standard deviations are also given for a comparison. The distribution of friction angle and cohesion is significantly modified via Bayesian updating. The standard deviation of friction angle is reduced from 0.15 to 0.114. The standard deviation of cohesion is reduced from 0.2 to 0.163. Compared to the COVs of the soil parameters of prior and posterior distribution, the COVs of both friction angle and cohesion are reduced with the application of the Bayesian framework. The posterior density will be more concentrated and more informative than the prior density. The COVs for both friction angle and cohesion are reduced, implying reduction in the uncertainty of strength parameters. The updated parameters show that the uncertainty of friction angle and cohesion is reduced, which reflects the measurement on the variable.

6.2. Predicted Limit Support Pressure. In this section, the probability distribution of the limit support stress is determined for a given condition. Using updated soil parameters of friction angle φ and cohesion c presented as posterior distributions, the limit support pressure on tunnel face can be estimated. The limit support pressures obtained by the posterior and prior distribution are shown in Figure 6, respectively. The outcomes resulting from prior and posterior parameters are demonstrated. The cumulative distributions are shown in Figure 7. The distributions have considerable effects on limit support pressure on tunnel face. With all the soil strength parameters updated via Bayesian framework, the three-dimensional model of tunnel face stability is able to compute probabilities more accurately and realistically.

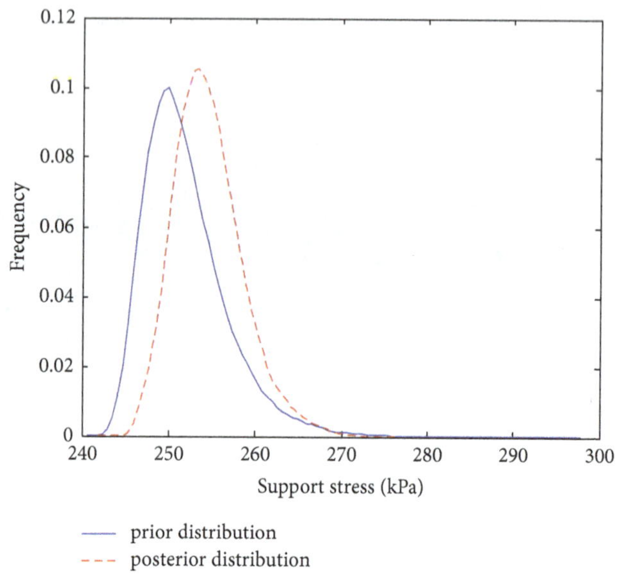

FIGURE 6: Posterior distribution of the limit support stress.

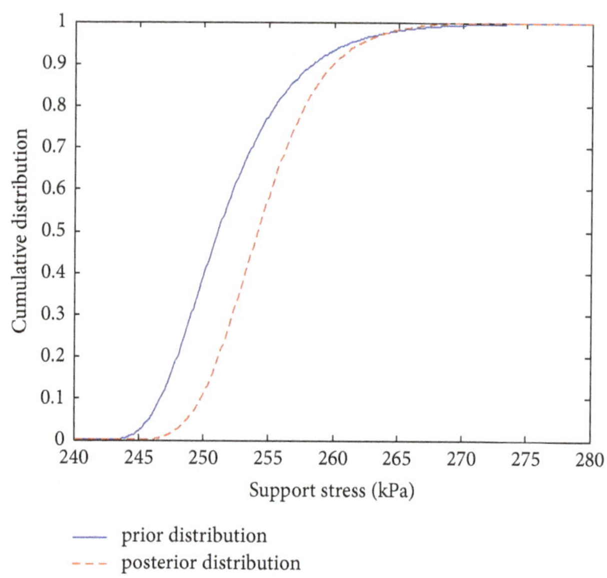

FIGURE 7: Cumulative distribution of the limit support stress.

7. Conclusions

In this paper, a Bayesian framework for updating soil strength parameters in tunnel excavation below river is presented. The method employs a MCMC simulation based approach to derive the posterior distribution of the parameters. The posterior distribution can be used for probabilistic analysis of the limit support stress on the circular tunnel face. The 3D analysis model with the limit equilibrium method is used as deterministic model. The uncertain parameters considered in the analysis are the soil strength parameters as the friction angle and cohesion. Comparing with the prior distribution, the means of prediction improve and the variation of prediction reduces by the proposed Bayesian framework. Therefore, the proposed method is effective in reducing the uncertainty of soil strength parameters, demonstrating its potential as a practical geotechnical engineering tool. The Bayesian framework with MCMC method might be more favorable in the uncertainty analysis and risk management.

Conflicts of Interest

The authors declare that they have no conflicts of interest.

Acknowledgments

The authors sincerely acknowledge the support from the National Natural Science Foundation of China (51468041), the Specialized Research Fund for the Doctoral Program of Higher Education (20123601110001), and Jiangxi Science Foundation (20161BAB203078).

References

[1] G. Mollon, K. K. Phoon, D. Dias, and A. H. Soubra, "A new 2D failure mechanism for face stability analysis of a pressurized tunnel in spatially variable sands," in *GeoFlorida 2010: Advances in Analysis, Modeling & Design*, pp. 2052–2061, 2010.

[2] A. H. Haghi, M. R. Asef, A. Taheri, and M. Mohkam, "Evaluation of the heading confinement pressure effect on ground settlement for EPBTBM Using full 3D numerical analysis," *International Journal of Mining & Geo-Engineering*, vol. 47, no. 1, pp. 13–32, 2013.

[3] N. Ebadati and M. Kaboli, "Modeling stress distribution and the change of ground behavior during excavation of Tehran city underground tunnel," *Journal of Geotechnical Geology*, vol. 9, no. 4, pp. 293–304, 2014.

[4] G. Idinger, P. Aklik, W. Wu, and R. I. Borja, "Centrifuge model test on the face stability of shallow tunnel," *Acta Geotechnica*, vol. 6, no. 2, pp. 105–117, 2011.

[5] I. M. Lee, J. S. Lee, and S. W. Nam, "Effect of seepage force on tunnel face stability reinforced with multi-step pipe grouting," *Tunnelling and Underground Space Technology*, vol. 19, no. 6, pp. 551–565, 2004.

[6] K.-K. Phoon and F. H. Kulhawy, "Characterization of geotechnical variability," *Canadian Geotechnical Journal*, vol. 36, no. 4, pp. 612–624, 1999.

[7] Z. Cao and Y. Wang, "Bayesian model comparison and selection of spatial correlation functions for soil parameters," *Structural Safety*, vol. 49, pp. 10–17, 2014.

[8] S. E. Cho, "Effects of spatial variability of soil properties on slope stability," *Engineering Geology*, vol. 92, no. 3-4, pp. 97–109, 2007.

[9] C.-P. Zhang, K.-H. Han, and D.-L. Zhang, "Face stability analysis of shallow circular tunnels in cohesive-frictional soils," *Tunnelling and Underground Space Technology*, vol. 50, pp. 345–357, 2015.

[10] G. Mollon, D. Dias, and A.-H. Soubra, "Probabilistic analysis of circular tunnels in homogeneous soil using response surface methodology," *Journal of Geotechnical and Geoenvironmental Engineering*, vol. 135, no. 9, pp. 1314–1325, 2009.

[11] G. Mollon, D. Dias, and A.-H. Soubra, "Probabilistic analysis of pressurized tunnels against face stability using collocation-based stochastic response surface method," *Journal of Geotechnical and Geoenvironmental Engineering*, vol. 137, no. 4, pp. 385–397, 2011.

[12] H.-Z. Li and B. K. Low, "Reliability analysis of circular tunnel under hydrostatic stress field," *Computers & Geosciences*, vol. 37, no. 1-2, pp. 50–58, 2010.

[13] P. Oreste, "A probabilistic design approach for tunnel supports," *Computers & Geosciences*, vol. 32, no. 7, pp. 520–534, 2005.

[14] T. Celestino, N. Aoki, R. Silva, R. Gomes, A. Bortolucci, and D. Ferreira, "Evaluation of tunnel support structure reliability," *Tunnelling and Underground Space Technology*, vol. 21, no. 3-4, pp. 311–322, 2006.

[15] J. Connor Langford and M. S. Diederichs, "Reliability based approach to tunnel lining design using a modified point estimate method," *International Journal of Rock Mechanics and Mining Sciences*, vol. 60, pp. 263–276, 2013.

[16] Q. Lü, C. L. Chan, and B. K. Low, "System reliability assessment for a rock tunnel with multiple failure modes," *Rock Mechanics and Rock Engineering*, vol. 46, no. 4, pp. 821–833, 2013.

[17] A. Eshraghi and S. Zare, "Face stability evaluation of a TBM-driven tunnel in heterogeneous soil using a probabilistic approach," *International Journal of Geomechanics*, vol. 15, no. 6, Article ID 04014095, 2015.

[18] M. Kattwinkel and P. Reichert, "Bayesian parameter inference for individual-based models using a Particle Markov Chain Monte Carlo method," *Environmental Modeling and Software*, vol. 87, pp. 110–119, 2017.

[19] J. A. Vrugt, H. V. Gupta, W. Bouten, and S. Sorooshian, "A shuffled complex evolution metropolis algorithm for optimization and uncertainty assessment of hydrologic model parameters," *Water Resources Research*, vol. 39, no. 8, pp. 113–117, 2003.

[20] W.-H. Zhou, K.-V. Yuen, and F. Tan, "Estimation of soil-water characteristic curve and relative permeability for granular soils with different initial dry densities," *Engineering Geology*, vol. 179, pp. 1–9, 2014.

[21] C. Hsein Juang, Z. Luo, S. Atamturktur, and H. Huang, "Bayesian updating of soil parameters for braced excavations using field observations," *Journal of Geotechnical and Geoenvironmental Engineering*, vol. 139, no. 3, pp. 395–406, 2013.

[22] L. L. Zhang, Z. B. Zuo, G. L. Ye, D. S. Jeng, and J. H. Wang, "Probabilistic parameter estimation and predictive uncertainty based on field measurements for unsaturated soil slope," *Computers & Geosciences*, vol. 48, pp. 72–81, 2013.

[23] H. Haario, M. Laine, A. Mira, and E. Saksman, "DRAM: efficient adaptive MCMC," *Statistics and Computing*, vol. 16, no. 4, pp. 339–354, 2006.

[24] G. Anagnostou and K. Kovári, "The face stability of slurry-shield-driven tunnels," *Tunnelling and Underground Space Technology*, vol. 9, no. 2, pp. 165–174, 1994.

[25] N. Horn, "Horizontaler erddruck auf senkrechte abschlussflächen von tunnelröhren," *Landeskonferenz Der Ungarischen Tiefbauindustrie*, pp. 7–16, 1961.

[26] K. Terzaghi, *Theoretical Soil Mechanics*, John Wiley & Sons, New York, NY, USA, 1943.

[27] J.-S. Xu, D.-C. Du, and Z.-H. Yang, "Upper bound analysis for deep tunnel face with joined failure mechanism of translation and rotation," *Journal of Central South University*, vol. 22, no. 11, pp. 4310–4317, 2015.

[28] A. Gelman, J. B. Carlin, H. S. Stern, and B. R. Rubin, *Bayesian Data Analysis*, CRC Press, Boca Raton, Fla, USA, 2nd edition, 2004.

[29] L. L. Zhang, J. Zhang, L. Zhang, and W. H. Tang, "Back analysis of slope failure with Markov chain Monte Carlo simulation," *Computers & Geosciences*, vol. 37, no. 7-8, pp. 905–912, 2010.

[30] R. Kelly and J. Huang, "Bayesian updating for one-dimensional consolidation measurements," *Canadian Geotechnical Journal*, vol. 52, no. 9, pp. 1318–1330, 2015.

[31] K.-K. Phoon and F. H. Kulhawy, "Evaluation of geotechnical property variability," *Canadian Geotechnical Journal*, vol. 36, no. 4, pp. 625–639, 1999.

[32] X. Pei, S. C. Wong, and N. N. Sze, "A joint-probability approach to crash prediction models," *Accident Analysis & Prevention*, vol. 43, no. 3, pp. 1160–1166, 2011.

[33] F. Korner-Nievergelt, T. Roth, S. von Felten, J. Guélat, B. Almasi, and P. Korner-Nievergelt, *Markov Chain Monte Carlo Simulation, Bayesian Data Analysis in Ecology Using Linear Models with R,BUGS, and STAN*, Chapter 12, Academic Press, Boston, Mass, USA, 2015.

2

Study and Application of Roof Cutting Pressure Releasing Technology in Retracement Channel Roof of Halagou 12201 Working Face

Ma Xingen,[1,2] **He Manchao**,[1,3] **Wang Yajun**,[3] **Zhang Yong**,[1,2] **Zhang Jiabin**,[1,2] **and Liu Yuxing**[1,2]

[1] *State Key Laboratory for Geomechanics and Deep Underground Engineering, China University of Mining and Technology (Beijing), Beijing 100083, China*
[2] *School of Mechanics and Civil Engineering, China University of Mining and Technology (Beijing), Beijing 100083, China*
[3] *Department of Geotechnical Engineering, College of Civil Engineering, Tongji University, Shanghai 200092, China*

Correspondence should be addressed to Wang Yajun; 464557359@qq.com

Academic Editor: Gisella Tomasini

The retracement channel roof cutting (RCRC) technology can change the overburden structure actively by cutting off the roof of channel along the direction of working face tendency and make use of the gangue collapsing from roof cutting range to fill the goaf and weaken the mining pressure during the retracement process of working face. In order to solve the problems of high stress in surrounding rock and serious deformation of retracement channel in Halagou coal mine, it is the first time that the pressure releasing test is carried out on the 12201 working face by the method of the directional presplitting roof cutting in retracement channel. First, according to statics theory and energy theory, the stress state of hydraulic support and roof deformation mechanism of retracement channel are analyzed. Then the roof cutting design of retracement channel is determined according to the geological conditions of 12201 working face, and the cutting effect is analyzed by numerical simulation. Finally, the field test is carried out on the 12201 working face to verify the effect of pressure releasing by roof cutting. The result shows that, with the roof cutting design including the roof cutting height being 8m and roof cutting angle being 45°, the roof subsidence of the 12201 working face retracement channel in Halagou mine is reduced to 132.5mm, and the hydraulic support resistance is maintained at 1361KN. And there is no hydraulic support crushed; the deformation of the retracement channel is also small; namely, the effect of roof cutting for pressure releasing is obvious.

1. Introduction

The coal mining method of wall system can realize continuous coal mining and has the advantages of high yield, high efficiency, high production efficiency, and strong adaptability, so it has been widely popularized all over the world [1]. In China, Russia, Poland, Ukraine, Britain, Germany, France, Japan, and other coal mining countries, the coal production with wall system all accounts for more than 80% of the whole underground mining production [2, 3]. But in the application of wall system coal mining method, the stopping and withdrawing process of working face is always the key link restricting its production succession and safety level [4]. There are two main ways regarding the retracement of fully mechanized mining working face in China. One is excavating the retracement channel in advance, that is, excavating 1 or 2 auxiliary lanes parallel to the mining face before the end of the mining work at the stopping line of the mining face, and is used for the retracement of the working face. The other is excavating the retracement channel by the coal mining machine; namely, when the mining face is pushed to the stopping line, the coal mining machine cuts coal wall to form the retracement channel by itself [5, 6]. No matter which method of retracement is adopted, the control of the working face surrounding rock at the end of the mining stage is an unavoidable problem. Particularly, the way of excavating the

TABLE 1: Parameters of each rock layer.

Lithology	Density/kN/m3	Tensile Strength/MPa	Internal Friction Angle/°	Cohesion/MPa	Bulk Modulus/GPa	Shear Modulus/GPa
Siltstone	22	4.3	28	0.8	2.11	1.86
Fine sandstone	23	7.3	30	1.0	3.81	3.05
Mudstone	20	3.2	28	0.2	0.20	0.15
Coal	13	3.3	29	0.2	0.35	0.18

retracement channel in advance has a higher demand of the roof support, because the channel will be influenced by the stress concentration caused by the advance of mining face [7, 8].

In view of the above problems, many scholars have carried out a large number of studies. Xi Long, taking the 2505 working face of Yongcheng coal mine as an example, studies the mechanical structure and motion state of the overlying rock mass upon the retracement channel and puts forward the reinforcement support measures of the retracement channel under different overburden fracture structures [9]. Kang Jichun points out that the instability process of retracement channel contains three stages, namely, the influence stage of the channel excavation, the influence stage of the mining work, and the influence stage of the main roof's instability. And, on the basis of this, the support design and process of the retracement channel are optimized according to the different stages of the instability [7, 10]. Through the deformation observation of the retracement channel of a broken roof working face, Zhang Jinhu finds out that the movement of its roof can be divided into three processes: relative stability, mutational motion, and significant motion. And, increasing the working resistance of the chock supports in the retracement channel under the relatively stable stage, the final deformation of channel can be controlled effectively [11]. The above research results provide good references for the mechanism exploration of the retracement channel's deformation and the design of the supporting countermeasures. However, in the current research, the deformation control of the retracement channel is still mainly realized by the passive reinforcement of support. While controlling the deformation of the surrounding rock, the support cost and workload also increase, and the retracement and succession of working face are also affected.

In 2009, academician He Manchao of the Chinese Academy of Sciences proposed the roof cutting pressure releasing gob-side entry retaining (RCPRGER) non-pillar mining technology [12]. The core of this technology is cutting off the stress transfer between goaf roof and entry roof through directional presplitting blasting implemented on the goaf side of entry roof. Then, after the working face mining, the roof of goaf area will collapse, the mined area adjacent to the retaining entry will be filled up with the broken fallen rocks, and the roadway will be retained automatically with no coal pillar or filling material [13, 14]. At present, the technology has been successfully carried out in many mines to retain the gob-side entry, and the effect of pressure release by roof cutting is obvious. Therefore, taking this as a reference, this paper studies the application of the roof cutting pressure releasing (RCPR) technology in the retracement channel firstly, namely, through the blasting cutting of the retracement channel roof to cut off the partial stress transfer in the roof and improving the stress environment of the channel surrounding rocks actively, relieving the roof support pressure.

2. Project General Situation

This paper takes the 12201 working face of Halagou coal mine as an engineering example to carry out the research of the retracement channel roof cutting (RCRC) technology. Halagou coal mine is located in the boundary region between Inner Mongolia and Shaanxi provinces in China (as shown in Figure 1(a)), and the 12201 working face is the first mining face of the No.2 panel in Shendong group. The working face is about 747m long in the strike length and 320m long in the tendency length. The thickness of the coal seam ranges from 1.6m to 2.4m, and the average thickness is 2.0m. The buried depth of the working face is about 60m to100m, and the dip angle of the coal seam is between 1° and 3°. The roof of this working face is a typical coal containing composite roof, its stratigraphic column is shown in Figure 1(b), and the relevant parameters of each rock layer are listed in Table 1.

The layout of 12201 working face is shown in Figure 1(c). The 560m length section before the stopping line of this working face's haulage entry is used to carry out the RCPRGER technology, and the retained entry can be used as the air return entry of 12202 working face. The 12201 working face takes the method of excavating two retracement channels in advance to realize the working face retracement. The length of the retracement channels is 320m, and the tunnel sections are rectangular. Among them, the main retracement channel is 5.2m wide and 2.0m high, and the auxiliary retracement channel is 5.5m wide and 2.0m high. In the working face retracement of adjacent panel, the maximum load of hydraulic support reached 60MPa, and the dynamic load coefficient reached 2.08. Some hydraulic supports were compressed severely, and a part of them were damaged (as shown in Figure 2). Besides, the reinforce support of main retracement channel was difficult, usually needing two rows of chock supports. Therefore, in order to alleviate this phenomenon, this paper takes the 12201 working face as an engineering example to carry out the research about the RCRC technology.

3. RCRC Pressure Releasing Technology

On the basis of the analysis of the overburden strata structure at the final mining stage, this paper puts forward the RCRC

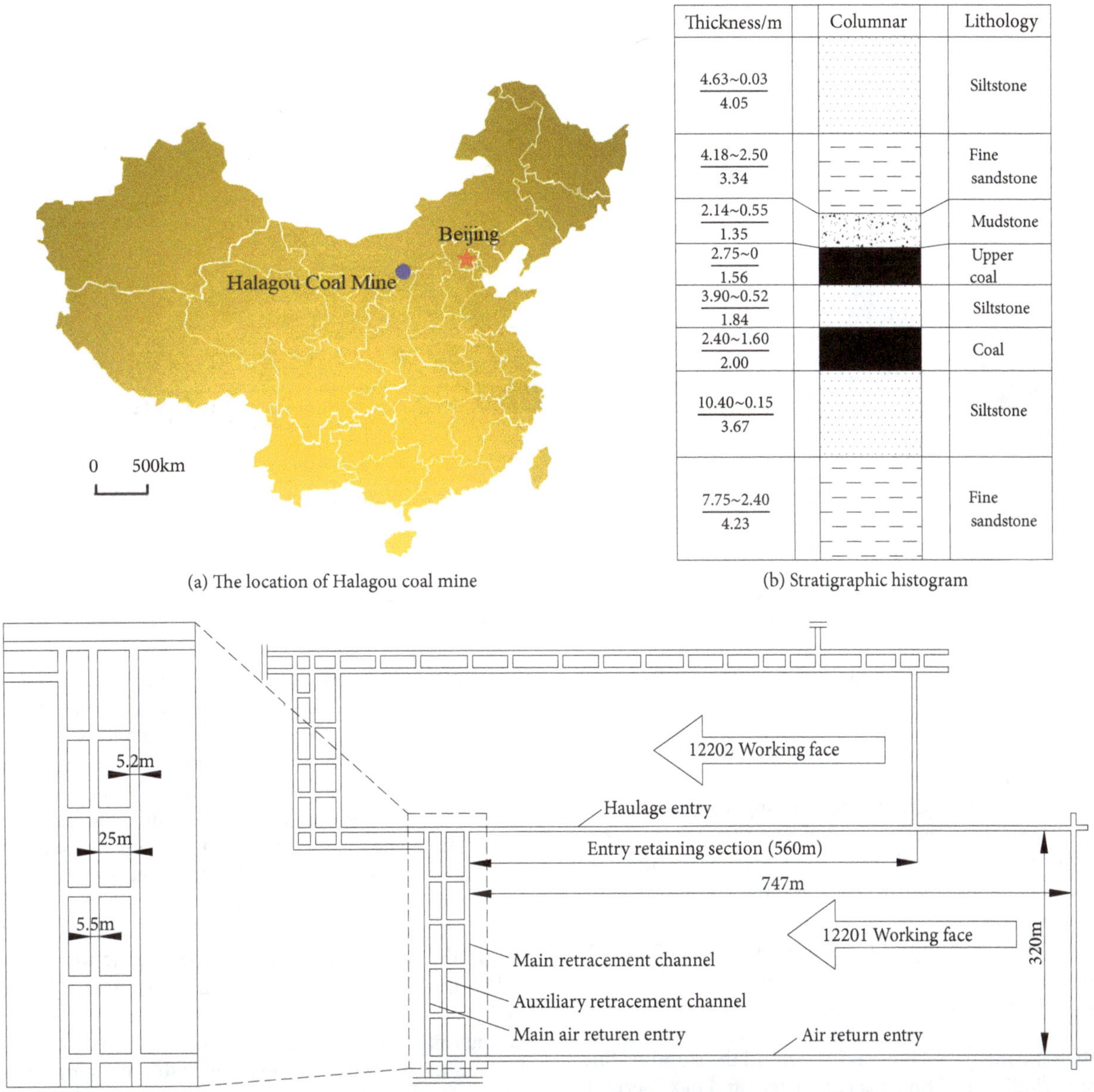

(a) The location of Halagou coal mine

(b) Stratigraphic histogram

(c) Layout of 12201 working face

FIGURE 1: Layout and roof lithology of 8304 working face.

(a) Overloaded

(b) Damaged

FIGURE 2: Overloaded and damaged hydraulic support.

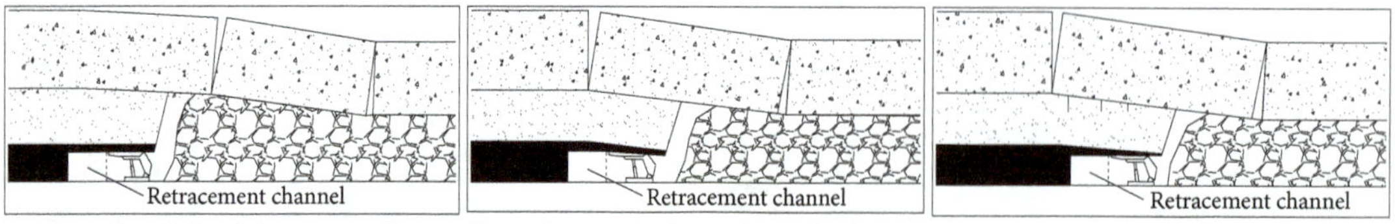

(a)Breaking behind the stopping line (b)Breaking upon the retracement channel (c)Breaking upon the coal wall

FIGURE 3: Structure diagram of overlying strata at stop line.

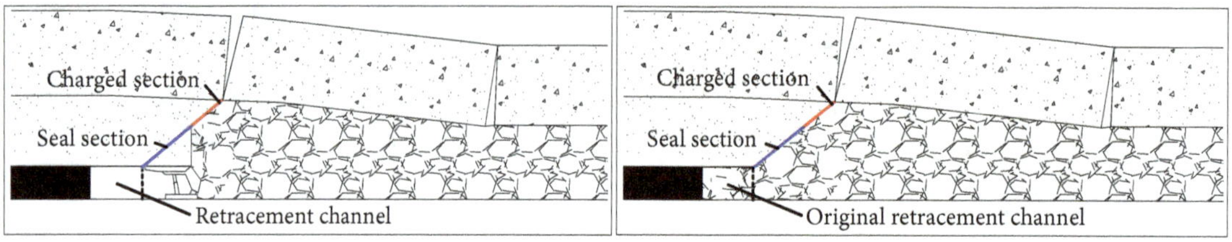

(a)First stage (b)Second stage

FIGURE 4: Breaking process under roof cutting condition.

technology and then combines it with the bidirectional concentrated tension blasting (BCTB) technology to change the structure of the overburden strata actively and improve the pressure environment of hydraulic supports in the retracement channel.

3.1. Analysis of Roof Structure at the Final Mining Stage. Firstly, the roof structure of the working face at the final mining stage under the traditional condition is analyzed. When the retracement channel is formed and connected with the working face, the overburden strata structure at the stopping line, according to the fault line position of the main roof, can be divided into three types: breaking behind the stopping line, breaking upon the retracement channel, and breaking upon the coal wall (as shown in Figures 3(a)–3(c)) [15].

When the main roof is breaking behind the stopping line, the last periodic weighting has completed, and next periodic weighting has not occurred. So the hydraulic supports in retracement channel are mainly subjected to the static load of the overlying rock mass, and the surrounding rock structure is more stable. When the main roof is breaking upon the retracement channel, the suspension length of main roof has reached the weighting step, and this periodic weighting is happening. So the hydraulic supports in retracement channel are affected by the rotation and sinking of main roof, and the surrounding rock structure is unstable. Under this condition, the large deformation of the retracement channel and the damage of hydraulic support are easy to occur during the retracement process of the working face. When the main roof is breaking upon the coal wall, although the suspension length of main roof has not reached the weighting step, the main roof has been breaking upon the coal wall crossing over the retracement channel and this periodic weighting has begun. So, under the disturbance of the retracement of hydraulic supports, the main roof is prone to further break and rotation. Then, during the retracement of the working face, there are also the risks of the retracement channel deformation and the hydraulic supports overload.

Through the above analysis, it can be found that the structure of the surrounding rock is relatively stable when the main roof is breaking behind the stopping line. But, under the traditional long wall mining condition, the three kinds of main roof breaking forms are random. Therefore, the RCRC technology is put forward. Namely, before the connection of retracement channel and working face, the BCTB technology is used to cut off the retracement channel roof along the direction of working face tendency and change the overburden structure actively. Then when the working face is connected with the retracement channel, a part of roof behind the working face can fall down along the cutting slit and fill the goaf firstly (as shown in Figure 4(a)). Next, when the hydraulic supports have been withdrawn, the roof in the cutting range can collapse and fill the goaf further to weaken the mining pressure and load of hydraulic support in the working face retracement process (as shown in Figure 4(b)).

3.2. BCTB Technology. In order to realize the directional presplitting cutting in the roof of retracement channel along the direction of working face tendency, the BCTB technology is used to carry out the roof cutting. As a kind of directional presplitting blasting, this technology can realize the directional transmission of explosive energy along the leading direction of the energy gathering pipe through the combined use of energy gathering pipe and explosive and then realize the directional breaking of the overburden strata along the direction of roof cutting. The schematic is as shown in Figure 5 [16].

The structure and blasting control mechanism of energy gathering pipe are shown in Figure 6. Some pipes can be connected and installed by connecting sleeves according to the requirement of blasting depth, and the locked groove can maintain the energy gathering pipe towards a certain angle at the time of installation. When cutting the roof, a certain

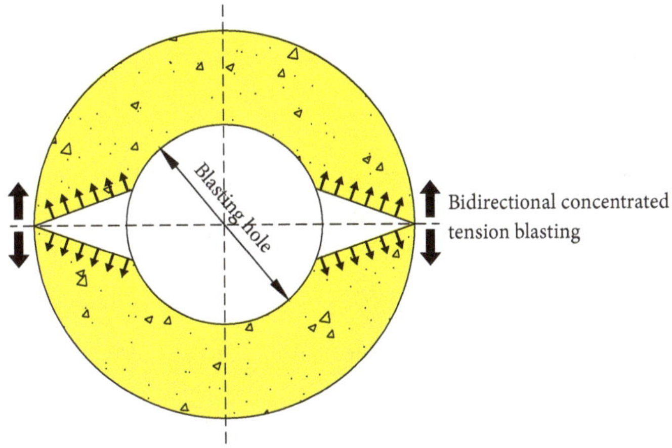

FIGURE 5: Mechanics model of BCTB.

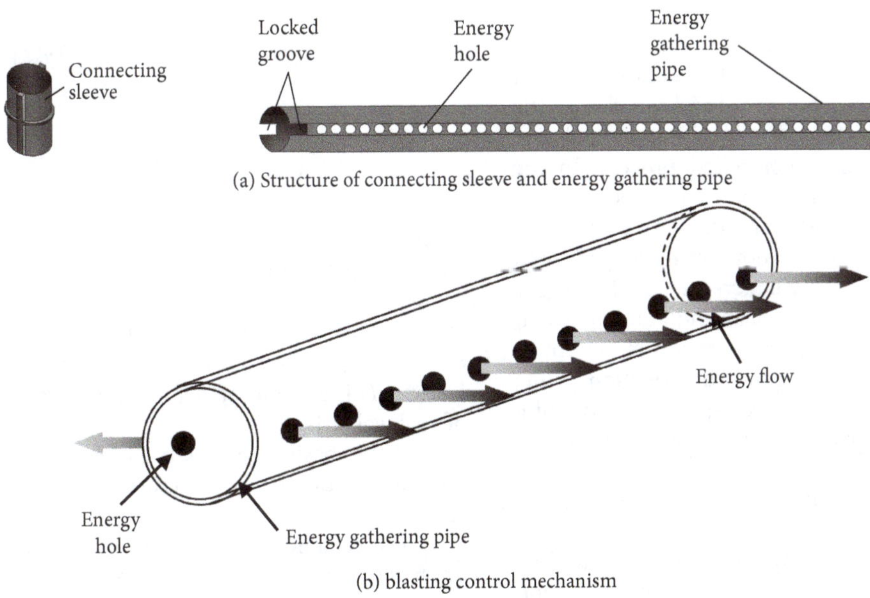

FIGURE 6: Device and principle of BCTB.

amount of explosive rolls is loaded into the energy gathering pipes, followed by putting them into the blasting hole in the roof and sealing the holes. The design of the energy hole can make the blasting energy produce the accumulation effect in a setting direction, which can form a tensile presplitting in the rock mass along the direction of energy gathering and protect the integrity of the rock mass in the energy nongathering area.

In order to achieve the ideal effect of presplitting cutting, the explosive charge structure and the blasting hole spacing should be rationally designed. In the application of the RCPRGER technology, the size of the energy gathering pipe is usually ϕ36.5mm×1500mm. Combined with the previous experience, the explosive charge structure of single hole should be determined by the field test according to the specific lithology of the roof. And a reasonable explosive charge structure of single hole can make two continuous cracks on the hole wall after blasting and cannot make the blasting hole collapse. Besides, the blasting hole spacing depends on the influence range of blasting. If the spacing is too large, the presplitting cracks cannot penetrate the roof rock completely; if the spacing is too small, some explosion energy will be wasted.

In addition, in the application of RCRC releasing technology, in order to avoid the roof leakage of the working face during the connection process of retracement channel and working face, only charging the explosive in the section of blasting hole which is behind the hydraulic supports when the working face stops mining and the other section of the hole is only treated as sealing section, that can make the roof upon the hydraulic supports have a certain integrity.

4. Analysis of the Roof Cutting Effect in Retracement Channel

The RCPRGER has been successfully tested in many mines, and its schematic is shown in Figure 7. According to the

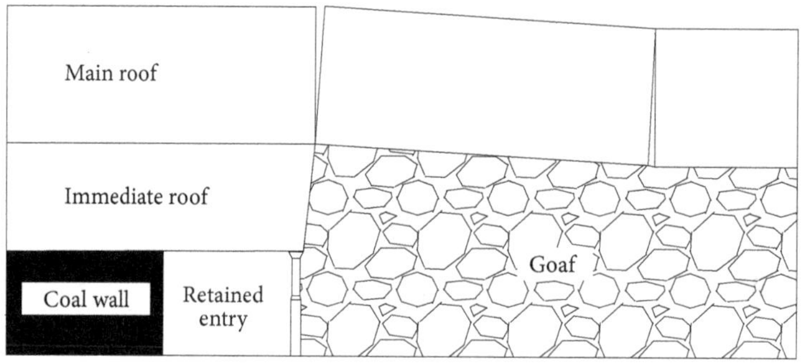

FIGURE 7: Simple diagram of roof cutting principle.

previous experiences about roof cutting, it can be seen that the roof cutting not only can affect the horizontal collapse of the roof but also has some influences on the roof vertical structure [17, 18].

In the horizontal direction, the roof presplitting cutting can cut off the stress transfer in the roof and can change the roof collapse form actively. Then the breaking and collapse of roof can turn to a human controlled process. In the vertical direction, part of main roof in the roof cutting affecting range can be converted to the immediate roof through the reasonable design of roof cutting height. Then the thickness of immediate roof will increase, and the gangue pile formed by immediate roof's collapsing and expansion can support main roof and other overlying strata much better. That can reduce the subsidence deformation of the overburden strata effectively. According to the structural evolution characteristics of the cutting roof, the roof cutting effect in retracement channel can be analyzed from two aspects of the hydraulic support pressure and channel deformation.

The working face is usually long under long wall mining, and the analysis emphasis is the working face advance section in this study. Based on the rock mass mechanical property of poor tensile capacity, the influence of both ends of the working face can be neglected, and the stress analysis of retracement channel can be simplified into 2D plane problem. Besides, during the retracement of working face, the retracement channel is supported well by various support types, so it can be regarded as a deformable body within a certain deformation range. Then, based on the classical assumptions in typical materials mechanics and elastic mechanics, such as continuity hypothesis, homogeneity hypothesis, small deformation hypothesis, the field problem can be solved by mathematical method in a certain content, and the analysis results can provide certain references to the engineering application.

4.1. Pressure Analysis of Hydraulic Support. Based on the stress balance theory of loose medium, Professor Hou Chaojiong and Professor Ma Nianjie combined the stress differential equilibrium equation to find the stress of the coal seam interface and the width of the limit equilibrium area of the coal stress. On this basis, the limit equilibrium area width and the support force of the plastic zone in the protective coal pillar at the stopping line can be solved as follows [19, 20]:

$$x_0 = \frac{mk_a}{x} \ln\left(\frac{k\gamma H + c'/\tan\varphi}{c'/\tan\varphi + p_x/k_a}\right), \tag{1}$$

$$\sigma = \left(\frac{c'}{\tan\varphi} + \frac{p_x}{k_a}\right)e^{2x\tan\varphi/mk_a} - \frac{c'}{\tan\varphi}, \tag{2}$$

where c' and φ are the cohesive force (MPa) and internal friction angle (°) of the interface between the coal seam and surrounding rock; m is the coal seam thickness, m; k_a is the lateral pressure coefficient; k is the maximum coefficient of stress concentration; γ is the average density of coal seam, KN/m^3; H is the mining depth, m; p_x is the support strength of coal wall, MPa.

On this basis, the pressure of hydraulic support under four overburden strata structures is calculated, respectively [21, 22].

The mechanical models of the overburden strata breaking upon retracement channel are shown in Figure 8, where q is the resultant force of main roof weight and pressure of other overburden strata; q_0 is the average pressure caused by immediate roof; T_A is the horizontal thrust to rock block A; N_A is the shear force to rock block A; M_A is the bending moment to rock block A at point A′; T_B, N_B, and M_B have the same meaning as T_A, N_A, and M_A; M_0 is the limit bending moment of direct roof; x_0 is the lateral width of limit equilibrium area in coal wall; σ is the support of coal wall plastic zone; F_G is the support of gangue wall; P_1, P_2, P_3, and P_4 are the support force of the hydraulic support under four conditions; a is the width of the retracement channel; b is the support length of the hydraulic support; c is the horizontal distance between retracement channel and point A′. Taking Figure 8(a) as an example, the stress state of main roof rocks can be analyzed by static equilibrium method:

To rock B,

$$\begin{aligned}&\sum F_x = 0\\&\sum F_y = 0\\&\sum M_{B'} = 0,\end{aligned} \tag{3}$$

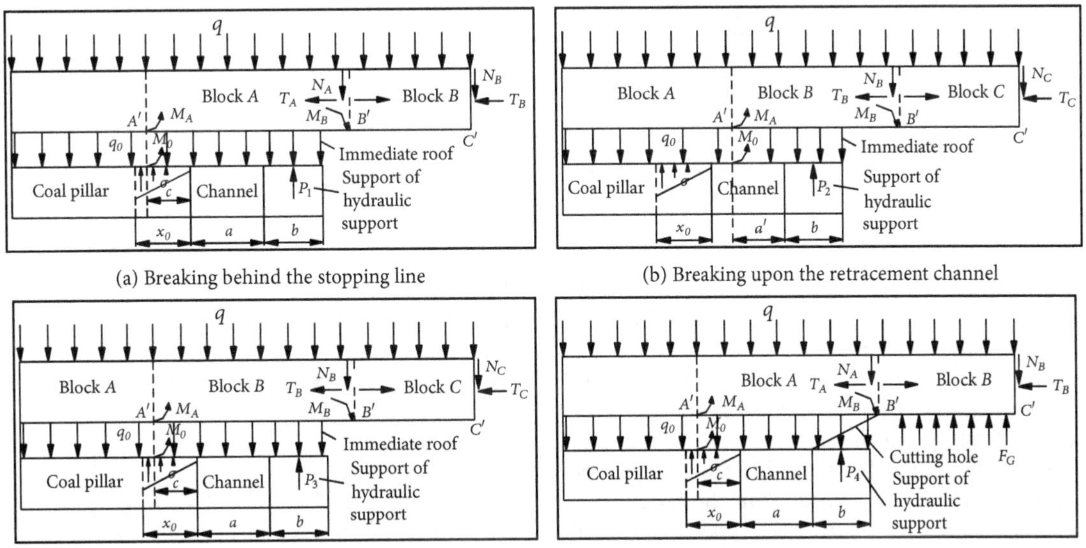

(a) Breaking behind the stopping line (b) Breaking upon the retracement channel

(c) Breaking upon the coal wall (d) Breaking with roof cutting

FIGURE 8: Mechanics model of main roof breaking upon retracement channel.

and then

$$T_A = T_D$$
$$N_A = N_B + qL \tag{4}$$
$$M_B + T_B(h - \Delta S_C) - N_B L - \frac{qL^2}{2} = 0,$$

and therefore

$$N_B = \frac{M_B + T_A(h - \Delta S_B) - qL^2/2}{L}, \tag{5}$$

$$N_A = \frac{M_B + T_A(h - \Delta S_B) + qL^2/2}{L}. \tag{6}$$

To rock A,

$$\sum M_A = 0, \tag{7}$$

and then

$$M_A + M_0 + P_1\left(c + a + \frac{b}{2}\right) + \int_0^c \sigma(c - x)\,dx + T_A(h - \Delta S_A) - M_B - \frac{qL^2}{2} - \frac{q_0(c + a + b)^2}{2} - N_A L = 0. \tag{8}$$

The horizontal thrust can be calculated as follows:

$$T_A = T_B = \frac{qL}{2(h - \Delta S_B)}, \tag{9}$$

where h is the thickness of main roof, m; L is the lateral breaking span of basic roof, m; ΔS_A is the subsidence of rock A at point B′; ΔS_B is the subsidence of rock B at point C′. Then P_1 can be solved as follows:

$$P_1 = \frac{\left[2M_B + qL^2 + q_0(c + a + b)^2/2 + qL(\Delta S_A - \Delta S_B)/[2(h - \Delta S_B)] - M_A - M_0 - \int_0^c \sigma(c - x)\,dx\right]}{(c + a + b/2)}. \tag{10}$$

In the same way, according to the mechanical models shown as Figures 8(b)–8(d), P_2, P_3, and P_4 can be solved as follows:

$$P_2 = \frac{\left[2M_B + qL^2 + q_0(a' + b)^2/2 + qL(h - \Delta S_C)/[2(h - \Delta S_B)] - M_A - M_0 - q'L/2\right]}{(a' + b/2)}, \tag{11}$$

$$P_3 = \frac{\left[2M_B + qL^2 + q_0(c+a+b)^2/2 + qL(\Delta S_B - \Delta S_C)/[2(h-\Delta S_B)] - M_A - M_0 - \int_0^c \sigma(c-x)\,dx\right]}{(c+a+b/2)}, \quad (12)$$

$$P_4 = \frac{\left[M_B(L+c+a+b) + q(c+a+b)^2/2 + qL(c+a+b)/2 + q_0(c+a+b)^2/2 + q(c+a+b)/2 - F_G(c+a+b) - M_A - M_0 - qL/4 - \int_0^c \sigma(c-x)\,dx\right]}{(c+a+b/2)}. \quad (13)$$

Comparing formula (11) and formula (12), $P_3 < P_2$ is easy to get. Because the residual bending moment at the fracture of rock beam is usually very small, which can be regarded as 0, M_B=0 in Formula (10) and M_A=M_B=0 in formulas (11) and (12). Then, through the subtraction between formula (12) and formula (10), it can be got that $P_1 < P_3 < P_2$. Generally, the roof suspension length behind the hydraulic supports under roof cutting condition is shorter than the no cutting condition, and $F_G \geq 0$. So $P_4 < P_1 < P_3 < P_2$ can be got through the subtraction between formula (13) and formula (10), which shows that the working pressure of hydraulic support decreases obviously after roof cutting.

4.2. Deformation Analysis of Roof. When the retracement channel is stable under various support types, the roof can be regarded as a deformable body that accumulates a certain amount of elastic energy, and the subsidence of it can be solved according to energy principle and variational method [23, 24]. The simplified mechanical model of retracement channel's immediate roof is shown in Figure 9. The left boundary is fixed; the right boundary is suffered pressure P_{z2}; the upper boundary is given certain deformation; the lower boundary is suffered support P_{z1} of anchor cables and support P of hydraulic supports; θ is the rotation of basic roof. The strain energy can be calculated as follows:

$$U = \iiint \Delta U \mathrm{d}x\,\mathrm{d}y\,\mathrm{d}z = \frac{1}{2}\iiint \sigma\varepsilon\,\mathrm{d}V, \quad (14)$$

where $\mathrm{d}V$ is an elastic element; σ and ε are the stress and strain of elastic body. The immediate roof deformation of retracement channel can be regarded as a plane strain problem, so the strain energy can be expressed by displacement component:

$$U = \frac{E}{2(1+\mu)}\iiint\left[\frac{\mu}{1-2\mu}\left(\frac{\partial\mu}{\partial x}+\frac{\partial\mu}{\partial y}\right)^2 + \left(\frac{\partial\mu}{\partial x}\right)^2 + \left(\frac{\partial\mu}{\partial y}\right)^2 + \frac{1}{2}\left(\frac{\partial\mu}{\partial x}+\frac{\partial\mu}{\partial y}\right)^2\right]\mathrm{d}V, \quad (15)$$

where u and v are the displacement components.

Assuming that the direct roof elastic body becomes stable after a slight deformation caused by external force, the Lagrange displacement variational equation can be obtained as follows:

$$\delta U = \iiint (X\delta u + Y\delta v)\,\mathrm{d}x\,\mathrm{d}y + \iint \left(\overline{X}\delta u + \overline{Y}\delta v\right)\mathrm{d}s, \quad (16)$$

where X and Y are the components of bulk force; $\overline{X}$ and $\overline{Y}$ are the components of surface force.

According to the boundary conditions of retracement channel, the displacement component on y-axis can be expressed as follows:

$$v = \frac{a}{l}\left[\frac{(4\mu-1)l\theta/16(1-2\mu)^2 - p_{z2}h'(1+\mu)/4E(1-2\mu)}{[(2l/3h')((1-\mu)/(1-2\mu)) + 2h'/3l][(2h'/3l)((1-\mu)/(1-2\mu)) + 2l/3h'] - 1/16(1-2\mu)^2} + \frac{\left[(2h'/3l)((1-\mu)/(1-2\mu)) + 2l/3h'\right]\left[\theta h'/6 - (2l^2\theta/3h')((1-\mu)/(1-2\mu)) + ((1+\mu)/E)\left[(p_{z1}a - h'\rho gl)/2 + (b/a+2b)p\right]\right]}{[(2l/3h')((1-\mu)/(1-2\mu)) + 2h'/3l][(2h'/3l)((1-\mu)/(1-2\mu)) + 2l/3h'] - 1/16(1-2\mu)^2}\right], \quad (17)$$

where l=a+b; E is the modulus of elasticity; μ is the Poisson ratio.

It can be seen that the main parameters affecting the roof deformation of retracement channel are the channel section size, the working resistance of hydraulic supports, the strength of channel support, the immediate roof rotation angle, and the immediate roof thickness. Under the same conditions of the channel section size, the working resistance of hydraulic supports, and the strength of channel support, the immediate roof thickness increases after the roof cutting. Then the gangue pile formed by immediate roof's collapsing and expansion can support main roof and other overlying strata much better, and the rotation angle of main roof also decreases. So the roof subsidence of the retracement channel should also be reduced.

5. Roof Cutting Design of Retracement Channel

5.1. Key Parameters of Roof Cutting. The key parameters of roof cutting mainly contain roof cutting height and roof cutting angle.

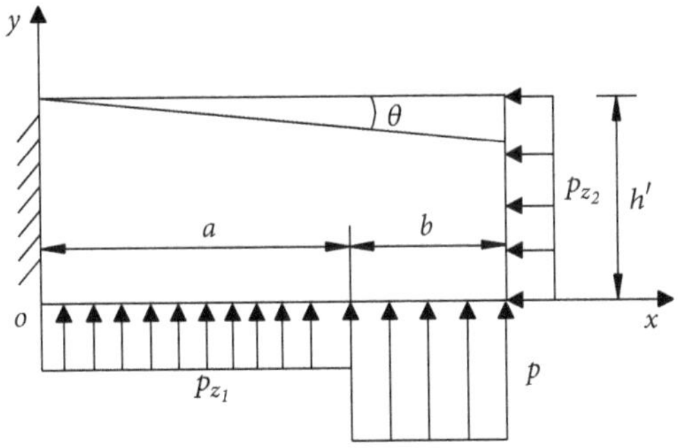

FIGURE 9: Mechanics model of retracement channel direct roof.

5.1.1. Roof Cutting Height. In the haulage entry retaining process of Halagou coal mine, the roof cutting height is designed by the roof rock's crushing expansion coefficient. In order to ensure the effective filling of the goaf by the gangue after the collapse of the roof in the roof cutting range, the minimum roof cutting height can be calculated as follows [25]:

$$H_F = \frac{(H_M - \Delta H_1 - \Delta H_2)}{(K-1)}, \tag{18}$$

where H_F is the minimum roof cutting height, m; H_M is the coal seam height, m; ΔH_1 is the roof subsidence volume, m; ΔH_2 is the floor heave volume, m; K is the roof rock's expansion coefficient. According to the weight calculation of the thickness of each rock layer in the roof cutting range (as shown in formula (19)), the roof rock's expansion coefficient K is 1.31. Then, ignoring the roof subsidence and floor heave, the minimum roof cutting height is calculated approximately to be 6.5m based on the coal seam height, which is 2m.

$$K = \sum_{i=1}^{n} K_i \frac{D_i}{H_M}, \tag{19}$$

where, assuming that the number of the rock layer in the roof cutting range is n ($n \geq 1$), the thickness of each rock layer is D_i ($1 \leq i \leq n$), and the rock's theoretical expansion coefficient Ki ($1 \leq i \leq n$) can be determined by indoor test.

In the practice of entry retaining test, the 6.5m roof cutting height can make the certain goaf zone adjacent to the retained entry be filled well and can form a gangue wall on one side of the retained entry, which can support upon main roof effectively. However, according to the roof lithologic histogram, the top of the roof cutting slit is located in the middle of the fine sand rock layer. When the entry is retained with the mining advance of working face, the immediate roof breaks along the roof cutting surface periodically with a certain suspension length affected by the dynamic mining pressure. But when the mining face reaches the stopping line after the roof cutting of retracement channel, the immediate roof of whole working face will break along the roof cutting surface basically at the same time under the affection of static pressure. So, in order to further reduce the tension of the roof breaking, the roof cutting height is raised to 8.0m in the roof cutting height design of retracement channel, namely, cutting off the fine sand rock layer completely as shown in Figure 12.

5.1.2. Roof Cutting Angle. The design of the roof cutting angle should meet two requirements. One is that the friction of the roof cutting surface should be small when the goaf roof collapses, and the other is that the goaf roof in the roof cutting range can collapse in time behind the hydraulic supports when the working face stops mining [26].

In general, the collapse of the goaf roof in the roof cutting range has two forms. One is the broken falling of lower immediate roof firstly, and the other is the rotary breaking of upper immediate roof then (as shown in Figure 10) [27]. At this condition, the rotary breaking process of upper immediate roof is the key to reduce the frictional effect to roof cutting surface caused by the collapse of goaf roof, and its geometric description is shown in Figure 11, where H_F is the roof cutting height and H_Z is the thickness of the lower immediate roof; the key rock block $OABN$ of upper immediate roof is subjected to a torque m with an axis O point. In order to ensure that the rotation with point O as axis of rock block $OABN$ has no extrusion and friction on the cutting face BM, the line segment OA needs to be longer than the line segment OB; namely, $L_{OA} \geq L_{OB}$.

According to the geometric relationship shown in Figure 11, the length of L_{OA} and L_{OB} can be calculated as follows [18]:

$$\begin{aligned} L_{OB} &= \sqrt{L_S{}^2 + (H_F - H_Z)^2} \\ \frac{L_{OA}}{\sin \angle 3} &= \frac{L_{OB}}{\sin \angle 4} \\ \angle 4 &= \angle 1 = \pi - \angle 2 - \angle 3 \\ \angle 2 &= \arctan \frac{H_F - H_Z}{L_S} \\ \frac{L_{OB}}{\sin \angle 5} &= \frac{L_{BM}}{\sin \angle 6} = \frac{L_{MO}}{\sin \angle 3} \\ L_{BM} &= H_F - \frac{K \cdot H_Z - H_M}{\cos \beta}. \end{aligned} \tag{20}$$

During the entry retaining process of 12201 working face in Halagou coal mine, the lower mudstone layer can collapse directly when it is behind the hydraulic supports and exposed in the goaf, but the upper layers in the roof cutting range usually break with rotation. According to the above formulas, the minimum roof cutting angles of upper coal layer, mudstone layer, and fine sandstone layer are calculated to be 23°, 17°, and 12°, respectively.

In addition, the hydraulic support type of 12201 working face is TAGOR10660/11.3/22.3, the top beam length of this support is 4.8m, and the support maximum length is 5.8m. And, through field observation, the maximum suspension length of the immediate roof behind the hydraulic supports is 2m in the mining process. So, in order to ensure that the roof in the roof cutting range can collapse in time when the working face stops mining, the roof cutting slit should extend 2m behind the hydraulic supports at least (as shown

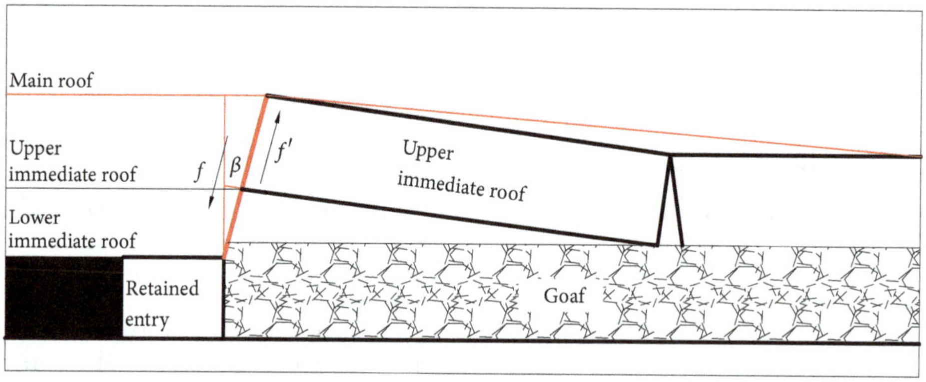

FIGURE 10: Breaking sketch map of upper immediate roof.

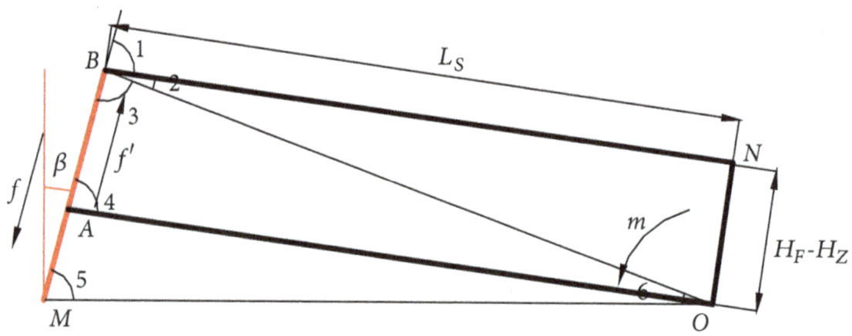

FIGURE 11: Breaking geometric description of upper direct roof.

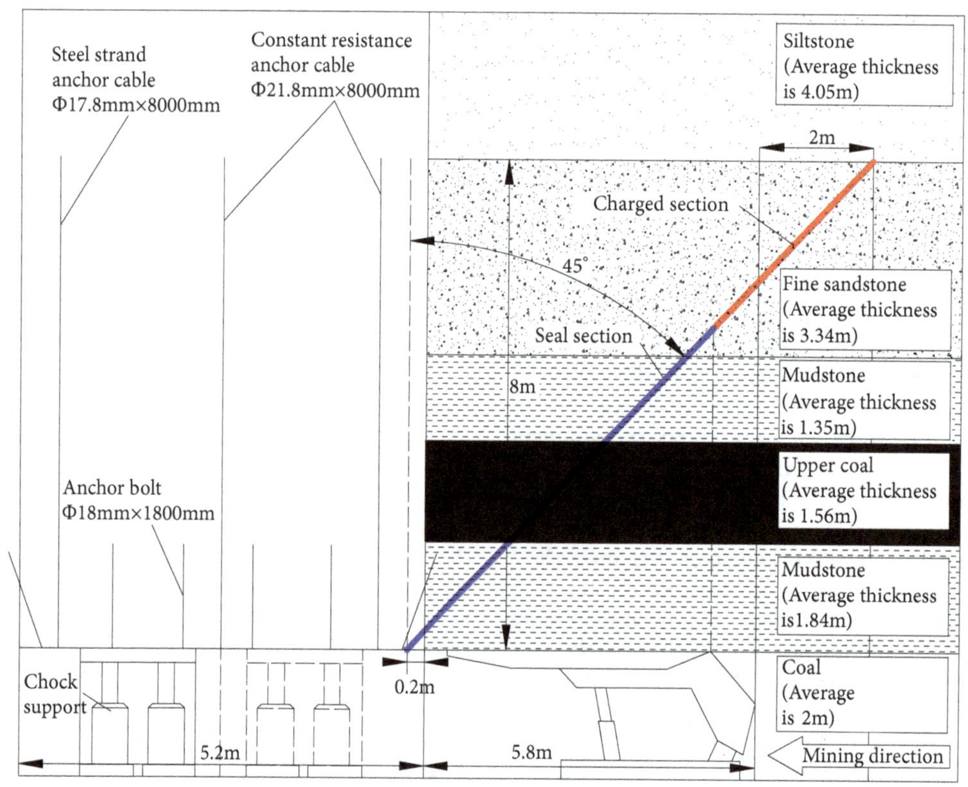

FIGURE 12: Roof cutting design of 12201 working face.

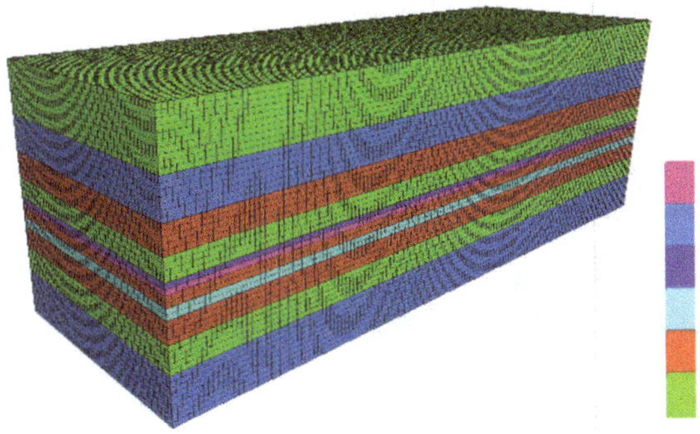

FIGURE 13: Numerical calculation model.

in Figure 12). Then the roof cutting angle can be calculated as follows:

$$\theta_F \geq \arctan \frac{L_D + L_J + d}{H_F}, \tag{21}$$

where L_D is the maximum suspension length of immediate roof, m; L_J is the maximum length of the hydraulic support, m; d is the distance between the cutting holes and the retracement channel principal wall, which is 200mm in this study. Taking relative parameters into formula (21), the result can be got as $\theta_F \geq 45^\circ$. At the same cutting height condition, the larger the cutting angle is, the deeper the cutting hole is and the larger the workload of roof cutting is. And if the roof cutting angle is too large, the roof leakage of working face during the connection process of retracement channel and working face may happen easily. So the roof cutting angle is designed as 45° finally, as shown in Figure 12.

5.2. Analysis of the Roof Cutting Effect. In view of the above roof cutting design, the relevant parameters can be taken into the formulas in Section 4 to verify the design effect. At the same time, according to the geological conditions of 12201 working face in Halagou coal mine, the design results can also be verified through the numerical simulation by the software FLAC 3D. So, in this study, we use two ways of mechanical analysis and numerical simulation to analyze the roof cutting effect in retracement channel.

5.2.1. Calculation Analysis. According to the field geological conditions of 12201 working face, the relevant parameters are assigned as follows: c=0.1MPa, φ=18°, k=2, γ=25KN/m^3, H=80m, p_X=0.04MPa, m=2m, and L=19m. Then the width of the limit equilibrium area of the mining stopping protected coal pillar is calculated to be 4.6m. Taking the relevant parameters about the size of the retracement channel and the specifications of hydraulic support into formulas (10)-(13) and in order to get the minimum value of P_1, P_2, P_3, and P_4, let a'=0 in the value range of $0 \leq a' \leq a$ and $c = x_0$ in the value range of $0 \leq c \leq x_0$. The calculation results of P_1, P_2, P_3, and P_4 are 2.284MN, 4.028MN, 3.146MN, and 1.594MN, respectively; it can be seen that pressure of hydraulic support reduces significantly after roof cutting. In addition, according to the support design of the retracement channel, the strength of the channel support P_{z1}=0.6MPa, and the strength of the hydraulic support p=1.2MPa. Taking the relevant parameters under no cutting condition into formula (17), the roof subsidence is calculated to be 0.26m. Meanwhile when considering the relevant parameters under roof cutting condition, the calculation result is 0.11m. Namely, the subsidence of roof also reduces obviously after roof cutting.

5.2.2. Numerical Simulation Analysis. The numerical model is built by the FLAC 3D software based on the geological conditions of 12201 working face in Halagou coal mine. The model size is 100m×50m×50m, which contains the 30m thick roof, the 18m thick floor, the 60750 grid units, and the 70677 nodes (as shown in Figure 13). After the completion of the model, the working face mining simulation is carried out, and the vertical stress distributions of the retracement channel surrounding rock under roof cutting and no cutting conditions are analyzed comparatively.

As shown in Figure 14, when the mining face is far away from the retracement channel, the stress distribution of the surrounding rock is symmetrical and the stress is relatively higher under no cutting condition. But the stress distribution is obviously changed after roof cutting, and the surrounding rock of the retracement channel presents the pressure relief state. In this condition, the stress distribution is asymmetrical, and the vertical stress of the channel vice wall is greater than that of the channel principal wall. With the advance of mining face, when the mining working face is closer to the retracement channel (5m), the channel principal wall and the mining wall are all in the high stress state under no cutting condition. So the rib spalling and large deformation phenomena can easily take place in the working face and retracement channel, which is bad to the working face retracement process. Meanwhile, under the roof cutting condition, the channel principal wall and the mining wall are all in the low stress state. Although there is a stress concentration area in the left pillar of the retracement channel, it is far from the channel vice wall. So the retracement channel is relatively stable under this condition. When the mining face is connected with the retracement channel, the channel

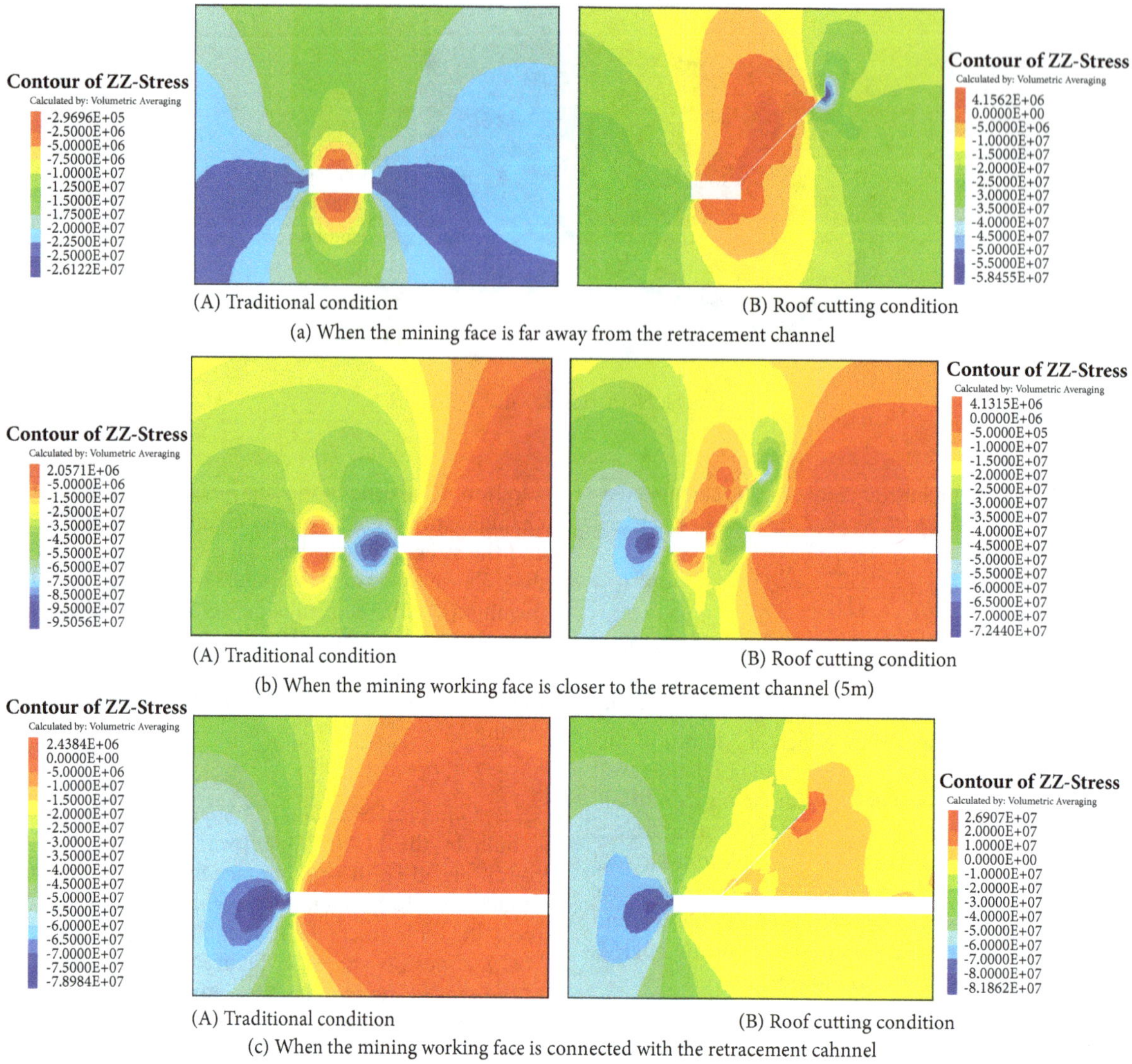

Figure 14: Numerical calculation results of vertical stress distributions.

vice wall is in the high stress state under either the no cutting condition or the roof cutting condition. But, under the roof cutting condition, the stress concentration area is smaller, the peak value is lower, and the distance between the stress peak and retracement channel is longer. So the retracement channel has a better stability under the roof cutting condition.

6. Field Engineering Test

According to the above analysis, the roof cutting angle is 45°, the roof cutting height is 8m, and the depth of roof cutting hole is 11.3m in the field test of RCRC about 12201 working face in Halagou coal mine. The roof of the retracement channel is supported by anchor bolts, anchor cables, and W steel strips. The specification of anchor bolt is ϕ16mm×1800mm, and the layout spacing is 900mm×1000mm.The specification of W steel strips is 3mm×230mm×4800mm. The anchor cables are installed as 3 columns in total on the roof. The column on the channel vice wall side is common steel strand anchor cable, whose specification is ϕ17.8mm×8000mm. The other two columns are constant resistance large deformation anchor cable, whose specification is ϕ21.8mm×8000mm. The row spacing of anchor cables is 1500mm. The original temporary reinforcement support of retracement channel needs two rows of chock supports, and only one row can be left along the channel vice wall under the roof cutting condition (as shown in Figure 12) [28].

As shown in Figure 12, in the roof cutting blasting of this test, only charging the explosive in the section of blasting hole which is behind the hydraulic supports when the working face stops mining (upper 4.2m), the other section of the hole is only treated as sealing section (lower 7.1m). That can ensure that the roof upon the hydraulic supports have a certain integrity. Through the field blasting test, the class 2 coal mine permissible emulsion explosive with the specification being ϕ32mm×200mm is chosen to carry out the roof presplitting cutting, the blasting spacing is 1000mm, and the explosive consumption of each blasting hole is 20 volumes.

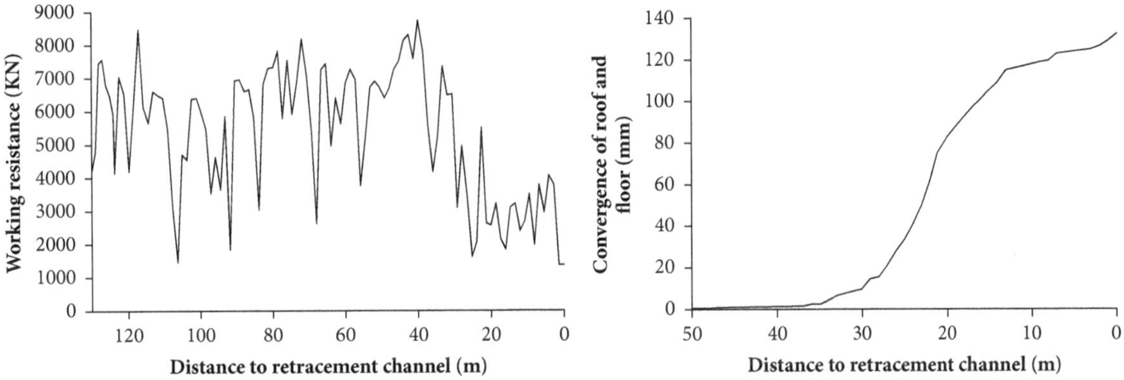

(a) Working resistance of hydraulic support

(b) Convergence of roof and floor

FIGURE 15: Field monitoring results.

In the connection process between working face and retracement channel, the working resistance of hydraulic support and the convergence of roof and floor at the middle of retracement channel, where the deformation is the most severe, are observed and recorded as Figure 15. It can be seen that the working resistance of the hydraulic support reaches a peak value of 8741KN when the distance between working face and retracement channel is 38m. Then the working resistance slumps during the distance range of 32m-15m, and the minimum value is 1612KN. In the distance range of 15m-4m subsequently, the working resistance rises slightly, and the maximum value reaches 3782KN. Finally, until the retracement of hydraulic support is completed, the working resistance is maintained at 1361KN. For the convergence of roof and floor, which begins to increase obviously at the distance of working face and retracement channel, is 37m, it increases sharply in the distance range of 28m-13m and then becomes stable. At the last stage, there are two slight rises of the convergence at the distance being 8m and 3m, and the final convergence of roof and floor is 132.5mm.

In addition, through the field observation, it is found that when the working face and the retracement channel are connected, the immediate roof behind the hydraulic support collapses along the roof cutting hole's explosive charged section firstly. Then, after the retracement of hydraulic support, the triangular remnant immediate roof upon the original hydraulic support collapses along the tendency direction of working face with the 2m-3m break step. Through the whole retracement process, the deformation of the retracement channel is small, and the working condition of hydraulic support is good with no overload phenomenon (as shown in Figure 16).

FIGURE 16: Scene of mining stopping.

7. Conclusion

(1) According to the fault line position of the main roof, the overburden strata structure at the stopping line can be divided into three types: breaking behind the stopping line, breaking upon the retracement channel, and breaking upon the coal wall, followed by putting forward the RCRC technology. Based on the statics theory and energy theory, the stress state of hydraulic support and roof deformation mechanism of retracement channel are analyzed to judge the stability of surrounding rock. According to the calculation and comparison analysis, under the roof presplitting cutting condition, the pressure of hydraulic support is lower than the other three traditional types. And the pressure of hydraulic support and the subsidence of the channel roof are all closely related to the thickness of the immediate roof and the weighting step of the working face.

(2) On the basis of the analysis about the overburden strata structure at stopping line and the roof cutting effect, the key parameters of the RCRC in 12201 working face of Halagou coal mine are designed as follows: the roof cutting height is 8m, the roof cutting angle is 45°, and the depth of the roof cutting blasting hole is 11.3m. Through the verified mechanical calculation and numerical simulation, the working resistance of hydraulic support and the subsidence of channel roof all reduce obviously through the channel roof's presplitting cutting, and the pressure release effect of surrounding rock is remarkable.

(3) According to the field supporting condition and the blasting test conclusion, the spacing of roof cutting blasting hole is 1000mm, the section of the hole which is charged with explosive is the upper 4.2m section, the lower 7.1m section of the hole is treated as seal section, the explosive specification is ϕ32mm×200mm, and the explosive consumption of single hole is 20 volumes. It is verified by the field engineering

test that the working resistance of hydraulic support at the middle of the working face finally remains at 1361KN, and the maximum roof subsidence of retracement channel is 132.5mm during the working face retracement process. Overall, the deformation of the retracement channel is small, and the working condition of hydraulic support is good with no overload phenomenon through the whole retracement process.

Conflicts of Interest

The authors declare that they have no conflicts of interest.

References

[1] B. H. G. Brady and E. T. Brown, *Rock Mechanics for Underground Mining*, Chapman and Hall, London, UK, 1992.

[2] G. D. Zhang and F. Yin, "Study on safety withdrawing technology of hydraulic powered supports from fully mechanized coal mining face," *Coal Science and Technology*, vol. 40, no. 8, pp. 14–17, 2012.

[3] W. Zhen, *Study on Surrounding Rock Control of Dismantling Tunnel in Fully-Mechanized Working Face*, Shandong University of Science and Technology, 2009.

[4] L. Wang, S. W. Jiang, and N. Liu, "Research on new technology of hydraulic support removal in fully mechanized mining face," *China Energy and Environment Protection*, vol. 39, no. 10, pp. 163–166, 2017.

[5] J. M. Ma, M. C. Shao, and J. S. Shi, "Research on supports retracement process of fully mechanized coal mining face," *Coal Science & Technology Magazine*, vol. 44, no. S1, pp. 21–23, 2016.

[6] Y. L. Yang, Z. H. Xu, and Y. M. Li, "Research on hydraulic supports withdrawing technique in fully mechanized mining face with sublevel caving with extra thick coal seam," *Coal Science and Technology*, vol. 44, no. 11, pp. 51–55, 2016.

[7] J. C. Kang, *Study on Key Technologies in Recovery Room of Large Mning Height Face*, Henan Polytechnic University, 2012.

[8] R. Varden, R. Lachenicht, A. G. P. Thompson Jr, and E. Villaescusa, "Development and implementation of the Garford Dynamic bolt at the Kanowna Belle mine," in *Proceedings of the 10th Underground operators conference*, pp. 95–104, AusIMM, Launceston, Melborune, Australia, 2008.

[9] L. Xi, D. N. Bian, and L. Li, "Practice and discussion about establishing retreat channel in fully mechanized coal mining face of chengjiao coal mine," *Shanxi Coking Coal Science and Technology*, vol. 6, no. 6, pp. 29–31, 2014.

[10] P. F. Gou, J. C. Tang, and D. Wang, "Research on supporting technology of fully excavated large section retractable channel in fully mechanized mining face," *Energy Technology and Management*, vol. 1, no. 1, pp. 19–21, 2012.

[11] J. H. Zhang, "Study on movement law of surrounding rock and support adaptability of the broken roof in return channel," *Coal Science & Technology Magazine*, vol. 43, no. 12, pp. 28–31, 2015.

[12] M. C. He, X. G. Ma, F. L. Niu et al., "Adaptability research and application of rapid gob-side entry retaining formed by roof cutting and pressure releasing with composite roof and medium thick coal seam," *Chinese Journal of Rock Mechanics and Engineering*, vol. 37, no. 12, pp. 1–15, 2018.

[13] M.-C. He, "Progress and challenges of soft rock engineering in depth," *Journal of the China Coal Society*, vol. 39, no. 8, pp. 1409–1417, 2014.

[14] B. S. Chi, K. F. Zhou, and M. C. He, "Optimization research on supporting parameters of roof cutting entry retaining with large mining height face," *Coal Science & Technology Magazine*, vol. 45, no. 8, pp. 128–133, 2017.

[15] Z. Ma and Y. F. Yang, "Study on rapid equipment withdrawing technology with hydraulic powered support of fully mechanized coal mining face as shield support," *Coal Engineering*, vol. 45, no. 4, pp. 38–40, 2013.

[16] H. Yao, "Study and application of new channel support and pressure relief cut-through technology for fully mechanized working face in shendong mining area," *Safety in Coal Mines*, vol. 48, no. 3, pp. 37–39, 2017.

[17] M. C. He, Z. Q. Song, and A. Wang, "Theory of longwall mining by using roof cutting shortwall team and 110 method," *Coal Science & Technology Magazine*, vol. 1, no. 1, pp. 1–9, 2017.

[18] X. G. Ma, M. C. He, J. D. Sun et al., "Research on the design of roof cutting parameters of non coal pillar gob-side entry retaining mining with roof cutting and pressure releasing," *Geotechnical and Geological Engineering*, no. 8, pp. 1–16, 2018.

[19] C. J. Hou and N. J. Ma, "Stress in in-seam roadway sides and limit equilibrium zone," *Journal of China Coal Sociery*, vol. 14, no. 4, pp. 21–29, 1989.

[20] X. Ma, M. He, J. Wang, Y. Gao, D. Zhu, and Y. Liu, "Mine strata pressure characteristics and mechanisms in gob-side entry retention by roof cutting under medium-thick coal seam and compound roof conditions," *Energies*, vol. 11, no. 10, 2018.

[21] P. Gong, Z. Ma, R. R. Zhang, X. Ni, F. Liu, and Z. Huang, "Surrounding rock deformation mechanism and control technology for gob-side entry retaining with fully mechanized gangue backfilling mining: A Case Study," *Shock and Vibration*, vol. 2017, Article ID 6085941, 15 pages, 2017.

[22] M. C. He, S. Y. Chen, and Z. B. Guo, "Control of surrounding rock structure for gob-side entry retaining by cutting roof to release pressure and its engineering application," *Journal of China University of Mining Technology*, vol. 46, no. 5, pp. 959–969, 2017.

[23] H. B. Yang, Z. G. Wang, Y. H. Zhao et al., "Study on roof control technology for withdrawal passageway in fully mechanized coal face with large mining height," *Mining Safety and Environmental Protection*, vol. 41, no. 5, pp. 59–61, 2014.

[24] L. St-Pierre, F. P. Hassani, P. H. Radziszewski, and J. Ouellet, "Development of a dynamic model for a cone bolt," *International Journal of Rock Mechanics and Mining Sciences*, vol. 46, no. 1, pp. 107–114, 2009.

[25] X. G. Ma, M. C. He, and J. Wang, "Study on design of chamber supporting behind cutting hole under condition of gob-side entry retaining in 8304 Face of Yanya coal mine," *Coal Technology*, vol. 36, no. 12, pp. 30–33, 2017.

[26] Z. Guo, J. Wang, T. Cao, L. Chen, and J. Wang, "Research on key parameters of gob-side entry retaining automatically formed by roof cutting and pressure release in thin coal seam mining," *Journal of China University of Mining and Technology*, vol. 45, no. 5, pp. 879–885, 2016.

3

Neural Networks Applied to the Wave-Induced Fatigue Analysis of Steel Risers

João P. R. Cortina, Fernando J. M. de Sousa, and Luis V. S. Sagrilo

Laboratory of Analysis and Reliability of Offshore Structures, Civil Engineering Program/COPPE, Federal University of Rio de Janeiro, Rio de Janeiro, Brazil

Correspondence should be addressed to João P. R. Cortina; jpr.cortina@gmail.com

Academic Editor: Lotfi Senhadji

Time domain stochastic wave dynamic analyses of offshore structures are computationally expensive. Considering the wave-induced fatigue assessment for such structures, the combination of many environmental loading cases and the need of long time-series responses make the computational cost even more critical. In order to reduce the computational burden related to the wave-induced fatigue analysis of Steel Catenary Risers (SCRs), this work presents the application of a recently developed hybrid methodology that combines dynamic Finite Element Analysis (FEA) and Artificial Neural Networks (ANN). The methodology is named hybrid once it requires short time series of structure responses (obtained by FEA) and imposed motions (evaluated analytically) to train an ANN. Subsequently, the ANN is employed to predict the remaining response time series using the prescribed motions imposed at the top of the structure by the floater unit. In this particular work, the methodology is applied aiming to predict the tension and bending moments' time series at structural elements located at the top region and at the touchdown zone (TDZ) of a metallic riser. With the predicted responses (tensions and moments), the stress time series are determined for eight points along the pipe cross sections, and stress cycles are identified using a Rainflow algorithm. Fatigue damage is then evaluated using SN curves and the Miner-Palmgren damage accumulation rule. The methodology is applied to a SCR connected to a semisubmersible platform in a water depth of 910 m. The obtained results are compared to those from a full FEA in order to evaluate the accuracy and computer efficiency of the hybrid methodology.

1. Introduction

As the oil industry faces the challenge of production on harsh environments, the design of offshore structures requires progressively a better understanding of the environment and the use of more sophisticated numerical methods, capable of representing the nonlinearities involved in the dynamic analysis of such structures. Both requirements, however, can lead to very high computational costs.

This is a consequence of the combination of refined numerical models and the need of several long stochastic time domain simulations that are needed to assure reliable statistics for the response parameters investigated. This aspect can be crucial in the wave-induced fatigue analysis of risers or in early design stages, where many structural configurations should be analyzed.

Depending on the water depth, geometric configuration, finite element mesh refinement, use of nonlinear physical properties, and nonlinear beam element formulation, a single 3-hour FEM-based stochastic dynamic analysis of a SCR can demand more than 3 hours of computational processing in a modern desktop computer. As, in the design practice, the wave-induced riser fatigue analysis requires the stochastic dynamic analysis for many load cases (sea state conditions), the computer burden of this analysis is tremendous.

One of the most promising alternatives to evaluate dynamic responses of marine structures with a significant reduction in computational time is the utilization of hybrid methods combining FEA with Artificial Neural Networks (ANNs) [1–7]. The basic idea of these methods is to employ the remarkable capacity of learning, generalization, and prediction of neural networks to replace the onerous numerical

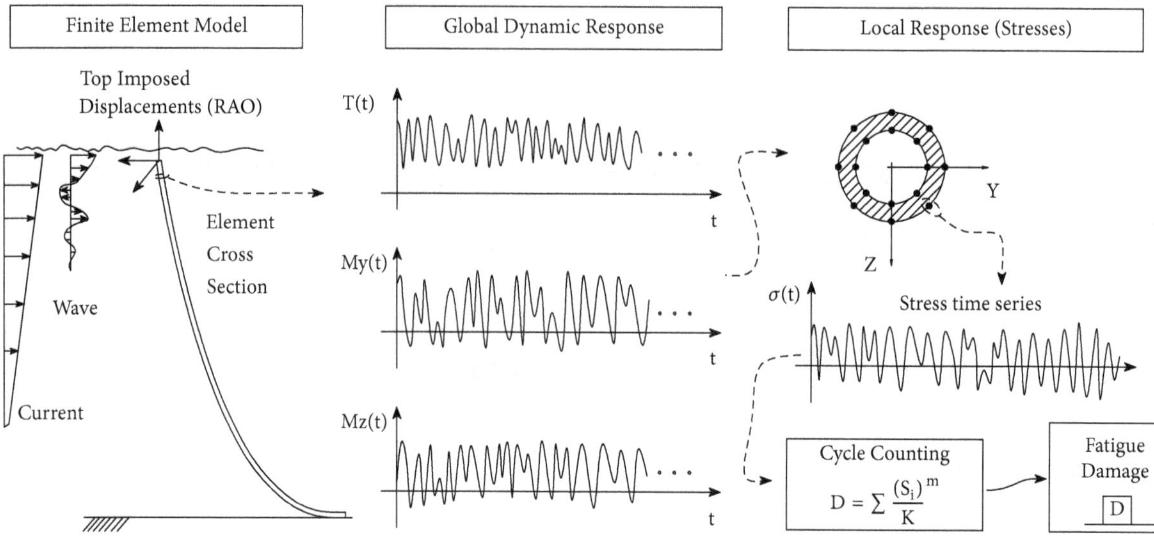

FIGURE 1: Steel riser fatigue procedure.

integration of a time domain dynamic analysis by finite elements method. In the most usual application, short time series of motions (imposed at the top of the structure by the floater unit) and the corresponding induced structure responses (provided by the FEA) can be used to train an ANN. Then, once the floater motions can be evaluated analytically, the remaining of the response time series is predicted by the ANN.

Guarize et al. [1] applied the hybrid ANN-FEA methodology in the prediction of the dynamic top tension of a mooring line and in the evaluation of the DNV-LRFD utilization factor [8] in the sagbend zone of a SLWR (Steel Lazy-Wave Riser). Both structures were connected to a FPSO (Floating, Production, Storage, and Offloading unit). The predictions showed very good agreement with the FEA results, and the computational time was reduced by a factor of about 20 times. Pina et al. [2] presented hybrid methods which use dynamic autoregressive models applied to the prediction of risers and mooring lines' top tensions. Christiansen et al. [3, 4] used a similar hybrid methodology in the prediction of top tensions for the fatigue assessment of a mooring line with a substantial reduction in the total simulation time. Chaves et al. [5] studied the application of the hybrid method in the prediction of tension and curvatures in the bend stiffener region of a flexible riser and used these results to perform a fatigue verification. Kim [6, 7] employed a NARX-based Volterra model for the prediction of dynamic response of slender marine structures, also obtaining good results.

This work presents an application of a hybrid ANN-FEA method for the prediction of axial tension and bending moments at the top region and at the touchdown zone of a SCR connected to a semisubmersible platform, in order to reduce computational costs of time domain stochastic simulations. To the authors' knowledge, the previous works did not employ a similar approach to predict the load effects for riser points near the TDZ nor to evaluate the total fatigue damage in this region.

The methodology is then used for all load cases (sea states) in order to evaluate fatigue damage in these critical regions. The fatigue results and also the computational costs of the hybrid method and those of the usual FEA approach are compared.

2. Stochastic Time Domain Fatigue Assessment of Metallic Risers

Steel risers are structures composed of welded segments of conventional steel pipes [9] and used for oil or gas production and for gas exportation. The most common configurations are the free hanging catenary (SCR) and the lazy-wave (SLWR, Steel Lazy-Wave Riser).

The wave-induced fatigue analysis of such structures is traditionally divided into two steps: the global and the local analysis. In the global analysis, a nonlinear FE-based time domain stochastic dynamic simulation is employed for each loading case to get the corresponding time series of axial tensions and bending moments along the riser. Usually, the global analysis is performed using a decoupled procedure, where the floater unit motions are computed separately in a first step and then imposed as prescribed displacements at the top of the riser finite element model. Subsequently, in the local analysis, the global responses (axial tension and bending moments) are used to compute the stress time series in some points of the riser cross sections. The procedure is illustrated in Figure 1. For a metallic riser, the normal stress time series $\sigma(t)$ in any point of a cross section can be determined by the following expression [10]:

$$\sigma(t) = \frac{T(t)}{A} + \frac{My(t)\, r \sin(\theta)}{I_y} + \frac{Mz(t)\, r \cos(\theta)}{I_z} \quad (1)$$

where T, My, and Mz are the axial tension and the bending moments, respectively, A and I are the cross sectional area and inertia moments, and r and θ are the radius and the

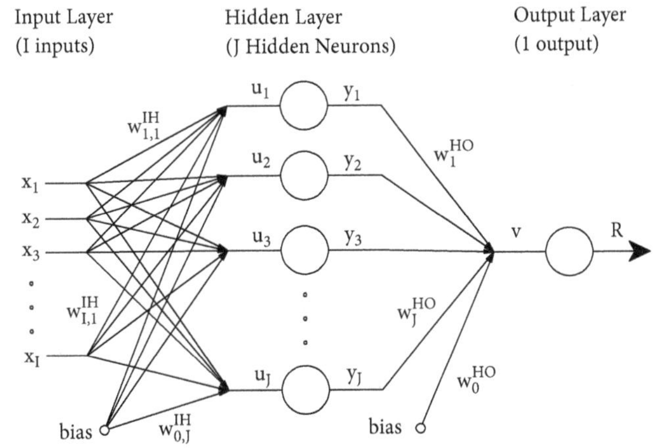

FIGURE 2: ANN architecture.

angle that determines the point where the stress is being computed on the cross section. For fatigue assessment, it is recommended that at least eight points in the internal and external circumferences should be analyzed [10].

After obtaining stresses time series, a cycle counting procedure, such as the Rainflow technique [11], is applied to identify and count the stress cycles for the fatigue damage assessment. The fatigue damage at any point of a cross section is evaluated by means of S-N curves and the Miner-Palmgren damage accumulation rule [12]. Through this procedure, the annual fatigue damage is computed as

$$D_{1-yr} = 2920 \sum_{j=1}^{N_{ST}} \sum_{i=1}^{N_C^j} \frac{\left(S_i^j\right)^m}{K} \cdot \gamma_j \tag{2}$$

where 2920 is the total number of 3-h sea states in one year, N_C^j is the number of stress cycles identified by the Rainflow algorithm in the jth 3h-loading case (sea state), N_{ST} is the total number of loading cases (sea states) considered in the fatigue analyses, S_i^j is the ith stress range of the jth loading case, m and K are the S-N curve parameters, and, finally, γ_j is the annual relative frequency of occurrence of the respective loading case ($\sum_j \gamma_j = 1$). Commonly, the number of loading cases is very large due to the wave scatter diagrams discretization (mainly for the waves periods) and when all wave directions are considered. The fatigue life (in years) is simply the inverse of the annual fatigue damage.

As mentioned before, the large amount of loading cases combined with the need of long response time series for statistical consistency in the identified stress cycles makes the global analysis the most time-consuming phase of riser fatigue assessment. By using ANNs to predict the global riser response, the procedure presented in this work aims to reduce the computational time associated with this specific phase of the metallic riser fatigue verification.

3. Artificial Neural Networks

Artificial Neural Networks (ANNs) constitute a form of mathematical processing of information with architecture and operation similar to the biological neural networks. The main feature of the ANNs is the learning ability, which in practical terms results in applications such as pattern recognition, clustering, classification, optimization, and function approximation, among others [13]. The use of ANNs has been increasing lately to help solving many engineering problems; for instance, see [14–16].

In this work, ANNs are employed to define a mathematical function relating the floating displacements imposed at the riser top with its global responses: axial tension T and bending moments My and Mz. For such application, the ANN must be trained, using the data and results of a very short simulation provided by the FEA. Then the ANN is used to predict longer loading responses time series.

3.1. ANN Architecture. The most usual ANN architecture applied to function approximation uses three layers: input layer, hidden layer, and output layer. The configuration is outlined on Figure 2. The first layer reads the inputs values of the network. The output layer delivers the responses of the network. The hidden and output layers are composed of neurons (numerical functions). Each connection between elements in neighboring layers contains a weight (synapse weighting). Linked to the hidden and output layers there are also bias parameters, with unitary values.

A neuron j in the hidden layer, for instance, receives an input u_j, given by

$$u_j = w_{0,j}^{IH} \cdot 1 + \sum_{i=1}^{I} \left(w_{i,j}^{IH} \cdot x_i\right) \tag{3}$$

where I is the number of network inputs, $w_{0,j}^{IH}$ is the bias weight of the input layer, $w_{i,j}^{IH}$ is the weight between the input i and the respective neuron j in the hidden layer ($j = 1 \ldots J$, J is the total number of hidden layer neurons), and x_i is the i network input. Each neuron returns an output y_j denoted by

$$y_j = \varphi_H\left(u_j\right) = \varphi_H\left(w_{0,j}^{IH} \cdot 1 + \sum_{i=1}^{I} \left(w_{i,j}^{IH} \cdot x_i\right)\right) \tag{4}$$

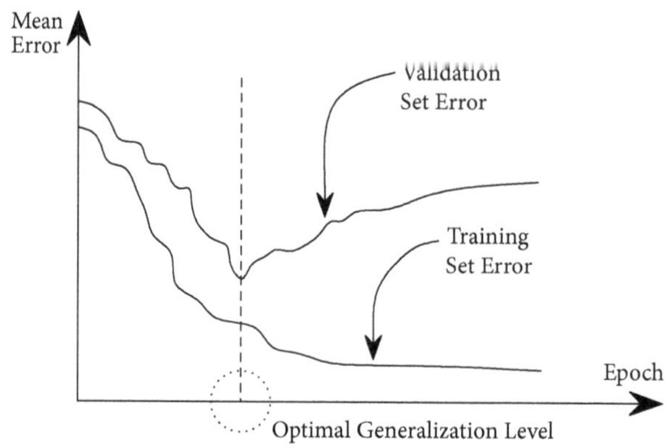

FIGURE 3: Usual behavior of the training and validation errors.

where $\varphi_H(.)$ is the so-called activation function, usually sigmoid functions like hyperbolic tangent and the logistic functions. For an ANN with a single output R, this parameter is determined by

$$R = \varphi_O(v) = \varphi_O\left(w_0^{HO} \cdot 1 + \sum_{j=1}^{J}\left(w_j^{HO} \cdot y_j\right)\right) \tag{5}$$

where $\varphi_O(.)$ is the activation function of the neuron in the output layer, usually a linear function. The remaining parameters are defined similarly to those in the hidden layer (see (3)).

3.2. Training. Considering the ANN previously presented and considering the inputs and outputs as time series, the relation between the inputs x_i and the output R can be written as

$$R(\mathbf{W}, t) = f(\mathbf{W}, \mathbf{x}(t)) \tag{6}$$

where $\mathbf{x}(t)$ is a vector containing the time series of the inputs and $\mathbf{W}$ is a matrix containing all ANN connecting weights. Given a sample set of inputs (time series) and the corresponding outputs, the network weights $\mathbf{W}$ can be obtained in such a way that the network output is in agreement to the given outputs sample. This weights adjustment is made by means of an optimization method and is called neural network training. Commonly the optimization problem is expressed by the minimization of the mean square error E, defined as

$$\begin{aligned} E(\mathbf{W}) &= \frac{1}{N}\sum_{t=1}^{N}(R(\mathbf{W}, t) - v(t))^2 \\ &= \frac{1}{N}\sum_{t=1}^{N}(f(\mathbf{W}, \mathbf{x}(t)) - v(t))^2 \end{aligned} \tag{7}$$

where N is the length (number of data points) of the training sample time series and v(t) is the desired output sample. An iterative optimization procedure, such as the steepest descend method [13], can be employed to obtain optimum values for the weights $\mathbf{W}$. Usually the weights are initialized using small random values. Each iteration in the optimization procedure is called epoch, and the input and output time-series are normalized to facilitate the training procedure [17, 18].

Generally, two types of errors can occur during the network training. One is related to the early training stopping, far from the optimum set of weights and originating high prediction errors. The other happens when the ANN is overtrained and it adapts too much to the training set (overfitting). Both situations lead to high prediction errors for new inputs (nontrained data) [13, 17, 18]. Thus, there is an optimum level of training to be achieved. The procedure applied to avoid these errors is to separate part of the training set for the training itself (weights update) and the other part for validation, with separated error functions. Commonly the network error follows the patterns exhibited in Figure 3. The training process must be stopped when the error for the validation set begins to increase [17]. This detail avoids the overfitting and ensures a good network generalization for new inputs.

4. ANN Model for Global Riser Analysis Responses

The goal of this work is to apply ANNs for predicting the global riser responses, using an ANN as a surrogate model for the computationally expensive FEA. For a given element of the riser finite element mesh, ANNs are trained to predict the axial tension T and bending moments My and Mz independently:

$$\begin{aligned} T(t) &= f_T(\mathbf{W}_T, x(t)) \\ My(t) &= f_{My}\left(\mathbf{W}_{My}, x(t)\right) \\ Mz(t) &= f_{Mz}(\mathbf{W}_{Mz}, x(t)) \end{aligned} \tag{8}$$

where $\mathbf{W}_T$,$\mathbf{W}_{My}$, and $\mathbf{W}_{Mz}$ are the optimal weights achieved during the ANN training phase for the three responses and $\mathbf{x}(t)$ are the ANN inputs, i.e., the prescribed translational and rotational floater motions at the riser top. Given the dynamic nature of the analysis, the memory effect of the system (riser) must be taken into account in the ANN model. This aspect can be accomplished in the model by

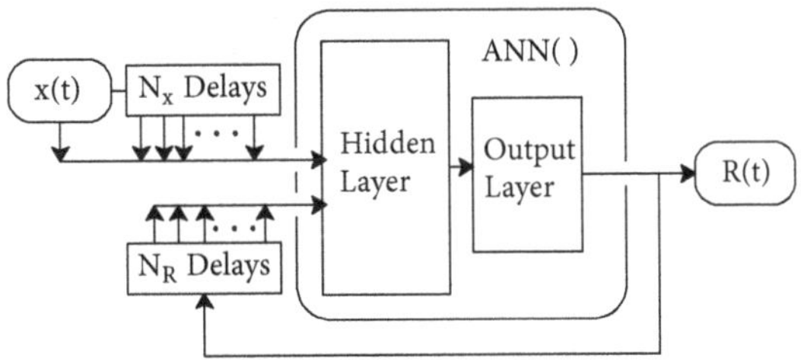

FIGURE 4: ANN-NARX model.

different approaches, e.g., AutoRegressive model (AR) and the EXogenous inputs model (EX). The model employed in this work is a Nonlinear AutoRegressive with eXogenous inputs (NARX) neural network [2]. Then, for the present work, the inputs of a ANN-NARX model consist of the top motions imposed by the floater in the considered time instant, the past values of these displacements, and also the corresponding past values of the respective global response (output). Considering that the time series are discretized at equal time steps Δt, then the ANN-NARX model is defined as

$$R(t) = f_R \begin{bmatrix} \{x(t), x(t-\Delta t), \ldots, x(t-N_x\Delta t)\}, \\ \{y(t), y(t-\Delta t), \ldots y(t-N_y\Delta t)\}, \\ \{z(t), z(t-\Delta t), \ldots, z(t-N_z\Delta t)\}, \\ \{rx(t), rx(t-\Delta t), \ldots, rx(t-N_{rx}\Delta t)\}, \\ \{ry(t), ry(t-\Delta t), \ldots, ry(t-N_{ry}\Delta t)\}, \\ \{rz(t), rz(t-\Delta t), \ldots, rz(t-N_{rz}\Delta t)\}, \\ \{R(t-\Delta t), \ldots R(t-N_R\Delta t)\} \end{bmatrix} \quad (9)$$

where R(t) represents either T(t), My(t), or Mz(t); x(t), y(t), and z(t) represent the global translational motions imposed by the floater; rx(t), ry(t), and rz(t) represent the rotational displacements; N_x, N_y, N_z, N_{rx}, N_{ry}, and N_{rz} are the number of time steps delays for the imposed displacements and N_R is the number of time steps delays of the response R(t). In Figure 4 an ANN implemented with the NARX model is outlined.

The ANNs developed in this work were trained and applied using the Neural Networks Toolbox™ in MATLAB™ [19]. A hyperbolic tangent function was used as activation function for the neurons in the hidden layer, a linear function was used for the neuron in the output layer, the time series were normalized to have zero mean and unitary standard deviation, the weights were initiated randomly, and the training algorithm utilized was the Levenberg-Marquardt [18, 19].

5. Hybrid ANN-FEA Methodology

The main goal of the hybrid ANN-FEA methodology is to be able to conciliate the dynamic model complexity with a reduction of its computational cost. The methodology is named hybrid since a FEA short simulation is used to determine the complex relation between the nonlinear riser response and the top end motions (ANN training). The hybrid ANN-FEA procedure is schematically presented in Figure 5.

Once the floater prescribed motions at the riser top are easily determined by means of the floater RAOs (Response Amplitude Operators), a short length FEA simulation is performed in order to obtain the riser global dynamic response time series. Then, these time series of inputs (top displacements) and outputs (riser responses) are used as training and validation sets for the ANN. After training, the ANN is used to predict the remaining global response, based on the complete time series of the top prescribed motions. In this way, the ANN replaces the onerous FEA numerical integration for a large part of the simulation length.

As usual, in the everyday riser design practice, the low-frequency floater motions were not considered in the present work (only static offsets were taken into account). However, if it is desired to include low-frequency motions in the analysis, the proposed hybrid FEM-ANN methodology can still be used. In this case, it is only necessary to provide time series of motions and responses used for training and verification, and time series of motions used for prediction, which include the low and wave frequency components. These motions can be obtained, for instance, from a simplified 3D floater motion dynamic analysis (see Guarize [1]).

In this work, the hybrid methodology is applied individually for some elements located at the top and TDZ regions of a SCR, which are the most critical points for fatigue. Individual ANNs are trained and used for each element, for each response parameter considered (T(t), My(t), and Mz(t)) and for each loading case. This choice of a specific ANN for each loading case takes into consideration that there is a different "catenary" configuration for each load condition and, as a consequence, the TDZ position can change substantially. In general, SCRs present nonlinear and remarkably high bending moments in the touchdown zone (TDZ), but the TDZ is not the same for all loading cases. For instance, a segment that belongs to the TDZ in a "far" loading case may not be affected by bending in a "near" loading case. Then, considering many loading conditions, the stress response for a given riser point in the touchdown zone is not well behaved as it almost is for the suspended riser segment. Therefore, an ANN trained for predicting the riser stresses at the TDZ, employing the combination of different signals from different sea state conditions, may not lead to good results for sea states not present in the training process.

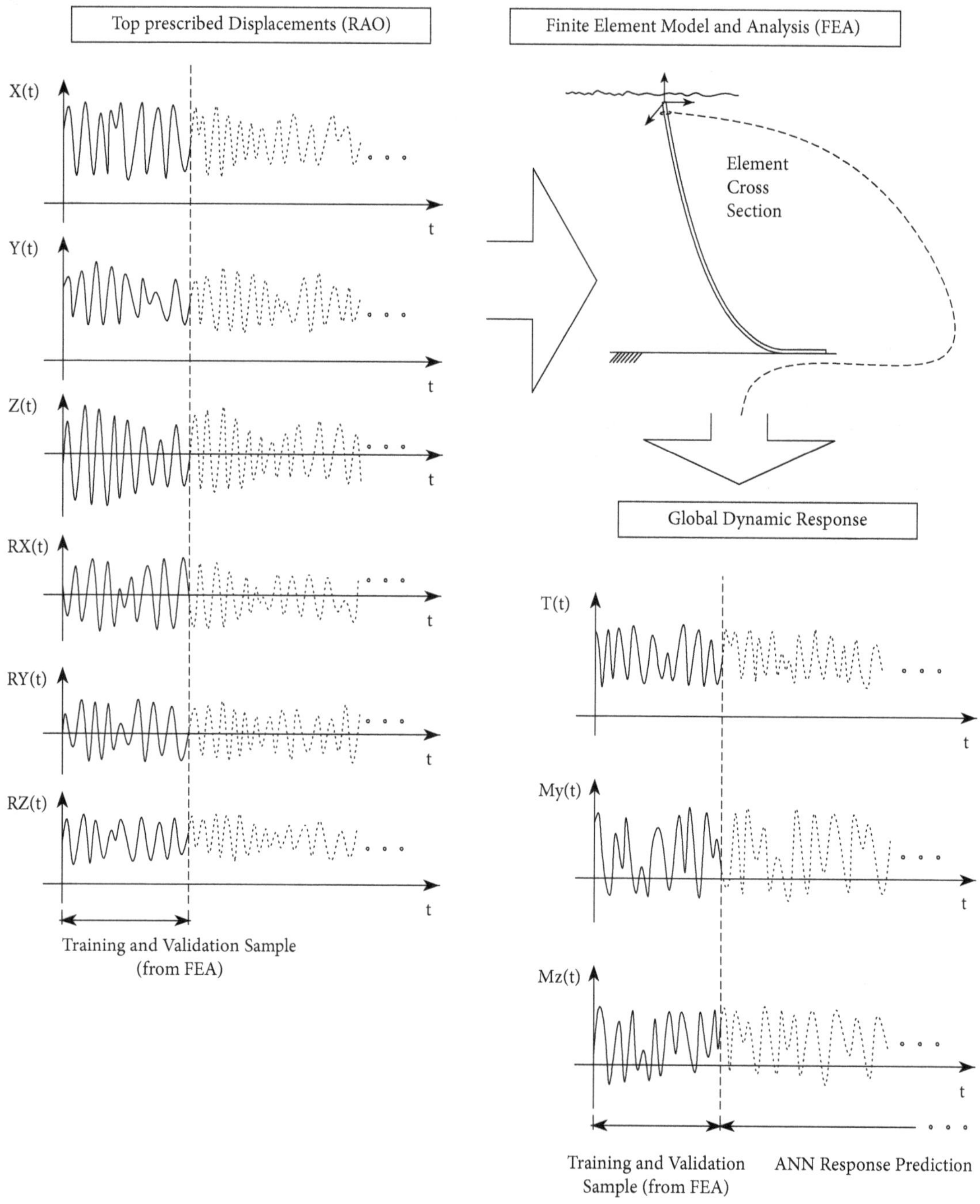

FIGURE 5: Hybrid ANN-FEA method.

6. Case Study

The hybrid ANN-FEA procedure is applied to compute the wave-induced fatigue life of some elements at the top and at the TDZ region of a steel catenary riser (SCR) (see Figure 6). This metallic riser is described by Senra [20] and consists of a 10.7$''$ gas export SCR connected to a semisubmersible platform in a water depth of 910m. The SCR model has a flexjoint in the connection. The physical and geometrical properties of the model are found in Table 1, while the information about the finite element mesh employed is presented in Table 2. The semisubmersible platform heading is 20° (in relation to true north) and the riser azimuth is 192.11° (see Figure 6).

The loading cases selected to perform the fatigue analysis are defined by static offsets and current and wave loadings. The same current profile was adopted for all loading cases considered in the fatigue analysis. This profile is triangular with a superficial velocity of 0.7 m/s, propagating to the south direction. The offset magnitude is a function of the significant wave height Hs of the respective load case, and it is proportional to 2.5% of the SWL (Still Water Level) for the

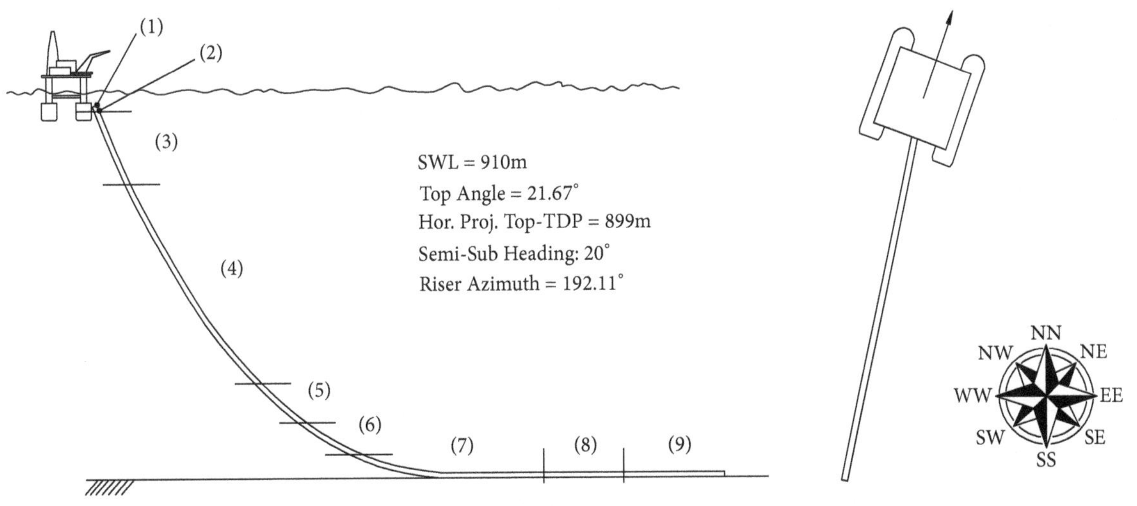

FIGURE 6: SCR model used in the study case.

TABLE 1: SCR properties.

Parameter	Value
External Diameter (m)	0.273 (10.7")
Internal Diameter (m)	0.232 (9.13")
Elasticity Module (kN/m^2)	2.07E+08
Specific Weight (kN/m^3)	77.00
Internal Fluid Weight (kN/m^3)	2.00 (gas)
Top Internal Pressure (MPa)	18.00
Soil vertical stiffness (kN/m)	1000.00
Riser Top Angle	21.67°
Riser Azimuth	192.11°

highest significant wave height of the wave scatter diagram. The offset direction is coincident with the wave propagation direction. The waves selected are divided into 8 incidence directions, with 11 waves (sea states) for each direction. The frequency of occurrence for each incidence direction is presented in Table 3, while the irregular wave parameters and their frequency of occurrence for each direction are found in Table 4. The JONSWAP spectrum was adopted to represent the sea surface elevation. Then, the fatigue analysis requires the irregular wave simulations for 88 loading cases. In fact, this is a simplified representation of wave data which was employed in order to make all comparisons performed in this work possible. Once the current is the same and the offset is defined by the significant wave height, the loading case is identified herein by the wave name. Specifically, a loading case identified as NE03 is a case where the wave has a northeast incidence direction, a significant wave height Hs of 1.25m, a spectral peak period Tp of 7.02s, and an annual frequency of occurrence of 2.17% (Tables 3 and 4).

A usual FEA-based fatigue analysis was initially performed for comparison purposes and it was also used to identify the elements with the lower fatigue life at the top and TDZ regions, ensuring that the hybrid methodology is applied (and evaluated) to the critical structure points. The fatigue assessment of the entire structure was based on the DNV E S-N curve [21].

6.1. ANN Architecture and Training. As it can be observed from the definition of an ANN, the architecture of the network is defined by the number of neurons in the hidden layer and the time-series steps delays. As said before, a specific ANN is established for each response parameter, for each loading case, and for each element along the structure. In this work an iterative process, described below, was employed to define the architecture of each ANN.

It is assumed that the number of delays is the same for all input and output time series, so the two open parameters to define the ANN architecture are the number of step delays and the number of neurons in the hidden layer. Based on initial tests and previous works [1–3, 5], a sequential trial set of numbers for these two parameters is specified. The sets employed were as follows (hidden neurons/delays): 10/10, 10/5, 10/2, 12/10, 12/5, 12/2, 15/10, 15/5, 15/2, 20/10, 20/5, 20/2, 5/10, 5/5, 5/2, 10/1, 12/1, 20/1, 3/10, 3/5, 3/2, 5/30, 5/40, and 5/50. The effectiveness of a given architecture (for each response loading case in a structure point) is assessed by comparing the ANN predicted response (trained with one of the trial sets) with the "true" FEA-based response used for training the network.

The assessment is performed based on statistical results, as described in Algorithm 1. If the trained ANN results satisfy the conditions specified in Algorithm 1, the ANN is taken as the surrogate model for the hybrid method. If not, other attempts are made using the same trial architecture, given the random weights initialization [17, 18]. If the maximum attempts per trial set are reached without complying with the conditions, the following trial set is evaluated, as outlined in Algorithm 1. The acceptable tolerances given in Algorithm 1 were determined by means of initial trial and error tests.

The time series were discretized in steps of 0.25s and the FEM-based simulation (used for training/validation) was

TABLE 2: SCR finite element mesh.

Seg	Length (m)	Number of elements	Element initial length (m)	Element final length (m)
1	1.102	14	0.08	0.08
2	10	112	0.08	0.10
3	200	364	0.10	1.00
4	600	600	1.00	1.00
5	150	171	1.00	0.75
6	150	240	0.75	0.50
7	500	1000	0.50	0.50
8	200	160	0.50	2.00
9	255.847	73	2.00	5.00

```
For each loading case response do
  nset = 1
  While [nt <= MaxTries and nset <= QttSet] do
    Train ANN
    Apply ANN
    if [Dif_μ <= Dmean and Dif_σ <= Dstd and Dif_max <= Dmax and FlagSp = 1] do
      TestFlag = 1
    else do
      TestFlag = 0
    end
    if [TestFlag = 1] do
      nt = 1000, nset = 1000 (successful training!)
    elseif [TestFlag = 0 and nt < MaxTries and nset <= QttSet] do
      nt = nt +1
    elseif [TesteFlag = 0 and nt = MaxTries and nset < QttSet] do
      nt = 1
      nset = nset + 1
    elseif [TesteFlag = 0 and nt = MaxTries and nset = QttSet] do
      nt = 1000, nset = 1000 (unsuccessful training!)
  end
end
```

nt = current number of attempt per trial set
nset = current trial set number
MaxTries = maximum number of attempt per trial set (5)
QttSet = trial sets quantity (24)
Dmean = acceptable difference for mean (5%)
Dstd = acceptable difference for standard deviation (8% for axial tension, 6% for moments)
Dmax = acceptable difference for maximum observed value (300% for axial tension, 200% for moments)
FlagSp = Energy spectrum flag, equal to 1 if no significant energy is observed in high frequencies, for ANN response

$$\text{Dif}_\mu = 100\%.\frac{|\mu_{\text{training}} - \mu_{\text{ANN}}|}{\mu_{\text{training}}}$$

$$\text{Dif}_\sigma = 100\%.\frac{|\sigma_{\text{training}} - \sigma_{\text{ANN}}|}{\sigma_{\text{training}}}$$

$$\text{Dif}_{\text{max}} = 100\%.\frac{|\text{Max}_{\text{training}} - \text{Max}_{\text{ANN}}|}{\text{Max}_{\text{training}}}$$

$\mu_{\text{training}}, \mu_{\text{ANN}}$ = mean values of the training and predicted time series
$\sigma_{\text{training}}, \sigma_{\text{ANN}}$ = standard deviation values of the training and predicted time series
$\text{Max}_{\text{training}}, \text{Max}_{\text{ANN}}$ = observed maximum values of the training and predicted time series

ALGORITHM 1: ANN training procedure.

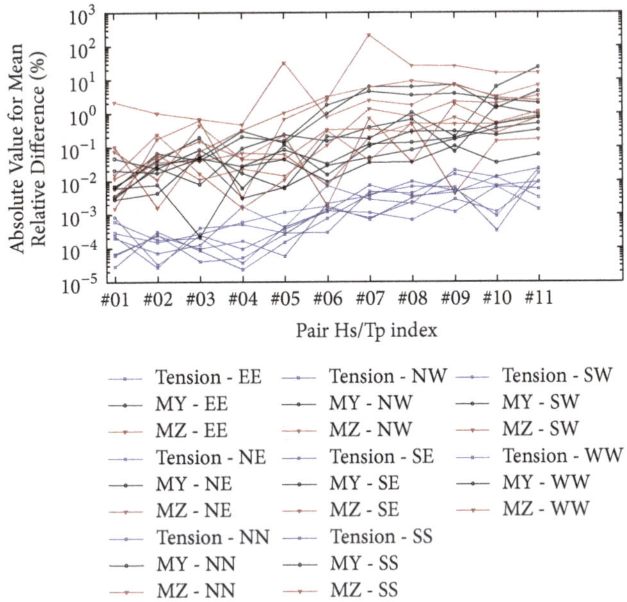

FIGURE 7: Relative differences for the response mean values (critical riser top element).

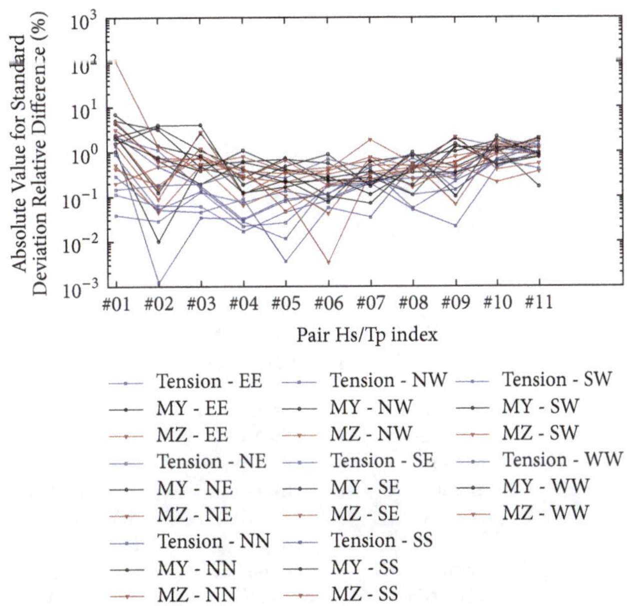

FIGURE 8: Relative differences for the response standard deviation values (critical riser top element).

500s long. The training is done using 80% of the sample, while the remaining 20% is used for validation. The training is stopped when the validation error starts to increase.

The time spent in this batch training procedure is quite low, mainly for those points located close to the riser top, where usually the first ANN architecture trial leads to very good results. For some TDZ response cases, the searching for a good configuration results in more verifications and tests of different ANN configurations; however the training time spent is very low when compared to the full-fatigue FEA (generally speaking, no more than 1 minute of CPU time is spent to train an ANN).

6.2. ANN Predicted Responses. With the trained ANNs available, the riser global responses (axial tension and bending moments) were predicted till 10800 seconds (3 hours), and the obtained results were compared with those obtained by the complete FEA, for the elements with the lowest fatigue lives at the top and TDZ regions, the critical ones. The results are compared through the relative difference for any response statistical parameter estimation or fatigue damage by the generic relationship: $((\text{Hybrid}_{\text{ANN-FEA}}\,\text{Result} - \text{FEA Result})/|\text{FEA Result}|) * 100\%$.

In Figures 7 and 8 the relative differences for the mean (μ) and standard deviation (σ) of the response parameters

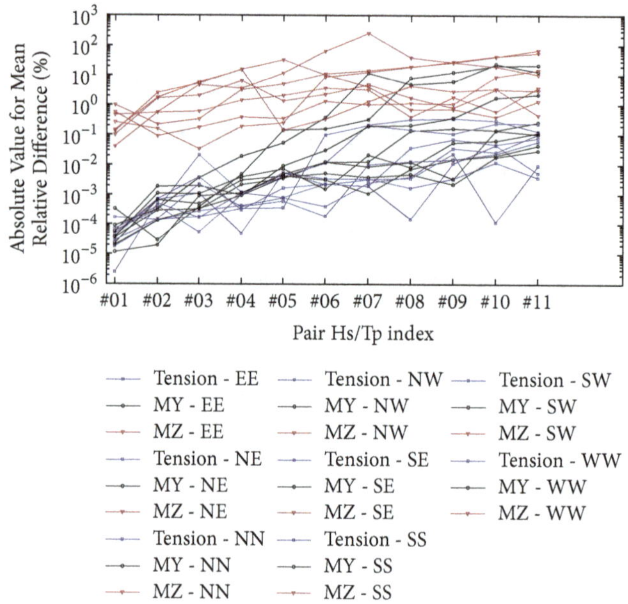

FIGURE 9: Relative differences for the response mean values (critical riser TDZ element).

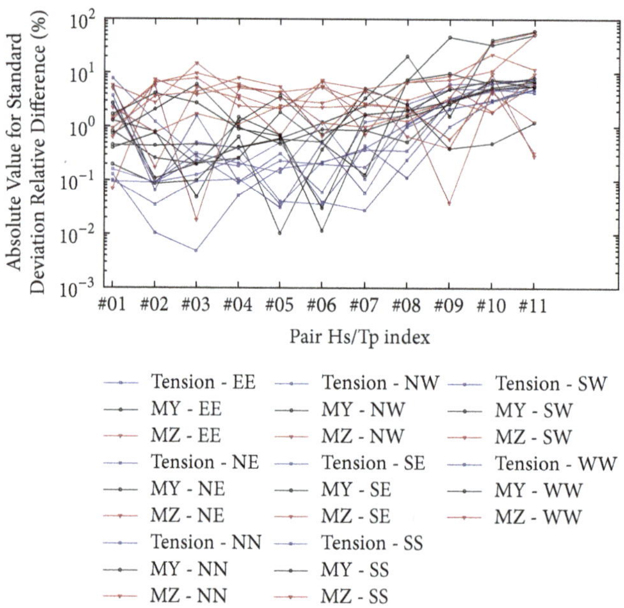

FIGURE 10: Relative differences for the response standard deviation values (critical riser TDZ element).

(absolute values) are presented, respectively, for each loading case, for the critical element cross section at the riser top. Figures 9 and 10 present the same results for the riser critical cross section located at the TDZ. As can be observed from the figures, the relative differences found for the axial tension are, in general, lower than those for the bending moments. The differences are larger for the critical point located at the TDZ than those for the riser top. This is expected since the TDZ is strongly influenced by nonlinearities due to the complex soil-structure interaction. However, for the majority of the cases, the differences can be considered low.

6.3. Fatigue Analysis Results. From the predicted global responses, the stress time series are determined for the selected SCR elements. Figure 11 shows a short window of a stress time series obtained by the hybrid ANN-FEA method and the corresponding one provided by the full FEA for the most critical point on the element with the lowest fatigue life at the top. The stress time series is taken from the loading case that contributes most to the total fatigue damage. Figure 12 shows similar comparison for the riser TDZ critical element cross section. As can be seen, the agreement between the ANN prediction and FEA results is remarkable.

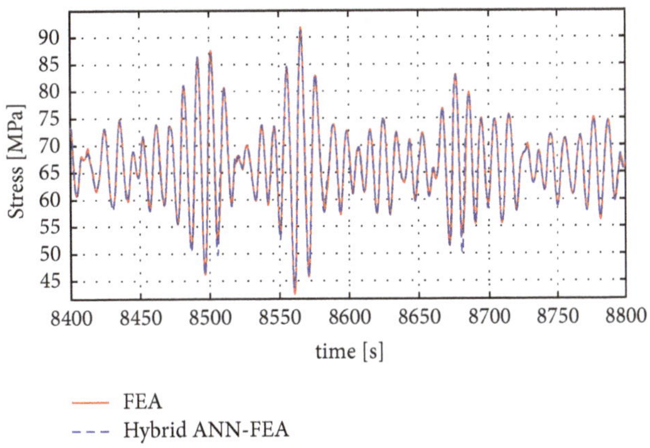

FIGURE 11: Stresses time series, critical point, and loading case in the riser critical element at top.

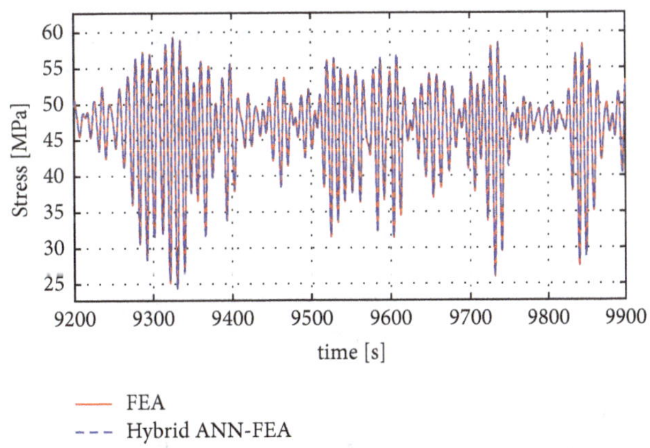

FIGURE 12: Stresses time series, critical point, and loading case in the riser critical element at TDZ.

TABLE 3: Waves incidence directions.

Incidence Direction		Frequency (%)
NN	North	26.0
NE	Northeast	9.0
EE	East	1.5
SE	Southeast	1.5
SS	South	10.0
SW	Southwest	23.0
WW	West	13.0
NW	Northwest	16.0
Sum:		100.0

Table 5 presents the correlation coefficients between ANN predicted and FEA stress time series for each loading case in the most critical point of the cross section at the riser top. Table 6 presents similar results for the riser critical element at the TDZ. For the riser top, in general, there is a very good correlation between both results. The lower correlation values are observed for loading cases with mild waves. For the TDZ region the results are also good, although some lower correlation values are observed for some scattered loading cases. In these cases, the combination of the top imposed displacements and the offset leads to considerable nonlinear responses for the specific element, especially for the My bending moment.

Figure 13 presents the annual fatigue damage for the loading cases related to waves coming from north direction (the direction with the highest accumulated damage) for the riser top critical element. Figure 14 shows similar results for the critical element located at the TDZ. These two figures show very good agreement between the hybrid methodology and the FEA results for the computed fatigue damage.

Finally, in Figure 15 the fatigue life computed by the hybrid ANN-FEA methodology is plotted together with the results given by the full FEA and also with the relative differences (%) for all cross sections elements evaluated on the top region. Figure 16 shows similar results for all cross sections located at the TDZ where fatigue results have been assessed. It is observed that the fatigue life differences are lower for the top region, where the maximum difference is 2.2%. For the elements located at the TDZ, higher differences are found (maximum difference is 11.0%). However, for the elements near the region with the lowest fatigue life in the TDZ, the maximum difference observed was lower, i.e., 3.2%.

The fatigue damage and life of the critical riser elements identified by both methods are summarized in Table 7. For the top zone, the critical riser point identified for both

TABLE 4: Wave parameters.

Index #	Hs (m)	Tz (s)	Tp (s)	Occurrence Frequency (%)
01	0.25	3.75	4.95	1.74E-03
02	0.75	4.75	6.33	3.40E+00
03	1.25	5.25	7.02	2.41E+01
04	1.75	6.25	8.41	3.33E+01
05	2.25	6.75	9.11	2.21E+01
06	2.75	7.25	9.81	1.05E+01
07	3.25	7.75	10.51	4.20E+00
08	3.75	8.25	11.21	1.60E+00
09	4.25	8.75	11.91	5.63E-01
10	4.75	9.25	12.62	2.01E-01
11	5.25	9.75	13.32	3.40E-02
		Sum:		100.0

TABLE 5: Stress correlation coefficients (riser top cross section).

#	Index Hs (m)	Tp (s)	EE	NE	NN	NW
01	0.25	4.95	0.9952	0.9332	0.8018	0.9759
02	0.75	6.33	0.9994	0.9787	0.9631	0.9977
03	1.25	7.02	0.9995	0.9846	0.9845	0.9833
04	1.75	8.41	0.9992	0.9977	0.9975	0.9984
05	2.25	9.11	0.9993	0.9982	0.9954	0.9986
06	2.75	9.81	0.9992	0.9985	0.9976	0.9990
07	3.25	10.51	0.9975	0.9990	0.9982	0.9992
08	3.75	11.21	0.9985	0.9986	0.9988	0.9979
09	4.25	11.91	0.9989	0.9965	0.9983	0.9992
10	4.75	12.62	0.9981	0.9984	0.9987	0.9978
11	5.25	13.32	0.9978	0.9987	0.9976	0.9986
#	Index Hs (m)	Tp (s)	SE	SS	SW	WW
01	0.25	4.95	0.9613	0.7846	0.7549	0.9841
02	0.75	6.33	0.9740	0.8935	0.9230	0.9991
03	1.25	7.02	0.9791	0.9328	0.9841	0.9992
04	1.75	8.41	0.9968	0.9939	0.9977	0.9977
05	2.25	9.11	0.9983	0.9964	0.9988	0.9988
06	2.75	9.81	0.9992	0.9988	0.9973	0.9989
07	3.25	10.51	0.9994	0.9981	0.9992	0.9980
08	3.75	11.21	0.9994	0.9988	0.9993	0.9982
09	4.25	11.91	0.9993	0.9991	0.9967	0.9976
10	4.75	12.62	0.9994	0.9984	0.9993	0.9980
11	5.25	13.32	0.9974	0.9989	0.9991	0.9786

methods is the same, but for the TDZ the hybrid ANN-FEA method identified a point which is 1.0 meter far apart from the critical one identified by FEA. Despite this, in practical terms, the distance is too short (given the SWL and the riser length) and the values estimated are in closer agreement.

7. Conclusions and Final Remarks

In this paper the use of a hybrid ANN-FEA methodology was investigated with the aim of speeding up the wave-induced fatigue analysis of Steel Catenary Risers. In general, at least for early design stages, this methodology predicted quite good global responses for elements evaluated on top and TDZ regions of the SCR investigated. Consequently, the estimated fatigue lives for these regions were also predicted with very good accuracy. The higher differences occur in some elements located at the TDZ. However, for the critical element in this region (lowest fatigue life), the predicted results were as good as the ones obtained for the SCR top.

TABLE 6: Stress correlation coefficients (riser TDZ cross section).

#	Index Hs (m)	Tp (s)	EE	NE	NN	NW
01	0.25	4.95	0.9922	0.9489	0.9253	0.9857
02	0.75	6.33	0.9985	0.9968	0.9983	0.9972
03	1.25	7.02	0.9970	0.9699	0.9986	0.9986
04	1.75	8.41	0.9953	0.9992	0.9992	0.9858
05	2.25	9.11	0.9971	0.9966	0.9990	0.9971
06	2.75	9.81	0.9963	0.9980	0.9979	0.9944
07	3.25	10.51	0.9940	0.9970	0.9893	0.9917
08	3.75	11.21	0.9744	0.9614	0.9520	0.9909
09	4.25	11.91	0.9597	0.8650	0.9805	0.9948
10	4.75	12.62	0.9803	0.9356	0.9851	0.9520
11	5.25	13.32	0.9776	0.9610	0.9954	0.9728
#	Index Hs (m)	Tp (s)	SE	SS	SW	WW
01	0.25	4.95	0.9906	0.9343	0.9438	0.9897
02	0.75	6.33	0.9806	0.8037	0.9570	0.9803
03	1.25	7.02	0.9722	0.9613	0.7965	0.9669
04	1.75	8.41	0.9742	0.9888	0.9808	0.9615
05	2.25	9.11	0.9843	0.9934	0.9892	0.9671
06	2.75	9.81	0.9890	0.9956	0.9930	0.9789
07	3.25	10.51	0.9907	0.9959	0.9940	0.9827
08	3.75	11.21	0.9934	0.9919	0.9952	0.9803
09	4.25	11.91	0.9902	0.9937	0.9955	0.9446
10	4.75	12.62	0.9458	0.9894	0.9750	0.9457
11	5.25	13.32	0.9533	0.9898	0.9729	0.9360

TABLE 7: Fatigue damage and life (riser top and TDZ critical points).

Region	Method	Dist. from top (m)	Annual Damage	Fatigue Life (yr)
Top	ANN-FEA	1.10	2.39E-02	41.8
	FEA	1.10	2.43E-02	41.1
TDZ	ANN-FEA	1341.60	6.68E-03	149.8
	FEA	1342.60	6.63E-03	150.9

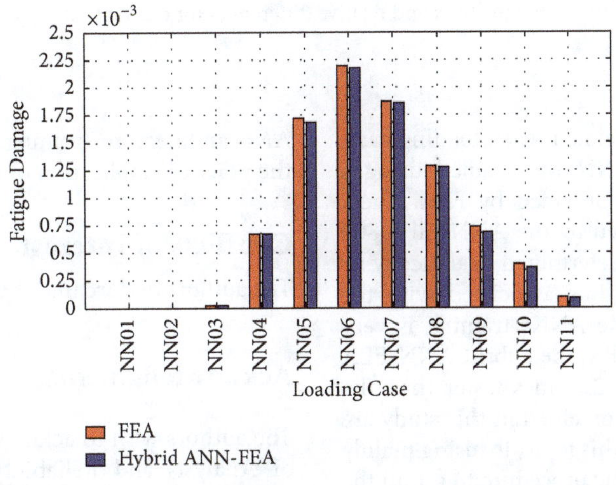

FIGURE 13: Fatigue damage for north direction waves. Riser top critical element.

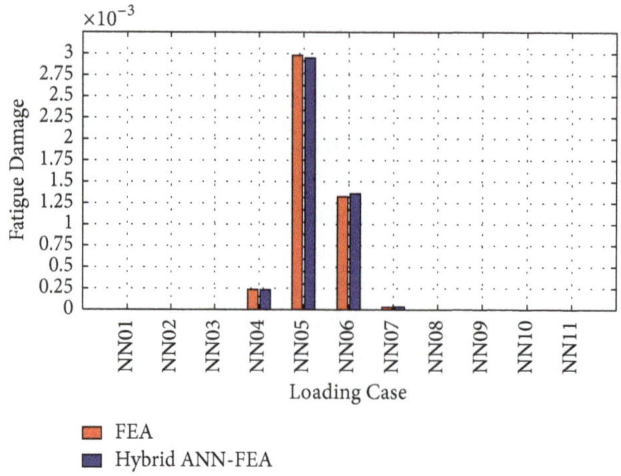

FIGURE 14: Fatigue damage for north direction waves. Riser TDZ critical element.

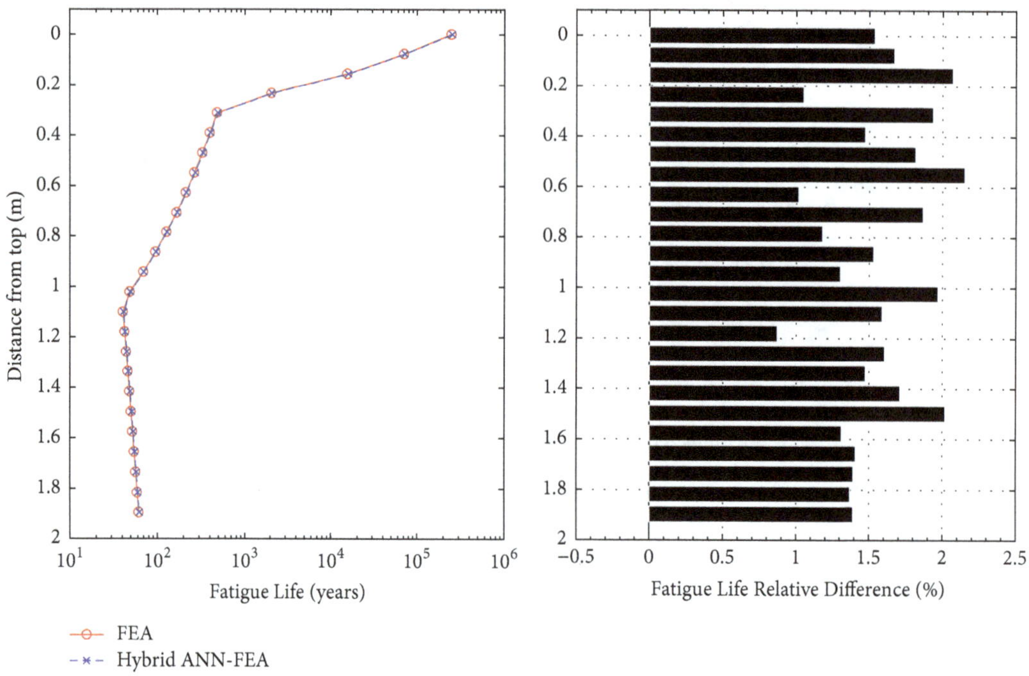

FIGURE 15: Fatigue lives and relative differences for elements at top riser.

In this work, for each riser element, each loading case, and each global response, an ANN was trained using a 500 s long response time series provided by FEA. Then, the predicted response is generated by the ANN till 10800 seconds using the displacements determined analytically by RAOs. Once the computational time associated with the fatigue damage evaluation and the ANNs training is very low compared to the full global FEA, the hybrid ANN-FEA method can be approximately 20-22 times faster than the usual FEA procedure. The results obtained in this study are very encouraging to go further in this topic, focusing mainly on the development of an automatic procedure to set up the ANN architecture, training, and evaluation, in order to make the practical application of the methodology easier.

Conflicts of Interest

The authors declare that they have no conflicts of interest.

Acknowledgments

The authors wish to acknowledge the support of Laboratory of Analysis and Reliability of Offshore Structures, Civil

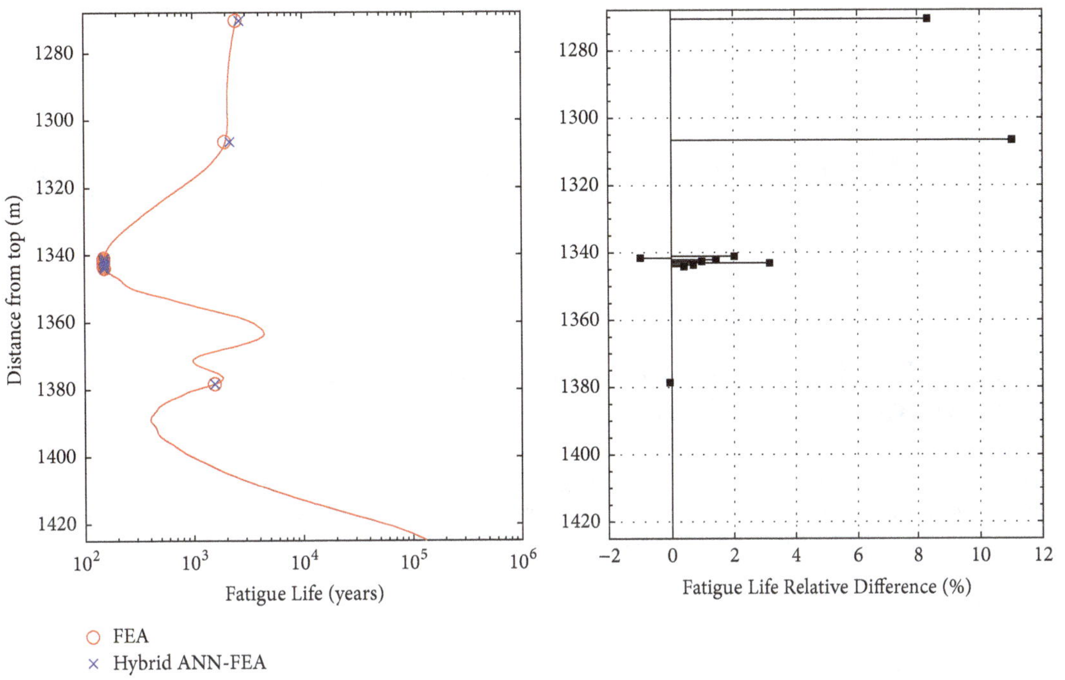

FIGURE 16: Fatigue lives and relative differences for elements at the TDZ.

Engineering Program, COPPE, Federal University of Rio de Janeiro, for the development of this work.

References

[1] R. Guarize, N. A. F. Matos, L. V. S. Sagrilo, and E. C. P. Lima, "Neural networks in the dynamic response analysis of slender marine structures," *Applied Ocean Research*, vol. 29, no. 4, pp. 191–198, 2007.

[2] A. C. de Pina, A. A. de Pina, C. H. Albrecht, B. S. Leite Pires de Lima, and B. P. Jacob, "ANN-based surrogate models for the analysis of mooring lines and risers," *Applied Ocean Research*, vol. 41, pp. 76–86, 2013.

[3] N. H. Christiansen, P. E. Torbergsen Voie, J. Hogsberg, and N. Sodahl, "Efficient mooring line fatigue analysis using a hybrid method time domain simulation scheme," in *Proceedings of the ASME 2013 32nd International Conference on Ocean, Offshore and Arctic Engineering*, Nantes, France.

[4] N. H. Christiansen, P. E. Voie, O. Winther, and J. Høgsberg, "Comparison of Neural Network Error Measures for Simulation of Slender Marine Structures," *Journal of Applied Mathematics*, vol. 2014, pp. 1-11, 2014.

[5] V. Chaves, L. V. Sagrilo, V. R. Silva, and M. A. Vignoles, "Artificial Neural Networks Applied to Flexible Pipes Fatigue Calculations," in *Proceedings of the ASME 2015 34th International Conference on Ocean, Offshore and Arctic Engineering*, p. V05BT04A022, St. John's, Newfoundland, Canada.

[6] Y. Kim, "Prediction of the dynamic response of a slender marine structure under an irregular ocean wave using the NARX-based quadratic Volterra series," *Applied Ocean Research*, vol. 49, pp. 42–56, 2015.

[7] Y. Kim, "Finite memory quadratic Volterra model for the response prediction of a slender marine structure under a Morison load," *Journal of Fluids and Structures*, vol. 56, pp. 75–88, 2015.

[8] "DNV-OS-F201 (Det Norsk Veritas)," Dynamic Risers, Hovik, Norway, 2010.

[9] API 5L (American Petroleum Insitute), *Specification for Line Pipe*, Washington, USA, 2000.

[10] DNV-RP-F204 (Det Norsk Veritas), *Riser Fatigue*, Norway, 2010.

[11] *Standard Practices for Cycle Counting in Fatigue Analsysis*, ASTM (American Society for Testing and Materials), West Conshohocken, USA, 2011.

[12] A. Almar-Naess, *Fatigue Handbook Offshore Steel Structures*, Tapir Publishers, Norway, 1985.

[13] S. Haykin, *Neural networks a comprehensive foundation*, Prentice Hall, 2005.

[14] S. N. Londhe, S. Shah, P. R. Dixit, T. M. B. Nair, P. Sirisha, and R. Jain, "A Coupled Numerical and Artificial Neural Network Model for Improving Location Specific Wave Forecast," *Applied Ocean Research*, vol. 59, pp. 483–491, 2016.

[15] I. López, L. Aragonés, Y. Villacampa, and J. C. Serra, "Neural network for determining the characteristic points of the bars," *Ocean Engineering*, vol. 136, pp. 141–151, 2017.

[16] I. Lazakis, Y. Raptodimos, and T. Varelas, "Predicting ship machinery system condition through analytical reliability tools and artificial neural networks," *Ocean Engineering*, vol. 152, pp. 404–415, 2018.

[17] M. Smith, *Neural Networks for Statistical Modeling*, vol. 10, Van Nostrand Reinhold, 1993.

[18] M. T. Hagan, H. B. Demuth, M. H. Beale, and O. de Jesús, Neural Network Design, ISBN 0-9717321-0-8.

4

Flow Partitioning in Rectangular Open Channel Flow

Yu Han,[1] Shu-Qing Yang,[2] Muttucumaru Sivakumar,[2] Liu-Chao Qiu,[1] and Jian Chen[3]

[1]*College of Water Resources & Civil Engineering, China Agricultural University, Beijing 100083, China*
[2]*School of Civil, Mining & Environmental Engineering, University of Wollongong, Wollongong, NSW 2522, Australia*
[3]*College of Engineering, China Agricultural University, Beijing 100083, China*

Correspondence should be addressed to Jian Chen; jchen@cau.edu.cn

Academic Editor: Marco Pizzarelli

Hydraulic engineers often divide a flow region into subregions to simplify calculations. However, the implementation of flow divisibility remains an open issue and has not yet been implemented as a fully developed mathematical tool for modeling complex channel flows independently of experimental verification. This paper addresses whether a three-dimensional flow is physically divisible, meaning that division lines with zero Reynolds shear stress exist. An intensive laboratory investigation was conducted to carefully measure the time-averaged velocity in a rectangular open channel flow using a laser Doppler anemometry system. Two innovative methods are employed to determine the locations of division lines based on the measured velocity profile. The results clearly reveal that lines with zero total shear stress are discernible, indicating that the flow is physically divisible. Moreover, the experimental data were employed to test previously proposed methods of calculating division lines, and the results show that Yang and Lim's method is the most reasonable predictor.

1. Introduction

A reliable assessment of the distribution of boundary shear stresses in turbulent open channel flow is of vital importance for analyzing various critical engineering problems such as channel design, sediment transport, environmental load, and wetland design [1]. Currently, researchers directly determine boundary shear stresses by measuring the turbulence structures in the viscous sublayer, where accurate data is very difficult to acquire [2, 3]. Thus, it is essential to develop a novel method for accurately determining the boundary shear stress using main flow data that is relatively easy to acquire.

From this perspective, the theory of flow partitioning, which is widely accepted to be an effective mathematical tool for simplifying hydraulic calculations, would be instrumental in the development of the proposed novel method. Leighly [4] was the first to propose that a flow region is divisible based on its isovels and orthogonal trajectories. Chiu and his colleagues [5–7] further developed this theory by proposing a mathematical model to describe isovel patterns. However, Nezu and Nakagawa [8] noted that the description of isovel curves by means of simple analytical functions introduces large errors. Hence, the implementation of flow divisibility remains an open issue and has not yet been implemented as a fully developed mathematical tool for modeling complex channel flows independently of experimental verification.

Since the theory of a divisible flow region was first proposed, efforts have been made to determine the location of division lines within the flow. The Keulegan method (KM; [9]) assumes that the division line is the bisector of the base angles of a polygonal channel and that it is applicable only to a channel with equal roughness on the sidewalls and bed. Einstein [10] also extended the theory of divisibility to river flows by assuming that the cross section of a river can be separated into three regions. While Chien and Wan [11] attempted to explain Einstein's hypothesis in terms of the energy transport mechanism, they did not provide a useful method of determining the division lines. Hence, this work has received little attention since its publication. In addition, a number of models have been proposed to express the division lines within a divisible flow, for example, Daido's method (DM; [12]), Guo and Julien's method (GJM; [13]), and Yang and Lim's method (YLM; [14]). These models were subjected to detailed reviews by Yang et al. [15] and Han et al. [16, 17], and these authors concluded that division lines could be located based on the existence of measured Reynolds shear stress values of zero below the free surface in rectangular channel flow. However, the Reynolds shear stress can only be measured using very sophisticated equipment, like laser Doppler anemometry (LDA), in controlled laboratory environments. Thus, researchers urgently require a novel method to determine the existence of division lines using a common parameter (e.g., the time-averaged velocity) that can be measured using relatively simple equipment. More importantly, practicing engineers can employ this new method to infer whether the flow region in their study is divisible.

Based on these considerations, the present study seeks to determine the existence of division lines by analyzing the mean velocity distribution in an open channel rectangular flume. In this study, two methods were developed to locate division lines: the first employs the condition of total zero-shear stress and the second employs the log-law. We experimentally verify the existence of division lines in an open rectangular channel and validate the two proposed methods.

2. Classic Log-Law and Implications of Shear Stress

2.1. Classic Log-Law and Empirical Determination of Its Parameters. The log-law is one of the most important and reliable laws in turbulent flow and has been demonstrated to accurately reflect the velocity of fluid flow near the boundary of the fluid region (i.e., the wall). In the wall region, the characteristic scales of length and velocity are represented using the inner variables ν/u_* and u_*, respectively, where ν is the kinematic viscosity and u_* is the shear velocity. In this region, the total shear stress layer is nearly constant and equal to ρu_*^2, where ρ is the fluid density. Based on this, Prandtl [18] expressed the momentum flux caused by secondary currents $\rho\overline{u'v'}_n$ in terms of the mixing length l_m, as follows:

$$-\rho\overline{u'v'} = \rho l_m^2 \left|\frac{du}{dy}\right| \frac{du}{dy} = \rho u_*^2, \quad (1)$$

where u is the time-averaged velocity along the wall and l_m is assumed to be proportional to the distance y to the boundary, that is, $l_m = \kappa y$. Here, $\kappa = 0.4$ is an empirical parameter denoted as the von Karman constant. Integration of (1) with the boundary condition that $u = 0$ at the wall position $y = y_0$, yields the well-known logarithmic velocity profile [19]:

$$\frac{u}{u_{*1}} = \frac{1}{\kappa} \ln \frac{y}{y_0}. \quad (2)$$

In the standard expression of the log-law, u_{*1} is the global shear velocity, and y_0 is empirically expressed by Nikuradse [20] as follows:

$$y_0 = \frac{\nu}{u_{*s}}, \quad (3)$$

where c is an experimentally determined coefficient and u_{*s} is the local shear velocity, which may differ from u_{*1}. For circular pipe flow under the force of gravity g, it is well known that $u_{*1} = u_{*s} = (gRS)^{0.5}$, where R is the hydraulic radius, and S is the channel slope. This empirical expression has been widely extended to channel flows. However, disagreements with experimental data have induced researchers to modify the log-law, and numerous alternatives have been proposed. For example, κ has been treated as a variable, or c has been made to vary according to flow conditions. Moreover, additional terms have been included, such as in the log-wake law. A comprehensive review of these modifications can be found in the literature (e.g., Guo [2] and Nezu and Nakagawa [8]).

In contrast to the empirical modifications discussed above, Tracy and Lester [21] first questioned the definition of shear velocities using experimental data obtained from central channel profiles. Their results showed that, for open channel flows, the value of u_{*1} employed on the left hand side of the log-law given by (2) is not identical to the value of u_{*s} by which y_0 is defined in (3), but, rather, $u_{*1} = (gRS)^{0.5}$ and $u_{*s} = (ghS)^{0.5}$, where h is the water depth. Surprisingly, the measured data agreed very well with the log-law without the introduction of any new terms, like the wake-law term. Furthermore, Yang et al. [22] reexamined experimentally measured data in the literature obtained for a rectangular duct flow and rectangular open channel flows with various aspect ratios. They verified that Tracy and Lester's [21] conclusion was valid even in the corner regions of channel flows when $u_{*s} = \sqrt{gL_nS}$, where L_n denotes the normal distance from the boundary to the division line. This establishes a relationship between the mean velocity and the division line, and a method to test whether a flow region is divisible can be thereby developed based on the local boundary shear stress according to a single parameter L_n. However, owing to the difficulties involved in the direct measurement of the local shear stress, it would be ideal if u_{*s} could be determined from the velocity distribution. Thus, it is worth investigating how to determine u_{*s} from a velocity profile.

As indicated by (2), u_{*1}/κ can be generally evaluated from the slope of a plot of the measured mean velocity against $\ln y_n$ (i.e., the distance to the boundary in the normal direction) based on the measured velocity profile. In addition, u_{*s} can be evaluated from its vertical axis intercept. Finally, the values of u_{*s} obtained from the velocity profiles can be employed to determine the characteristic distance L_n in an open channel flow as $L_n = u_{*s}^2/(gS)$. Therefore, the upper boundary of the subregion or division line can be inferred from the location of L_n. This forms the basis of the first simple method proposed in the present study to determine division lines in a channel flow.

2.2. Inferred Division Lines from the Zero Total Shear Stress (ZTSS) Condition. The total shear stress τ along a division line is expressed in the following form:

$$\tau = \mu \frac{\partial u}{\partial n} - \rho u v_n - \rho\overline{u'v'}_n, \quad (4)$$

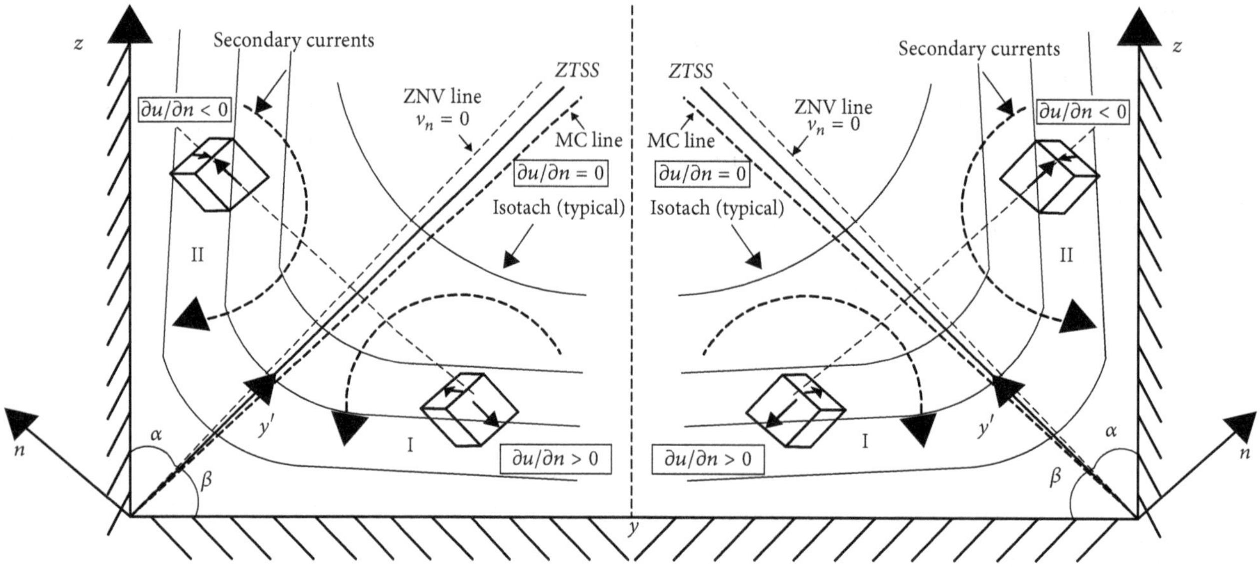

FIGURE 1: Relationships between mean velocity contours, secondary currents, and division lines in a rectangular open channel flow.

where μ is the dynamic viscosity, n represents the direction normal to the division line interface, and v_n is the velocity component of secondary flow normal to the division line. The Reynolds shear stress can be modeled by drawing an analogy with the coefficient of molecular viscosity in Stokes' law in the following form:

$$-\overline{u'v'_n} = v_t \frac{\partial u}{\partial n}, \tag{5}$$

where v_t is the turbulent kinematic viscosity. Thus, (4) can be rewritten as

$$\tau = (\mu + v_t) \frac{\partial u}{\partial n} - \rho u v_n. \tag{6}$$

Hence, we can infer based on (6) that meeting the ZTSS condition of $\tau = 0$ requires that the following conditions are simultaneously satisfied:

(1) $\partial u/\partial n = 0$.

(2) $v_n = 0$.

The first condition states that either the gradient of u along the division line defined in (6) is approximately equal to zero or that u is at a maximum at the line. The second condition states that no secondary fluid flow penetrates the division line defined in (6) and is therefore parallel to the division line.

It is therefore necessary to identify the locations where $\partial u/\partial n = 0$ and $v_n = 0$ in the flow region. This can be addressed by plotting the isovels in the flow region according to the y and z directions, as shown in Figure 1. Experimental research conducted by Melling and Whitelaw [23] indicates that the isovel lines in open rectangular channel flow bulge markedly towards the corners. Therefore, we can calculate the curvature k_c of every point along an isovel and determine the points of maximum curvature (MC). We then plotted a bold dashed line linking the MC points on the isovels in Figure 1, which is denoted as the MC line. In addition, the secondary current vectors were determined, and a fine dashed line indicating the line along which $v_n = 0$, denoted as the ZNV line, was plotted in Figure 1 accordingly. These lines divide the flow region into regions I and II. We note that $\partial u/\partial n > 0$ in flow region I and $\partial u/\partial n < 0$ in region II.

The procedure for determining the locations of division lines in an open channel flow by considering the conditions of ZTSS based on the simultaneous satisfaction of conditions 1 and 2 can be summarized as follows:

(i) First, plot an isovel graph u/U_m, where U_m is maximum velocity in flow. Next, fit the isovel lines to polynomials as

$$y = f(z). \tag{7}$$

(ii) Calculate k_c for every point along an isovel as follows:

$$k_c = \frac{|y''|}{\left(1 + y'^2\right)^{3/2}}. \tag{8}$$

(iii) Select the MC points for each isovel, and obtain the MC line linking all MC points on the isovels, which approximately satisfies the condition $\partial u/\partial n = 0$.

(iv) The secondary current vectors are plotted on the same graph and are partitioned according to the ZNV line.

(v) The ZTSS division line is plotted between the ZNV and MC lines. Thus, both $\partial u/\partial n$ and v_n are approximately equal to zero along the ZTSS line, which provides a good approximation of the physical division line.

2.3. Experimental Verification of Division Lines in an Open Rectangular Channel. The experiment was conducted in

TABLE 1: Experimental condition.

Flow Parameter	Length L (m)	Width b (m)	Depth h (m)	Aspect ratio, b/h	$\sqrt{gRS}$ (m/s)	$\sqrt{ghS}$ (m/s)
UOW data	11	0.3	0.11	2.73	0.025	0.033
Nezu et al. [24] data (A1)	6	0.04	0.04	1	0.031	0.039

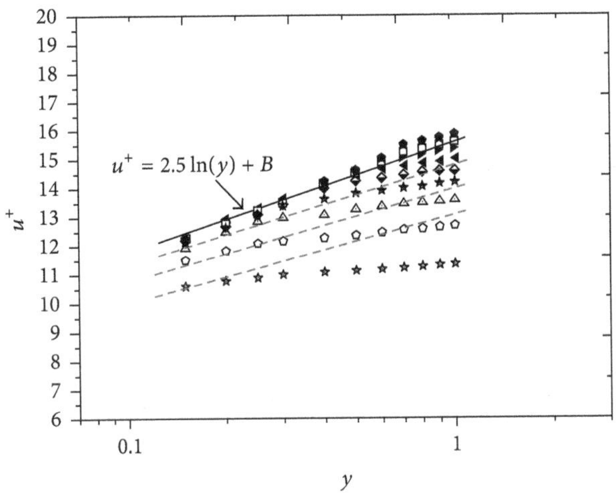

FIGURE 2: Measured velocity distributions along different layers z (where $z = 0$ represents the channel bottom) determined relative to the channel depth b plotted in the form of $u^+ = u/(gRS)^{0.5}$ versus the natural logarithm of the distance from a wall (y) in a smooth rectangular channel (data from Nezu et al. [24]).

a smooth rectangular open channel at the University of Wollongong (UOW). The flume was 10.5 m long, $h = 0.3$ m wide, and $b = 0.45$ m deep, and a fully developed uniform flow was established at the test section 6.0 m downstream from the entrance of channel. A series of velocity profiles were measured using a Dantec two component LDA system. A hot-wire system was used to measure the flow velocity. The detailed experimental conditions are presented in Table 1. The experimental data obtained by Nezu et al. [24] were also examined for comparison.

First, the velocity distributions measured by Nezu et al. [24] and at the UOW along different layers z (where $z = 0$ represents the channel bottom) determined relative to the channel depth b were plotted in the form $u^+ = u/\sqrt{gRS}$ versus the natural logarithm of the distance from a wall ($\ln y$), as shown in Figures 2 and 3, respectively. As predicted by (2), the results show that, to a certain extent, the measured velocity in the near wall region follows the solid straight line given in the figures. It is interesting to note that the other dashed lines have an equivalent slope with that of the solid line for data points near the wall, but involve different intercepts. This is particularly evident for the UOW data presented in Figure 3. The identical slope implies that u_{*1} in (2) is a constant, while the different intercepts imply that u_{*s} in (3) is a variable. Thus, according to the intercepts of each straight line, the results show that u_{*s} increases with increasing z/b.

As discussed in Section 2.1, the distance L_n along the normal direction from the division line to the boundary can be determined from the calculated values of u_{*s} as $L_n = u_{*s}^2/(gS)$. Therefore, the upper boundary of the subregion or division line can be obtained, and the results based on the values of L_n determined from the intercepts of the straight lines obtained from Figures 2 and 3 are shown in Figures 4(a) and 4(b), respectively, and are plotted relative to the channel width h with respect to z/b. Yang and Lim [14] hypothesized that turbulent energy must be transported through the minimum relative distance towards the boundary for dissipation. Thus, the distance L_n can also be considered as the energy transfer path. For comparison, the division lines based upon models proposed by different authors are also included in Figure 4. In addition, the ZTSS line is also plotted in Figure 4. We obtained the ZTSS line according to the procedure presented in Section 2.2 for both the data obtained from Nezu et al. [24] and UOW. According to the outlined procedure, we plotted the isovel graphs u/U_m, as shown in Figures 5(a) and 5(b) for the data obtained from Nezu et al. [24] and UOW, respectively. All MC points along each isovel were computed according to Eq. (8) using MATLAB, and the MC lines were obtained by linking all the MC points. The MC line is presented in Figure 5 as a red dashed line (long dashes) running from corner to corner. Meanwhile, the ZNV line, which clearly partitions the two rotating secondary cells, is also plotted in Figure 5 as a red dashed line (short dashes) running from corner to corner. Accordingly, the ZTSS division line lies between the ZNV and MC lines, where both $\partial u/\partial n$ and v_n are approximately zero.

As shown in Figure 4, the locations of division lines obtained according to L_n and the ZTSS line are quite similar. Such similarity in results obtained from very different methods suggests that division lines exist physically and that the zero-shear stresses can be determined by the velocity contours. A comparison of the various model results indicates that the division line provided by YLM appears to be the most reasonable predictor of division lines relative to the ZTSS and L_n results, with a minimum averaged deviation of 18% and 16%, respectively.

3. Conclusion

It is generally assumed that the division of flow regions in a rectangular open channel flow is only a mathematical concept, and no consensus as to whether division lines physically exist has developed. To clarify this, a specially

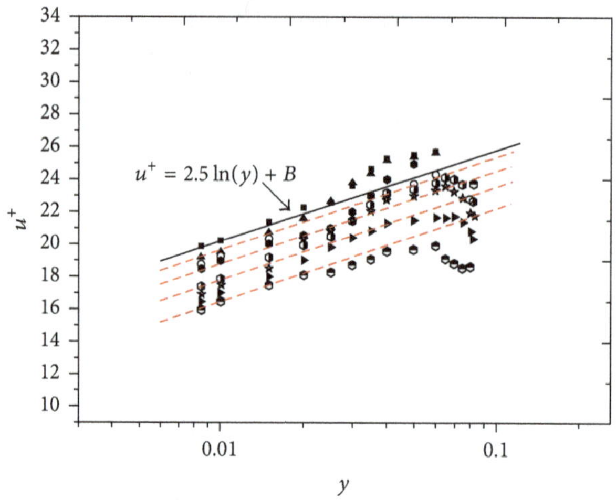

FIGURE 3: Measured velocity distributions along different layers z (where $z = 0$ represents the channel bottom) determined relative to the channel depth b plotted in the form of $u^+ = u/(gRS)^{0.5}$ versus the natural logarithm of the distance from a wall (y) in a smooth rectangular channel (data from UOW).

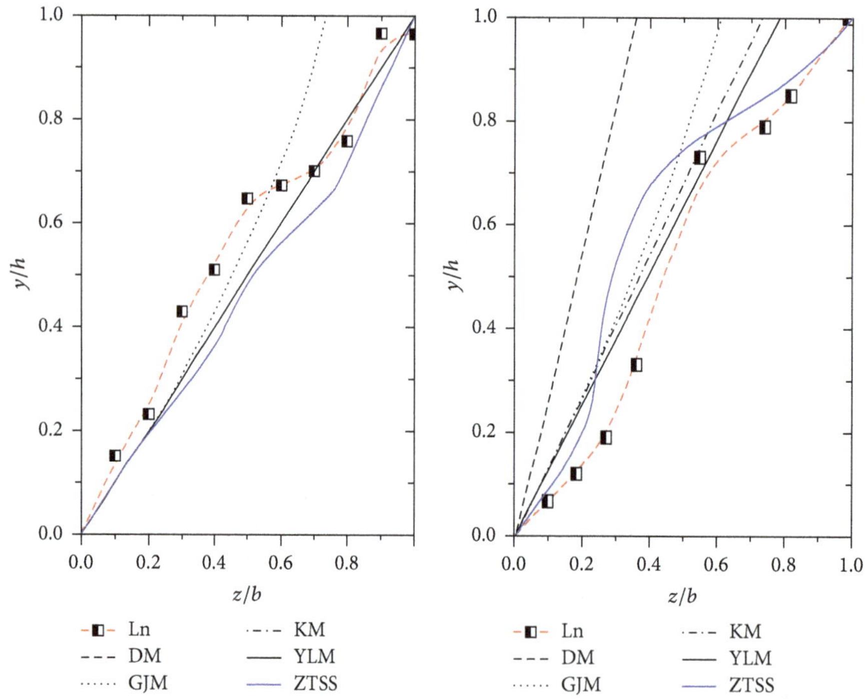

FIGURE 4: Comparison between measured division lines established according to the ZTSS line and the calculated values of L_n, and the division lines predicted by DM, GJM, KM, and YLM (data from Nezu et al. [24] (a) and from UOW (b)).

designed apparatus was employed to infer the zero-shear lines from mean velocity distributions in a three-dimensional channel. This investigation made the following contributions, and various conclusions can be drawn.

(1) A clear method was presented to infer the existence of ZTSS lines in a rectangular open channel flow based on velocity contours and secondary current data. This method is a direct and straightforward means of identifying the existence of division lines using the criteria suggested.

(2) The division line can also be determined from the velocity distribution according to the distance L_n

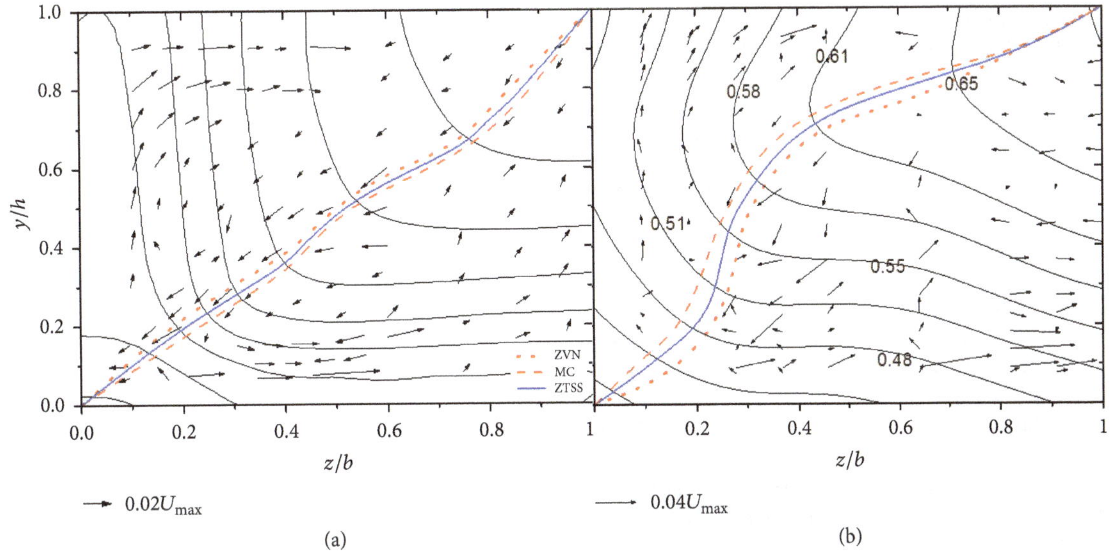

FIGURE 5: Measured ZTSS division lines obtained according to the locations of the MC and ZVN lines (data from Nezu et al. [24] (a) and from UOW (b)).

along the normal direction from the division line to the boundary. The experimental division lines obtained from the L_n results were very similar to those obtained according to the ZTSS line. Thus, the division line can be determined by identifying the ZTSS and L_n locations for a channel flow.

(3) The division lines obtained from the proposed ZTSS and L_n methods were compared with those obtained using existing models: KM, DM, YLM, and GJM. The lines obtained by the YLM demonstrated the closest agreement with the division lines obtained from the proposed methods.

Conflicts of Interest

The authors declare that there are no conflicts of interest regarding the publication of this paper

Acknowledgments

The authors express gratitude for the financial support from the National Natural Science Foundation of China (Grant no. 51509248) and National Key R&D Program of China (Grants nos. 2017YFC0403203, 2016YFC0400207, 2017YFD0701000 and 2016YFD200700).

References

[1] S. Dey, "Sediment pick-up for evolving scour near circular cylinders," *Applied Mathematical Modelling*, vol. 20, no. 7, pp. 534–539, 1996.

[2] J. Guo, "Modified log-wake-law for smooth rectangular open channel flow," *Journal of Hydraulic Research*, vol. 52, no. 1, pp. 121–128, 2014.

[3] J. Guo, "Sidewall and non-uniformity corrections for flume experiments," *Journal of Hydraulic Research*, vol. 53, no. 2, pp. 218–229, 2015.

[4] J. B. Leighly, "Toward a theory of the morphologic significance of turbulence in flow of water in streams," *University of California Publications in Geography*, vol. 6, no. 1, pp. 1–22, 1932.

[5] C. L. Chiu and J. D. Chiou, "Structure of 3-D flow in rectangular open channels," *Journal of Hydraulic Engineering*, vol. 112, no. 11, pp. 1050–1068, 1986.

[6] C.-L. Chiu and D. W. Murray, "Variation of velocity distribution along nonuniform open-channel flow," *Journal of Hydraulic Engineering*, vol. 118, no. 7, pp. 989–1001, 1992.

[7] C.-L. Chiu and C. A. A. Said, "Maximum and mean velocities and entropy in open-channel flow," *Journal of Hydraulic Engineering*, vol. 121, no. 1, pp. 26–35, 1995.

[8] I. Nezu and H. E. D. Nakagawa, *Turbulence in open-channel flows*, Balkema, Rotterdam, Netherlands, 1993.

[9] G. H. Keulegan, "Laws of turbulent flow in open channels," *Journal of Research of the National Bureau of Standards*, vol. 21, no. 6, pp. 707–741, 1938.

[10] H. A. Einstein, *Formulas for the transportation of bed load*, vol. 107, American Society of Civil Engineers, 1942.

[11] N. Chien and Z. Wan, *Mechanics of Sediment Transport*, American Society of Civil Engineers, Reston, VA, USA, 1999.

[12] A. Daido, "Effect of aspect ratio of channel and sediment density on transport phenomena," in *Proceedings of the Int. Symp. On Hydraulic Research in Nature*, pp. 5–7, Chinese Hydraulic Engineering Society, Beijing, China, 1992.

[13] J. Guo and P. Y. Julien, "Shear stress in smooth rectangular open-channel flows," *Journal of Hydraulic Engineering*, vol. 131, no. 1, pp. 30–37, 2005.

[14] S.-Q. Yang and S.-Y. Lim, "Mechanism of energy transportation and turbulent flow in a 3D channel," *Journal of Hydraulic Engineering*, vol. 123, no. 8, pp. 684–692, 1997.

[15] S.-Q. Yang, Y. Han, P. Lin, C. Jiang, and R. Walker, "Experimental study on the validity of flow region division," *Journal of Hydro-environment Research*, vol. 8, no. 4, pp. 421–427, 2014.

[16] Y. Han, S.-Q. Yang, N. Dharmasiri, and M. Sivakumar, "Effects of sample size and concentration of seeding in LDA measurements on the velocity bias in open channel flow," *Flow Measurement and Instrumentation*, vol. 38, pp. 92–97, 2014.

[17] Y. Han, S.-Q. Yang, N. Dharmasiri, and M. Sivakumar, "Experimental study of smooth channel flow division based on velocity distribution," *Journal of Hydraulic Engineering*, vol. 141, no. 4, Article ID 06014025, 2015.

[18] L. Prandtl, "Bericht uber untersuchungen zur aubgebildeten turbulenz," *Zeitschrift für Angewandte Mathematik und Mechanik*, vol. 5, no. 2, pp. 136–169, 1925.

[19] I. Marusic, J. P. Monty, M. Hultmark, and A. J. Smits, "On the logarithmic region in wall turbulence," *Journal of Fluid Mechanics*, vol. 716, pp. 1–11, 2013.

[20] J. Nikuradse, "Stromungsgesetze in rauhen rohren," *VDI-Forsch*, p. 361, 1933.

[21] H. J. Tracy and C. M. Lester, *Resistance Coefficients and Velocity Distribution in Smooth Rectangular Channel*, USGS Water-supply paper, Washington, DC, USA, 1961.

[22] S.-Q. Yang, S.-Y. Lim, and J. A. McCorquodale, "Investigation of near wall velocity in 3-D smooth channel flows," *Journal of Hydraulic Research*, vol. 43, no. 2, pp. 149–157, 2005.

[23] A. Melling and J. H. Whitelaw, "Turbulent flow in a rectangular duct," *Journal of Fluid Mechanics*, vol. 78, no. 2, pp. 289–315, 1976.

[24] I. Nezu, H. Nakagawa, and A. Tominaga, "Secondary currents in a straight channel flow and the relation to its aspect ratio," *Turbulent Shear Flows*, vol. 4, no. 2, pp. 246–260, 1985.

5

Analytical Solution and Application for One-Dimensional Consolidation of Tailings Dam

Hai-ming Liu, [1,2] Gan Nan, [1] Wei Guo, [1] Chun-he Yang, [2] and Chao Zhang [2]

[1] *Faculty of Civil Engineering and Mechanics, Kunming University of Science and Technology, Kunming, Yunnan 650224, China*
[2] *State Key Laboratory of Geomechanics and Geotechnical Engineering, Institute of Rock and Soil Mechanics, Chinese Academy of Sciences, Wuhan 430071, China*

Correspondence should be addressed to Hai-ming Liu; haiming0871@163.com

Academic Editor: Shuo Wang

The pore water pressure of tailings dam has a very great influence on the stability of tailings dam. Based on the assumption of one-dimensional consolidation and small strain, the partial differential equation of pore water pressure is deduced. The obtained differential equation can be simplified based on the parameters which are constants. According to the characteristics of the tailings dam, the pore water pressure of the tailings dam can be divided into the slope dam segment, dry beach segment, and artificial lake segment. The pore water pressure is obtained through solving the partial differential equation by separation variable method. On this basis, the dissipation and accumulation of pore water pressure of the upstream tailings dam are analyzed. The example of typical tailings is introduced to elaborate the applicability of the analytic solution. What is more, the application of pore water pressure in tailings dam is discussed. The research results have important scientific and engineering application value for the stability of tailings dam.

1. Introduction

Based on the assumption that the soil is isotropic and uniform, an external surface load is instantaneously applied and is held constant; a classical one-dimensional (1D) consolidation theory was proposed by Terzaghi [1]. In order to analyze time-dependent loading, an analytical solution to the layered consolidation problem for a general set of boundary conditions and an arbitrary load history was presented by Schiffman and Stein [2]. The 1D consolidation analytical solutions considering ramp loading were presented by Olson [3]. A simple semianalytical method to solve the problem of 1D consolidation by taking into account the varied compressibility of soil under cyclic loadings was brought up by Cai et al. [4]. A rigorous solution of the conventional Terzaghi 1D consolidation under haversine cyclic loading with any rest period was proposed by Müthing et al. [5], which is achieved using Fourier harmonic analysis for the periodic function representing the rate of imposition of excess pore water pressure. A semianalytical solution to 1D consolidation of viscoelastic unsaturated soils with a finite thickness under oedometric conditions and subjected to a sudden loading was put forward by Qin et al. [6]. A semianalytical solution to 1D consolidation equation of fractional derivative Kelvin-Voigt viscoelastic saturated soils subjected to different time-dependent loadings was presented by Wang et al. [7]. Under the condition of the increasing weight of superincumbent material and the length of the drainage path varies, a solution for the 1D consolidation of a clay layer whose thickness increases with time was proposed by Gibson [8]. An exact analytical solution of the nonhomogeneous partial differential equation governing the conventional 1D consolidation under haversine repeated loading was derived and discussed by Razouki et al. [9, 10].

In order to analyze different boundary conditions, single drainage solutions for several specific variations of the permeability and shear modulus were given by Mahmoud and Deresiewicz [11]. Several analytical solutions for the consolidation analysis of a soil layer with fairly general laws of variation of permeability and compressibility for both a single-drained condition and a double-drained condition were proposed by Zhu and Yin [12]. Tang et al. [13] propose a closed-form solution for consolidation of three-layered soil with a vertical drain system and a method to solve a convergence solution by enlarging the precision for the entire thickness of the foundation and adding controlling precision for the overall average degree of consolidation of each soil layer. The solution of Terzaghi 1D consolidation equation with general boundary conditions is proposed by Mei and Chen [14], whose solution is validated by comparing it to the classical solution. Hawlader et al. [15] develop a new constitutive model for the compressibility behavior of soft clay sediments at low effective stress level and the model is used to solve finite strain 1D consolidation with pertinent initial and boundary conditions. A solution to the consolidation equation with boundary conditions that are cyclic with time is given by Rahalt and Vuez [16]. A nonlinear theory of consolidation has been developed for an ideal normally consolidated soil by Davis and Raymond [17]. A simple calculation procedure to analyze the one-dimensional response of saturated soil layers to pore pressure variations at the boundary described by a general time-dependent function is developed by Conte and Troncone [18].

The tailings dam is an important geotechnical structure in mining engineering. For a long time, the theory of reservoir dam is applied to tailings dam without any modification. However, there are many differences between tailings dam and reservoir dam, which lead to inaccurate calculation results of the pore water pressure. According to the classical Terzaghi consolidation theory, the analytical solution of the pore water pressure is discussed in this paper.

2. The General Equation of 1D Consolidation

According to Darcy's law, the flow drag resistance of the z-direction is as follows:

$$F_z = -i_z \gamma_w = -n \frac{\gamma_w}{k} v_z. \tag{1}$$

The negative sign on the right side of (1) indicates that the drag resistance is opposite to the direction of flow velocity.

According to the mechanical equilibrium conditions, the flow drag resistance of the z-direction can also be expressed as follows:

$$F_z = \frac{\partial p_w}{\partial z} + \gamma_w. \tag{2}$$

Equations (1) and (2) can be combined as follows:

$$\frac{\partial p_w}{\partial z} + \gamma_w + n \frac{\gamma_w}{k} v_z = 0. \tag{3}$$

It is assumed that the water in porosity is incompressible under the consolidation process. Thus, γ_w is a constant. During the process of soil consolidation, because the porosity of soil particles is continuously compressed, the permeability coefficient of soil is decreased continuously and the permeability coefficient k of soil is changed with depth, that is, $k = f(z)$. Taking the derivative of (3), one can have

$$\frac{\partial^2 p_w}{\partial z^2} + n\gamma_w \frac{\partial}{\partial z}\left(\frac{v_z}{k}\right) = 0. \tag{4}$$

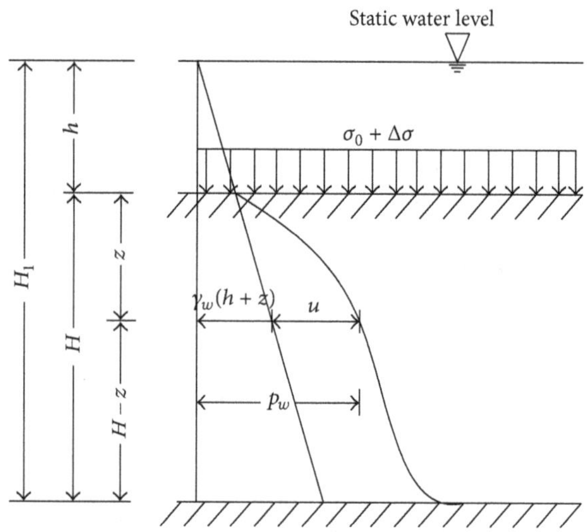

FIGURE 1: Schematic diagram of stress in soil.

According to fractional derivative rule, (4) can be expanded as follows:

$$\frac{\partial^2 p_w}{\partial z^2} + n\gamma_w \frac{1}{k}\frac{\partial v_z}{\partial z} - n\gamma_w v_z \frac{1}{k^2}\frac{dk}{dz} = 0. \tag{5}$$

Assume that the horizontal direction consolidation of the soil can be neglected; that is, the consolidation is under confined compression condition. According to the 1D consolidation condition, (6) can be deduced

$$n\frac{\partial v_z}{\partial z} = \frac{\partial \varepsilon_z}{\partial t}. \tag{6}$$

Under the condition of 1D consolidation, the axial strain ε_v is equal to the volume strain ε_v of the soil. According to Pane and Schiffman [19] research, (4) and (5) can be rewritten as follows:

$$\frac{\partial^2 p_w}{\partial z^2} + \frac{\gamma_w}{k}\frac{\partial \varepsilon_z}{\partial t} = 0, \tag{7}$$

$$\frac{\partial^2 p_w}{\partial z^2} + \frac{\gamma_w}{k}\frac{\partial \varepsilon_z}{\partial t} - n\gamma_w v_z \frac{1}{k^2}\frac{dk}{dz} = 0. \tag{8}$$

In the consolidation problem, it is of great significance to study the change of pore water pressure with time and position under external load. Therefore, (8) can be expanded according to pore water pressure.

Assuming that the soil profile is shown as Figure 1, the pore water pressure of a point located z under load σ_0 and $\Delta\sigma_0$ is u. According to the Terzaghi effective stress principle, the effective stress σ' could be given as follows under consolidation:

$$\sigma' = \sigma - p_w = \sigma - \left[\gamma_w (h+z) + u\right]. \tag{9}$$

According to Figure 1, the total stress σ of a point located z is as follows:

$$\sigma = (\sigma_0 + \Delta\sigma) + \gamma_s z + \gamma_w h. \tag{10}$$

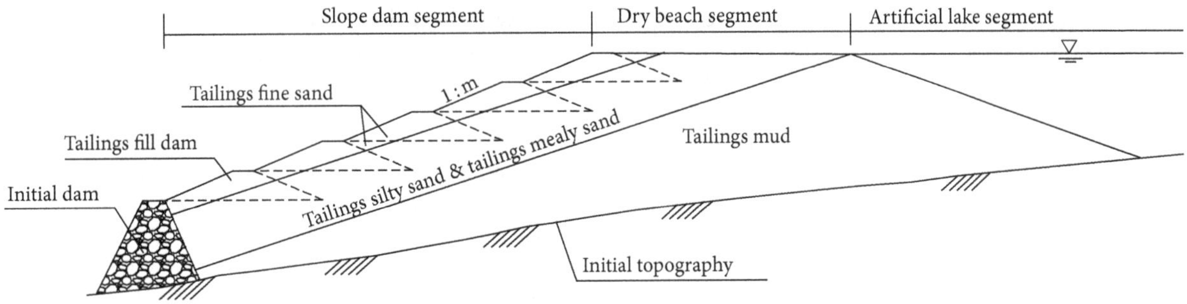

FIGURE 2: Schematic diagram of the upstream tailings dam.

Substituting (10) into (9) yields

$$\sigma' = (\sigma_0 + \Delta\sigma) + \gamma' z - u. \quad (11)$$

Take the derivative of (11) with respect to t and consider the relationship $\partial z/\partial t$ equals $\partial H/\partial t$

$$\frac{\partial \sigma'}{\partial t} = \frac{\partial}{\partial t}(\Delta\sigma) + \gamma' \frac{\partial H}{\partial t} - \frac{\partial u}{\partial t}. \quad (12)$$

According to the compression curve of 1D consolidation test, the volume compressibility factor m_v can be defined as follows:

$$m_v = \frac{1}{1+e}\frac{\partial e}{\partial \sigma'}, \quad (13)$$

$$\frac{\partial \varepsilon_z}{\partial t} = \frac{1}{1+e}\frac{\partial e}{\partial t}. \quad (14)$$

Substituting (13) into (14) gives

$$\frac{\partial \varepsilon_z}{\partial t} = m_v \frac{\partial \sigma'}{\partial t}. \quad (15)$$

Substituting (15) into (8), it is noted that the last term of (8) has the following relation:

$$n\gamma_w v_z \frac{1}{k^2}\frac{dk}{dz} = -F_z \frac{1}{k}\frac{dk}{dz} = -\left(\frac{\partial p_w}{\partial z} + \gamma_w\right)\frac{1}{k}\frac{dk}{dz}. \quad (16)$$

Equation (8) can be derived as

$$\frac{\partial^2 p_w}{\partial z^2} + \frac{m_v \gamma_w}{k}\frac{\partial \sigma'}{\partial t} + \left(\frac{\partial p_w}{\partial z} + \gamma_w\right)\frac{1}{k}\frac{dk}{dz} = 0. \quad (17)$$

Substituting (15) into (17), (17) can be expressed as pore water pressure. Equation (18) can be obtained

$$\frac{\partial^2 u}{\partial z^2} + \frac{m_v \gamma_w}{k}\left[\frac{\partial \Delta\sigma}{\partial t} + \gamma' \frac{\partial H}{\partial t} - \frac{\partial u}{\partial t}\right] + \frac{1}{k}\frac{dk}{dz}\frac{\partial u}{\partial z} = 0. \quad (18)$$

Equation (18) is a general equation which reflects the 1D consolidation process. It takes into account the situation where the applied load changes with time, the soil thickness changes with time, the soil permeability changes with depth, and so on.

3. Accumulation and Dissipation of Pore Water Pressure in Tailings Dam

Tailings are mine wastes produced in the mining engineering, which are sent to tailings reservoir by pipe or flume. The tailings dam is an important part in mining engineering, which is consisted by the initial dam and fill dam. The initial dam is made by permeable rockfill generally, and the fill dam is formed by tailings. In general, the construction of a tailing dam takes many decades or even a century. According to construction method, the tailings dam can be divided into upstream tailings dam, downstream tailings dam, midline method tailings dam, and so on. The number of tailings ponds in China has reached more than 12,000 by statistics. Because the downstream of the tailings dam is residents living area or mining production area generally the social and people's property damage a huge impact if the tailings dam is failed.

Because the upstream tailings dam has advantages of simple operation, low construction costs, the less need for coarse particles, and so on, according to statistics, 95% of the tailings dams are adopting the construction of the upstream tailings dam in China. On the other hand, the construction process of the upstream tailings dam cannot precisely control the shape of the tailings dam. The deposition structure of the tailings dam is very complex. The upstream tailings have shortcoming of long infiltration distant and poor stability. In view of the tailings slurry discharge and tailings particle deposition following the sediment mechanics, the section profile of the tailings dam has obvious regularity. The coarser the particle size is, the shorter the average distance migrates and vice versa. Generally, the deposition order of the tailings dam along the dry beach face is tailings fine sand, tailings silt, tailings sand, and tailings mud.

If particle size distribution of the tailings and the length of the dry beach face are kept constant, the interface of the tailings material should be substantially parallel to the surface of the dam slope, which is shown in Figure 2. For most of the tailings dams, because the tailings water contains more heavy metals, the initial topography and the side of tailings dam are treated as an impermeable boundary. On the other hand, the tailings dam extends very long in the direction of the reservoir area. Therefore, the consolidation of the tailings dam is a 1D problem.

As can be seen from Figure 2, the closer to inside the reservoir area is, the thicker the tailings mud is and the worse the average consolidation is. In other words, the closer to inside the reservoir area is, the greater pore water pressure of the tailings dam is and the lower shear strength of the tailings material is. Generally speaking, the construction of a tailings dam needs several decades to reach the design elevation. With the continuous production of mining, the height of the tailings dam is increasing. During this process, the accumulation and dissipation of the excess pore water pressure of the tailings material experience several different stages of development.

3.1. Tailings Thickness Increases with Time. In this stage, the thickness of the tailings increases with time. The tailings material produces great compressive deformation under self-weight effect. The pore water pressure produced by the self-weight pressure at the early stage is partially dissipated. Nevertheless, the pore water pressure caused by self-weight pressure is increased with the tailings thickness at the later stage, which leads to the increase of the pore water pressure with time. Therefore, the net pore water pressure is accumulated during this process.

3.2. Load on the Tailings Layer Increases with Time. The self-deposition of the tailings mud continues to extend to interior of the tailings reservoir area in this stage. The self-deposition of the tailings layer has ended where the end of the tailings reservoir is near. It overlapped with particles coarse tailings and tailings mud. The tailings mud has to bear a growing load with time. The tailings mud continues to consolidate under the combined action of self-weight and additional load. At this stage, the pore pressure produced by the self-weight pressure dissipates a little part, and the more pore water pressure produced by additional load accumulates. The net pore water pressure tends to accumulate in this process.

3.3. The Additional Load of the Tailings Material Remains Constant. For a specific part of the tailings dam, the upper boundary of the tailings dam has reached the designed elevation at a certain moment. Therefore, the additional load will remain unchanged from this moment. During this stage, the tailings mud will continue to consolidate under the combined action of self-weight and constant additional load. Thus, the pore water pressure will dissipate within the tailings dam. Obviously, the different parts of tailings mud have different consolidation stages. There is no additional load at the thickest tailings mud in the tailings reservoir until it is closed. Therefore, the tailings mud is always in the first stage and will never enter the second and third stages. From the above qualitative analysis, the pore water pressure, which is located at the slope dam segment, reaches the maximum value when the additional load just stops growing. The pore water pressure, which is located at dry beach segment and artificial lake segment, achieves the maximum value when the tailings dam reaches the maximum height. The numerical values of the pore water pressure at different locations and moments can be obtained by solving the partial differential equations of consolidation problems.

4. Analytical Solutions of the Pore Water Pressure

It is necessary to accurately calculate the accumulation and dissipation of the pore water pressure in tailings dam, of which a solution of 1D consolidation problem can be simplified. During the consolidation process, the soil parameters of tailings material (such as the change of bulk density, permeability coefficient, and consolidation coefficient) are changed with the consolidation process. The change law can be determined through many experiments. Under normal circumstances, it is difficult to obtain the analytical solution when the change law of soil parameters is considered. Therefore, the exact solution of the problem can only be depended on numerical calculation method.

As the horizontal length of the tailings reservoir is usually far greater than the vertical thickness (i.e., the horizontal length is 10 times more than the vertical thickness), the drainage consolidation effect of the horizontal direction can be ignored. Therefore, the vertical direction of drainage is needed to consider. Then, it is only a 1D consolidation problem. Based on the assumption of small-strain and constant of soil parameters, such as m_v, k, $\partial\Delta\sigma/\partial t$, γ', and $\partial H/\partial t$ are constant, (18) is degenerated into a constant coefficient nonhomogeneous parabolic partial differential equation

$$\frac{\partial^2 u}{\partial z^2} + \frac{1}{c_v}\left(\alpha + \gamma' Q - \frac{\partial u}{\partial t}\right) = 0, \tag{19}$$

where

$$\alpha = \frac{\partial \Delta\sigma}{\partial t}, \quad Q = \frac{\partial H}{\partial t}, \quad c_v = \frac{k}{m_v \gamma_w}. \tag{20}$$

The above three different stages are specifically discussed as follows.

4.1. The Thickness Mud of Tailings Mud Increases with Time. The overburden load of the tailings mud at this stage is zero, that is, $\alpha = 0$. The consolidation equation yields

$$\frac{\partial^2 u}{\partial z^2} + \frac{1}{c_v}\left(\gamma' Q - \frac{\partial u}{\partial t}\right) = 0. \tag{21}$$

If the bottom of the tailings mud is impermeable, the coordinates origin is taken as the impermeable bottom. The boundary conditions of the problem can be given as follows:

$$z = 0, \quad \frac{\partial u}{\partial z} = 0, \tag{22}$$

$$z = H, \quad u = 0. \tag{23}$$

The initial condition of the problem is given by

$$t = 0,\quad u = 0. \tag{24}$$

According to Gibson's [20] study, (25) is chosen as the solution of (21)

$$u = \gamma' h - t^{-1/2} \int_0^\infty g(\xi)\left(e^{-(x-\xi)^2/4c_v t} + e^{-(x+\xi)^2/4c_v t}\right) d\xi. \tag{25}$$

It can be verified that (25) satisfies the boundary condition (22). The choice of function $g(\xi)$ needs to satisfy (23). Substituting (25) into (23) gives

$$\frac{1}{2}\gamma' h t^{1/2} e^{h^2/4c_v t} = \int_0^\infty g(\xi) e^{-(x+\xi)^2/4c_v t} \cosh \frac{h\xi}{2c_v t} d\xi. \tag{26}$$

For any function $h(t)$, the partial differential equation (26) has no numerical solution. With regard to $h(t) = mt$, the following parameters can be introduced to transform the equation

$$\xi^2 = \tau,\quad p = \frac{1}{4}c_v t,\quad F(\tau) = g(\tau)\tau^{-1/2} \cosh \frac{k\tau^{1/2}}{2c_v}. \tag{27}$$

Equation (26) can be transformed into

$$\frac{1}{8} m\gamma' c_v^{-3/2} p^{-3/2} e^{m^2/16 c_v^2 p} = \int_0^\infty F(\tau) e^{-p\tau} d\tau. \tag{28}$$

Equation (28) can be obtained through the Laplace change:

$$F(\tau) = \frac{\gamma'}{2\pi^{1/2} c_v^{1/2}} \sinh \frac{m\tau^{1/2}}{c_v}. \tag{29}$$

Thence

$$g(\xi) = \frac{\gamma'\xi}{2\pi^{1/2} c_v^{1/2}} \tanh \frac{m\xi}{c_v}. \tag{30}$$

Combining the solutions of (25) and (30), the solution of (21) is obtained

$$u = \gamma' Qt - \gamma' (\pi c_v t)^{-1/2} \exp\left(\frac{-z^2}{4c_v t}\right) \int_0^\infty \xi \tanh \frac{Q\xi}{2c_v} \cdot \cosh \frac{z\xi}{2c_v t} \exp\left(\frac{-\xi^2}{2c_v t}\right) d\xi. \tag{31}$$

It can be verified that (31) satisfies the initial condition of (24).

4.2. The Additional Load of Tailings Mud Increases with Time. During this stage, the thickness of tailings mud not only does not increase with time but also gradually decreases with the increase of effective stress. For the sake of simplicity, assuming that the thickness of tailings mud H is constant, the basic consolidation differential (19) can be simplified as

$$c_v \frac{\partial^2 u}{\partial z^2} = \frac{\partial u}{\partial t} - \frac{\partial \Delta\sigma}{\partial t}. \tag{32}$$

The initial condition of the problem is given by

$$t = t_1,\quad u = u_0(z). \tag{33}$$

The boundary conditions of the problem can be given as

$$z = 0,\quad u = 0;\quad z = H,\quad \frac{\partial u}{\partial z} = 0. \tag{34}$$

The following three questions are called problem A, problem B, and problem C, respectively:

A:

$$\frac{\partial u}{\partial t} = c_v \frac{\partial^2 u}{\partial z^2}$$

$$\text{Initial conditions: } t = t_1,\quad u = u_0(z)$$

$$\text{Boundary conditions: } z = 0,\quad u = 0;\quad z = H,\quad \frac{\partial u}{\partial z} = 0. \tag{35}$$

B:

$$\frac{\partial u}{\partial t} = c_v \frac{\partial^2 u}{\partial z^2} + \frac{\partial \Delta\sigma}{\partial t}$$

$$\text{Initial conditions: } t = t_1,\quad u = 0$$

$$\text{Boundary conditions: } z = 0,\quad u = 0;\quad z = H,\quad \frac{\partial u}{\partial z} = 0. \tag{36}$$

C:

$$\frac{\partial u}{\partial t} = c_v \frac{\partial^2 u}{\partial z^2} + \frac{\partial \Delta\sigma}{\partial t}$$

$$\begin{aligned} &\text{Initial conditions:} \quad t = t_1, \\ &u_0(z) \\ &\text{Boundary conditions:} \quad z = 0, \\ &u = 0 \\ &z = H, \\ &\frac{\partial u}{\partial z} = 0. \end{aligned} \tag{37}$$

If the solution of problem A, problem B, and problem C is u_A, u_B, and u_C respectively, it can be directly proved through substitution method:

$$u_C = u_A + u_B. \tag{38}$$

The solution of u_A belongs to the Terzaghi classical 1D consolidation problem. The analytical solution can be obtained.

$$u_A = \sum_{j=1}^{\infty} A_n \sin\frac{j\pi z}{2H} \exp\left[\frac{-j^2\pi^2 c_v (t - t_1)}{4H^2}\right], \tag{39}$$

where

$$A_n = \frac{1}{H}\int_0^{2H} u_0(z) \sin\frac{j\pi z}{2H} dz. \tag{40}$$

$u_0(z)$ is the solution of (31). Substituting (31) into (40) by Matlab software, A_n can be obtained. Substituting it into (39), the numerical solution u_A can be given.

For solution to the problem u_B, Schiffman [21] acquires the following series solutions:

$$u_B = \frac{16\alpha H^2}{\pi^3 c_v} \sum_{1,3,5}^{\infty} \frac{1}{j^3} \sin\frac{j\pi z}{2H} \cdot \left\{1 - \exp\left[\frac{-j^2\pi^2 c_v (t - t_1)}{4H^2}\right]\right\}. \tag{41}$$

When the tailings mud is below the groundwater level, the total stress $\sigma(z)$ of the tailings mud at the depth z can be expressed as follows:

$$\sigma(z) = \gamma' z + \frac{\partial \Delta\sigma}{\partial t}(t - t_1). \tag{42}$$

The ratio of pore water pressure to total stress can be easily determined as follows:

$$U_z = \frac{u_A + u_B}{\sigma(z)}. \tag{43}$$

4.3. The Additional Load on the Tailings Mud Remains Constant. In this stage, the basic differential equation and the definite condition are problem D:

D:

$$\begin{aligned} &\frac{\partial u}{\partial t} = c_v \frac{\partial^2 u}{\partial z^2} \\ &t = t_2, \\ &u = (u_A + u_B)\big|_{t=t_2} \\ &z = 0, \\ &u = 0 \\ &z = H, \\ &\frac{\partial u}{\partial z} = 0. \end{aligned} \tag{44}$$

The following questions can be called problem E:

E:

$$\begin{aligned} &\frac{\partial u}{\partial t} = c_v \frac{\partial^2 u}{\partial z^2} \\ &t = t_2, \\ &u = u_B\big|_{t=t_2} = u_1(z) \\ &z = 0, \\ &u = 0 \\ &z = H, \\ &\frac{\partial u}{\partial z} = 0. \end{aligned} \tag{45}$$

If the solution of problem A is u_A and the solution of problem E is u_E, the solution of problem D can be easily demonstrated is u_D as follows:

$$u_D = u_A + u_E. \tag{46}$$

The definite condition of problem E and problem A is exactly the same in form, except that the function $u_1(z)$ is different from the function $u_0(z)$.

Suppose that

$$u_E = \sum_{j=1}^{\infty} B_n \sin\frac{j\pi z}{2H} \exp\left[\frac{-j^2\pi^2 c_v (t - t_2)}{4H^2}\right]. \tag{47}$$

Then

$$B_n = \frac{1}{H}\int_0^{2H} u_1(z) \sin\frac{j\pi z}{2H} dz, \tag{48}$$

where

$$u_1(z) = \frac{16\alpha H^2}{\pi^3 c_v} \cdot \sum_{1,2,3}^{\infty} \frac{1}{j^3} \sin\frac{j\pi z}{2H} \left\{1 - \exp\left[\frac{-j^2\pi^2 c_v (t_2 - t_1)}{4H^2}\right]\right\}. \tag{49}$$

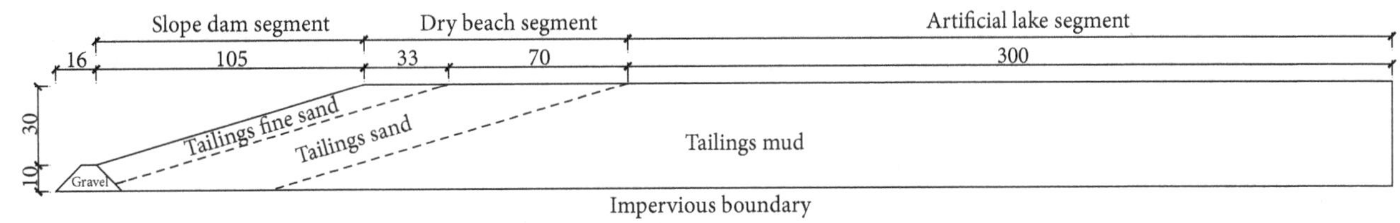

FIGURE 3: Simplified section for calculation of pore water pressure in a tailings dam.

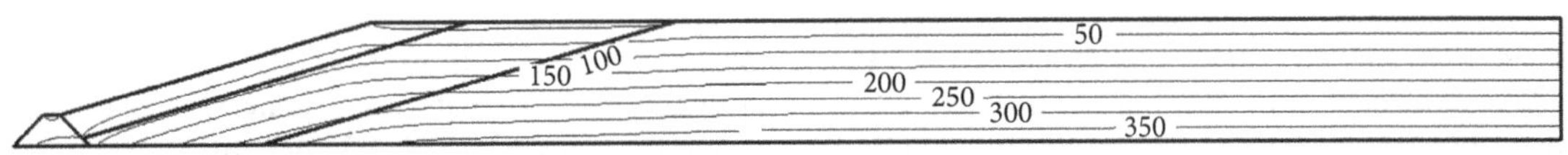

FIGURE 4: Distribution of pore water pressure at the moment when the dam rises to its maximum height.

According to the orthogonal rule of the solution, the following relationship can be obtained:

$$\frac{1}{H}\int_0^{2H}\sin\frac{m\pi z}{2H}\sin\frac{j\pi z}{2H}dz=\begin{cases}0, & m\neq j\\ 1, & m=j.\end{cases} \quad (50)$$

Equation (48) can be simplified through (50)

$$B_n = \begin{cases}\dfrac{16\alpha H^2}{\pi^3 c_v}\dfrac{1}{j^3}\left\{1-\exp\left[\dfrac{-j^2\pi^2 c_v(t_2-t_1)}{4H^2}\right]\right\} & j=1,3,5\\ 0 & j=2,4,6.\end{cases} \quad (51)$$

Substituting (51) into (47) yields

$$u_E = \frac{16\alpha H^2}{\pi^3 c_v}\sum_{1,3,5}^{\infty}\frac{1}{j^3}\left\{1-\exp\left[\frac{-j^2\pi^2 c_v(t_2-t_1)}{4H^2}\right]\right\} \cdot \sin\frac{j\pi z}{2H}\exp\left[\frac{-j^2\pi^2 c_v(t-t_2)}{4H^2}\right]. \quad (52)$$

For the third stage, the analytical solution of pore water pressure is obtained.

Similar to (42), the ratio of pore water pressure to total stress at a certain depth of tailings mud at this stage is as follows:

$$U_z = \frac{u_A+u_E}{\sigma(z)}. \quad (53)$$

During this stage, $\sigma(z)$ does not change with time, and it can take the following value:

$$\sigma(z) = \gamma' z + \alpha(t_2-t_1). \quad (54)$$

5. Example

The simplified calculation section of the representative section of a tailings dam is shown in Figure 3. The length of the tailings dam is 368 m. Therefore, the problem can be simplified as 1D consolidation. The slime layer below the level is assumed for impervious boundary. The rising speed of the dam height is 1.75 m/y. The saturation density of gravel is 22.0 kN/m^3, the saturation density of the tailings fine sand is 19.6 kN/m^3, the saturation density of the tailings sand is 19.1 kN/m^3, and the saturation density of the tailings mud is 18.8 kN/m^3. The permeability coefficient of gravel is 3×10^{-3} m/s, the permeability coefficient of the tailings fine sand is 5×10^{-6} m/s, the permeability coefficient of the tailings sand is 2×10^{-6} m/s, and the permeability coefficient of the tailings mud is 1.5×10^{-8} m/s. The consolidation coefficient of gravel is 3.8×10^{-1} cm^2/s, the consolidation coefficient of the tailings fine sand is 5.4×10^{-3} cm^2/s, the consolidation coefficient of the tailings sand is 2.8×10^{-3} cm^2/s, and the consolidation coefficient of the tailings mud is 1.2×10^{-5} cm^2/s. Please determine the distribution of pore water pressure when the tailings dam reaches maximum height.

The above problem can be solved through (31), (46), and (52). The calculation can be accomplished through the MATLAB software. Firstly, a coordinate system can be set up. Then, the corresponding coordinates are taken out of the interval 1 m. The region is determined according to the coordinates. Then, the pore water pressure can be calculated through the corresponding equation. Finally, the contour map can be drawn through a set of array. The results of the calculation are shown in Figure 4.

6. Discussion

Due to the similarity between the tailings dam and the reservoir dam in the geotechnical structure, many scholars directly introduce the calculation methods of the reservoir dam, whose theory is relatively mature, into the tailings dam. Considering the difference on construction cycle, construction materials between the tailings dam and reservoir dam, it makes the calculation results inconsistent with the actual.

Based on the Terzaghi consolidation theory of 1D, the tailings dam is divided into the slope dam segment, the dry beach segment, and the artificial lake segment. The solutions of the pore water pressure are derived, respectively. The analysis shows that the additional load of the slope dam segment is unchanged, which can be calculated using (46). The analysis indicates that tailings thickness of the artificial lake segment increases with time during construction period, which can be calculated by (31). The situation of the dry beach section is slightly complicated. The additional load of the dry beach section which is under the slope dam segment is basically the same, which can be calculated by (52). The additional load on the other sections of the dry beach segment increases with time, which can be calculated using (31).

The theoretical derivation is based on 1D consolidation theory. It is only considering consolidation in the vertical direction. Since the horizontal scale of most tailings dams is much larger than the vertical direction, it has little influence to ignore the drainage of horizontal. From the point of engineering view, it is conservative to the stability of the tailings dam. Considering drainage of the horizontal, the pore water pressure will be lessened and the safety factor of the tailings dam will be greater.

It is assumed that the deformation of the tailings is small deformation during the consolidation process. If the actual tailings are loose relatively, the deformation of the tailings is large deformation. He et al. [22] studied the pore water pressure of saturated soils using the updated Lagrangian formulation of large strain method (ULM), total Lagrangian formulation of large strain method (TLM), and small strain method (SSM). The results point out that the pore water pressure of the above three methods is exactly the same when the strain is less than 2.4%. When the strain is greater than 2.4%, the pore water pressure of the ULM and SSM still keeps the same, and the pore water pressure of the TLM is obviously less than the ULM and SSM. When the strain reaches 7.5%, the pore water pressure of the ULM and SMM began to bifurcate. When the deformation of the tailings dam is examined for a relatively short time, the consolidation of prophase tailings dam has been completed under previous load. The consolidation deformation of the whole tailings dam is relatively small, and its deformation conforms to the small deformation assumption. On the other hand, it is difficult to calculate and solve partial differential equations using large deformation theory. The differences between large deformation and small deformation need to be further studied.

It is supposed that the mechanical parameters such as permeability coefficient and consolidation coefficient are constant during the consolidation process. Previous studies [23] have shown that the permeability coefficient of rock and soil decreases with the increase of deformation during the consolidation process. It is believed that the value of C_c/C_k decides whether it is necessary to take under consideration the effect of nonlinear property by Zhuang et al. [24]. The results show that pore water pressure calculated depends on the ratio of loading intensity to the initial effective vertical stress subjected to time-dependent loading by Conte and Troncone [25]. How the permeability coefficient of tailings material changes during the consolidation process is less studied by relevant scholars. How the permeability coefficient of tailings material changes during the consolidation process is less studied by relevant scholars. From the aspect of engineering view, the larger the pore water pressure of actual is, the smaller the safety factor of tailings dam is. Therefore, it is dangerous for the tailings dam. The law of permeability coefficient of tailing material during consolidation process needs to be further studied.

7. Conclusion

Tailings dam is a very important geotechnical structure of mine engineering. The calculation of pore water pressure has a great impact on the safety factor of tailings dam slope. How to accurately estimate pore water pressure is very difficult. Based on the assumption of 1D consolidation and small strain of tailings material, a general equation of the pore water pressure is proposed. According to dissipation and accumulation characteristics of the pore water pressure in the tailings dam, the tailings dam can be divided into the slope dam segment, the dry beach segment, and the artificial lake segment. The analytic solutions of the corresponding segment are obtained through solving the partial differential equation, which has great significance to the stability of the tailings dam.

Notations

τ: Shear strength of the tailings material
$\sigma(z)$: Maximum increases in vertical total stress as a function of depth z
ξ: A parameter introduced to transform an equation
α: The rate of the additional load on the tailings mud, $\alpha = \partial\Delta\sigma/\partial t$
x: Coordinate of x direction
p: A parameter introduced to transform an equation
n: Porosity of the tailings material
m, j: Counters, 1, 2, 3,. . .
$g(\tau)$: A function of variable τ
$g(\xi)$: A function introduced into solving partial differential equation
Q: The rate of the thickness of the tailings mud, $Q = \partial H/\partial t$
$F(\tau)$: A function introduced to transform an equation
σ: Stress of the tailings material
z: Coordinate of z direction
u: The pore water pressure
k: Coefficient of coefficient
h: The height of static water level
H: Thickness of tailings mud
e: Porosity ratio
$\Delta\sigma$: Increment of stress
$u_1(z)$: Termination pore water pressure as a function of depth z
$u_0(z)$: Initial pore water pressure as a function of depth z
t_2: Termination time
t_1: Initial time
c_v: Coefficient of consolidation
U_z: Saturation
B_n: Coefficients to be determined

A_n: Coefficients to be determined
σ_0: Initial total stress
σ': Effective vertical stress
ε_z: Strain in the z direction
γ_w: Bulk density of water
γ_s: Saturated bulk density of tailings dam
γ': Buoyancy unit weight of tailings material
v_z: The actual velocity along the flow direction in the tailings dam
p_w: Water pressure
m_v: Coefficient of volume compressibility
i_z: Hydraulic gradient in the z direction
H_1: The total height of tailings mud plus static water level
F_z: The drag resistance force on the pore wall of a unit volume in the z direction.

Conflicts of Interest

The authors declare that they have no conflicts of interest.

Acknowledgments

This paper is supported by the Open Research Fund of State Key Laboratory of Geomechanics and Geotechnical Engineering, Institute of Rock and Soil Mechanics, Chinese Academy of Sciences, (Grant no. Z013009), the National Key Research and Development Program of China (Project no. 2017YFC0804601), the National Natural Science Foundation of China (Grant nos. 51764020; 51741410; 51234004), and the Natural Science Foundation of Yunnan Province (Grant no. 2015FB130). The authors would like to thank them for providing the financial support for conducting this research.

References

[1] K. Terzaghi, *Theoretical Soil Mechanics*, John Wiley & Sons, New York, NY, USA, 1943.

[2] R. L. Schiffman and J. R. Stein, "One-dimensional consolidation of layered systems," *Journal of the Soil Mechanics and Foundations Division*, vol. 96, no. 4, pp. 1499–1504, 1970.

[3] R. E. Olson, "Consolidation under time-dependent loading," *Journal of the Geotechnical Engineering Division*, vol. 103, no. 1, pp. 55–60, 1977.

[4] Y.-Q. Cai, X.-Y. Geng, and C.-J. Xu, "Solution of one-dimensional finite-strain consolidation of soil with variable compressibility under cyclic loadings," *Computers & Geosciences*, vol. 34, no. 1, pp. 31–40, 2007.

[5] N. Müthing, S. S. Razouki, M. Datcheva, and T. Schanz, "Rigorous solution for 1-D consolidation of a clay layer under haversine cyclic loading with rest period," *SpringerPlus*, vol. 5, no. 1, article no. 1987, 2016.

[6] A. Qin, D. Sun, and J. Zhang, "Semi-analytical solution to one-dimensional consolidation for viscoelastic unsaturated soils," *Computers & Geosciences*, vol. 62, pp. 110–117, 2014.

[7] L. Wang, D. Sun, P. Li, and Y. Xie, "Semi-analytical solution for one-dimensional consolidation of fractional derivative viscoelastic saturated soils," *Computers & Geosciences*, vol. 83, pp. 30–39, 2017.

[8] R. E. Gibson, "The Progress of Consolidation in a Clay Layer Increasing in Thickness with Time," *Géotechnique*, vol. 8, no. 4, pp. 171–182, 1958.

[9] S. S. Razouki and T. Schanz, "One-dimensional consolidation under haversine repeated loading with rest period," *Acta Geotechnica*, vol. 6, no. 1, pp. 13–20, 2011.

[10] S. S. Razouki, P. Bonnier, M. Datcheva, and T. Schanz, "Analytical solution for 1D consolidation under haversine cyclic loading," *International Journal for Numerical and Analytical Methods in Geomechanics*, vol. 37, no. 14, pp. 2367–2372, 2013.

[11] M. S. Mahmoud and H. Deresiewicz, "Settlement of inhomogeneous consolidating soils—I: The single-drained layer under confined compression," *International Journal for Numerical and Analytical Methods in Geomechanics*, vol. 4, no. 1, pp. 57–72, 1980.

[12] G. Zhu and J.-H. Yin, "Analysis and mathematical solutions for consolidation of a soil layer with depth-dependent parameters under confined compression," *International Journal of Geomechanics*, vol. 12, no. 4, pp. 451–461, 2012.

[13] X. W. Tang, B. Niu, G. C. Cheng, and H. Shen, "Closed-form solution for consolidation of three-layer soil with a vertical drain system," *Geotextiles and Geomembranes*, vol. 36, pp. 81–91, 2013.

[14] G.-X. Mei and Q.-M. Chen, "Solution of Terzaghi one-dimensional consolidation equation with general boundary conditions," *Journal of Central South University*, vol. 20, no. 8, pp. 2239–2244, 2013.

[15] B. C. Hawlader, B. Muhunthan, and G. Imai, "State-dependent constitutive model and numerical solution of self-weight consolidation," *Géotechnique*, vol. 58, no. 2, pp. 133–141, 2008.

[16] M. A. Rahalt and A. R. Vuez, "Analysis of settlement and pore pressure induced by cyclic loading of silo," *Journal of Geotechnical and Geoenvironmental Engineering*, vol. 124, no. 12, pp. 1208–1210, 1998.

[17] E. H. Davis and G. P. Raymond, "A non-linear theory of consolidation," *Géotechnique*, vol. 15, no. 2, pp. 161–173, 1965.

[18] E. Conte and A. Troncone, "Soil layer response to pore pressure variations at the boundary," *Géotechnique*, vol. 58, no. 1, pp. 37–44, 2008.

[19] V. Pane and R. L. Schiffman, "A note on sedimentation and consolidation," *Géotechnique*, vol. 35, no. 1, pp. 69–72, 1985.

[20] R. E. Gibson, "A heat conduction problem involving a specified moving boundary," *Quarterly of Applied Mathematics*, vol. 16, no. 4, pp. 426–430, 1959.

[21] R. L. Schiffman, "Consolidation of soil under time-dependent loading and varying permeability," in *Proceedings of the Thirty-Seventh Annual Meeting of the Highway Research Board*, vol. 37, pp. 584–617, 1958.

[22] K. S. He, Z. J. Shen, and X. X. Peng, "The comparison of large strain method using total and updated Lagrangian finite element formulation and small strain method," *Chinese Journal of Geotechnical Engineering*, vol. 22, no. 1, pp. 30–34, 2000.

[23] G. Mesri and R. E. Olson, "Mechanisms controlling the permeability of clays," *Clays and Clay Minerals*, vol. 19, no. 3, pp. 151–158, 1971.

[24] Y.-C. Zhuang, K.-H. Xie, and X.-B. Li, "Nonlinear analysis of consolidation with variable compressibility and permeability," *Journal of Zhejiang University (Engineering Science)*, vol. 6, no. 3, pp. 181–187, 2005.

[25] E. Conte and A. Troncone, "Nonlinear consolidation of thin layers subjected to time-dependent loading," *Canadian Geotechnical Journal*, vol. 44, no. 6, pp. 717–725, 2007.

6

Displacement and Stress Analysis of Thin Plate for Cement Concrete Pavement

Liang Jia ,[1] Li Zhang,[1] Jian Guo,[1] Kai Yao ,[2] and Shikai He[1]

[1]*College of Civil Engineering, Lanzhou University of Technology, Lanzhou 730050, China*
[2]*Department of Civil & Environmental Engineering, National University of Singapore, Singapore 117576*

Correspondence should be addressed to Kai Yao; yaokai@u.nus.edu

Academic Editor: Fazal M. Mahomed

In order to analyze the displacement and stress of thin plate for cement concrete pavement with rectangle shape and resting on Winkler soil foundation, an analysis model was set up based on the theory of elastic thin plate on Winkler foundation. According to elasticity Kirchhoff theory of thin plates and Winkler soil foundation model, the expressions of displacement and stress were yielded by using the inverse Hankel transform. Numerical method was employed to calculate displacement and stress of the thin plate. Results showed that displacement and stress corresponded with using the method from *Code for Highway Design-Pavement*. Results showed that the method can provide a new computation method for displacement and stress of the thin plate about cement concrete pavement.

1. Introduction

The model of a thin plate on elastic foundation was mainly used in structural engineering in last years. Currently, thin plates or films of metal, ceramic, or synthetic materials were bonded in the surface of machine structural parts or electronic devices to improve their mechanical, thermal, electronic, or tribological properties [1]. At these applications, the subgrade of the thin plate can be simulated as a Winkler foundation, which reacted with pressure proportional to the deflection of the plate at each point. The plate in the above applications was loaded by vertical loads or bending moments. For cement concrete pavement, the applied load is mainly from the vehicles, which is vertically applied on pavement. So that the width and thickness of the pavement are usually determined by the traffic volume. And the thickness will increase with the designed traffic volume. A large number of research works have been published to solve problem of a classical thin plate [2–4] or a thin plate on an elastic foundation [5–9]. Fisher investigated the elastic impact of a sphere on a thin plate which is continuous contact with a foundation [10]. Jędrysiak investigated the dynamic response of thin elastic plates in Winkler foundation [11]. Some researches introduced an efficient boundary element approach for the analysis of thin plate resting on an elastic Winkler foundation with different boundary conditions [12–15]. Shao combined the Fourier spectral method and differential quadrature method for solving problems for thin plates resting on Winkler foundations with irregular domains [16].

However, few research works have been published concerning the stress and displacement of a thin plate on Winkler foundation. The present work is aimed to obtain an analytical solution of a finite rectangle thin plate on an elastic foundation under vertical forces. For this purpose, both the Hankel transform and inverse Hankel transform were adopted.

2. Basic Assumption and Governing Equations

According to the theory of elastic thin plate on Winkler soil foundation, a rectangle thin plate with thickness h was considered to rest on a Winkler soil foundation (Figure 1), the thickness of plate was less than one-fifth the size of plane, the plate was loaded by the vertical pressure $q(x, y)$ and the foundation reaction pressure p assumed to be proportional

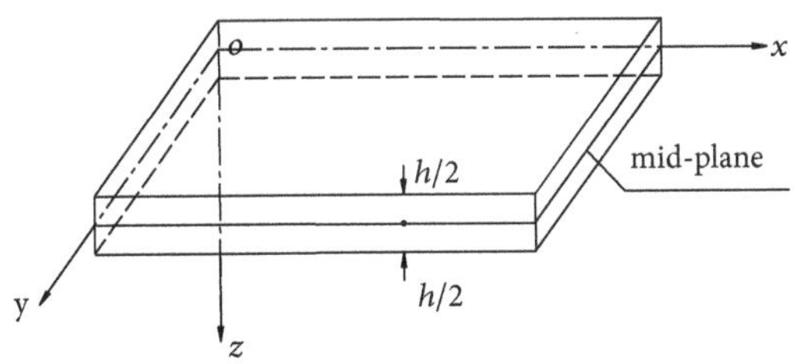

FIGURE 1: Model of the thin plate.

to the vertical deflection $\omega(x, y)$ of the plate [17], and the rectangular coordinate system was placed on midplane.

Kirchhoff proposed the basic assumption of the thin plate as follows [17].

(1) Length of normal line of the plate is unchanged before and after deformation, that is,

$$\varepsilon_z = 0 \quad (1)$$

where ε_z is normal strain.

(2) Before deformation, the normal line of the plate is perpendicular to midplane; after deformation, the normal line is still perpendicular to midplane, that is,

$$\gamma_{zx} = 0, \quad \gamma_{zy} = 0 \quad (2)$$

where γ_{zx}, γ_{zy} are shear strain.

(3) The plane of parallel to midplane does not squeeze, that is,

$$\sigma_z = 0 \quad (3)$$

where σ_z is stress.

(4) Horizontal displacement of the midplane does not occur during the process thin plate bending, which are

$$u_0 = 0, \quad v_0 = 0 \quad (4)$$

where u_0, v_0 are displacement.

According to the hypothesis (2), it yields
$\gamma_{zx} = 0$, namely, $\partial w/\partial x + \partial u/\partial z = 0$,
$\gamma_{zy} = 0$, namely, $\partial w/\partial y + \partial v/\partial z = 0$.
Thus,

$$u = -z\frac{\partial\omega}{\partial x} + u_0, \quad v = -z\frac{\partial\omega}{\partial y} + v_0 \quad (5)$$

According to the hypothesis (4), the displacement of thin plate can be obtained as follows:

$$u = -z\frac{\partial\omega}{\partial x}, \quad v = -z\frac{\partial\omega}{\partial y} \quad (6)$$

According to (6), the strain of thin plate can be obtained as follows:

$$\varepsilon_x = \frac{\partial u}{\partial x} = -z\frac{\partial^2\omega}{\partial x^2}, \quad \varepsilon_y = \frac{\partial v}{\partial y} = -z\frac{\partial^2\omega}{\partial y^2}, \quad \gamma_{xy} = \frac{\partial u}{\partial y} + \frac{\partial v}{\partial x} = -2z\frac{\partial^2\omega}{\partial x\partial y} \quad (7)$$

According to general Hook's law, we can obtain

$$\varepsilon_x = \frac{1}{E}\left[\sigma_x - \mu\left(\sigma_y + \sigma_z\right)\right], \quad \varepsilon_y = \frac{1}{E}\left[\sigma_y - \mu\left(\sigma_x + \sigma_z\right)\right], \quad \gamma_{xy} = \frac{1}{E}\left[\sigma_z - \mu\left(\sigma_x + \sigma_y\right)\right] \quad (8)$$

Substituting (7) into (8), the stress of thin plate can be obtained as follows:

$$\sigma_x = -\frac{Ez}{1-\mu^2}\left(\frac{\partial^2\omega}{\partial x^2} + \mu\frac{\partial^2\omega}{\partial y^2}\right), \quad \sigma_y = -\frac{Ez}{1-\mu^2}\left(\frac{\partial^2\omega}{\partial y^2} + \mu\frac{\partial^2\omega}{\partial x^2}\right), \quad \tau_{xy} = -\frac{Ez}{1+\mu}\frac{\partial^2\omega}{\partial x\partial y} \quad (9)$$

where E is elasticity modulus, μ is Poisson's ratio, u and v are displacement, ω is vertical deflection of the plate, and $\omega = \omega(x, y)$.

3. Stress Fields of the Elastic Thin Plate

A stress field is a region in a body for which the stress is defined at every point [18]. Stress fields were widely used in materials science. The stress field, which was composed of these generalized stresses, was called the generalized stress field.

A hexahedron element with thickness h was taken out from the thin plate (Figure 2); stress components acting on the four sides of hexahedron element were $M_x dy$, $M_y dx$, $M_{xy} dy$, $M_{yx} dx$, $Q_x dy$, and $Q_y dx$, where M_x, M_y, M_{xy}, M_{yx}, Q_x, and Q_y are called the generalized stress. Generalized stress can be expressed as follows:

$$M_x = -\int_{-h/2}^{h/2}\sigma_x z\,dz$$

$$M_y = -\int_{-h/2}^{h/2}\sigma_y z\,dz$$

$$M_{xy} = M_{yx} = -\int_{-h/2}^{h/2}\tau_{xy} z\,dz$$

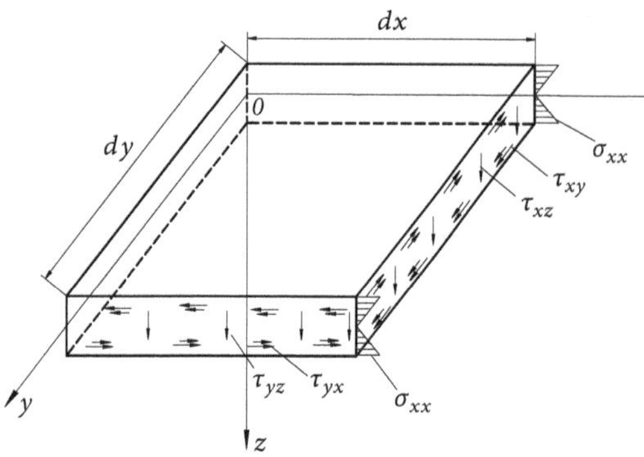

FIGURE 2: Stress of the plane.

$$Q_x = -\int_{-h/2}^{h/2} \tau_{zx}\, dz$$

$$Q_y = -\int_{-h/2}^{h/2} \tau_{zy}\, dz \tag{10}$$

Substituting (9) into (10) yields

$$M_x = -D\left(\frac{\partial^2\omega}{\partial x^2} + \mu\frac{\partial^2\omega}{\partial y^2}\right)$$

$$M_y = -D\left(\frac{\partial^2\omega}{\partial y^2} + \mu\frac{\partial^2\omega}{\partial x^2}\right)$$

$$M_{xy} = -D(1-\mu)\frac{\partial^2\omega}{\partial x\partial y} \tag{11}$$

$$Q_x = -D\frac{\partial(\nabla^2\omega)}{\partial x}$$

$$Q_y = -D\frac{\partial(\nabla^2\omega)}{\partial y}$$

where $\nabla^2 = \partial^2/\partial x^2 + \partial^2/\partial y^2, \nabla^2$ is the Laplacian operator, $D = Eh^3/12(1-\mu^2)$, and D is relative stiffness of the plate.

Combining (9) and (11) gives

$$\sigma_x = \frac{12M_x z}{h^3}$$

$$\sigma_y = \frac{12M_y z}{h^3}$$

$$\tau_{xy} = \frac{12M_{xy} z}{h^3} \tag{12}$$

$$\tau_{zx} = -\frac{6Q_x}{h^3}\left(\frac{h^2}{4} - z^2\right)$$

$$\tau_{zy} = -\frac{6Q_y}{h^3}\left(\frac{h^2}{4} - z^2\right)$$

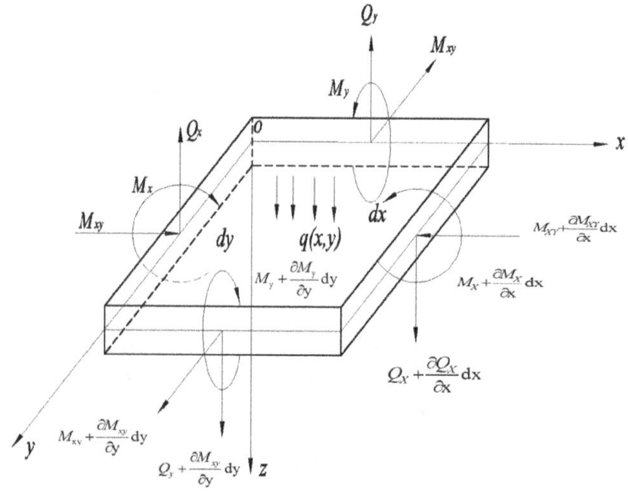

FIGURE 3: Internal force of the plane.

Under the condition of axisymmetric loading, stress of the plane can be obtained

$$M_r = -D\left(\frac{d^2w}{dr^2} + \frac{\mu}{r}\frac{dw}{dr}\right)$$

$$M_\theta = -D\left(\mu\frac{d^2w}{dr^2} + \frac{1}{r}\frac{dw}{dr}\right) \tag{13}$$

4. Bending Equations of the Thin Plate

As illustrated in Figure 3, uniform load $q(x,y)$ is applied on hexahedron element, while M_x, M_y, M_{xy}, M_{yx}, Q_x, and Q_y are generalized stress on the four sides of the element.

The loading added up to zero in the direction of the z axis, and solving bending moments of parallel to the x axis and the y axis yields

$$Q_x = \frac{\partial M_x}{\partial x} + \frac{\partial M_{xy}}{\partial y} \tag{14}$$

$$Q_y = \frac{\partial M_y}{\partial y} + \frac{\partial M_{xy}}{\partial x} \tag{15}$$

$$\frac{\partial Q_x}{\partial x} + \frac{\partial Q_y}{\partial y} + q = 0 \tag{16}$$

Substituting (14) and (15) into (16) yields

$$\frac{\partial^2 M_x}{\partial x^2} + 2\frac{\partial^2 M_{xy}}{\partial x\partial y} + \frac{\partial^2 M_y}{\partial y^2} + q = 0 \tag{17}$$

Substituting (11) into (17) yields

$$\mathrm{D}\nabla^2\nabla^2\omega(\mathrm{x},\mathrm{y}) = q(x,y) \tag{18}$$

where $\nabla^2 = \partial^2/\partial x^2 + \partial^2/\partial y^2$, ∇^2 is the Laplace operator.

According to the view of Winkler [1], a rectangle thin plate was considered to rest on a Winkler soil foundation, and the foundation reaction pressure p was only related to the vertical deflection ω of the plate.

$$p = k\omega \tag{19}$$

where k is reaction modulus of the foundation.

The plate was loaded as follows:

$$q(x, y) - k\omega(x, y) \tag{20}$$

A finite elastic thin plate on Winkler soil foundation; the bending equation was obtained as follows:

$$D\nabla^2\nabla^2\omega(x, y) = q(x, y) - k\omega(x, y) \tag{21}$$

5. The Equations of Stress and Displacement

Under the axisymmetric loading, the bending equation of the thin plate on Winkler soil foundation in cylindrical can be obtained as follows:

$$D\nabla^2\nabla^2\omega(r) = q(r) - kw(r) \tag{22}$$

where $\nabla^2 = (d^2/dr^2 + (1/r)(d/dr))$.

The application of the first class of zero order Hankel transform to equation (22) yields

$$D\xi^4\overline{\omega}(\xi) + k\overline{\omega}(\xi) = \overline{q}(\xi) \tag{23}$$

where $\overline{\omega}(\xi) = \int_0^\infty \omega(r)J_0(\xi r)r\,dr$, $\overline{q}(\xi) = \int_0^\infty q(r)J_0(\xi r)r\,dr$.

Solution for (23) yields [17]

$$\overline{\omega}(\xi) = \frac{\overline{q}(\xi)}{D\xi^4 + k} \tag{24}$$

where $D = Eh^3/12(1-\mu^2)$.

Set $l^4 = D/k$; l is the relative stiffness radius of the plate on Winkler soil foundation. The application of Hankel transform to (24) and expression of deflection can be obtained as follows.

$$\omega(r) = \frac{1}{k}\int_0^\infty \frac{\overline{q}(\xi)J_0(\xi r)}{1 + l^4\xi^4}\xi\,d\xi \tag{25}$$

Substituting (25) into (13) yields

$$\begin{aligned} M_r &= -D\left(\frac{d^2\omega(r)}{dr^2} + \frac{\mu}{r}\frac{d\omega(r)}{dr}\right) \\ &= l^4\int_0^\infty \frac{\overline{q}(\xi)J_0(\xi r)}{1 + l^4\xi^4}\left[\xi J_0(\xi r) - \frac{1-\mu}{r}J_1(\xi r)\right]\xi^2 d\xi \end{aligned} \tag{26}$$

$$\begin{aligned} M_\theta &= -D\left(\mu\frac{d^2\omega}{dr^2} + \frac{1}{r}\frac{d\omega}{dr}\right) \\ &= l^4\int_0^\infty \frac{\overline{q}(\xi)J_0(\xi r)}{1 + l^4\xi^4}\left[\mu\xi J_0(\xi r) + \frac{1-\mu}{r}J_1(\xi r)\right]\xi^2 d\xi \end{aligned} \tag{27}$$

The maximum radial stress σ_r and tangential stress σ_θ can be obtained as follows:

$$\sigma_r = \frac{6M_r}{h^2} \tag{28}$$

$$\sigma_\theta = \frac{6M_\theta}{h^2} \tag{29}$$

On the position r=0, (25)-(27)can be transformed as follows:

$$\omega(r) = \frac{1}{k}\int_0^\infty \frac{\overline{q}(\xi)}{1 + l^4\xi^4}\xi\,d\xi \tag{30}$$

$$M_r = l^4\int_0^\infty \frac{\overline{q}(\xi)}{1 + l^4\xi^4}\xi^3 d\xi \tag{31}$$

$$M_\theta = l^4\int_0^\infty \frac{\overline{q}(\xi)}{1 + l^4\xi^4}\mu\xi^3 d\xi \tag{32}$$

Equations (30)-(32) can be solved by numerical method. Combining equations (31) and (32) gives

$$M_\theta = \mu M_r \tag{33}$$

The thin plate for cement concrete pavement, μ, was less than 1. Therefore, σ_r was greater than σ_θ on the position $r = 0$. The maximum stress σ_r was calculated in the following example.

6. Calculation Example

In order to check the validity of the method presented in the paper, an example of thin plate for cement concrete pavement was given. The material parameters were chosen as follows: h = 0.18 m and h = 0.22 m, E = 11000 MPa, μ = 0.15, k = 1.4x10^7 N/m^3, δ = 0.151 m $q(x,y)$ = 0.7 MPa. We selected the integral interval changed from 0-10, 0-20, 0-30, 0-40, 0-50, and 0-1000, respectively, the displacement and stress were calculated, and the results were summarized in Tables 1, 2, 3, and 4.

According to the result of Table 1, integral interval tends to 0-20; when the thickness h of the thin plate is from 0.18 m to h = 0.22 m, displacement of the thin plate on the position r = 0 was 0.615 mm and 0.595 mm, respectively. According to the results of Table 2, integral interval tends to 0-40; when the thickness h of the thin plate is from 0.18 m to h = 0.22 m, stress of the thin plate on the position r = 0 was 1.53 MPa and 1.41 MPa, respectively.

7. Validation of the Method

The material parameters were chosen as the calculation example; according to the literature [19], calculation procedure was as follows.

Relative stiffness radius l of the pavement is

$$\begin{aligned} l_{h=0.18m} &= \sqrt[4]{\frac{Eh^3}{12(1-\mu^2)k}} = 0.25\text{ m} \\ l_{h=0.22m} &= \sqrt[4]{\frac{Eh^3}{12(1-\mu^2)k}} = 0.42\text{ m} \end{aligned} \tag{34}$$

Based on this, we can obtain coefficient α:

$$\begin{aligned} \alpha_{h=0.18m} &= \frac{\delta}{l_{h=0.18m}} = 0.604 \\ \alpha_{h=0.22m} &= \frac{\delta}{l_{h=0.18m}} = 0.356 \end{aligned} \tag{35}$$

TABLE 1: Displacement of plate in the middle position ($h = 0.18$ m).

integration interval	0-10	0-20	0-30	0-40	0-50	0-1000
displacement of the plate in the middle position/mm	0.614963	0.615458	0.615507	0.615473	0.615469	0.615433

TABLE 2: Stresses of plate in the middle position ($h = 0.18$ m).

integration interval	0-10	0-20	0-30	0-40	0-50	0-1000
stress of the plate in the middle position /MPa	1.46	1.53	1.54	1.53	1.53	1.53

TABLE 3: Displacement of plate in the middle position ($h = 0.22$ m).

integration interval	0-10	0-20	0-30	0-40	0-50	0-1000
displacement of the plate in the middle position/mm	0.594542	0.595469	0. 595564	0. 595535	0. 595531	0. 595529

TABLE 4: Stresses of plate in the middle position ($h = 0.22$ m).

integration interval	0-10	0-20	0-30	0-40	0-50	0-1000
stress of the plate in the middle position /MPa	1.35	1.41	1.42	1.41	1.41	1.41

The deflection of the thin plate on the position $r = 0$ can be found in the table [19].

$$\begin{aligned} \alpha &= 0.604, \\ \varpi &= 1.23 \times 10^{-2} \\ \alpha &= 0.356, \\ \varpi &= 1.19 \times 10^{-2} \end{aligned} \tag{36}$$

Deflection of the plate on the position $r = 0$ can be obtained as follows:

$$\begin{aligned} \omega_{h=0.18m} &= \frac{p}{k}\varpi = \frac{0.7}{14} \times 1.23 \times 10^{-2} = 6.15 \times 10^{-4}\ \text{m} \\ \omega_{h=0.22m} &= \frac{p}{k}\varpi = \frac{0.7}{14} \times 1.19 \times 10^{-2} = 5.95 \times 10^{-4}\ \text{m} \end{aligned} \tag{37}$$

The bending moment coefficient on the position $r = 0$ can be found in the table [19].

$$\begin{aligned} \alpha &= 0.604, \\ \overline{M_r} &= 0.172 \\ \alpha &= 0.356, \\ \overline{M_r} &= 0.237 \end{aligned} \tag{38}$$

Bending moment of the plate on the position $r = 0$ can be obtained as follows:

$$\begin{aligned} M_{rh=0.18m} &= p\overline{M_r} = 48 \times 10^3 \times 0.172 = 8.26 \times 10^3\ \text{N} \\ M_{rh=0.22m} &= p\overline{M_r} = 48 \times 10^3 \times 0.237 \\ &= 11.37 \times 10^3\ \text{N} \end{aligned} \tag{39}$$

Radial stress of the plate on the position $r = 0$ can be obtained as follows:

$$\begin{aligned} \sigma_{rh=0.18m} &= \frac{6M_r}{h^2} = \frac{6 \times 8.26 \times 10^3}{0.18^2} = 1.53\ \text{MPa} \\ \sigma_{rh=0.22m} &= \frac{6M_r}{h^2} = \frac{6 \times 11.37 \times 10^3}{0.22^2} = 1.41\ \text{MPa} \end{aligned} \tag{40}$$

From the above analysis, the stress and deflection based on the method were equal to using the method from *Code for Highway Design-Pavement* [19]. It also indicated that the present method was reasonable in the paper.

8. Conclusions

(1) The expressions of displacement and stress for cement concrete pavement on elastic foundation were derived.

(2) In order to verify the theoretical results, an example of thin plate for cement concrete pavement was given; the stress and deflection correspond with using the method from *Code for Highway Design-Pavement*.

(3) Future experiment research is needed to be carried out to verify the theoretical results present in this paper.

Conflicts of Interest

The authors declare that they have no conflicts of interest.

Acknowledgments

This work was supported by the National Natural Science Foundation of China (Grant no. 51568044).

References

[1] D. G. Pavlou, "Green's function for a pre-stressed thin plate on an elastic foundation under axisymmetric loading," *Engineering Analysis with Boundary Elements*, vol. 29, no. 5, pp. 428–434, 2005.

[2] M. V. Delyavs'kyi, N. V. Zdolbits'ka, L. I. Onyshko, and A. P. Zdolbits'kyi, "Determination of the stress-strain state in thin orthotropic plates on Winkler's elastic foundations," *Journal of Materials Science*, vol. 50, no. 6, pp. 782–791, 2015.

[3] L. De Oliveira Neto and J. B. De Paiva, "Cubic approximation for the transverse displacement in BEM for elastic plates analysis," *Engineering Analysis with Boundary Elements*, vol. 28, no. 7, pp. 869–880, 2004.

[4] L. P. Jr, "Plate bending analysis using the classical or the Reissner-Mindlin models," *Engineering Analysis with Boundary Elements*, vol. 27, no. 6, pp. 603–609, 2003.

[5] W. Yu and F. Chen, "Global bifurcations and homoclinic trees in motion of a thin rectangular plate on a nonlinear elastic foundation," *Meccanica*, vol. 48, no. 5, pp. 1251–1261, 2013.

[6] M. Utku, E. Çitipitioglu, and I. Inceleme, "Circular plates on elastic foundations modelled with annular plates," *Computers & Structures*, vol. 78, no. 1, pp. 365–374, 2000.

[7] K. Chandrashekhara and S. J. Antony, "Elastic analysis of an annular slab-soil interaction problem using a hybrid method," *Computers & Geosciences*, vol. 20, no. 2, pp. 161–176, 1997.

[8] T. Morimoto and Y. Tanigawa, "Elastic stability of inhomogeneous thin plates on an elastic foundation," *Archive of Applied Mechanics*, vol. 77, no. 9, pp. 653–674, 2007.

[9] F. Cafarella, "Thin plates on elastic foundation with unilateral behaviour," *Meccanica*, vol. 7, no. 1, pp. 39-39, 1972.

[10] H. D. Fisher, "The impact of an elastic sphere on a thin elastic plate supported by a winkler foundation," *Journal of Applied Mechanics*, vol. 42, no. 1, pp. 133–135, 1975.

[11] J. Jedrysiak, "Dynamics of thin periodic plates resting on a periodically inhomogeneous Winkler foundation," *Archive of Applied Mechanics*, vol. 69, no. 5, pp. 345–356, 1999.

[12] A. Joodaky, I. Joodaky, M. Hedayati, R. Masoomi, and E. B. Farahani, "Corrigendum to "Deflection and stress analysis of thin FGM skew plates on Winkler foundation with various boundary conditions using extended Kantorovich method," *Composites*, vol. 51, no. 3, pp. 191–196, 2013.

[13] A. El-Zafrany, S. Fadhil, and K. Al-Hosani, "A new fundamental solution for boundary element analysis of thin plates on Winkler foundation," *International Journal for Numerical Methods in Engineering*, vol. 38, no. 6, pp. 887–903, 1995.

[14] Z. Yang and W. Li, "Subharmonic resonance of rectangular thin plate with four sides free on Winkler foundation and its singularity analysis," *Journal of Mechanical Strength*, vol. 29, no. 2, pp. 329–333, 2007.

[15] A. Joodaky, I. Joodaky, M. Hedayati, R. Masoomi, and E. B. Farahani, "Deflection and stress analysis of thin FGM skew plates on Winkler foundation with various boundary conditions using extended Kantorovich method," *Composites Part B: Engineering*, vol. 51, pp. 191–196, 2013.

[16] W. Shao and X. Wu, "Fourier differential quadrature method for irregular thin plate bending problems on Winkler foundation," *Engineering Analysis with Boundary Elements*, vol. 35, no. 3, pp. 389–394, 2011.

[17] Z. Chuanchao and W. Binggang, *Mechanical analysis of road structures*, China Communications Press, Beijing, 2003.

[18] E. Lomakin, S. Alexandrov, and Y.-R. Jeng, "Stress and strain fields in rotating elastic/plastic annular discs," *Archive of Applied Mechanics*, vol. 86, no. 1-2, pp. 235–244, 2016.

[19] Y. Zukuang, *Code for highway design-pavement*, China Communications Press, Beijin, China, 3rd edition, 2003.

7

Dynamic Analysis of a Plate on the Generalized Foundation with Fractional Damping Subjected to Random Excitation

Ali Hosseinkhani,[1] Davood Younesian,[1] and Saman Farhangdoust[2]

[1] *Center of Excellence in Railway Transportation, School of Railway Engineering, Iran University of Science and Technology, Tehran 16846-13114, Iran*
[2] *Department of Civil and Environmental Engineering, Florida International University, Miami, FL 33174, USA*

Correspondence should be addressed to Saman Farhangdoust; sfarh006@fiu.edu

Academic Editor: Giovanni Falsone

Stochastic response of a plate on the generalized foundation driven by random excitation is solved in this paper. Governing differential equation is obtained by employing the Galerkin method. The generalized harmonic function technique is applied to the governing equation of motion. Using the stochastic averaging method (SAM), the system is approximated by the time homogeneous diffusive Markov process. Corresponding approximate stationary probability function is achieved by solving associated Fokker-Plank-Kolmogorov (FPK). An analytical solution is presented for the stationary probability of the amplitude and velocity. Validity of the stationary probability is verified by Monte-Carlo simulation. Parametric study is carried out to investigate effects of foundation parameters and excitation intensity on the stationary probability function. It is found that the fractional properties act similar to the foundation stiffness and damping and can be employed as a new control parameter for the support design.

1. Introduction

Fractional calculus has ancient history over 300 years but has been limited to the pure mathematics until new decades. In recent years an increasing interest to the fractional calculus is observed, because of its various applications in physics and engineering [1, 2]. One of the active branches of fractional calculus is addressed to the material science. Fractional calculus is widely used nowadays in modeling of the viscoelastic fractional material applicable in damping elements in structural and bio applications [3, 4].

A plate resting on a viscoelastic foundation can physically model a broad range of engineering problems from runways and slab/panel elements in structural and vehicle engineering to tissue modeling in biomechanics. Free vibration of plates on the generalized nonlinear viscoelastic foundation was analytically studied by Younesian et al. [5]. However, in many real cases, the plate is subject to harmonic or random excitation. In such cases, any kind of geometrical or material nonlinearity in presence of random excitation imposes high complexity to obtain analytical or even numerical solutions. Xu and Cheung [6] have extended averaging method to strongly nonlinear systems. Stochastic averaging of a SDOF system under combined harmonic and noise excitation was analyzed by Lal and Singh [7]. Response and stability of a SDOF system subjected to wide-band random excitation were studied by Huang et al. [8]. Zhu et al. introduced a modified stochastic averaging method by use of the energy balance method [9]. Li et al. [10] numerically studied dynamic stability of a column with fractional derivative by use of the averaging method. Spanous et al. presented a standard formulation for modeling vibration of systems with frequency-dependent parameters and fractional properties [11]. They developed a linearization-based solution to a stochastic nonlinear fractional equation to analyze nonlinear random vibration of a beam with fractional properties [12, 13]. Wahi and Chatterjee studied effect of the delay and fractional derivative terms in oscillations of a resonator [14]. Yang and Xu used stochastic averaging to investigate the effects of fractional parameters on the output voltage of an energy harvester with fractional damping [15].

In another parallel research line, other researchers have concentrated on the fractional systems under random excitations. Huang and Jin [16] employed the stochastic averaging and generalized harmonic functions procedure for a SDOF strongly nonlinear oscillator under Gaussian white noise excitations; they found that the order of fractional derivative has a significant effect on the response of system. Response of a strong nonlinear oscillator [17], its reliability function [18], and its stochastic/asymptotic stability [19, 20] have been studied by Chen et al. They could separate fractional derivative into the equivalent quasilinear dissipative force and quasilinear restoring force [21]. Yang et al. studied the stationary and stochastic response of nonlinear system with fractional derivative under white Gaussian noise input [22, 23]. Failla and Pirrotta presented a numerical method to calculate the response of a system under stochastic excitation based on the discretization of fractional derivative [24].

Noise-induced bifurcation is an undesired phenomenon in stochastic system; In this regard, Sun et al. performed a stochastic sensitivity analysis for the probability response of a nonautonomous system and estimated the critical intensity of noise which leads to explosive bifurcation [25]. Chen et al. derived the amplitude probability density of a Duffing oscillator by the aid of stochastic averaging and by employing the corresponding FPK's equation. Then, they designed a fractional order PID controller in order to eliminate stochastic bifurcations, while the integer-order controller was not able to provide stable response [26]. Chen and Sun presented exact solution for FPK equation of nonlinear systems; they considered an exponential polynomial as the basic function for the response and by an iteration scheme, the basic solution was converged to the exact solution [27]. Yurchenko et al. introduced a suitable change of variable to a vibro-impact system with fractional damping under random excitation; and by use of stochastic averaging, amplitude of mean squared response was presented analytically [28]. Yang et al. obtained an approximate-analytical solution for a noisy vibro-impact system by employing nonsmooth transformation and stochastic averaging and studied stochastic bifurcation phenomena for such systems [29].

In other side, remarkable research has been devoted to identify dynamic characteristics of the conventional plates under random loading. Lal and Singh studied nonlinear free vibration of laminated composite plates resting on elastic foundation with random system properties in thermal environment [7]. Free and forced vibration analysis for the Reissner-Mindlin plates on Pasternak foundation and Winkler elastic foundation were presented by Shen et al. [30]. Dynamic response for a shear deformable plate on two-parameter elastic foundation under transverse dynamic load has been solved by use of perturbation technique by Huang and Zheng [31]. Kostantakopoulos et al. used the mode summation approach and solved dynamic response of a plate on the Winkler foundation under the action of an impact load [32]. Dang-Trung et al. used the Cell-Based Smoothed Triangular Plate Element and analyzed dynamic response of a visco-elastically supported plate under moving load [33]. More recently, Cai et al. used the Galerkin's approach to study dynamic characteristic behavior of a thin plate on fractional Pasternak-type foundation [34]. Reutskiy [35, 36] proposed a new methodology based on the combination of the two techniques: the method of external excitation (MEE) and the backward substitution method (BSM) to solve higher order fractional eigenvalue problems. Using statistical linearization-based approach, Malara and Spanos presented large displacement response of a plate with fractional damping nature subjected to random load [37].

According to the reviewed literature, a plate on a foundation can be considered as the mathematical model of several practical problems. In this study, the model of foundation is considered to be the most general form (including fractional, general nonlinearity, and general damping). In such cases, due to mathematical complexity, there are very few reported publications on the context particularly when we have geometrical nonlinearity, fractional properties, and random excitation all together. In many practical applications, properties of the materials used as a foundation or vibration isolator are frequency-dependent. Therefore, conventional models such as Maxwell or Kelvin-Voigt are not accurate in such cases. We in this study show that how fractional characteristics can be applied for this purpose. The objective here is to provide a straightforward methodology to obtain a closed-form solution for a plate resting on one of the most generalized forms of viscoelastic fractional foundation. Presenting the integral-based closed-form solution, in this study, provides the computational platform for design engineers to calculate design parameters of the foundations based on the statistical pattern of the loading condition.

Apart from the methodology, questions about the effects of different material and geometrical properties of the foundation on the probability responses are answered. To this end, the averaging method and harmonic function are employed to solve the governing equation. Then the corresponding Fokker-Planck-Kolmogorov equation is derived and solved to achieve the probability function of the response. The drift and diffusion coefficients are found in a series form for the generalized foundation with fractional damping. Monte-Carlo simulation is applied for a plate subjected to a random excitation and validity of the procedure is verified. Effects of the random excitation intensity, the fractional order and foundation parameters on the deflection response probability are examined.

2. Mathematical Modeling

The generalized form of nonlinear Duffing stiffness is represented by the following [38]:

$$g(X) = \sum_{\substack{n=2k+1 \\ k=1,2,\cdots,m}}^{n} k_n X^n \tag{1}$$

where k_n includes the linear and nonlinear stiffness coefficients. Corresponding governing equation of a plate motion on the foundation with generalized stiffness and damping is derived as

$$D\nabla^4 w + \rho h\ddot{w} + \varepsilon c_1 \dot{w} + \varepsilon c_2 \dot{w}^3 + \varepsilon a D^\alpha w + \sum_{\substack{n=2k+1 \\ k=1,2,\cdots,m}}^{n} k_n w^n = \xi(t) \tag{2}$$

where $w = w(x, y, t)$ is the transverse displacement, $D = Eh^3/12(1 - \vartheta^2)$ denotes the flexural rigidity of the plate and E, h and ϑ are the Young's modulus, thickness, and Poisson's ratio of the plate, respectively. Also $\nabla^4 = (\partial^2/\partial x^2 + \partial^2/\partial y^2)^2$ and ρ is the mass density of the plate material.

At this stage, the Galerkin method is employed to solve the motion equation which suggests the following solution for a simple-supported plate [39]:

$$w(x, y, t) = \sum_{i=1}^{N}\sum_{j=1}^{M} \sin\left(\frac{i\pi x}{a}\right)\sin\left(\frac{j\pi y}{b}\right) q_{ij}(t) \tag{3}$$

Substituting (3) into (2) and taking the main mode into account, the following equation is obtained:

$$\ddot{q}(t) + \beta_1 \dot{q}(t) + \beta_2 \dot{q}^3(t) + \beta_3 D^\alpha q(t) + \sum_{\substack{n=2k+1 \\ k=1,2,\ldots,m}}^{n} \lambda_i q^n(t) = \xi_1(t) \tag{4}$$

in which

$$\begin{aligned} \beta_1 &= \frac{\varepsilon c_1}{\rho h}, \\ \beta_2 &= \frac{9\varepsilon c_2}{16\rho h}, \\ \beta_3 &= \frac{\varepsilon a}{\rho h}, \\ \xi_1(t) &= \frac{16}{\pi^2}\xi(t) \\ \lambda_1 &= \frac{D\left((\pi/a)^2 + (\pi/b)^2\right)^2 + k_1}{\rho h}, \\ \lambda_i &= \frac{4k_i}{ab\rho h}\int_0^a\int_0^b \sin^{i+1}\left(\frac{\pi x}{a}\right)\sin^{i+1}\left(\frac{\pi y}{b}\right) dy\, dx, \\ & i = 3, 5, \ldots \end{aligned} \tag{5}$$

3. Solution Method

Back to (4), we consider the corresponding nonlinear system with fractional derivative damping subjected to stochastic excitation:

$$\ddot{X}(t) + \varepsilon h(\dot{X}) + \varepsilon a D^\alpha X(t) + g(X) = \xi(t) \tag{6}$$

where

$$\varepsilon h(\dot{X}) = \varepsilon c_1 \dot{X}(t) + \varepsilon c_2 \dot{X}^3(t) \tag{7}$$

in which εc_1, εc_2, and εa are the light damping coefficients, $g(X)$ consists of the linear and nonlinear restoring force, and $\xi(t)$ is Gaussian white noise with the following correlation function:

$$E[\xi(t)\xi(t+\tau)] = 2D_1\delta(\tau) \tag{8}$$

D^α donates the fractional derivative operator and according to Riemann-Liouville's definition is represented by the following [40, 41]:

$$D^\alpha X(t) = \frac{1}{\Gamma(1-\alpha)}\frac{d}{dt}\int_0^t \frac{X(t-\tau)}{\tau^\alpha} d\tau, \quad 0 < \alpha < 1 \tag{9}$$

The most well-known, practical representations for fractional derivative operators are Riemann-Liouville (RL), Caputo, and Grunwald-Letnikov (GL) definitions. For many cases, it has been proved that these fractional derivative definitions are equivalent. Grunwald-Letnikov definition presents the relation in the form of a series and is more appropriate for discrete systems. Riemann-Liouville definition is widely used for calculations of analytical solutions in continues systems. Caputo definition can include the initial conditions of a problem in the same form of nonfractional problems; so it can provide a physical meaning for the initial conditions. In the present study, according to the nature of stochastic analysis and continuous modeling hypothesis, definition of Riemann-Liouville is taken into account.

The free conservative vibration associated with the system (6) is therefore represented by the following:

$$\ddot{X}(t) + g(X) = 0 \tag{10}$$

The response of system (6) is a nearly random spread of the system (10): therefore, by use of generalized harmonic function, the following form of response is assumed for the following solution:

$$\begin{aligned} X(t) &= A\cos\theta(t) \\ \dot{X}(t) &= -Av\sin\theta(t) \\ \theta(t) &= \Phi(t) + \Gamma(t) \end{aligned} \tag{11a}$$

$$v(A, \theta) = \frac{d\Phi}{dt} = \sqrt{\frac{2(U(A) - U(A\cos\theta))}{A^2\sin^2\theta}} \tag{11b}$$

A, Φ, and Γ are the random process functions. By applying the generalized Van der Pol transformation from X, $\dot{X}$ to A and Γ, (11a) and (11b) yield to

$$\frac{dA}{dt} = \varepsilon F_1 + G_1\xi(t) \tag{12a}$$

$$\frac{d\Gamma}{dt} = \varepsilon F_2 + G_2\xi(t) \tag{12b}$$

in which

$$
\begin{aligned}
F_1 &= F_{11} + F_{12} \\
F_{11} &= \frac{Av\sin\theta}{g(A)} aD^{\alpha}(A\cos\theta) \\
F_{12} &= \frac{Av\sin\theta}{g(A)} h(-Av\sin\theta) \\
F_2 &= \frac{v\cos\theta}{g(A)} aD^{\alpha}(A\cos\theta) + \frac{v\cos\theta}{g(A)} h(-Av\sin\theta) \\
G_1 &= \frac{-Av\sin\theta}{g(A)} \\
G_2 &= \frac{-v\cos\theta}{g(A)}
\end{aligned} \tag{13}
$$

Since (12a) is independent from Γ, the Itô equation related to A should be satisfied:

$$dA = m(A)\,dt + \sigma(A)\,dB(t) \tag{14}$$

where $m(A)$ and $\sigma(A)$ are the draft and diffusion coefficients:

$$
\begin{aligned}
m(A) &= \varepsilon \left\langle F_{11} + F_{12} + D_1 \frac{dG_1}{dA} G_1 + D_1 \frac{dG_1}{d\Gamma} G_2 \right\rangle_\theta \\
\sigma^2(A) &= \varepsilon D_1 \left\langle 2G_1^2 \right\rangle_\theta
\end{aligned} \tag{15}
$$

in which $\langle . \rangle_\theta$ represents the average operator with respect to θ:

$$\langle . \rangle_\theta = \lim_{T\to\infty} \frac{1}{T} \int_0^T \langle . \rangle\, dt = \frac{1}{2\pi} \int_0^{2\pi} \langle . \rangle\, d\theta \tag{16}$$

Since A and Γ are slowly varying with time, $\theta(t)$ could be approximated by the instantaneous frequency:

$$\theta(t-\tau) \approx \theta(t) - w(A)\tau \tag{17}$$

By use of (17) and averaging over the first term of (15), $\langle F_{11}\rangle_\theta$ is reduced to

$$
\begin{aligned}
\langle F_{11}\rangle_\theta &= \frac{a}{g(A)} \lim_{T\to\infty} \frac{1}{T} \int_0^T D^{\alpha}(A\cos\theta) \times Av\sin\theta dt \\
&= \frac{a}{g(A)\Gamma(1-\alpha)} \lim_{T\to\infty} \frac{1}{T} \\
&\cdot \int_0^T Av\sin\theta d\left(\int_0^t \frac{A\cos\theta(t-\tau)}{\tau^{\alpha}} d\tau\right) \\
&= \frac{a}{g(A)\Gamma(1-\alpha)} \\
&\cdot \lim_{T\to\infty} \left\{ \frac{1}{T} \left(Av\sin\theta \int_0^t \frac{A\cos\theta(t-\tau)}{\tau^{\alpha}} d\tau \right)\Bigg|_0^T \right. \\
&\left. - \frac{1}{T} \int_0^T \left(\int_0^t \frac{\cos\theta(t-\tau)}{\tau^{\alpha}} d\tau \right) \frac{d}{dt}(Av\sin\theta)\, dt \right\}
\end{aligned} \tag{18}
$$

Using the following asymptotic integrals

$$
\begin{aligned}
&\int_0^t \frac{\cos(\omega\tau)}{\tau^q} d\tau = \omega^{(q-1)} \int_0^s \frac{\cos(u)}{u^q} du \\
&= \omega^{(q-1)} \left(\Gamma(1-q)\sin\left(\frac{q\pi}{2}\right) + \frac{\sin(s)}{s^q} + O\left(s^{-q-1}\right) \right) \\
&\int_0^t \frac{\sin(\omega\tau)}{\tau^q} d\tau = \omega^{(q-1)} \int_0^s \frac{\sin(u)}{u^q} du \\
&= \omega^{(q-1)} \left(\Gamma(1-q)\cos\left(\frac{q\pi}{2}\right) - \frac{\cos(s)}{s^q} + O\left(s^{-q-1}\right) \right)
\end{aligned} \tag{19}
$$

We can simplify (18) to

$$
\begin{aligned}
\langle F_{11}\rangle_\theta &\approx \frac{-a}{g(A)\Gamma(1-\alpha)} \lim_{T\to\infty} \frac{1}{T} \int_0^T Ag(A\cos\theta) \\
&\cdot \left[\cos\theta \int_0^t \frac{\cos(\omega(A)\tau)}{\tau^q} d\tau \right. \\
&\left. + \sin\theta \int_0^t \frac{\sin(\omega(A)\tau)}{\tau^q} d\tau \right] \\
&\approx \frac{-Aa}{2\pi g(A)\omega^{1-\alpha}} \int_0^{2\pi} g(A\cos\theta) \left[\cos\theta\sin\left(\frac{\alpha\pi}{2}\right) \right. \\
&\left. + \sin\theta\cos\left(\frac{\alpha\pi}{2}\right) \right] d\theta
\end{aligned} \tag{20}
$$

In the following step, G_i is expanded to the Fourier series with respect to θ:

$$G_i = G_{i0}(A) + \sum_{n=1}^{\infty} G_{in}^{(c)}(A)\cos n\theta + G_{in}^{(s)}(A)\sin n\theta, \quad i = 1,2 \tag{21}$$

By substituting G_i and F_{11} into (13) the averaged drift and diffusion coefficients are obtained as

$$
\begin{aligned}
\overline{m}(A) &= \varepsilon\langle F_{11}\rangle_\theta + \frac{\varepsilon A}{g(A)} \int_0^{2\pi} v\sin\theta h(-Av\sin\theta)\, d\theta \\
&+ \varepsilon D_1 \left[\frac{dG_{10}}{dA} G_{10} + \frac{1}{2} \sum_{n=1}^{\infty} \frac{dG_{1n}^{(c)}}{dA} G_{1n}^{(c)} + \frac{dG_{1n}^{(s)}}{dA} G_{1n}^{(s)} \right. \\
&\left. + nG_{1n}^{(s)} G_{2n}^{(c)} - nG_{1n}^{(c)} G_{2n}^{(s)} \right] \\
\overline{\sigma}^2(A) &= \varepsilon D_1 \left[2G_{10}^2 + \sum_{n=1}^{\infty} G_{1n}^{(c)^2} + G_{1n}^{(s)^2} \right]
\end{aligned} \tag{22}
$$

The FPK's equation associated with (14) is therefore derived as

$$\frac{\partial p}{\partial t} = -\frac{\partial}{\partial A}\left[\overline{m}(A)p\right] + \frac{1}{2}\frac{\partial^2}{\partial A^2}\left[\overline{\sigma}^2(A)p\right] \tag{23}$$

Since $g(X)$ is a monotonic increasing function, the boundary condition for (23) should satisfy

$$\begin{aligned} &if \ : \ A = 0 \Longrightarrow p = finite, \\ &if \ : \ A \longrightarrow \infty \Longrightarrow p, \ \frac{\partial p}{\partial A} \longrightarrow 0 \end{aligned} \tag{24}$$

Therefore, the stationary response of FPK's equation is

$$p(A) = \frac{C}{\sigma^2(A)} \exp\left[\left[\int_0^A \frac{2m(u)}{\sigma^2(u)} du\right]\right] \tag{25}$$

where C is the normalization constant. Back to the main problem, corresponding potential energy for the generalized restoring force is given by

$$U(X) = \int_0^X g(x)\,dx = \sum_{\substack{n=2k+1 \\ k=1,2,\cdots,m}}^{n} \frac{k_n X^{n+1}}{n+1} \tag{26}$$

At this step, drift and diffusion coefficients can be extracted for the original system with (4). Using (1), the integral of (20) can be expressed as

$$\langle F_{11} \rangle_\theta = -\frac{A\beta_3 \sum^n_{\substack{n=2k+1 \\ k=1,2,\dots,m}} \left(\lambda_i A^n \int_0^{2\pi} \cos^{n+1} u\, du\right)}{2\pi \lambda_1^{(1-\alpha)/2} \sum^n_{\substack{n=2k+1 \\ k=1,2,\dots,m}} \lambda_i A^n} \tag{27}$$

where $\lambda_1 = \omega^2$. Using definition of h, i.e., (7) with the coefficients defined by (5), we have

$$\begin{aligned} &\frac{A}{g(A)} \int_0^{2\pi} v \sin\theta h(-Av\sin\theta)\, d\theta \\ &= \frac{-A^2 \sum^n_{\substack{n=2k+1 \\ k=1,2,\dots,m}} \left\{\beta_1 \int_0^{2\pi} (2\lambda_n/(n+1))\left(1-\cos^{n+1} u\right) du + \beta_2 \left[\int_0^{2\pi} (2\lambda_n/(n+1))\left(1-\cos^{n+1} u\right) du\right]^2\right\}}{2\pi \sum^n_{\substack{n=2k+1 \\ k=1,2,\dots,m}} \lambda_i A^n} \end{aligned} \tag{28}$$

By substituting (27), (28) into (22), the drift and diffusion coefficients for the main equation are consequently obtained as ($\varepsilon = 1$):

$$\begin{aligned} m(A) &= \frac{-A^2 \sum^n_{\substack{n=2k+1 \\ k=1,2,\dots,m}} \left\{\beta_1 \int_0^{2\pi} (2\lambda_n/(n+1))\left(1-\cos^{n+1} u\right) du + \beta_2 \left[\int_0^{2\pi} (2\lambda_n/(n+1))\left(1-\cos^{n+1} u\right) du\right]^2\right\}}{2\pi \sum^n_{\substack{n=2k+1 \\ k=1,2,\dots,m}} \lambda_i A^n} \\ &\quad - \frac{A\beta_3 \sum^n_{\substack{n=2k+1 \\ k=1,2,\dots,m}} \left(\lambda_i A^n \int_0^{2\pi} \cos^{n+1} u\, du\right)}{2\pi \lambda_1^{(1-\alpha)/2} \sum^n_{\substack{n=2k+1 \\ k=1,2,\dots,m}} \lambda_i A^n} + \frac{D_1}{2}\left[\frac{dG_{11}^{(s)}}{dA} G_{11}^{(s)} + G_{11}^{(s)} G_{21}^{(c)}\right] \\ \sigma^2(A) &= D_1 \left[G_{11}^{(s)}\right]^2 \end{aligned} \tag{29}$$

where G_{i1} is obtained according to (21):

$$\begin{aligned} G_{11}^{(s)} &= \frac{-A(b_0 - b_2/2)}{\sum^n_{\substack{n=2k+1 \\ k=1,2,\dots,m}} \lambda_i A^n}, \\ G_{21}^{(c)} &= \frac{-(b_0 + b_2/2)}{\sum^n_{\substack{n=2k+1 \\ k=1,2,\dots,m}} \lambda_i A^n} \end{aligned} \tag{30}$$

b_0 and b_2 are calculated from the following approximation:

$$v(A,\theta) = b_0 + b_2\cos 2\theta + b_4 \cos 4\theta + b_6 \cos 6\theta \tag{31}$$

Using (11b) and (26), we have the following relation for $v(A,\theta)$:

$$v(A,\theta) = (\eta(1+u))^{0.5} \tag{32}$$

where

$$\eta = \sum_{\substack{n=2k+1 \\ k=1,2,\cdots m}}^{n} \frac{2\lambda_n A^{n-1}}{n+1},$$

$$u = \sum_{\substack{n=2k+1 \\ k=1,2,\cdots m}}^{n} \eta_n \cos^{n-1}\theta,$$

$$\eta_j = \frac{1}{\eta}\sum_{i=j}^{n-1}\frac{2\lambda_i A^{i-1}}{i+1}, \quad j = 3, 5, \ldots, n \tag{33}$$

b_0 and b_2 are then obtained from (31) and (33):

$$b_0 = \frac{\sqrt{\eta}}{2\pi}\int_0^{2\pi}\sqrt{1-u}\,d\theta$$

$$b_0 = \frac{\sqrt{\eta}}{\pi}\int_0^{2\pi}\sqrt{1-u}\cos 2\theta\,d\theta, \tag{34}$$

$$\sqrt{1-u} \approx 1 + \frac{u}{2} - \frac{u^2}{8} + \frac{u^3}{16}$$

The stationary probability density function is obtained by substituting $A = U^{-1}(H)$ in the following relation:

$$p(H) = p(A)\left|\frac{dA}{dH}\right| = \left.\frac{p(A)}{g(A)}\right|_{A=U^{-1}(H)} \tag{35}$$

in which U^{-1} is the inverse of potential function. The joint stationary probability of displacement and velocity can also be obtained as

$$p(x_1, x_2) = \left.\frac{p(H)}{T(H)}\right|_{H=x_2^2+U(x_1)} \tag{36}$$

in which

$$T(H) = \left.\frac{2\pi}{w(H)}\right|_{H=U^{-1}(A)} \tag{37}$$

Finally, the stationary probability density function of displacement and velocity are obtained to be

$$p(x_1) = \int_{-\infty}^{\infty} p(x_1, x_2)\,dx_2$$

$$p(x_2) = \int_{-\infty}^{\infty} p(x_1, x_2)\,dx_1 \tag{38}$$

and the stationary mean amplitude can be obtained by

$$E[A] = \int_0^{\infty} Ap(A)\,dA \tag{39}$$

4. Numerical Results

In this section, numerical results are presented and discussed for a case study. The aim is to show validity of the solution procedure and to provide a parametric study. The validation is performed by the comparison between the results from the present formulation and the results from Monte-Carlo simulation. Monte-Carlo simulation takes all of the parameters with uncertainty into account and presents the probability distribution for a parameter by performing the calculation over and over. By taking all of the calculated results for a desired parameter into account, the corresponding probability density or probability distribution can be achieved.

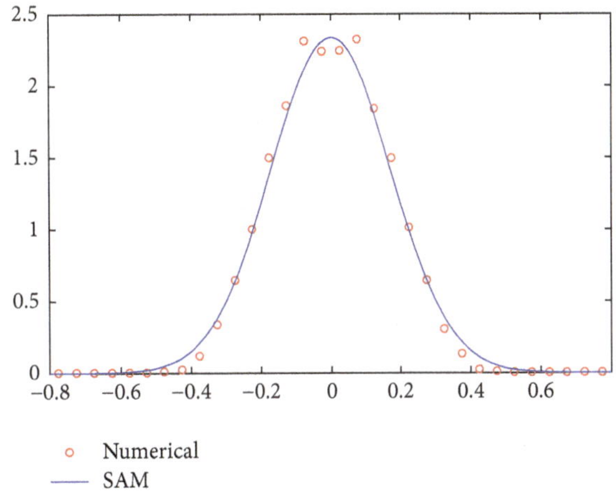

FIGURE 1: The stationary probability density of the displacement.

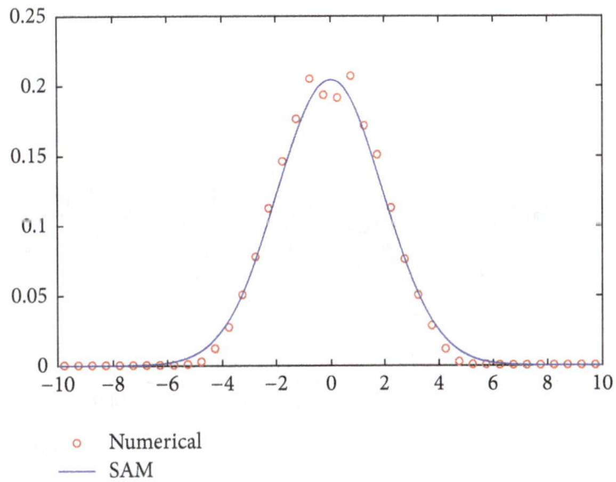

FIGURE 2: The stationary probability density of the velocity.

Parameters of the plate are set as: a = 4 m, b = 2 m, h = 2 mm, E = 207 Gpa, $\vartheta = 0.3$, $\rho = 7800\ kg/m^3$. Properties of the foundation are $c_1 = 0.4\ Ns/m$, $c_2 = 0$, $a = 1\ Ns^{\alpha}/m$, $\alpha = 0.55$, $D_1 = 0.1$, $k_1 = 600\ N/m$, $k_3 = 12\ kN/m^3$, $k_5 = 25\ kN/m^5$, $k_7 = 40\ kN/m^7$, $k_9 = k_{11} = \ldots = 0$. Drift and diffusion coefficients are obtained for such a system according to the previous section. Stationary probability and expected value can be calculated from (38) and (39). Figures 1 and 2 show the stationary probability density function of amplitude and velocity, respectively. The solid lines are correlated with the SAM and dots are outcomes of the Monte-Carlo numerical simulation obtained from Figures 3 and 4.

The displacement response and velocity response samples from Monte-Carlo numerical simulation are illustrated in Figures 3 and 4. The joint probability density function between the deflection and velocity is illustrated in Figure 5. A good agreement between the analytical proposed solution and those from numerical simulation is observed in these figures.

In the following parts, effects of the foundation properties on the stationary responses are investigated. Figure 6 illustrates influence of the power order of the stiffness on

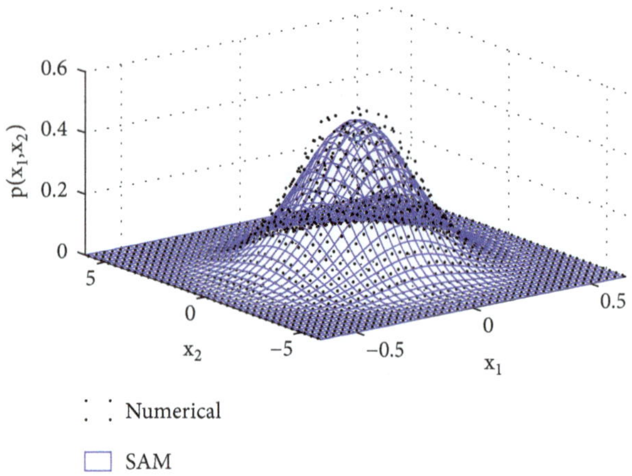

FIGURE 3: The joint stationary probability density between the displacement and velocity.

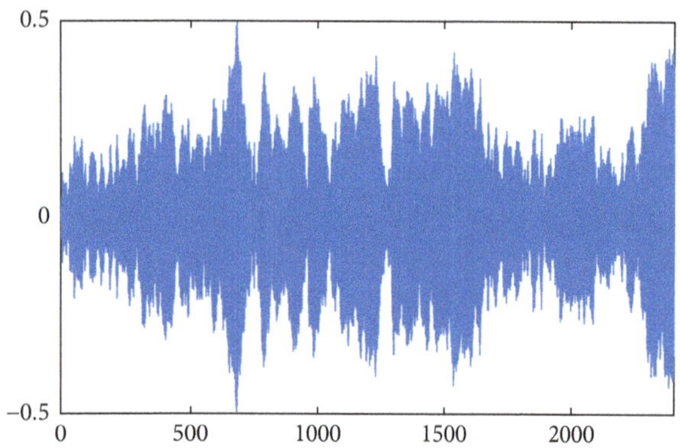

FIGURE 4: Displacement response from Monte-Carlo simulation.

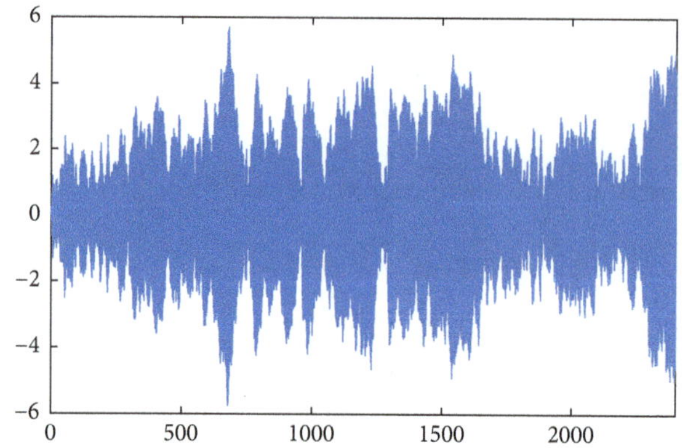

FIGURE 5: Velocity response from Monte-Carlo simulation.

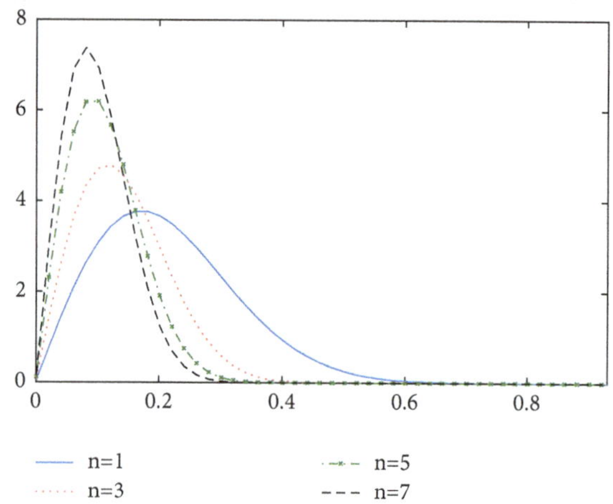

FIGURE 6: Effect of the power order of stiffness on the probability function of amplitude.

the stationary probability amplitude. It is seen that for larger powers we get more concentrated probability function around a smaller value of amplitude.

The effect of fractional order derivative on stationary probability amplitude is illustrated in Figure 7. It can be seen that the fractional derivative order has a remarkable effect on the responses. It acts similar to the power order of the stiffness; i.e., larger fractional order gives a more condensed probability function centered in smaller amplitudes.

Figure 8 shows variation of the amplitude probability function for four different values of fractional damping coefficient. It is observed that having smaller fractional damping coefficient gives a broadly spread probability. It is also seen that the sensitivity of the system is quite higher with respect to this parameter at large values of fractional damping. Figure 9 illustrates the effect of nonlinear damping coefficient c_2 on the stationary probability density of the amplitude. It is seen that sensitivity of the responses with respect c_2 decreases at high values of the nonlinear damping.

Variation of the expected value versus the fractional order derivative is illustrated for $n = 1$ and $n = 3$ in Figure 10. It is seen that the expected value decreases with increasing α and its decreasing rate enhances as α increases. It is also seen that the sensitivity with respect to α is higher when $n = 1$.

5. Conclusions

A straightforward and closed-form solution was proposed for nonlinear vibration analysis of a supported plate subjected to stochastic excitation. The support was modeled by a generalized form of stiffness and damping with fractional properties. Generalized harmonic function in conjunction with the stochastic averaging method was used to extract the averaged *Itô* and FPK's equation. Validity of the solution procedure was numerically verified by use of Monte-Carlo simulation. It was shown that the stationary probability density function of the responses is remarkably influenced by the fractional parameters. This means that instead of changing the stiffness and damping of the foundation which is costly and hard to implement, designers can use a fractional layer. Similar to the order of foundation stiffness, it was found that larger fractional order corresponds to more condensed probability function and to smaller amplitudes. It was proved that the fractional damping coefficient acts similar to the nonlinear damping but with an extra benefit. Having more sensitivity at high fractional damping reduces the risk of saturation at high material damping. It was also shown that for a linear foundation we can have desirable sensitivity with respect to fractional order so that instead of adding extra

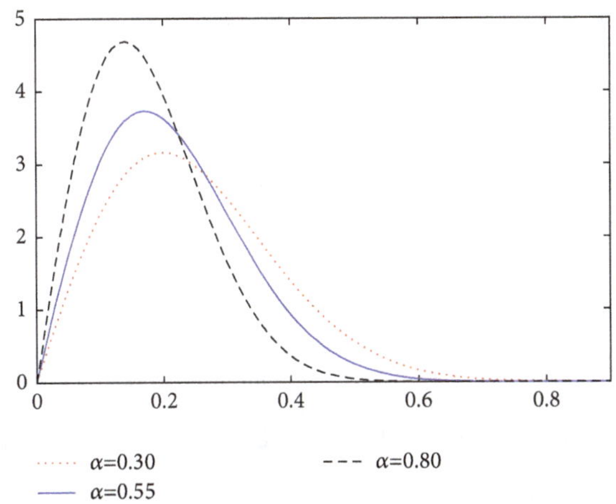

FIGURE 7: Effect of fractional order derivative on the probability function of amplitude.

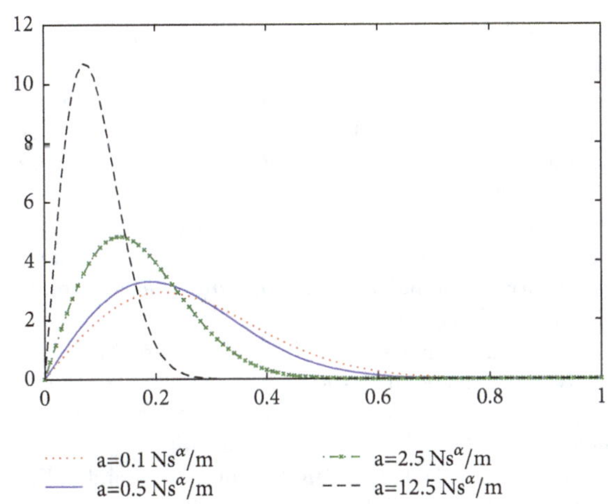

FIGURE 8: Effect of fractional damping on the probability function of amplitude.

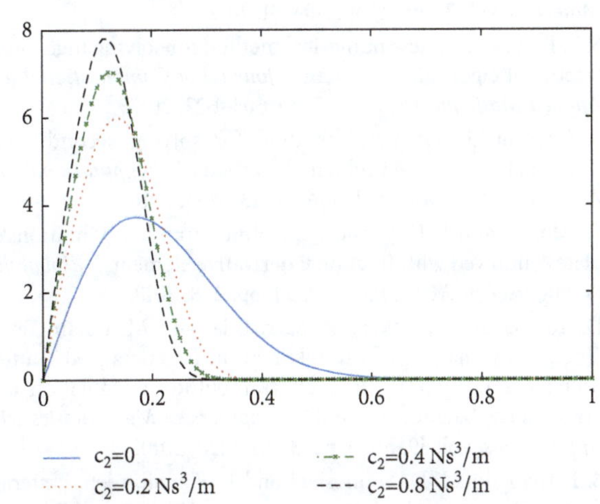

FIGURE 9: Effect of fractional damping on the probability function of amplitude.

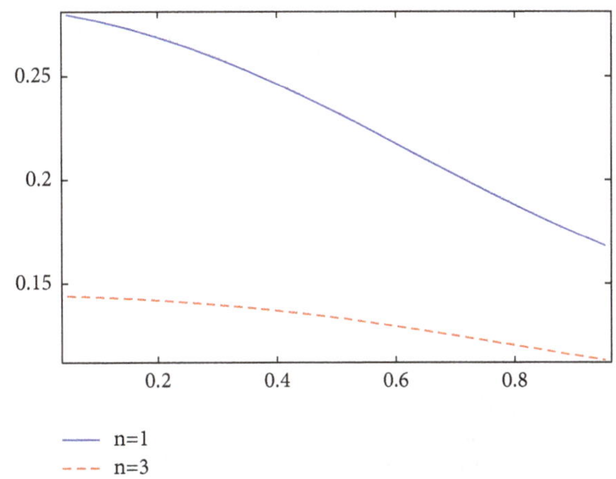

FIGURE 10: The mean amplitude E[A] as a function of α.

nonlinear damping we can employ a material with a larger fractional order.

Conflicts of Interest

The authors declare that they have no conflicts of interest.

References

[1] L. Debnath, "Recent applications of fractional calculus to science and engineering," *International Journal of Mathematics and Mathematical Sciences*, pp. 3413–3442, 2013.

[2] M. Dalir and M. Bashour, "Applications of fractional calculus," *Applied Mathematical Sciences*, vol. 4, pp. 1021–1032, 2010.

[3] Y. A. Rossikhin and M. V. Shitikova, "Application of fractional calculus for dynamic problems of solid mechanics: novel trends and recent results," *Applied Mechanics Reviews*, vol. 63, no. 1, Article ID 010801, 2010.

[4] F. C. Meral, T. J. Royston, and R. Magin, "Fractional calculus in viscoelasticity: an experimental study," *Communications in Nonlinear Science and Numerical Simulation*, vol. 15, no. 4, pp. 939–945, 2010.

[5] D. Younesian, Z. Saadatnia, and H. Askari, "Analytical solutions for free oscillations of beams on nonlinear elastic foundations using the variational iteration method," *Journal of Theoretical and Applied Mechanics*, vol. 50, no. 2, pp. 639–652, 2012.

[6] Z. Xu and Y. K. Cheung, "Averaging method using generalized harmonic functions for strongly non-linear oscillators," *Journal of Sound and Vibration*, vol. 174, no. 4, pp. 563–576, 1994.

[7] A. Lal and B. N. Singh, "Stochastic nonlinear free vibration of laminated composite plates resting on elastic foundation in thermal environments," *Computational Mechanics*, vol. 44, no. 1, pp. 15–29, 2009.

[8] Z. L. Huang, W. Q. Zhu, and Y. Suzuki, "Stochastic averaging of strongly non-linear oscillators under combined harmonic and white-noise excitations," *Journal of Sound and Vibration*, vol. 238, no. 2, pp. 233–256, 2000.

[9] W. Q. Zhu, Z. L. Huang, and Y. Suzuki, "Response and stability of strongly non-linear oscillators under wide-band random excitation," *International Journal of Non-Linear Mechanics*, vol. 36, no. 8, pp. 1235–1250, 2001.

[10] G.-G. Li, Z. Zheng-you, and C. Chang-jun, "Dynamical stability of viscoelastic column with fractional derivative constitutive relation," *Applied Mathematics and Mechanics-English Edition*, vol. 22, no. 3, pp. 294–303, 2001.

[11] P. D. Spanos and B. A. Zeldin, "Random vibration of systems with frequency-dependent parameters or fractional derivatives," *Journal of Engineering Mechanics*, vol. 123, no. 3, pp. 290–292, 1997.

[12] P. D. Spanos and G. Malara, "Nonlinear Random Vibrations of Beams with Fractional Derivative Elements," *Journal of Engineering Mechanics*, vol. 140, no. 9, p. 04014069, 2014.

[13] P. D. Spanos and G. I. Evangelatos, "Response of a non-linear system with restoring forces governed by fractional derivatives-Time domain simulation and statistical linearization solution," *Soil Dynamics and Earthquake Engineering*, vol. 30, no. 9, pp. 811–821, 2010.

[14] P. Wahi and A. Chatterjee, "Averaging oscillations with small fractional damping and delayed terms," *Nonlinear Dynamics*, vol. 38, no. 1–4, pp. 3–22, 2004.

[15] Y. Yang and W. Xu, "Stochastic analysis of monostable vibration energy harvesters with fractional derivative damping under Gaussian white noise excitation," *Nonlinear Dynamics*.

[16] Z. L. Huang and X. L. Jin, "Response and stability of a SDOF strongly nonlinear stochastic system with light damping modeled by a fractional derivative," *Journal of Sound and Vibration*, vol. 319, no. 3–5, pp. 1121–1135, 2009.

[17] L. C. Chen, Q. Q. Zhuang, and W. Q. Zhu, "Response of SDOF nonlinear oscillators with lightly fractional derivative damping under real noise excitations," *The European Physical Journal Special Topics*, vol. 193, no. 1, pp. 81–92, 2011.

[18] L. Chen and W. Zhu, "First passage failure of SDOF nonlinear oscillator with lightly fractional derivative damping under real noise excitations," *Probabilistic Engineering Mechanics*, vol. 26, no. 2, pp. 208–214, 2011.

[19] L. Chen, Q. Lou, Z. Li, and W. Zhu, "Stochastic stability of the harmonically and randomly excited Duffing oscillator with damping modeled by a fractional derivative," *Science China Physics, Mechanics & Astronomy*, vol. 55, no. 12, pp. 2284–2289, 2012.

[20] L. Chen, H. Li, Z. Li, and W. Zhu, "Asymptotic stability with probability one of MDOF nonlinear oscillators with fractional derivative damping," *Science China Physics, Mechanics & Astronomy*, vol. 56, no. 11, pp. 2200–2207, 2013.

[21] L. Chen, W. Wang, Z. Li, and W. Zhu, "Stationary response of Duffing oscillator with hardening stiffness and fractional derivative," *International Journal of Non-Linear Mechanics*, vol. 48, pp. 44–50, 2013.

[22] Y. Yang, W. Xu, W. Jia, and Q. Han, "Stationary response of nonlinear system with Caputo-type fractional derivative damping under Gaussian white noise excitation," *Nonlinear Dynamics*, vol. 79, pp. 139–146, 2014.

[23] Y. G. Yang, W. Xu, X. D. Gu, and Y. H. Sun, "Stochastic response of a class of self-excited systems with Caputo-type fractional derivative driven by Gaussian white noise," *Chaos, Solitons & Fractals*, vol. 77, pp. 190–204, 2015.

[24] G. Failla and A. Pirrotta, "On the stochastic response of a fractionally-damped Duffing oscillator," *Communications in Nonlinear Science and Numerical Simulation*, vol. 17, no. 12, pp. 5131–5142, 2012.

[25] Y. Sun, L. Hong, J. Jiang, and Z. Li, "Estimation of critical conditions for noise-induced bifurcation in nonautonomous nonlinear systems by stochastic sensitivity function," *International Journal of Bifurcation and Chaos*, vol. 26, no. 11, 1650184, 12 pages, 2016.

[26] L. Chen, T. Zhao, W. Li, and J. Zhao, "Bifurcation control of bounded noise excited Duffing oscillator by a weakly fractional-order $PI_\lambda D_\mu$ feedback controller," *Nonlinear Dynamics*, vol. 83, no. 1-2, pp. 529–539, 2016.

[27] L. Chen and J.-Q. Sun, "The closed-form solution of the reduced Fokker-Planck-Kolmogorov equation for nonlinear systems," *Communications in Nonlinear Science and Numerical Simulation*, vol. 41, pp. 1–10, 2016.

[28] D. Yurchenko, A. Burlon, M. Di Paola, G. Failla, and A. Pirrotta, "Approximate Analytical Mean-Square Response of an Impacting Stochastic System Oscillator With Fractional Damping," *ASCE-ASME J. Risk and Uncert. in Engrg. Sys., Part B: Mech. Engrg.*, vol. 3, no. 3, p. 030903, 2017.

[29] Y. Yang, W. Xu, and G. Yang, "Bifurcation analysis of a noisy vibro-impact oscillator with two kinds of fractional derivative elements," *Chaos: An Interdisciplinary Journal of Nonlinear Science*, vol. 28, no. 4, 043106, 8 pages, 2018.

[30] H. S. Shen, J. Yang, and L. Zhang, "Free and forced vibration of Reissner-Mindlin plates with free edges resting on elastic foundations," *Journal of Sound and Vibration*, vol. 244, no. 2, pp. 299–320, 2001.

[31] X.-L. Huang and J.-J. Zheng, "Nonlinear vibration and dynamic response of simply supported shear deformable laminated plates on elastic foundations," *Engineering Structures*, vol. 25, no. 8, pp. 1107–1119, 2003.

[32] T. G. Kostantakopoulos and G. T. Michaltsos, "Modeling and analysis of a plate on elastic foundation subjected to landing airplanes' forces," *International Journal of Structural Stability and Dynamics*, vol. 10, no. 1, pp. 37–54, 2010.

[33] H. Dang-Trung, H. Luong-Van, T. Nguyen-Thoi, and K. K. Ang, "Analyses of stiffened plates resting on viscoelastic foundation subjected to a moving load by a cell-based smoothed triangular plate element," *International Journal of Structural Stability and Dynamics*, vol. 17, no. 1, 1750011, 28 pages, 2017.

[34] W. Cai, W. Chen, and W. Xu, "Fractional modeling of Pasternak-type viscoelastic foundation," *Mechanics of Time-Dependent Materials*, vol. 21, no. 1, pp. 119–131, 2017.

[35] S. Y. Reutskiy, "A new numerical method for solving high-order fractional eigenvalue problems," *Journal of Computational and Applied Mathematics*, vol. 317, pp. 603–623, 2017.

[36] S. Y. Reutskiy, "A novel method for solving second order fractional eigenvalue problems," *Journal of Computational and Applied Mathematics*, vol. 306, pp. 133–153, 2016.

[37] G. Malara and P. D. Spanos, "Nonlinear random vibrations of plates endowed with fractional derivative elements," *Probabilistic Engineering Mechanics*, vol. 54, pp. 2–8, 2018.

[38] D. Younesian, H. Askari, Z. Saadatnia, and M. KalamiYazdi, "Frequency analysis of strongly nonlinear generalized Duffing oscillators using HE's frequency-amplitude formulation and HE's energy balance method," *Computers & Mathematics with Applications*, vol. 59, no. 9, pp. 3222–3228, 2010.

[39] S. Farhangdoust, D. Younesian, and E. Esmailzadeh, "Interaction of Higher Modes in Nonlinear Free Vibration of Stiffened Rectangular Plates," in *Proceedings of the ASME 2017 International Design Engineering Technical Conferences and Computers*

and Information in Engineering Conference, p. V008T12A043, Cleveland, Ohio, USA.

[40] Haiwu Rong and Xiangdong Wang, "Subharmonic Response of Linear Vibroimpact System with Fractional Derivative Damping to a Randomly Disrobed Periodic Excitation," *Mathematical Problems in Engineering*, vol. 2015, pp. 1–10, 2015.

[41] Ying Li and Ye Tang, "Analytical Analysis on Nonlinear Parametric Vibration of an Axially Moving String with Fractional Viscoelastic Damping," *Mathematical Problems in Engineering*, vol. 2017, pp. 1–9, 2017.

8

Key Characteristics of a Floating Slab Track Based on Longitudinal Interaction Analysis

Goo Sam Jee,[1] Huan Ha Nguyen,[1] Seung Yup Jang,[2] Heeyoung Lee,[1,3] and Wonseok Chung[1]

[1]*Department of Civil Engineering, Kyung Hee University, Yongin, Gyeonggi-do 17104, Republic of Korea*
[2]*Department of Transportation System Engineering, Graduate School of Transportation, Korea National University of Transportation, Gyeonggi-do 16106, Republic of Korea*
[3]*Department of Civil Engineering, University of Colorado Denver, 1200 Larimer Street, Denver, CO 80217, USA*

Correspondence should be addressed to Heeyoung Lee; heeyoung0908@khu.ac.kr and Wonseok Chung; wschung@khu.ac.kr

Academic Editor: Francesco Pellicano

The passage of railway vehicles produces mechanical vibrations across an extensive range of frequencies. A floating slab track (FST) has recently been proposed and constructed in Korea to reduce the vibrations and noise that result from typical railway operation. The main objective of this study is to investigate the behavior of an FST by means of an interaction analysis of a track-slab-isolator. Key FST parameters considered in this study are additional rail stress, slab stress and displacement, and stopper displacement. The FST parameters include slab length and thickness, isolator stiffness, and longitudinal stiffness. The sensitivity of the key characteristics of each FST parameter is examined. This study also determines the maximum length of an FST based on the maximum additional rail stress criterion. Finally, a formula to predict maximum additional rail stress is proposed for the design of FST systems.

1. Introduction

Numerous high-speed railway tracks have recently been developed and constructed in Korea. The passage of railway vehicles produces mechanical vibrations across an extensive range of frequencies. Thus, the placement of structural vibration isolation systems is required to decrease vibrations and noise. Floating slab track (FST) systems are the most effective for prevention of mechanical vibrations and ground-borne noise generated by passing trains [1, 2]. FST systems have been successfully installed in various civil structures such as tunnels, high-speed rail lines, bridges, and viaducts. Various different types of FST systems have been improved and constructed around the world in places such as Japan, Europe, and North America [3, 4]. In Korea, an FST system, such as the one shown in Figure 1, has been successfully developed and tested, and one is currently being constructed in the test-bed of a commercial railway line.

FST systems typically consist of simple masses and springs that can isolate vibrations due to wheel-rail interactions. In an FST system, a continuous welded rail (CWR) is fixed to a massive concrete slab through rail pads with isolators under the concrete track. Although an FST is expensive and requires greater section heights because of the isolator, this relatively high cost is more than compensated for by the reduced noise and vibration.

Grootenhuis studied the effects of an FST on noise and vibrations [6]. Dynamic analysis was used, and the results show that the FST system can not only eliminate noise and vibrations but also avoid resonance. The effects of noise and vibration on railway tracks were later studied by several researchers with various two- and three-dimensional finite element models [7–9]. Again, significant reductions in noise and vibration were found in various conditions. However, the effect of slab length is an important factor in FST development which has not yet been fully researched [10]. Slab length

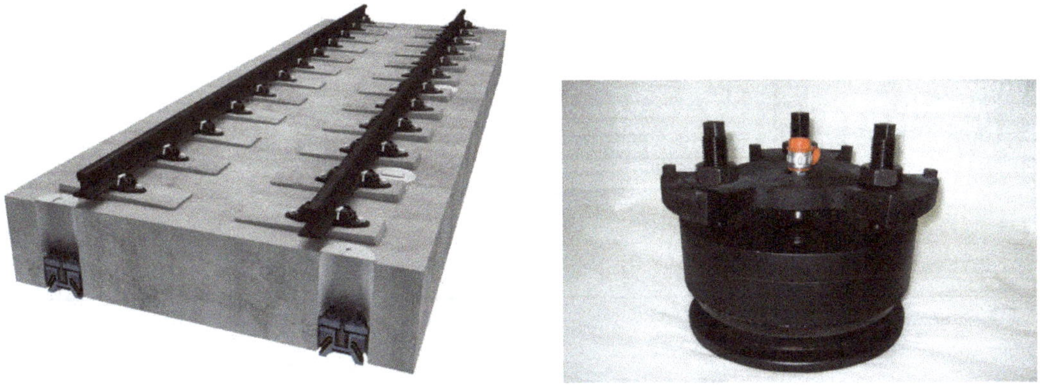

(a) An FST system (b) Isolator

FIGURE 1: Typical configuration of an FST system and isolator.

influences many aspects of an FST system, such as additional rail stress, tensile slab stress, and slab displacement. The relationships between additional rail stress and other FST parameters should be considered as particularly significant associations in the design of an FST system [11].

The key purpose of this study is to investigate the behavior of an FST by a track-slab-isolator interaction analysis. The key behaviors of an FST considered in this study are additional rail stress, slab stress and displacement, and stopper displacement. FST parameters include slab length and thickness, isolator stiffness, and longitudinal stiffness. The sensitivity of each parameter to the key response of an FST was examined based on interaction analysis. This study also determines the maximum length of an FST based on the maximum additional rail stress criterion. There are no fully accepted design criteria in FST systems for railway tracks, especially pertaining to recommendations on the length of the slab. In this study, the maximum additional tensile and compressive rail stress limit is 92 MPa, as described in the International Union of Railways (UIC) code [11]. Finally, a formula is proposed to predict the maximum additional rail stress for the design of an FST system.

2. Interaction Analysis Model

2.1. Research Methodology. The approach adopted in this work includes the development of a reliable track-slab-isolator interaction analysis model and the postulation and verification of a formula for additional rail stress. First, the dimensions of a standard FST were chosen based on experimental tests on a currently under-development FST [12]. Additionally, based on previous research [10], the dimensions of a standard FST were chosen, with a length of 34.5 m and a thickness of 0.54 m. Second, the applicable range for each parameter was selected to fully investigate the behavior of FST systems under various conditions.

In this study, the standard values of an FST system were modeled and fully populated with key variables such as slab length and thickness, vertical isolator stiffness, and Young's modulus. In an FST system, the longitudinal displacements are essential variables because they are generated by additional rail stresses; this necessitates the evaluation of these variables. Three stopper configurations are considered for analysis; these slab track configurations with and without stoppers are illustrated in Figure 2.

At both ends of the concrete slab track, bed blocks were installed, as shown in Figure 3, because the additional rail stresses change rapidly in this region. When the train operates over the transition between the embankment and the FST (at both ends of the concrete slab track), a vehicle-track force is created at the transition due to an abrupt change in track stiffness. As a result, rail stresses of increased magnitude are generated in this area [13].

2.2. Finite Element Model. In this study, numerical simulations were implemented using the ABAQUS package [14]. ABAQUS has an extensive range of abilities for simulation of linear and nonlinear functions, and the elements for the FST system model were carefully selected from the various elements the software is capable of modeling. In this study, an FST system was modeled by the connections between the rail, slab, isolator, bed block, and stopper (if any). All analysis elements were considered in a simplified two-dimensional (2D) finite element model (FEM). Figure 3 illustrates a typical detailed model used in this study for visualization of the FST system. CWRs and concrete slabs were modeled with Timoshenko beam elements that allow for transverse shear deformation. The length of a CWR can be quite long, up to several kilometers, but the floating slab is many times shorter. Therefore, an infinite Timoshenko beam (B21) was selected for the rail and a finite one was selected for the floating slab. The rail and concrete slab were assumed to be linear, elastic, and isotropic.

The upper beam elements (rail) are fixed at both ends and interact with the lower beam element (floating slab) by means of the track stiffness. The connection was modeled as a two-node spring element. The spring exhibits nonlinear behavior in the longitudinal direction and linear behavior in the vertical direction. The longitudinal stiffness of the track is defined in UIC code 774-3R [11] and is a function of the

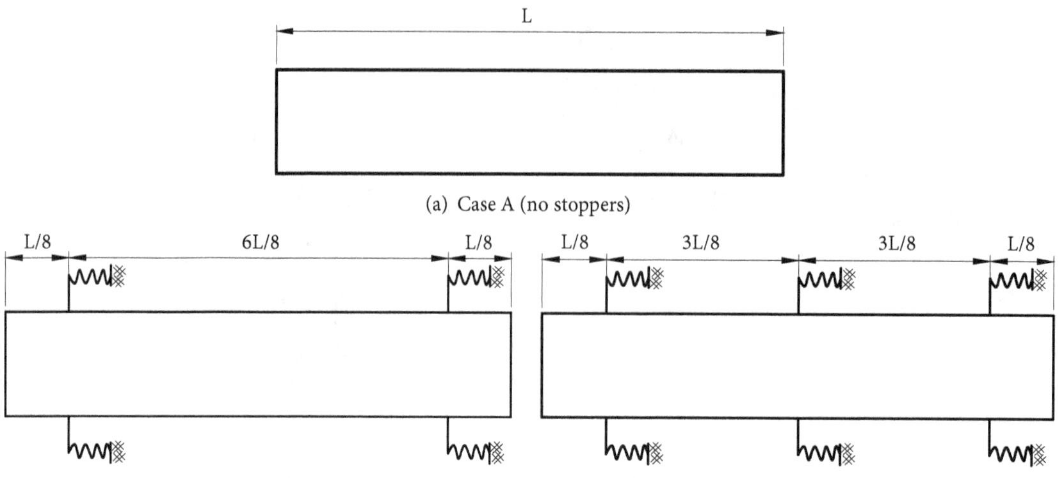

(a) Case A (no stoppers)

(b) Case B (2 stoppers)

(c) Case C (3 stoppers)

FIGURE 2: Stopper configurations of an FST system (plan view).

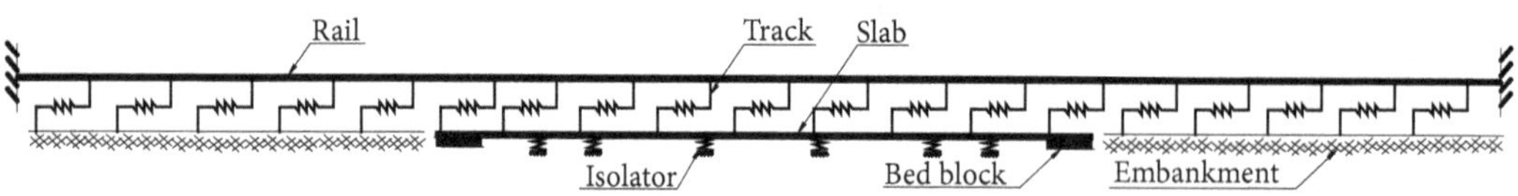

FIGURE 3: Analysis model of an FST system.

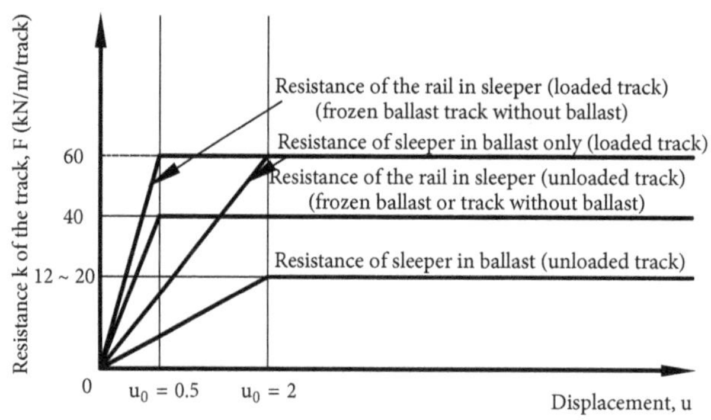

FIGURE 4: Track stiffness.

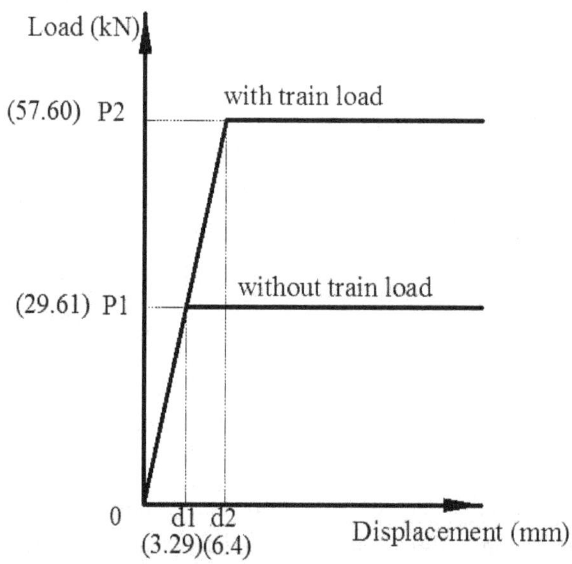

FIGURE 5: Longitudinal isolator stiffness.

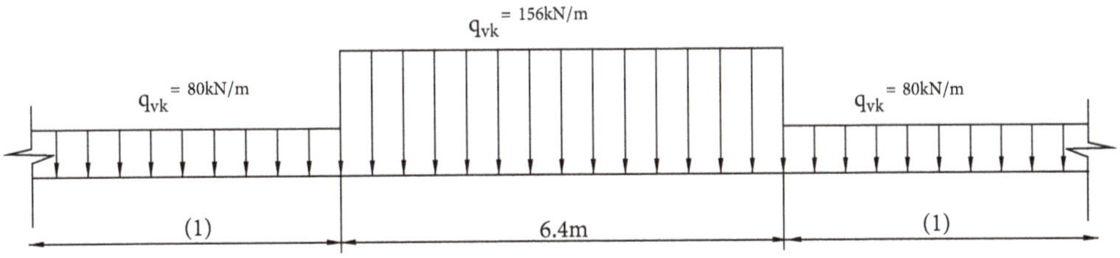

(1) No limitation

FIGURE 6: Load model and characteristic values for vertical train loads [5].

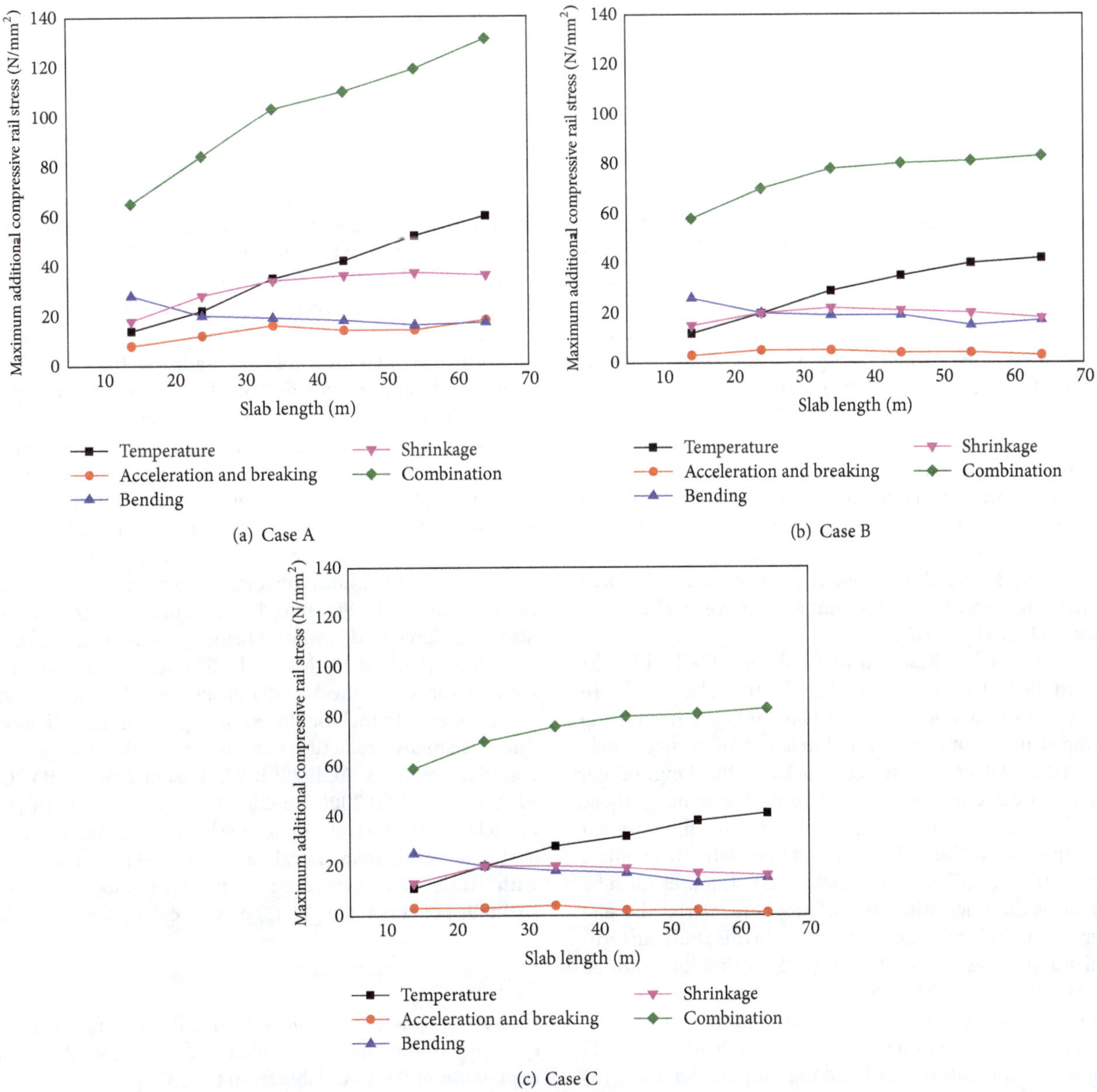

FIGURE 7: Results of additional compressive rail stress with increasing slab length.

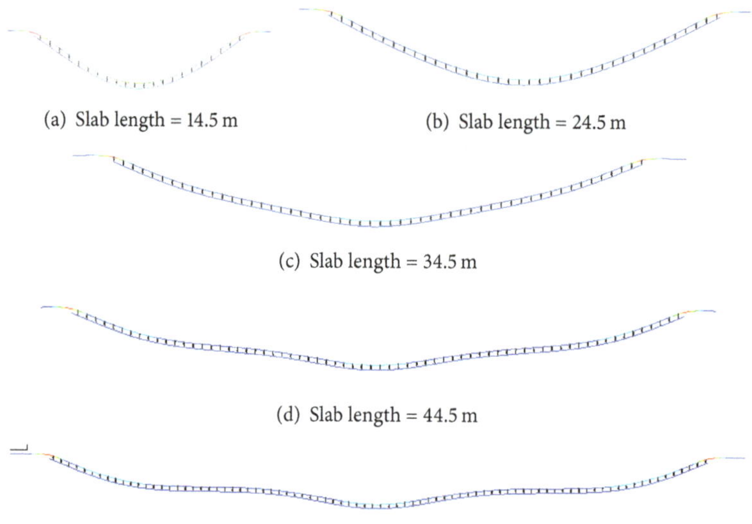

(a) Slab length = 14.5 m
(b) Slab length = 24.5 m
(c) Slab length = 34.5 m
(d) Slab length = 44.5 m
(e) Slab length = 54.5 m

FIGURE 8: Bending behaviors of the rail and the slab.

displacement of the rail relative to its supporting structure. To simplify calculations, the stiffness is stated as a bilinear function of the rail displacement, as shown in Figure 4. The vertical track stiffness was estimated as 47 kN/mm based on the tests carried out by Pandrol [15].

The floating slab was connected to an isolator, and its rigid foundation was modeled by a spring element whose stiffness acts in a fixed direction. The isolator spring stiffness was identified by experimental tests [12]. Since the isolator is made of a rubber-based material, it shows hyperelastic and frequency-dependent behavior. In this study, the isolator's static stiffness was chosen for further analysis. The isolator's longitudinal stiffness value was taken from a previous study [12] and was used to simplify the behavior, as shown in Figure 5. The longitudinal isolator stiffness was calculated from these test values as 9 kN/mm and the vertical isolator stiffness as 10 kN/mm [12].

As mentioned in Research Methodology, the bed blocks were installed at both ends of the floating slab track. To simplify the simulation, each bed block was modeled using 11 spring elements over the entire 1 m length of the bed block. The stiffness values of the bed blocks in the longitudinal and vertical directions were taken to be the same as those of the isolator. As shown in Figure 3, an embankment was connected to the rail of the floating slab. To simulate continuity in the FST system, the stiffness of this area must be the same as the track stiffness. The stoppers restrict the slab movement in the longitudinal direction. In this study, an individual stopper was modeled by a spring with a longitudinal stiffness value of 1200 kN/mm.

Several load cases can be considered in an FST system. This study considered the following four load cases: (1) changes in temperature, (2) breaking and acceleration, (3) bending, and (4) shrinkage. As the temperature changes, the slab expands and contracts. Assuming that the floating slab is subjected to uniform temperature changes, the temperature of the floating slab deviates from the reference temperature, as indicated in the UIC code. Breaking and acceleration forces applied at the top of the rail are considered to be distributed evenly over the length. These loads were chosen from the UIC code [11] as an acceleration force of 33 kN/m per track over a load length less than 33 m and a breaking force of 20 kN/m per track over a load length less than 400 m. The results of the acceleration and breaking forces were compared to each other, and the maximum values were chosen for the final analysis. Vertical train loads produce bending of the floating slab, leading to rotation of the end sections and vertical displacement of the upper edge of the floating slab end. According to railroad-bridge longitudinal interaction analysis guidelines in Korea [5], the load arrangement and the characteristic values for vertical train loads were taken to be those shown in Figure 6. Finally, concrete shrinkage was also considered in this study. Shrinkage is not included in the UIC code [11], but in this study shrinkage is an important contributor to the mechanical properties of a concrete slab. It is necessary to include these parameters in the FST model. There are many prediction models for shrinkage in hardened concrete, such as ACI 209R-92, Bazant-Baweja B3, CEB MC90-99, and GL2000 [16–20]. This study adopts the CEB model [19]. The total strain due to shrinkage can be calculated as the sum of shrinkage strain and the product of compliance with stress [16]. According to the CEB model, the total shrinkage of concrete $\varepsilon_{sh}(t,t_c)$ may be calculated as

$$\varepsilon_{sh}(t,t_c) = \varepsilon_{cas}(t) + \varepsilon_{cds}(t,t_c) \qquad (1)$$

where $\varepsilon_{cas}(t)$ is the autogenous shrinkage component and $\varepsilon_{cds}(t,t_c)$ is the drying shrinkage after 1 year. A detailed expression of (1) is available from the ACI [16].

In this study, finite element analysis was chosen to solve the problem; for this reason, the effect of shrinkage was converted to a thermal problem by the effective modulus method.

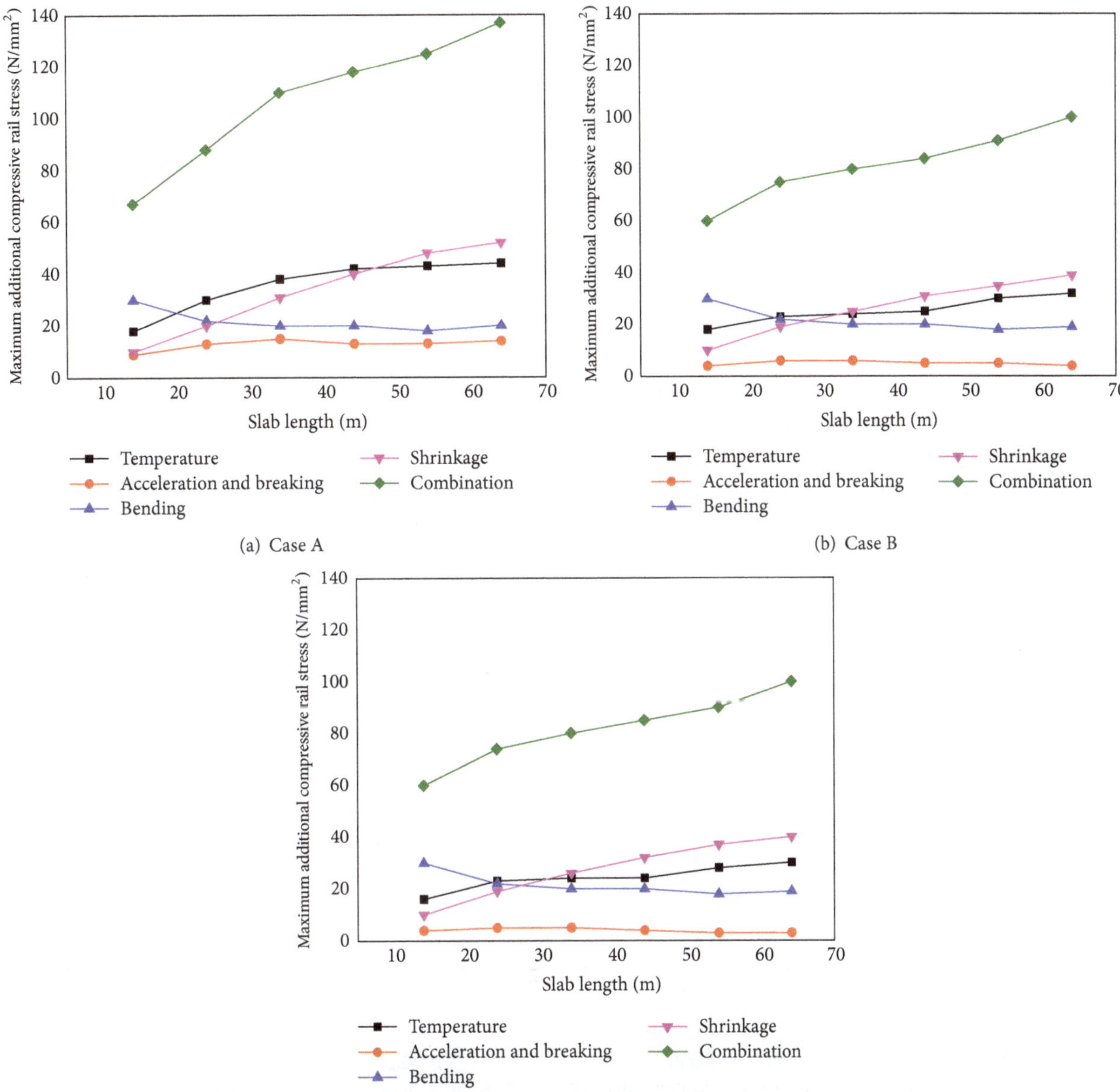

(a) Case A

(b) Case B

(c) Case C

FIGURE 9: Results of additional tensile rail stress with increasing slab length.

By this method, temperature changes (ΔT) were applied to the FST system to simulate shrinkage. The temperature changes can be estimated from the total strain $\varepsilon_{sh}(t, t_c)$ due to the shrinkage of the FST and were calculated to be âĂŠ255 $\mu\varepsilon$, which can be converted into a temperature of âĂŠ21.25°C. In finite element analysis, material properties must be given to each structural element. The element material properties of the rail are given in Table 1.

TABLE 1: Material constants.

Material property	Rail	Floating slab
Young's modulus of elasticity, E (MPa)	200000	28000
Shear modulus, G (MPa)	79300	21000
Coefficient of thermal expansion α (/°C)	1.2×10^{-5}	1.2×10^{-5}
Poisson's ratio	0.3	0.2

3. Numerical Results and Discussions

Finite element analysis of a standard FST was performed to understand the interaction behaviors of the FST system. The selected results of this study are the rail stresses, the slab tensile stresses and longitudinal displacement, and the stoppers' longitudinal displacements. Figure 7(a) shows the maximum additional compressive rail stress in Case A (with no stoppers). As the slab length increases, the overall rail

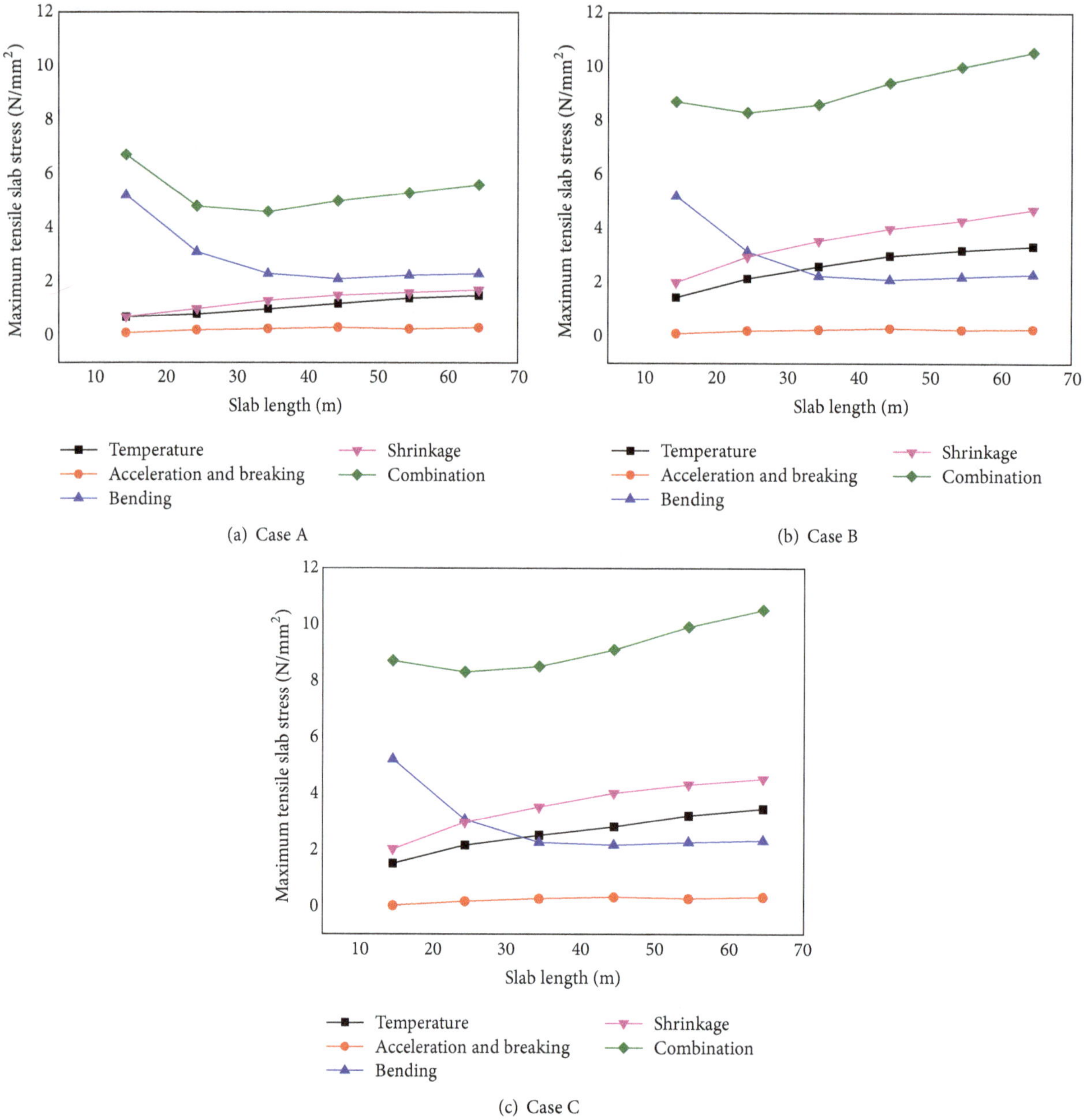

FIGURE 10: Results of additional tensile slab stress with increasing slab length.

stress has a similar growth rate to the stress due to changes in temperature; the stresses due to other loads do not show any consistent trends. The deformed configuration of the FST due to bending is shown in Figure 8. The maximum rail stresses due to the FST bending always occur at the transition between the embankments and the floating slab track. This can be explained by the sudden change in stiffness from the embankment to the isolator. Figure 8 shows that the shape of the deformed track changes as the slab length changes. A shape with a single curve was observed in shorter slab lengths, while multiple curvatures were observed in longer slab lengths. This is a result of the bilinear behavior of the isolator stiffness. When the load increases to a limit value, the isolator stiffness value approaches infinity. At that time, it has a maximum reaction force and acts in the opposite direction, moving the slab in the upward direction. The variations in these values are small compared to the permissible maximum additional stresses in the CWR in the UIC code [11], indicating that the contributions of variation in maximum additional stresses are relatively nonsignificant. The total maximum additional rail stresses due to load combinations monotonically increase as the slab length increases. Actions due to changes in temperature are the major influencing factors in the combined maximum rail stress.

Figure 7(b) shows the maximum additional compressive rail stresses in Case B (with 2 stoppers), and Figure 7(c) shows

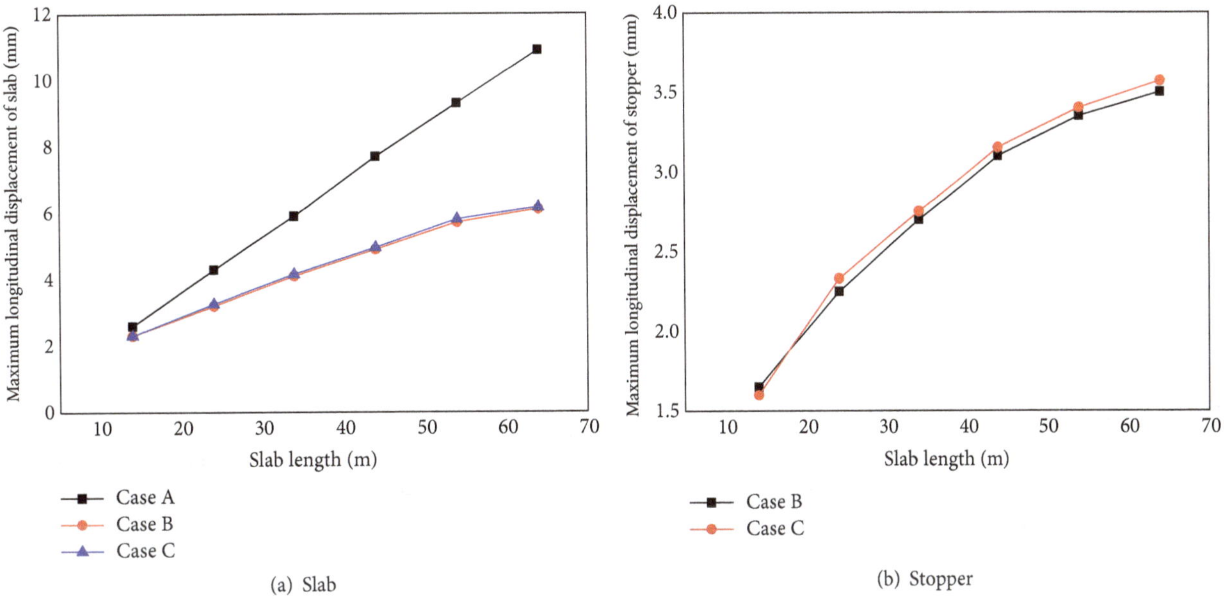

FIGURE 11: Longitudinal displacement.

the maximum additional compressive rail stresses in Case C (with 3 stoppers). The overall stress trends are similar to those of Case A (no stoppers); however, the magnitudes of the stresses in cases with stoppers (Cases B and C) are much lower compared to that with no stoppers (Case A). Specifically, the maximum stresses due to slab bending are the same for all cases because the stoppers do not restrict the vertical movement of the slab in a 2D FST system model; hence, additional compressive rail stresses have no effect on the bending. Particularly, in Cases B and C, the maximum stress values due to temperature and shrinkage loads have equal values. Because the center stopper in Case C is positioned at the axis of symmetry, the forces here due to temperature change and due to shrinkage act only in the longitudinal direction through the length of floating slab. Consequently, the stopper in the middle of the slab has no influence on the temperature and shrinkage forces. This observation indicates that the location of a stopper significantly affects the results of the analysis. Also, it is clear that the stresses increase as the slab length increases. In Case A (no stoppers), the maximum additional compressive rail stress increases rapidly over the allowable additional compressive rail stress with slab lengths over 30 m. In Cases B and C, the maximum additional compressive rail stresses satisfy the permissible additional compressive rail stress requirement for all slab lengths. The cumulative additional tensile rail stresses are shown in Figure 9. Similar to the cumulative additional compressive rail stresses shown in Figure 7, the total combined stresses increase with slab length.

The maximum slab length is defined by the permissible additional stresses on the slab in the CWR. As mentioned in the Korean version of the UIC code [5], the maximum additional rail stress of UIC60 CWR is 92 N/mm^2 for additional compressive and tensile rail stresses. In Cases B and C, the tensile rail stress reaches the maximum value at slab lengths of 54.5 m, while the maximum length for Case A is 24.5 m, a 30 m difference. This result highlights the importance of the stopper as a factor in the additional rail stress of an FST system.

Figure 10 shows the maximum tensile slab stresses with increasing slab lengths. The maximum stress due to bending occurs in the middle span of the floating slab and varies quite substantially with slab lengths between 14.5 and 34.5 m. This can be explained as a result of the slab behavior, illustrated in Figure 8. Additional isolators are added when the slab length is increased; the slab is lifted under vertical loading when the slab length is between 14.5 and 34.5 m. Length values exceeding this range will gradually increase stress. Figure 10(a) illustrates the increasing maximum tensile slab stresses due to shrinkage and temperature change; their values are nearly the same in Case A. However, Figures 10(b) and 10(c) show rapidly increasing stress due to temperature when stoppers were added. This occurred because the stoppers act in the longitudinal direction, the same direction in which the temperature and shrinkage loads act. Figure 10 shows that the effects of acceleration and braking on the tensile slab stress are small because the load is applied to the rail along the track direction, not directly to the slab. Figure 10 also illustrates the cumulative tensile slab stress, which, in contrast to the cumulative tensile rail stress, is lower when there are no stoppers (Case A) than when stoppers are present (Cases B and C).

Figure 11(a) shows the maximum longitudinal displacement of the slab. In Figure 11(a), the slab displacement in Case A increases more quickly than in the other cases, which again illustrates the significant influence of stoppers on longitudinal displacement. When stoppers are installed, the maximum longitudinal displacement increases with an increase in slab length, as can be inferred from the results for Cases B and C. With the exception of acceleration and breaking

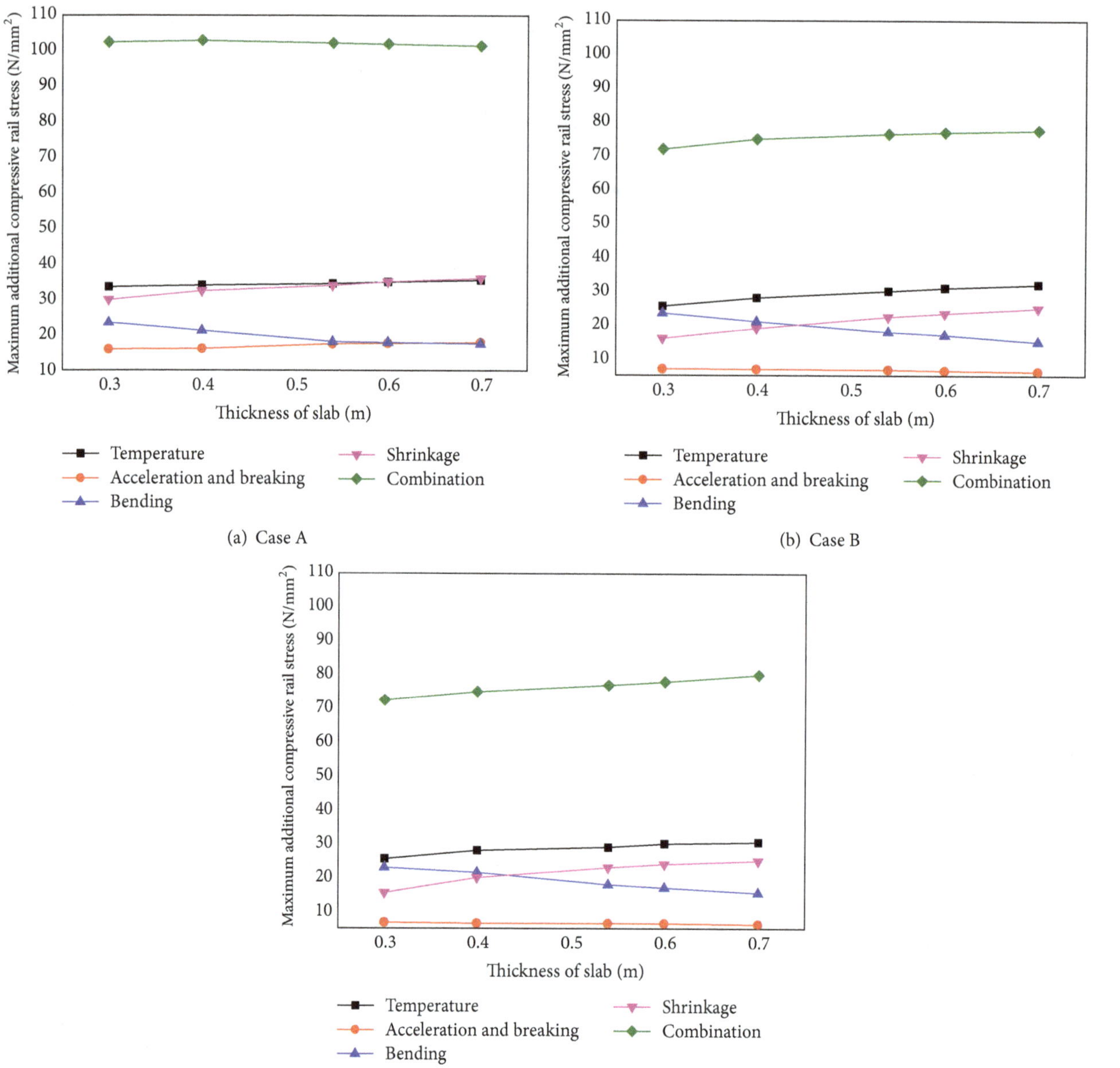

(a) Case A

(b) Case B

(c) Case C

FIGURE 12: Results of additional compressive rail stress with increasing slab thickness.

loading, the position of a stopper in the mid-span region has no influence on the rail stresses and longitudinal slab displacement. Figure 11(b) shows the maximum longitudinal displacement of the stopper. As the slab length increases, the maximum longitudinal displacement of both the slab and stopper increases in all three cases. The maximum longitudinal displacement of the stopper increases as the slab length increases.

The maximum additional compressive rail stresses with increasing slab thickness are shown in Figure 12. As can be seen in Figure 12(a), when no stoppers are used (Case A), the increases in maximum stress due to both temperature and acceleration and breaking are small when the slab thickness increases. Under vertical loading, the compressive rail stress has an inverse ratio to slab thickness, a trivially obvious result. Compressive rail stress due to shrinkage increases with slab thickness. The total additional compressive rail stress of a 0.6 m thick slab in Case A is 92.17 N/mm^2, over the maximum permissible additional stress from the UIC code [11]. The overall trends in Figures 12(b) and 12(c) are the same as for Case A but with lower magnitudes. The additional compressive stress trends in Cases B and C can be seen more clearly than in Case A, especially those due to temperature changes. The cumulative loads in Figure 12 clearly show the beneficial effect of stoppers in an FST system with varying slab thicknesses. A reasonable slab thickness should be chosen and combined with other appropriately chosen variables such as slab length and isolator stiffness. The maximum additional tensile rail stresses with increasing slab thickness are plotted in Figure 13. In general, the trends in this case are

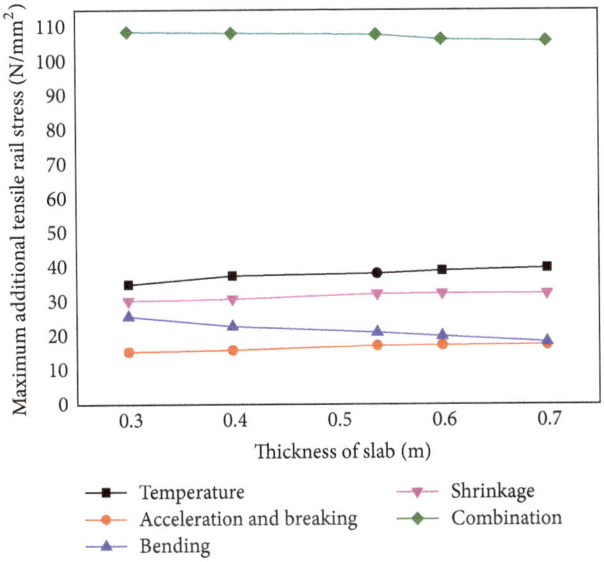

(a) Case A

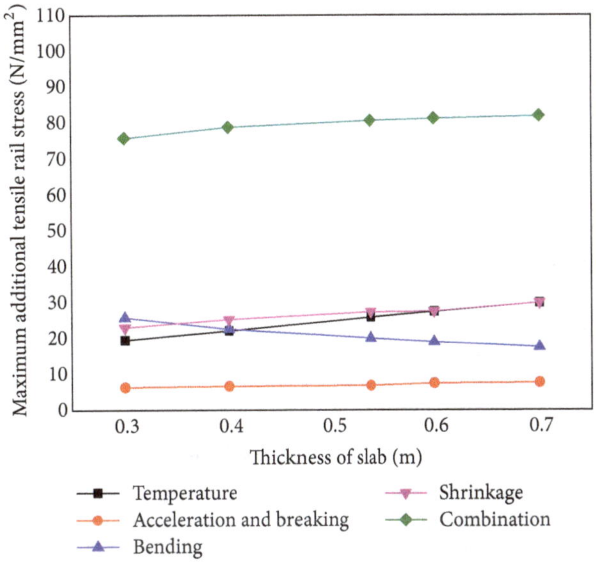

(b) Case B

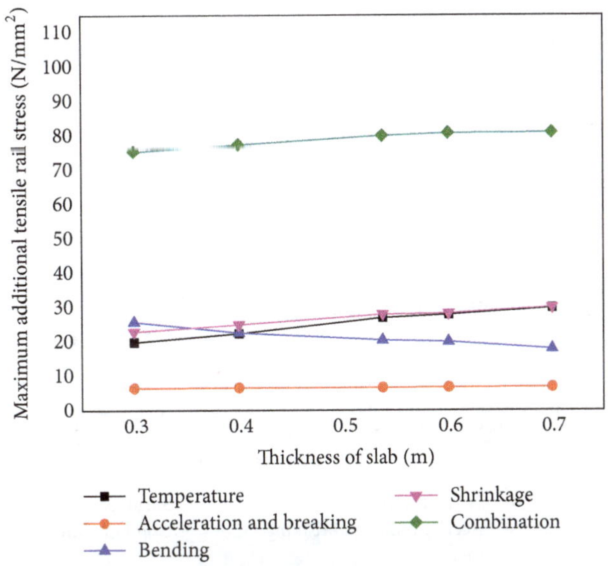

(c) Case C

FIGURE 13: Results of additional tensile rail stress with increasing slab thickness.

the same as those for maximum additional compressive rail stresses, but the tensile stresses for Cases B and C increase faster with increasing slab thickness than for Case A. This is again explained as an effect of the stoppers, which restrict the longitudinal displacement of the slab, causing the stresses to increase faster with increasing slab thickness. Figure 14 shows the maximum tensile slab stress with increasing slab thickness. In Case A, the concrete flexural stiffness increases as the slab thickness increases, which resists the generation of tensile slab stress over the entire slab. Thus, the stress decreases with increasing slab thickness for all loading types. These decreasing stresses also occur in Cases B and C. The cumulative maximum tensile slab stress is significantly lower when no stoppers are installed (Case A) than in the other cases in which they are used (Cases B and C).

Figure 15 shows the maximum additional compressive rail stress with changes in the isolator stiffness. The main contributor to the cumulative load combination is the bending load, which is an effect of the behavior of isolator stiffness, which only acts in the vertical direction in an FST system. In this simulation, loads due to temperature, shrinkage, and acceleration and breaking act only in the longitudinal direction and do not cause stress in the rail. The three stopper cases (Cases A, B, and C) show the same stress trend but with different magnitudes. Figure 16 shows the maximum additional tensile rail stress with changing isolator stiffness. For the same

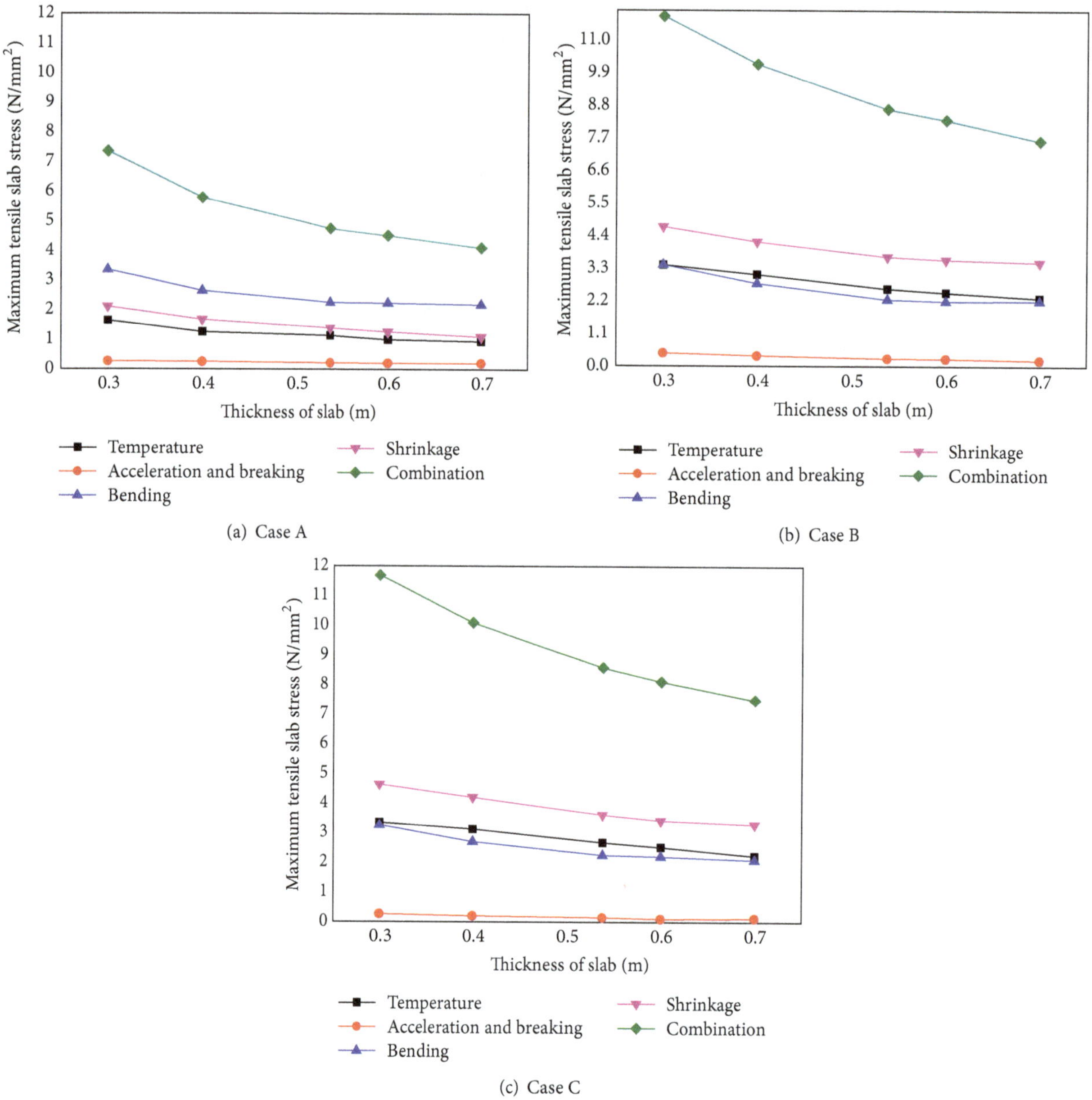

FIGURE 14: Results of tensile slab stress with increasing slab thickness.

reasons as the compressive rail stress, the maximum tensile rail stresses decrease as the isolator stiffness increases. Here, the only factor causing changes in the stress is the slab bending. The cumulative load combinations in Figure 16 show only a small difference between Cases B and C due to the stopper placement but show much higher stresses when no stoppers are installed, as in Case A. Figure 17 shows the maximum tensile slab stress with changing isolator stiffness. The tensile slab stress decreases as the isolator stiffness increases. The stress shows the same trends in all three cases, but the magnitude of the cumulative load combination stresses in Cases B and C is larger than that of Case A. This result is the opposite of that of rail stress because of the installation of stoppers.

In general, the results show that temperature change, bending, and shrinkage result in greater stresses on the FST system than acceleration and breaking. In almost all cases, the stresses due to acceleration and breaking have very small values. Thus, these behaviors have no significant impact on the load combination case.

4. Formula for Maximum Additional Rail Stress

The maximum additional rail stress of an FST system is the most important factor in the design of a CWR system, since the slab length is limited based on the maximum additional rail stress; however, modeling and analysis of an FST system with many parameters such as slab length and thickness and isolator stiffness are time-consuming. Thus, knowledge of the relationships between the parameters is needed to reduce the

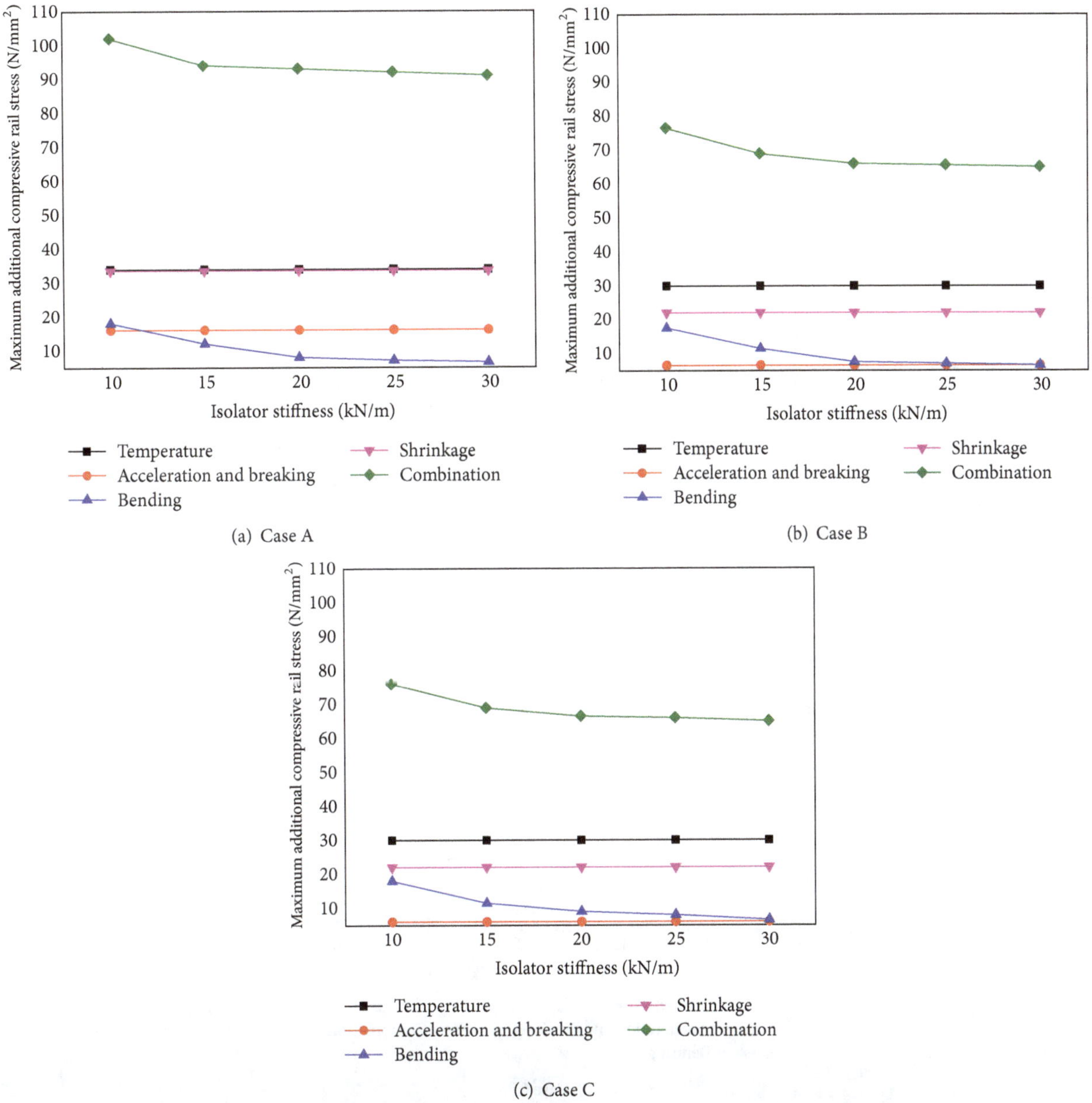

FIGURE 15: Results of compressive rail stress with increasing isolator stiffness.

calculation volume. The purpose of this section is to find that relationship based on previously analyzed data.

This study proposed a formula for maximum additional rail stress based on a nonlinear regression analysis of the interaction analysis results. The formula is proposed to confirm the sensitivities of each parameter. In this study, the interaction analysis was carried out up to the maximum slab length for FSTs. Thus, the applicable range for the slab length was chosen as 14.5 to 54.5 m. In the same manner, the applicable slab thickness range was 0.3 to 0.7 m. The isolator stiffness was fixed at 9 kN/mm. Young's modulus of the material used to make the slab track is between 20 and 35 GPa.

Within the chosen applicable ranges, the additional compressive rail stress is chosen to be a dependent variable, and the slab length and thickness are chosen as two independent variables; the formula obtained based on a nonlinear regression analysis was

$$\sigma_c = a + bL + cL^2 + dL^3 + eL^4 + \frac{f}{t} + \frac{g}{t^2} + \frac{h}{t^3} + \frac{i}{t^4} \tag{2}$$

where σ_c is the additional compressive rail stress, L is the slab length, t is the slab thickness, and the coefficients a, b, c, d, e, f, g, h, and i have the values shown in Table 2.

In the same manner, the formula for additional tensile rail stress based on interaction analysis is proposed to be

$$\sigma_t = a + bL + cL^2 + dL^3 + eL^4 + fL^5 + g\ln(t) + h\ln(t)^2 + i\ln(t)^3 \tag{3}$$

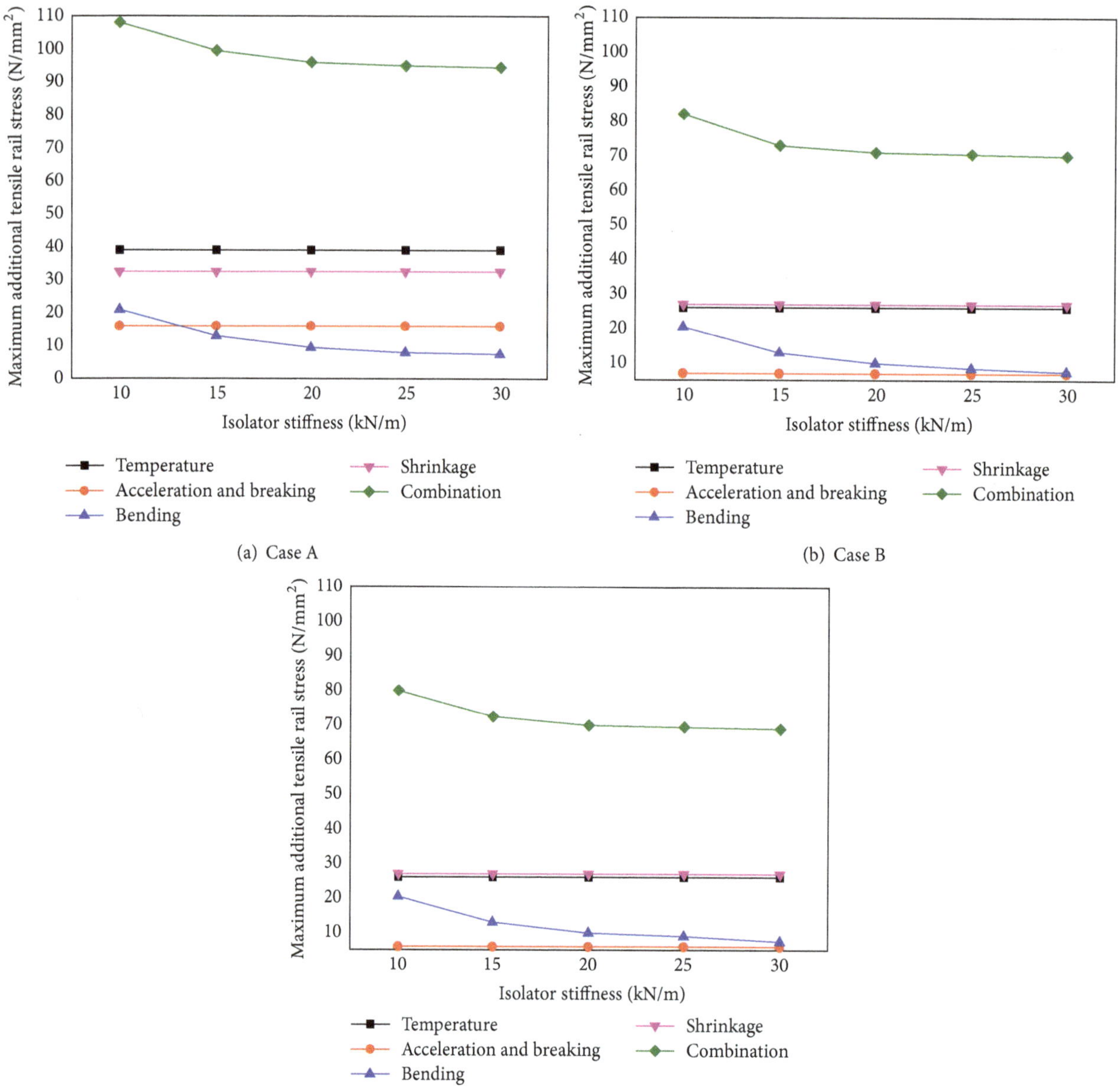

(a) Case A

(b) Case B

(c) Case C

Figure 16: Results of tensile rail stress with increasing isolator stiffness.

where σ_t is the additional tensile rail stress, L is the slab track length, t is the slab track thickness, and the coefficients a, b, c, d, e, f, g, h, and i have the values shown in Table 2.

The additional rail stresses from the proposed formula are compared with those obtained from finite element analysis in the applicable slab length and thickness ranges, as shown in Figures 18 and 19, respectively. These figures plot the ratio of the additional rail stresses obtained from the interaction analysis to those from the proposed formula, showing that the additional rail stress from the design equation is always greater than the finite element results. Within the applicable range of an FST system, the formula always produces additional rail stresses that are conservative by up to 10 %; because of this, the proposed formula is expected to become a safety margin in the design of an FST.

5. Conclusions

This paper examines the behavior of an FST by means of a track-slab-isolator interaction analysis. Parametric studies were conducted on the geometry and material properties of an FST to investigate the key behaviors of an FST. Many FST system cases were modeled and analyzed from a standard case to investigate the response behavior of an FST. The results from the analysis were used in a proposed formula for additional rail stress of an FST

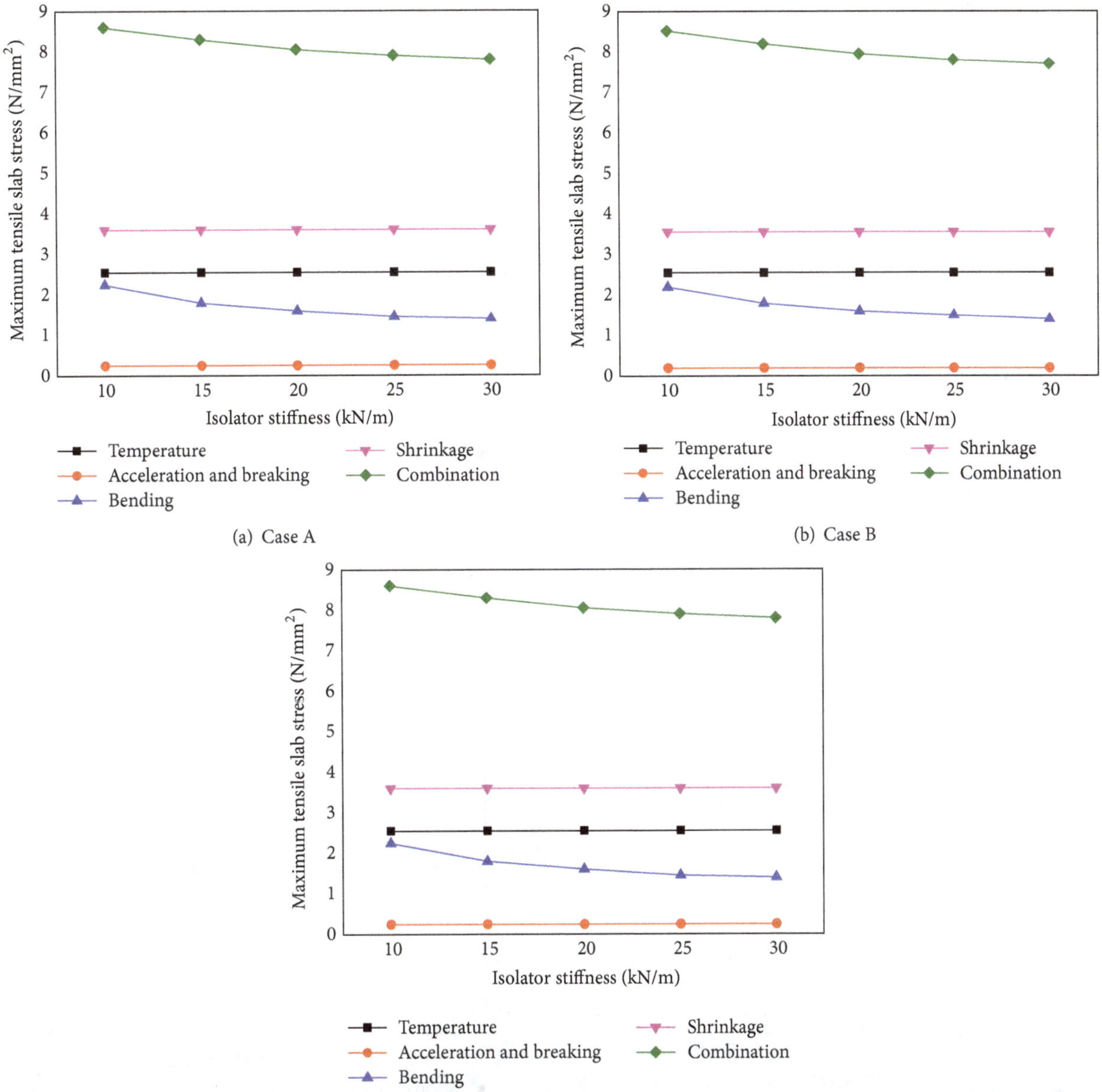

(a) Case A

(b) Case B

(c) Case C

FIGURE 17: Results of tensile slab stress with increasing isolator stiffness.

TABLE 2: Coefficients in the rail stress formula.

Additional compressive rail stress		Additional tensile rail stress	
Coefficient	Value	Coefficient	Value
a	-32.494	a	11.952
b	-1.659	b	6.233
c	-3.12×10^{-3}	c	-0.225
d	7.389×10^{-4}	d	4.36×10^{-3}
e	-7.333×10^{-6}	e	-4.3×10^{-5}
f	-17.448	f	1.824×10^{-7}
g	15.753	g	3.416
h	-5.057	h	-3.56
i	0.581	i	-0.521

system. The following conclusions were derived from the research.

The results of additional rail stresses demonstrate that temperature, bending, and shrinkage produce greater stresses on the FST system than acceleration and breaking. It has been shown that slab length is the most sensitive parameter influencing the additional rail stress; the maximum slab length for a standard FST is 54.5 m with stoppers and 24.5 m without stoppers, a significant difference. For this reason, stoppers should be used and carefully installed to optimize the overall function of the FST system.

Finally, the formula proposed for additional rail stress always produces results up to a 10% conservative safety margin, within the applicable parameter range of the FST system, and can be effectively used in the design phase.

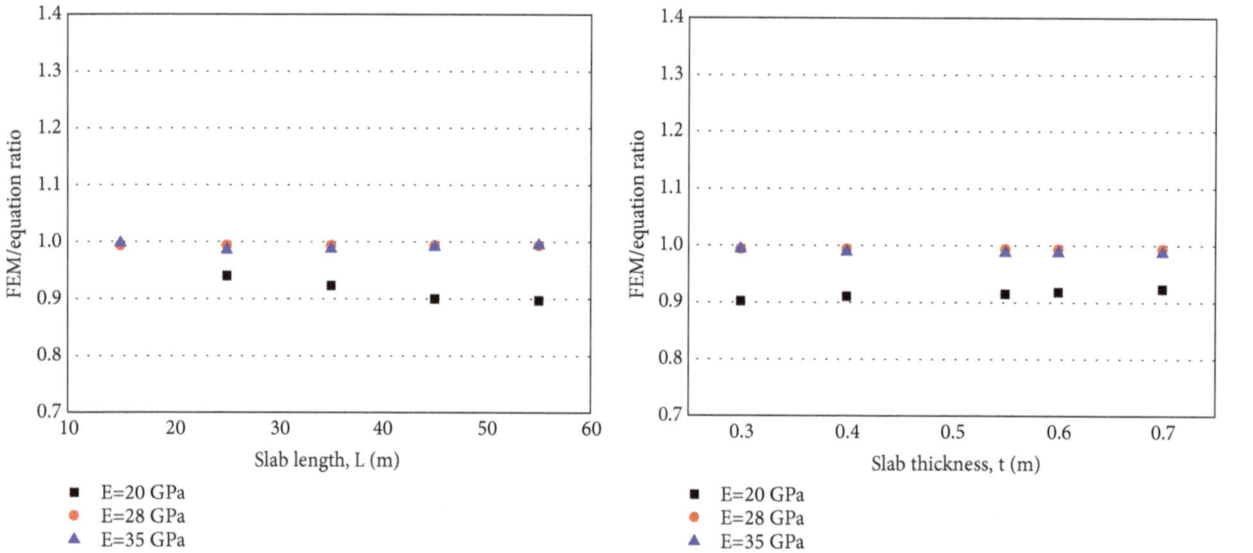

(a) Slab length (b) Slab thickness

FIGURE 18: Additional compressive rail stress comparisons between the design equation and the FEM.

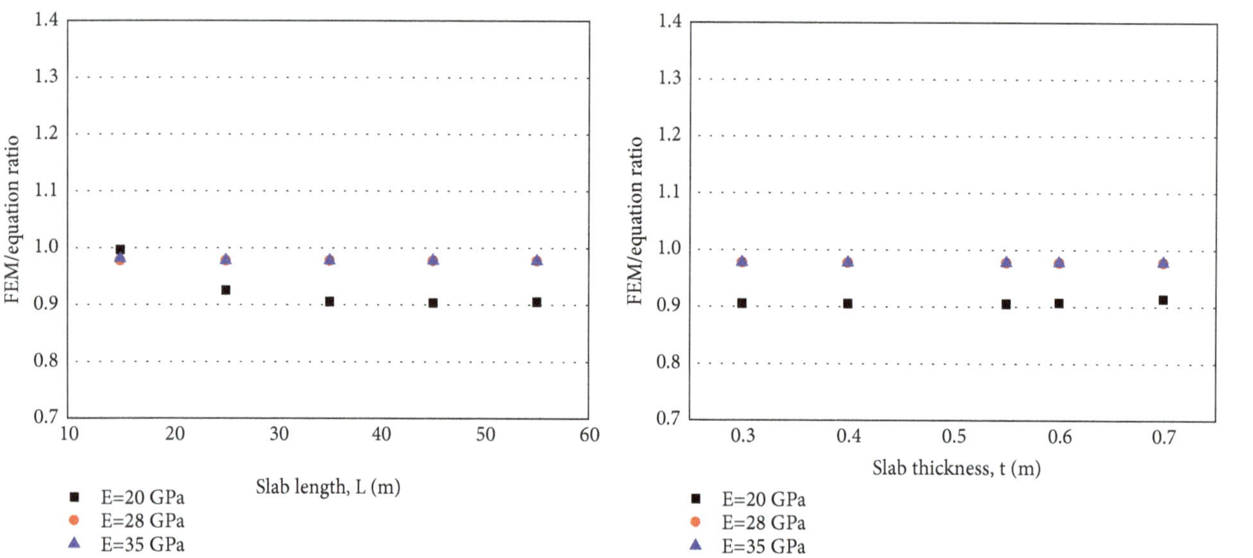

(a) Slab length (b) Slab thickness

FIGURE 19: Additional tensile rail stress comparisons between the design equation and the FEM.

Conflicts of Interest

The authors declare that they have no conflicts of interest.

Acknowledgments

This study is a basic research project conducted with support from the National Research Foundation of Korea through government funds (Ministry of Science, ICT, and Future Planning, Republic of Korea). The project numbers are 2017R1A2B4010467 and 2017R1C1B1006732.

References

[1] J. T. Nelson, "Recent developments in ground-borne noise and vibration control," *Journal of Sound and Vibration*, vol. 193, no. 1, pp. 367–376, 1996.

[2] G. Lombaert, G. Degrande, B. Vanhauwere, B. Vandeborght, and S. François, "The control of ground-borne vibrations from railway traffic by means of continuous floating slabs," *Journal of Sound and Vibration*, vol. 297, no. 3-5, pp. 946–961, 2006.

[3] W. D. Henn, "System comparison: ballasted track - slab track," *Rail Engineering International*, vol. 22, no. 2, 1993.

[4] D. N. Bilow and G. M. Randich, "Slab track for the next 100 years. Presented at the AREMA," in *Proceedings of the 2000 Annual Conference*, 2000.

[5] Korea Rail Network Authority, *Guideline of Railway Design*, 2011.

[6] P. Grootenhuis, "Floating track slab isolation for railways," *Journal of Sound and Vibration*, vol. 51, no. 3, pp. 443–448, 1977.

[7] S.-T. Yen and Y.-H. Lee, "Parameter identification and analysis of a slab track system using 3D ABAQUS program," *Journal of Transportation Engineering*, vol. 133, no. 5, pp. 288–297, 2007.

[8] G. Degrande, D. Clouteau, R. Othman et al., "A numerical model for ground-borne vibrations from underground railway traffic based on a periodic finite element-boundary element formulation," *Journal of Sound and Vibration*, vol. 293, no. 3-5, pp. 645–666, 2006.

[9] H. G. Wagner and A. Herrmann, "Floating slab track above ground for turnouts in tram lines," *Noise and Vibration Mitigation for Rail Transportation Systems*, vol. 99, pp. 86–93, 2008.

[10] J. Yuan, Y. Chang, Z. Meng, and L. Song, "Modal analysis and parameter assessment of floating slab track," in *Proceedings of the 2nd International Conference on Transportation Engineering, ICTE 2009*, pp. 3294–3299, July 2009.

[11] International Union of Railways, *Track/Bridge Interaction. Recommendations for Calculations*, CODE 774 3R, UIC, 2nd edition, 2001.

[12] Land Transport and Maritime R&D Report, *Development of Low Vibration Track (Floating Slab Track) And Technology for Improvement of Performance of Long-Span Bridges*, 2011.

[13] Z. G. Li and T. X. Wu, "On vehicle/track impact at connection between a floating slab and ballasted track and floating slab track's effectiveness of force isolation," *Vehicle System Dynamics*, vol. 47, no. 5, pp. 513–531, 2009.

[14] ABAQUS Inc, *ABAQUS/Standard User's Manual*, ABAQUS Inc., Pawtucket, RI, USA, 2007.

[15] Pandrol, "Testing of in-line SFC on Rheda 2000 to KHSRC specification," Tech. Rep. 46011-9, 2004.

[16] ACI Committee 209, *209.2R-08 Guide for Modeling and Calculating Shrinkage and Creep in Hardened Concrete*, American Concrete Institute, 2008.

[17] Z. P. Bažant and S. Baweja, "Justification and refinements of model B3 for concrete creep and shrinkage 1. statistics and sensitivity," *Materials and Structures*, vol. 28, no. 7, pp. 415–430, 1995.

[18] Z. P. Bazant, "Criteria for rational prediction of creep and shrinkage of concrete," *ACI Special Publications*, vol. 194, pp. 237–260, 2000.

[19] C. F. M. Code, *CEB-FIP Model Code for Concrete Structures, Euro-International Committe for Concrete*, Bulletin, 1990.

[20] N. J. Gardner and M. J. Lockman, "Design provisions for drying shrinkage and creep of normal strength concrete," *ACI Materials Journal*, vol. 98, no. 2, pp. 159–167, 2001.

Soil-Rock Slope Stability Analysis by Considering the Nonuniformity of Rocks

Shunqing Liu,[1] **Xianwen Huang**,[1] **Aizhao Zhou**,[1] **Jun Hu**,[2] **and Wei Wang**[3]

[1]*Department of Civil and Architecture Engineering, Jiangsu University of Science and Technology, Zhenjiang, Jiangsu 212003, China*
[2]*College of Civil Engineering and Architecture, Hainan University, Haikou, Hainan 570228, China*
[3]*School of Civil Engineering, Shaoxing University, Shaoxing, Zhejiang 312000, China*

Correspondence should be addressed to Wei Wang; wangwells@qq.com

Academic Editor: Payman Jalali

Soil-rock slopes are widely distributed around the world, while the commonly adopted method by simplifying it as a uniform media tends to be excessively conservative. In this study, a slope stability analysis method considering the nonuniform characteristics of rocks was proposed. It was found that the distribution, relative position, and shape of rock have significant effect on slope stability. For the influence of distribution, large rocks at the foot of slope have the most significant effect on slope stability while the effect is insignificant when the rocks are on the slope surface. In terms of the relative position of rocks, four plastic expansion modes of bypass, diversion, inclusion, and penetration were put forward through the analysis on the expansion mode of the plastic zone. Moreover, rock shape also has influence on slope stability.

1. Introduction

Soil-rock slopes are present in most mountainous areas all over the world, especially in the western China region [1, 2]. According to statistics, six of every seven landslide accidents that occurred on the Sichuan-Tibet Highway were associated with soil-rock slopes, along which the earthwork prone to landslides amounted to 300 million m^3. In addition, according to Liao's statistical data on the 816 historical landslide accidents reported in the Panxi area by the end of 2006, 500 landslide accidents, about 61.3% were associated with soil-rock slopes, indicating that soil-rock slopes account for a large proportion of slope accidents [3]. In July 2010, 29 people were killed and several others missing in the landslide of the soil-rock slopes in Zhaizigou, Shanxi Province. In May 2011, a large landslide occurred in Luojiang Village, Guangxi Province, in which the quantity of earth and rock that was dislocated amounted to 200,000 m^3, and caused economic losses of millions of dollars. In June 2017, 10 people were killed and 73 went missing in the massive landslide that occurred in Maoxian County, Sichuan Province, and the length of the river blockage area caused by the landslide reached 2 km. From these, it can be seen that landslide hazards are great threats to the life and property of humans and so it is necessary to prevent them [4]. Compared to homogeneous soil slope and cement-based material-treated soil slope, the nonhomogeneous soil-rock slope may have long and twisted developing path of plastic area because of the rock existing, which produces bigger safety factor [5–7]. Traditionally, soil-rock slope is often considered as homogeneous like soil slope [8–10]. This method of simplifying the soil-rock slope as a homogeneous soil slope neglects the enhancement of rock distribution on slope stability, and the over-conservative assumption is likely to cause wastage of resources. Therefore, it is imperative to analyze the stability and failure characteristics of soil-rock slopes based on the rock distribution, adopting an economical and reasonable support scheme for soil-rock slopes. Soil-rock slopes are mainly composed of soil-rock mixture, which is an extremely uneven multi-phase rock mass material [11, 12].

The mechanical properties of the soil-rock mixture from different aspects, such as stone content, rock shape, rock distribution, position, and bonding strength of the soil-rock interface, have been studied by many scholars. Sonmez

[13] adopted a serious triaxial tests without lateral constraints, proving that the rock content and size can have great influence on the compressing strength, and putting forward prediction formula whose versatility about all kinds of mixtures is not confirmed till now. Gong Jian [14] adopted discrete element method to make research on the effect of rock content to shear strength of soil-rock mixtures including cohesion and friction angel, and then based on the theory of Taylor, explaining the greater shear strength of mixtures which is owing to the contribution of interlocking of rocks. Zhang [15] according to the rock characteristic which is extracted from the slope photos in view, setting many models about soil-rock mixtures with many different sizes. Through analysis, it had be proved that the shape of modulus have influence on the strength characteristic of mixtures and put forward that the arrangement and distribution of rocks are the main factors, which is good agreement with the conclusion of Meng [16]. Fragaszy [17] presented a numerical homogenization study of the elastic property of a soil-rock mixture using random mesostructure generation. Through analysis, it can be found that the connection and broken of bigger rocks are the two main controlling factors to the strength of mixtures. For better evaluating the mechanical characteristic of soil-rock mixtures in engineering, Zhang [18] made a serious in situ test of the soil-rock mixtures in Nuozhadu embankment dam. The results showed that the elastic modulus and friction angel of soil-rock mixtures is bigger than uniform soils obviously, but the cohesion is smaller than soils. There have been extensive studies and research advances regarding the mechanical properties of soil-rock mixtures, while research progress on soil-rock slopes is relatively slow.

Ohene [19] and Hencher [20] put forward a series of stability analysis methods and the concept of slope treatment on slide rock and colluvial landslides based on a large number of site surveys. Liu [21] adopted two different media to simulate the soil and rock and set up corresponding soil-rock slope in random and then studied the the destructive mode and safety factor of slopes. Through analysis, it can be found that the destructive mode of slopes is complicated but the safety factor of it is predictable. Wu [22], based on the elastic-plastic modulus of soil, put forward an calculation modulus of soil-rock slopes considering the interaction of soil and rocks and obtained the conclusion that the soil-rock interaction has a great influence on the landslide shape in terms of rock size and slope angle and the deformation characteristics of landslide at the slope toe is slightly different with that at the slope top. Xu [23] designed a program about polygonal creation which is used to set up soil-rock slopes, and discrete element method is used to simulate the destructive progress and shear band of soil-rock mixtures which were expended in the safe analysis of soil-rock slope finally. Through analysis, it can be proved that the rock shape have a great influence on the shear band developing of soil-rock slope. Maria [5] based on the different rock content and set up a serious 2D slope modulus randomly and then adopted finite element and limit equilibrium methods to study the stability of slope and plastic zone expansion mode. So the conclusion can be achieved that the distribution of rocks can have influence on the stability and safety factor of slopes, but the size of influence is not been discussed and the mechanism of how the rocks affect the plastic zone expansion mode detailed. So it is necessary to study the stability of slopes under the effect of rocks in the slope and put forward a reliable way to analyze the safety of soil-rock slope.

Based on the above reasons, a method for the stability analysis of soil-rock slopes based on the rock distribution and positions is proposed in this paper. The main ideas are as follows. (1) extraction of the original rock position and grading parameters, (2) generating the corresponding random rock contour lines according to rock parameters, (3) using the finite difference finite element method to establish and analyze the corresponding model, and (4) evaluating the stability of the soil-rock slopes through a large number of stability analysis results combined with engineering experience.

2. Random Model Establishment

2.1. Acquisition of Rock Properties. There are two common methods for obtaining the structural parameters of the slope model, namely, CT scanning of unimpaired slopes, obtaining the distribution, location, and rock shape in the slope by nondestructive testing. Li [24] reconstructed the original model by CT scanning, and then compared the results of the numerical simulation with experimental results, obtaining good consistency between the results and thereby establishing the reliability of the model reconstruction for the original slope simulation. Although this method could accurately reflect the distribution characteristics of rock in the original slope, it is the complex preliminary work required limited its applicability for actual projects. Another method is to randomly analyze the slope parameters, such as rock gradation, stone content and distribution location, in the original slope through analogy, reasoning, small-scale in-situ tests, and other methods [25, 26]. Slope safety was evaluated on the basis of a large number of numerical analysis results combined with engineering experience. This method has the advantages of convenient operation and accurately calculated results compared with CT scanning and other in situ tests. Therefore, in this paper, the stability of soil-rock slopes was mainly analyzed by the second method, and this method for obtaining the rock outline would be described in detail as follows.

(1) The method of analogy analysis was adopted to accurately reflect the structural characteristics of the regenerated soil-rock slopes, and a photograph of the slope section adjacent to which a landslide had occurred was captured. Then, using the software Adobe Photoshop, rocks and soil were identified on the basis of the RGB color difference between them, and the corresponding binary image was generated as shown in Figure 1.

(2) Considering the large number of rocks, the artificial identification of the rock area, stone content, distortion coefficient, and other characteristics was a complex process, and the feasibility was not high. Therefore, the relevant computer code was written, and a program was used to identify them. First, the preprocessed image was meshed

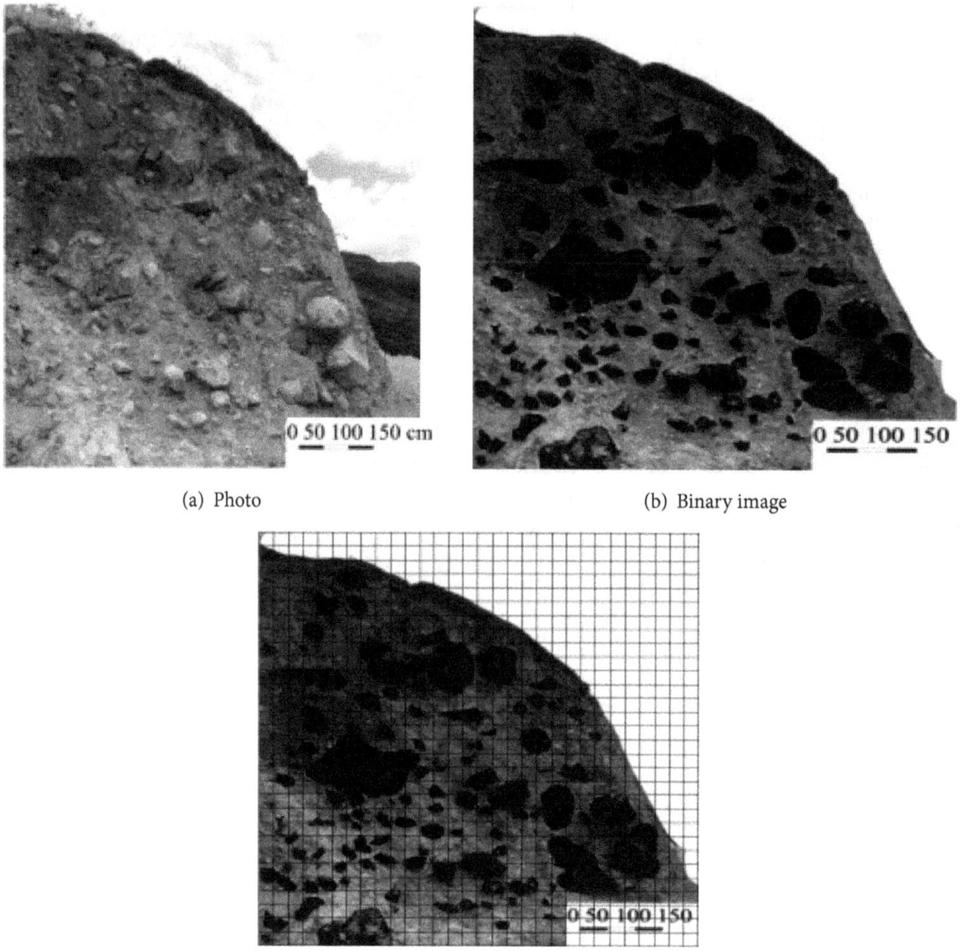

(a) Photo

(b) Binary image

(c) Grid partition

FIGURE 1: Imaging processing.

finely and evenly, and ten thousand grid points with definite coordinate positions were generated and written into the array. Then, the RGB colors at each grid point were identified, and they were written in the corresponding grid array, as shown in Figure 1(c).

(3) This step involved the identification of the entire rock. The expansion calculation was conducted on the defined rock unit according to the order of the first X axis and the rear Y axis. As shown in the Figure 2, unit a with the position coordinates of (X_a, Y_a) was judged as rock, and a number designating it as stone or rock was assigned to the unit according to the sequence. The coordinates of the eight units in the periphery of a were $(X_a \pm 1, Y_a \pm 1)$, $(X_a, Y_a \pm 1)$, and $(X_a \pm 1, Y_a)$, respectively, and if they contained rock units, the whole unit was assigned the same rock number as unit a. All the rocks were thus merged and identified, and assigned individual codes as the process proceeded, and the result of this process is shown in Figure 2(a).

(4) The final process was the calculation of the rock denaturation. Assuming that an entire rock is assigned the same rock number, the maximum diameter $d_{\max}$, minimum diameter $d_{\min}$, and distortion coefficient $\delta = d_{\max}/d_{\min}$ of the corresponding rock were calculated according to the formula $\sqrt{(X_i - X_j)^2 + (Y_i - Y_j)^2}$. Finally, the corresponding rock area rate and distortion coefficient were obtained through the comprehensive analysis of all the grids, and the result of the calculation was shown in Figure 2(b). In this figure, the minimum diameter is the distance from the center of area with maximum diameter to the outside edge of nearest rock.

2.2. Random Rock Contour Generation. After the image processing of the soil-rock slopes, the corresponding rock area rate and distortion coefficient of the gradation were obtained, and then the rocks in the corresponding gradation section were generated. The method used in this study was to generate a random circle corresponding with the gradation, using it to control the position and maximum diameter of the aggregates, and the corresponding alphabetic strings were x_o, y_o, d. The maximum diameter angle was set as $\theta(i_{\max}) = 2\pi\text{rand}$ (where rand indicates a random value between 0 and

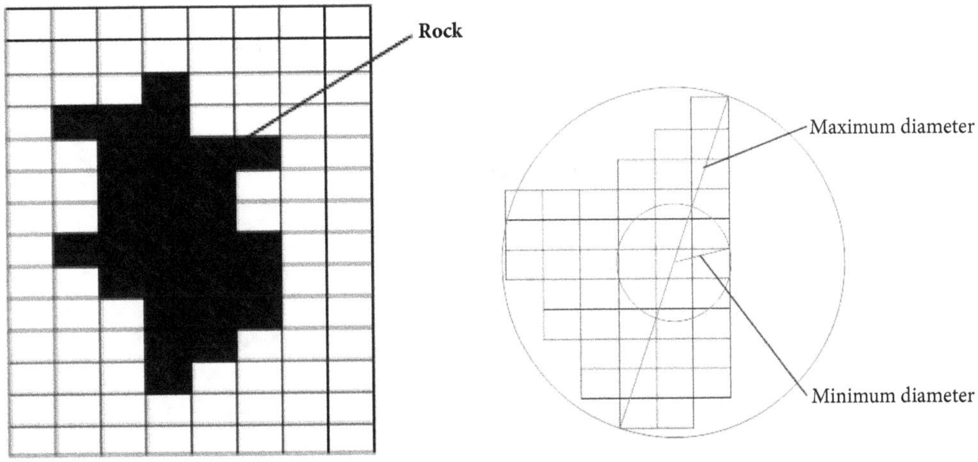

(a) Rock integrity

(b) Distortion coefficient of rock

FIGURE 2: Calculation effect.

1), and the corresponding extreme point coordinates were as follows:

$$(x_o \pm 0.5d\cos(\theta_{\max}), y_o \pm 0.5d\sin(\theta_{\max})) \tag{1}$$

$$(x_o \pm 0.5\delta\cos(\theta_{\min}), y_o \pm 0.5\delta\sin(\theta_{\min})) \tag{2}$$

where δ is the random side length under the influence of the distortion coefficient and the calculation formula is $\delta = (1 + k(2\text{rand} - 1))\mu d$, where k represents the variation threshold value of the distortion coefficient (between 0 and 0.1) and μ represents the distortion coefficient (between 0 and 0.9).

After the control points were generated, the assistant random boundary points of the rocks were generated, and the random functions of diameters and angles were $d_{\text{rand}} = \delta + (d - \delta)\text{rand}$ and $\theta_{\text{rand}} = 2\pi\text{rand}$, respectively. Thus, the boundary point coordinates of different rocks could be calculated according to the coordinates of the center point of the circle, and the set maximum polygon edge number (n-4) was taken as the control standard of the calculation number. As the boundary points were of randomly selected angles, all the boundary points of a single rock need to be sorted counterclockwise.

The rock area with rationalization of the boundary point was calculated using the triangle summation method. For any polygon A1, A2,..., with the circle (x_o, y_o) as the triangle vertex and l_{ab} as the bottom edge, the formula for calculating the area of any triangle is

$$S_i = \frac{1}{2}\begin{vmatrix} x_o & y_o & 1 \\ x_n & y_n & 1 \\ x_{n+1} & y_{n+1} & 1 \end{vmatrix} \tag{3}$$

The area of a single rock is obtained as $S = \sum S_i$.

2.3. Release of Rocks. To judge the possibility of invasion between polygonal rocks, the vector area method, and the included angle sum test method were proposed by previous works [27, 28], while DE and Taerwe [29] used the space partition-filling method. Although all the three methods can be used to determine whether or not rocks invade, the calculations are all rather complicated. In this study, the area determination method was adopted to directly judge whether or not the rocks invade by optimizing the release, hence simplifying the complex invasion judgment process. The probability of the random circle inside of the rock was avoided on the basis of the descending order of release gradation, greatly improving the rock release efficiency.

The release process of the rocks was as follows (shown in Figure 3): ① inputting initial parameters, including slope profile function, rock gradation parameters, stone content, and distortion parameters; ② generating the random circle based on the prescribed gradation; ③ determining whether there were rocks invading into the random circle; ④ generating polygonal rocks according to the distortion coefficient, and making node data rationalizations of a single rock; ⑤ drawing the corresponding rock outline map and then repeating steps ②③④ and ⑤ until the system generates an outline map meeting the requirements of the stone content, gradation curve, and distortion coefficient.

2.4. Finite Element Modeling

2.4.1. Finite Difference Strength Reduction Theory. The basis for calculating the slope safety factor in the strength reduction method [30] was that the shear strength parameters of soil were regularly reduced, as shown below:

$$c' = \frac{c}{F_s}$$
$$\varphi' = \arctan\left(\frac{\varphi}{F_s}\right) \tag{4}$$

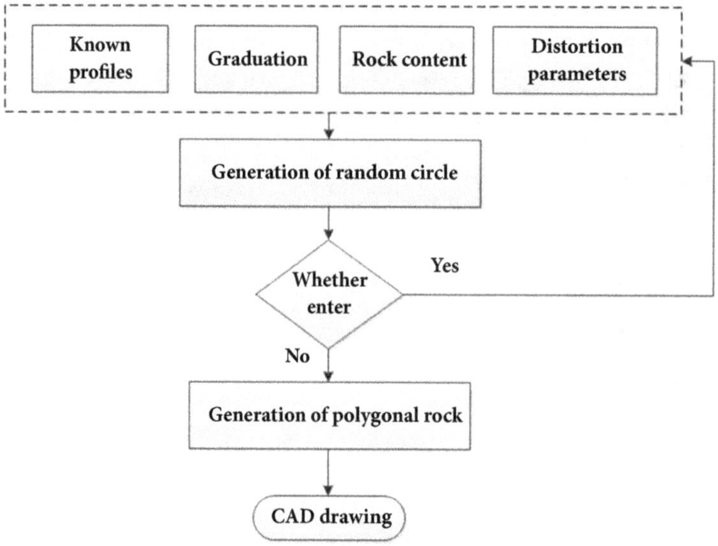

FIGURE 3: Flow chart of rock generation.

TABLE 1: Calculation model parameters of soil-rocks slopes.

Constituent	Density (kN/m^3)	Modulus of elasticity (kPa)	Poisson's ratio	Cohesion (kPa)	Internal friction angle (°)
Soil	1.80	50.0	0.35	30	24
Rocks	2.41	2e4	0.2	900	42

When the slope happened to be in the critical damage state, the reduction factor was the safety factor of the slope and the critical failure criterion mainly included

① whether the calculations made by the nonlinear program were convergent

② whether the plastic zone was in state of transfixion and formed a complete plastic zone

③ whether the displacement had a mutation.

In the MIDAS GTS software, the main basis for judging slope instability was whether the nonlinear calculation on energy, displacement, and internal force was convergent. It had the advantages that there was no need to presume the location and geometry of the sliding surface in advance, the influence of the soil deformation on slope stability could be taken into account, the sliding zone could be calculated more accurately, and the safety coefficient could be calculated directly.

2.4.2. Finite Element Model. The randomly generated rock model of the soil-rock slope was imported into the finite element software, and the model was discretized using a triangular mesh, with the maximum unit size being 0.5 times the minimum rock size. The total number of mesh elements was 32550. Horizontal constraints were set on the left and right boundaries of the model, and horizontal and vertical constraints were set at the bottom boundary; in addition, the vertical load due to the self-weight was added to the soil. The physical parameters of rock and soil were shown in Table 1.

2.4.3. Homogeneous Slope Calculation Results. In order to establish the influence relationship of rock locations on the original plastic zone of the slope, the corresponding model of the homogeneous soil slope was established and the sliding zone location was extracted.

As can be seen from Figure 4, the plastic zone started from the foot of the slope and extended to the top of slope. Finally, a perforative sliding zone was formed, and a wedge-shaped slider with a width of 8.34 m was generated in the slope.

Random rock generation software was used to simulate the soil-rock slopes and the strength reduction method was adopted to calculate the slope stability.

It can be seen from Figure 5 that the rock distribution affected the smoothness of the plastic zone. Rocks with high strength blocked the normal development path of the plastic zone, with the slope stability factor increasing to 1.180.

3. Influence of Larger Rock Position on Slope Stability

3.1. Model Feature Description. Four distribution models were established in this paper depending on the relative positions of the rocks and the plastic zone of the homogeneous soil slope. The heights of the areas II, III, and IV were taken based on the trisection of the sliding zone length, and the width of the top of the wedge-shaped slide was considered for the calculations. The area I was the undivided area of the slide, as shown in Figure 6.

An examination of the influence of the rock distribution locations on slope stability indicated that the slope stability

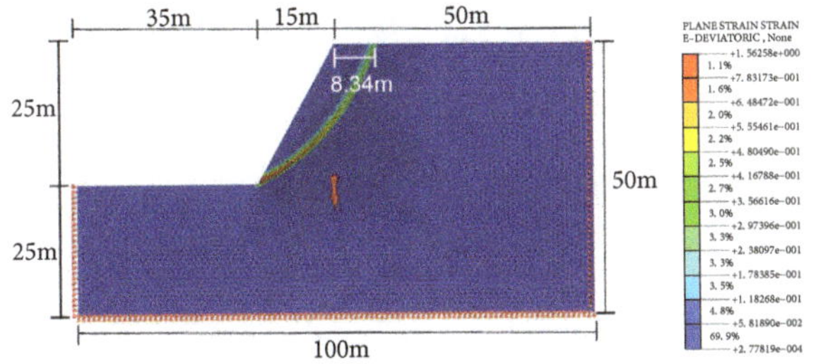

FIGURE 4: Calculation results of plastic zone in homogeneous soil slope (FOS: 1.01).

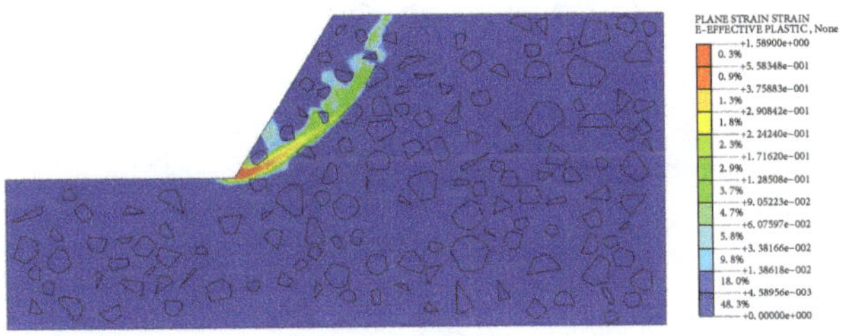

FIGURE 5: Calculation results of soil-rocks slopes (FOS: 1.18).

could be greatly influenced when the rock diameter was 0.3 times or more the area width by calculating rock models with different sizes. Therefore, the positions of the larger rocks in the slope were given more attention, and the control range of random rock sizes used in calculation was 0.3~0.5 times the width of the entire slope. The four distribution models are shown in Figure 7 were established by adjusting only the positions of the larger rocks, while the stone content and particle shapes were maintained the same.

3.2. Analysis of Calculation Results. As can be seen from Figure 8, the existence of larger rocks caused the originally smooth plastic zone to develop by flocculation, bifurcation and burr, and the plastic zone expansion exhibited an obvious "bypass" effect. The different distribution positions of larger rocks in the slope had different effects on the expansion mode and stability of the plastic zone.

As shown in Figure 8(a), when the larger rocks distributed in area I, the slope stability coefficient increased from 1.01 for the homogeneous soil slope to 1.188. It was assumed that the sliding portion of the slope was a massive block, and the larger rocks were located in the position indicated by area I, that is, the surface of the slope. The larger mass of the rocks made the barycenter of the sliding body deviate to the slope surface. Under the premise that the sliding zone was fixed, the angle of gravity direction of the barycenter of the sliding body and tangential angle of the sliding zone increased. The stress σ_x in the soil in the sliding zone along its normal direction increased, while the stress σ_y in the tangential direction and the shear stress τ decreased, which finally led to the decrease of the sliding force generated by the sliding body. Thus, the slope exhibited a greater safety factor. On the other hand, the smaller rocks in some slopes caused the development

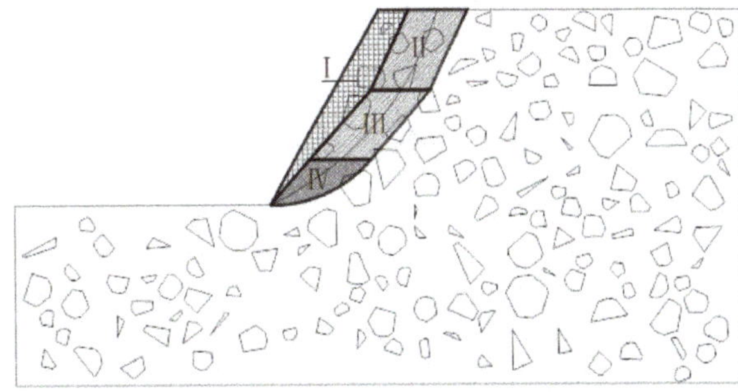

FIGURE 6: Location division of rocks.

route of the plastic zone in the slope to zigzag, increasing the development length of the plastic zone and improving the slope stability.

The larger rocks were stacked at the upper part of the plastic zone (area II), forming a plastic development area as shown in Figure 8(b). The plasticity development route of the slope was more complicated compared to the original homogeneous soil slope. The distribution pattern of the large rocks directly hindered the development route of the plastic zone at the top of slope, increasing the slope safety factor from 1.010 for homogenous soil to 1.316. Compared with the barycenter offset of the sliding mass in Figure 8(a), the larger rock distribution in this area directly affected the plasticity development route of the slope, making it impossible to form a penetrating and smooth primary plastic zone.

When the larger rocks were predominantly distributed in area III, the slope safety factor was 1.437, which was larger compared to the rock distribution in area II. The distribution of larger rocks in the middle of the slope plasticity development area hindered the development route of the plastic zone, making the formation of a penetrating and smooth primary plastic zone impossible. When the spacing of rocks was smaller, two front and back plastic zones around whole rocks formed. When rocks were spaced closely together, the phenomenon of "inclusion" can be observed, in which the plastic zone is developed first among the rocks and then continues onward to form the main plastic zone and develop further.

As can be seen from Figure 8(d), when larger rocks were distributed at the foot of the slope, there was a significant effect on the slope stability. And the stability coefficient increased from 1.01 for the homogeneous soil slope to 1.492 for the soil-rock mixture. When rocks with higher strength are distributed at the foot of the slope, the plasticity development route gets deviated from the foot of slope to the top of the rock accumulation. The movement of the plastic zone reduced the effective height of the slope in a relative sense and improved the slope stability.

Through the stability analysis of slopes with four different types of large-rock distributions, it was found that their magnitudes and manner of their influence on the slope stability were different. The obtained relationship with slope stability followed the relation Area IV (accumulation of larger

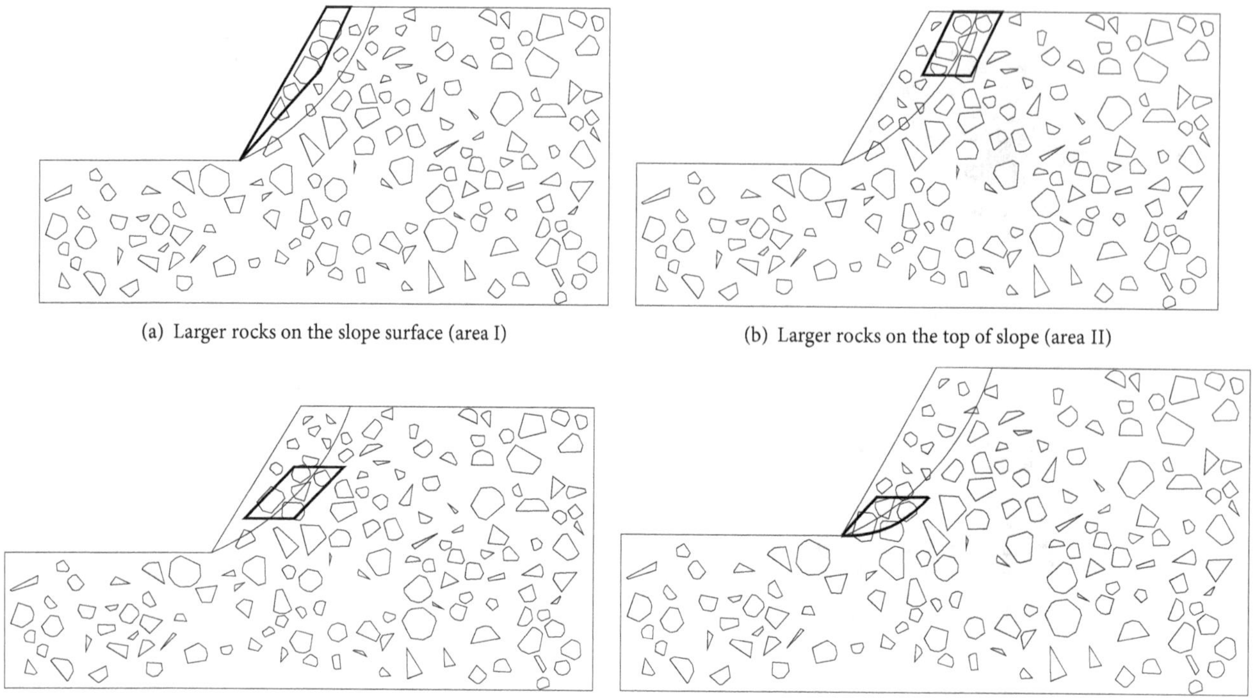

(a) Larger rocks on the slope surface (area I)

(b) Larger rocks on the top of slope (area II)

(c) Larger rocks on the slope surface (area III)

(d) Larger rocks on the top of slope (area IV)

FIGURE 7: Distribution of larger rocks with four kinds of positions.

rocks at the foot of the slope) > Area III (accumulation at the middle part of the slope) > Area II (accumulation at the top of the slope) > Area I (accumulation at the slope surface).

4. Influence of Relative Spacing of Rocks on Slope Stability

4.1. Model Feature Description. Although the distribution position of larger rocks had a greater impact on the overall slope stability, the impact of the relative position of rocks on slope stability could not be ignored. Even when the larger rocks were distributed within the same area, different relative distances and relative positions caused the plastic zone development to exhibit different patterns, thereby affecting the slope safety factor. According to the relative positions of the rocks, four main modes of plastic zone development were put forward. The condition of crack development in soil-rock mixtures was difficult to measure directly with existing technologies. Therefore, the authors took advantage of the characteristic that crack pairs would form along weak areas when the soil-rock mixture is air-dried. Using air-drying to disturb the soil-rock mixture, the method of replacing the sliding zone of the soil-rock mixture by an air shrinkage crack was used to represent the plasticity extension path in the soil-rock mixture, and Figure 9 shows the four different development modes that were observed using this method.

Mode I: As shown in Figure 9(a1), the larger rocks were distributed in the plastic zone extension path of the original homogeneous soil, causing an offset in the extension path, expanding along the weaker side of the whole rocks, and exhibiting "bypass." As shown in Figure 9(a2), the rocks blocked the primitive crack development of the soil, causing them to develop along the rock edges and changing the path of crack development.

Mode II: As shown in Figure 9(b1), the plastic zone expanded inside the rocks. Due to the greater strength of the rocks, the plastic zone expanded along the larger rocks creating a "tree root" pattern. As shown in Figure 9(b2), the original, single plasticity extension path turned into multiple plasticity extension paths after passing through larger rocks, resulting in the formation of "diversions." It also shows that a single crack divided into two cracks after passing through a block of rocks, which was different from the theoretical trajectory.

Mode III: As shown in Figure 9(c1), a single plasticity expansion path continued to develop in the same manner after entering the larger rock area. The "inclusion" phenomenon of the plastic zone appeared in the rock group; that is, the hitherto separate plastic zones merged into a single main plastic zone and continued to develop after passing around rocks. As shown in Figure 9(c2), obvious cracks appeared around the rocks and the presence of the rocks did not affect the crack development path.

Mode IV: As shown in Figure 9(d1), the plastic zone developed just along the cracks between the rocks, resulting in a "penetration" phenomenon relative to the rock group. As shown in Figure 9(d2), cracks in the soil-rock mixture developed in the interstices between two rocks, giving rise to "penetration."

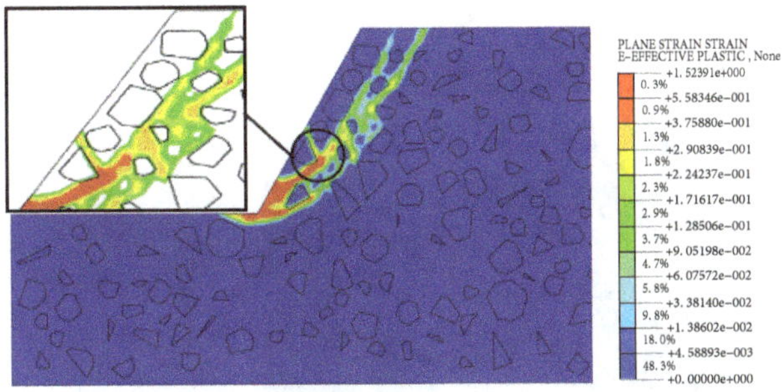

(a) Larger rocks on the slope surface (area I) (FOS=1.188)

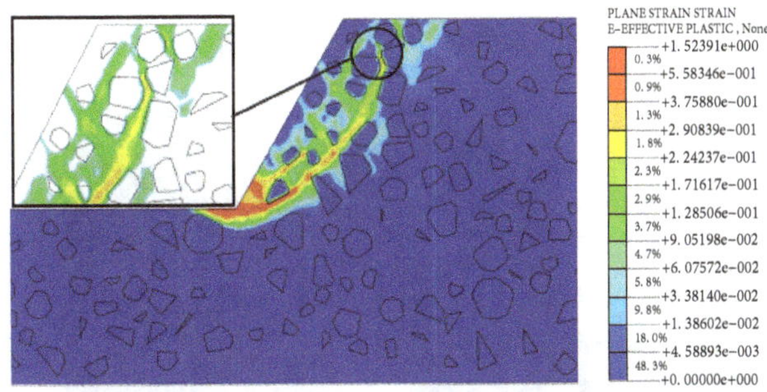

(b) Larger rocks on the top of slope (area II) (FOS=1.316)

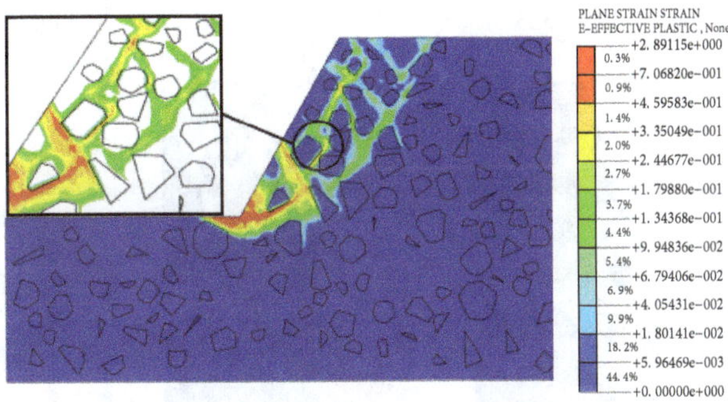

(c) Larger rocks on the slope surface (area III) (FOS=1.437)

(d) Larger rocks on the top of slope (area IV) (FOS=1.492)

FIGURE 8: Slope stability calculation results.

4.2. Results of the Analysis

4.2.1. Analysis of Plastic Zone Expansion Mode. By adjusting the relative position of the larger rocks, it was found that the relative position of the rocks had little effect on the plastic zone development path of the slope and the effect on the slope safety factor was smaller when the larger rocks were stacked in area I. Therefore, the slope stability corresponding to larger rocks in the area II, III, and IV alone was analyzed, as shown in Table 2. It can be seen from the table that the slope safety factor manifested certain discreteness under conditions of same rock distribution positions and different relative positions and intervals of rocks.

Regarding the distribution of larger rocks in area II, the slope safety coefficients under the four modes were 1.341, 1.325, 1.316, and 1.275, respectively. As can be seen from Figure A, the close-packed rocks made the plastic zone expand along the edge of the rock stack toward its top. On the one hand, the existence of rock edges and corners made the plastic zone develop along a zigzag path. On the other hand, the higher strength of the rock increased the antislip force of the slope and improved the slope stability. As can be seen from Figure B, the plastic zone was diverted under the action of larger rocks at the top of the slope, and one of the small plastic zones continued to develop, eventually producing a perforative plastic zone on the slope surface. Figure C reflected the development of the plastic zone when it contained rocks, and the existence of larger rocks did not have a significant impact on the plastic zone development. As can be seen in Figure D, the rock distribution did not

Table 2: The results of soil-rocks slopes.

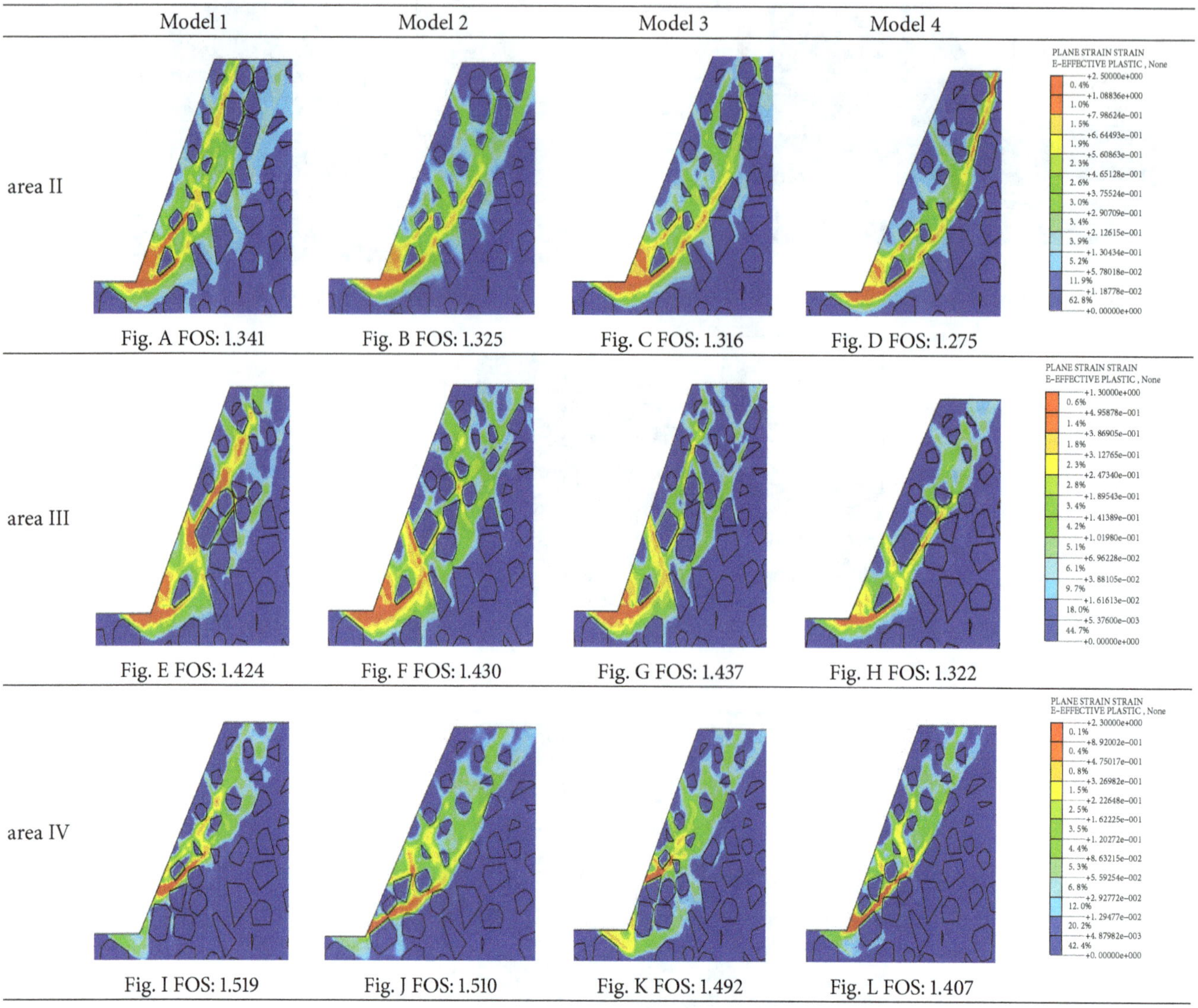

affect the route of the plastic zone development. Through comprehensive analysis, in this area, the impact of each mode on slope stability follows the order Mode I > Mode II > Mode III > Mode IV.

When the larger rocks were distributed in area III, the slope safety coefficients under the four kinds of modes were 1.424, 1.43, 1.437, and 1.322, respectively. When the spacing between rocks was relatively small, the plastic zone moved along the edge of whole rocks to the slope surface and continued to develop as shown in Figure E. The plastic zone extension lines in Figure F and Figure G are similar. The plastic zone in the area of rock accumulation area in the slope developed along the rock edges and then gradually aggregated into the main plastic zone toward the top of the slope. The rocks in Figure H are distributed around the original plastic zone, and they did not have a significant influence on the path of the plastic zone development. It was determined from the analysis that, in this area, the influence relationship of each mode on slope stability followed the order Mode III > Mode II > Mode I > Mode IV.

The larger rocks distributed at the foot of the slope had the greatest impact on the increase in slope stability, with the corresponding safety factors being 1.519, 1.510, 1.492, and 1.407, respectively. Larger rocks directly obstructed the plastic zone development, making the plastic zone at the foot of the slope extend along the top of the rocks (Figure I). As the relative distance between the rocks increased and the relative position was adjusted, larger rocks at the foot of slope still blocked the onset of the plastic zone (Figure J and K). This indicates that the initial position of the plastic zone would be affected when larger rocks were distributed at the foot of the slope. When rocks were distributed along the plastic zone consisting of homogeneous soil, the influence on the plastic zone development was of a lesser degree and the increase in the slope safety factor was smaller. The results indicate that, in

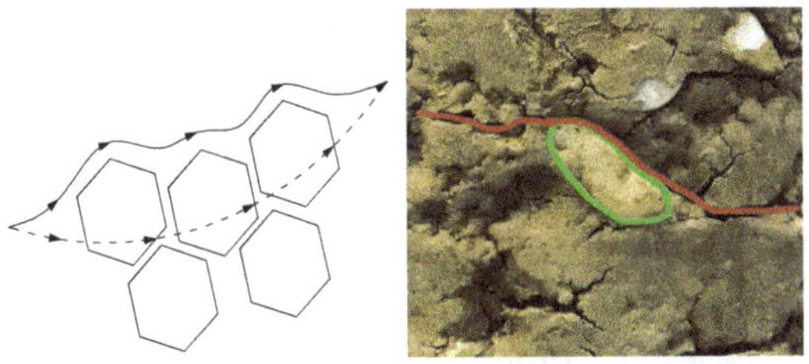

(a1) "Bypass" model (a2) "Bypass" example

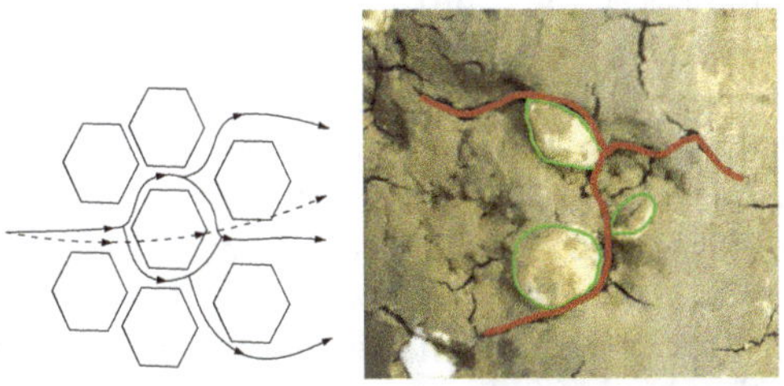

(b1) "Diversion" model (b2) "Diversion" example

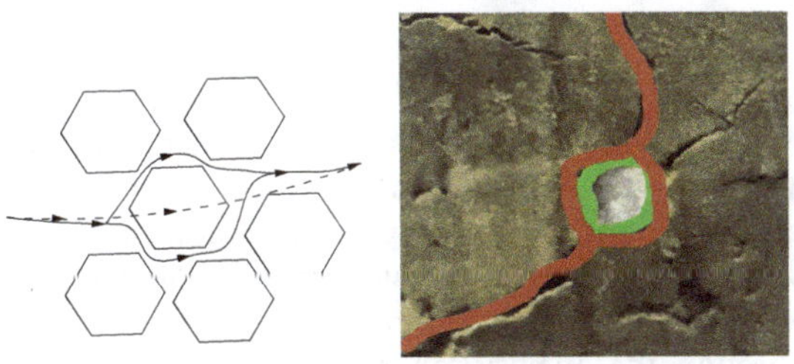

(c1) "Inclusion" model (c2) "Inclusion" example

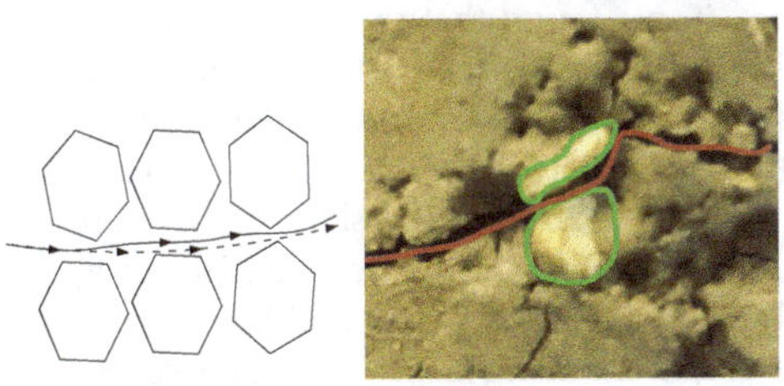

(d1)"Penetration" model (d2)"Penetration" example

FIGURE 9: Four plasticity expansion modes.

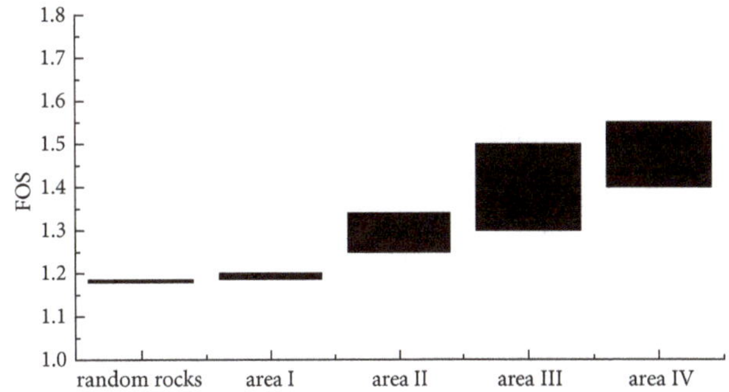

FIGURE 10: Slope safety factor.

this area, the impact of each mode on the slope stability was Mode III > Mode II > Mode I > Mode IV.

4.2.2. Slope Stability Analysis. For a given area of accumulation, the relative position of the larger rocks was continuously adjusted to obtain the slope safety factor corresponding to different positions of large-rock accumulation as per the statistical method, as shown in Figure 10.

As can be seen from Figure 10, the slope safety factor varied with the distribution position of larger rocks, showing great discreteness. By comparing the values of the slope safety factor under different large-rock accumulation conditions in each area, it was found that the discreteness of larger rock accumulation in the middle part of slope was the largest. The reason was that the plastic zone in the slope existed mainly in three modes, namely, the "bypass," "inclusion," and "penetration." The three modes influence the expansion path of the plastic zone to different extents and in different manners, giving rise to greater discreteness. When the larger rocks were accumulated at the foot of the slope, the discrete value of the slope safety factor was 0.15. Here, the plastic zone generally passed through the rock accumulation area in the "bypass" and "penetration" modes, with the former having an impact on the starting position of the plastic zone and the latter increasing the slip resistance of the slope, so that the rock accumulation in this position exhibited a certain degree of discreteness. When the larger rocks were accumulated at the top of the slope, the discrete value of the slope safety factor was 0.11. Here, the plastic zone generally passed through the rock area in the modes of "diversion," "bypass," and "penetration." At the top of the slope where the plastic zone development ended, the slope safety factor exhibited smaller discreteness as long as the plastic zone developed along the weakest path to form a penetrating sliding zone. When the larger rocks were accumulated at the slope surface, the changes in their relative positions did not affect the overall development of the plastic zone; there was a phenomenon of "bypass," and the discrete value of the slope safety factor was only 0.04. When the distribution of the larger rocks was not considered, the discrete value of the slope safety factor was the least, only 0.02.

5. Influence of Stone Shape on Slope Stability

5.1. Description of Model Features. In nature, different geological causes cause the soil-rock mixture to have different shapes [31]. For example, rocks in sedimentary soil-rock mixtures are scoured and transported by water flow for long periods of time, making them relatively smooth. On the other hand, rocks in colluvial soil-rock mixtures exhibit strong angularity, taking on needle-plate like shapes. The difference in rock shapes has a huge impact on the mechanical properties of the soil-rock mixture [32–35]. Nouguier-Lehon [36] based on different rock shapes adopted granular flow software to simulate mechanical properties of soil-rock mixtures and achieved that the rock shape has great influence on the strength of soil-rock mixtures. After that, Xu [37], owing to the use of CT method, adopted the greater shear testing instrument to study the strength characteristic and destructive path of soil-rock mixtures, but they ignore the rock shape to the strength characteristic and destructive path of soil-rock mixtures.

In this work, based on the study of dry shrinkage cracks in the soil-rock mixture, it was found that there were differences in the crack width and extension for pebbles and gravels, and the development of cracks in pebbles and gravels was analyzed according to the pattern differences, as shown in Figure 11.

Through the comparative analysis of the cracks in pebbles and gravels in the four modes, it was found that cracks usually developed at the ends of the rocks in the case of pebbles, while cracks usually develop along the rock edges and corners in gravels. Combined with the crack numbers and widths shown in Figures 11(e2) and 11(e3), it was found that the cracks were wider for pebbles and there were few microcracks in the periphery, while the cracks associated with the crushed stone were finer, and comparatively more microcracks appeared in the periphery. Therefore, it was believed that pebbles had a certain concentration effect on the crack development of the soil-rock mixture; that is, the cracks were concentrated along the main cracks. Meanwhile, the gravels did not have a concentration effect and the soil cracks were more chaotic. Therefore, it was necessary to analyze the influence of larger rock shapes on slope stability.

In this study, circular and triangular rocks were used to simulate pebbles and gravels respectively. In order to avoid the impact of the rock area ratio on slope stability, the triangle side length was maintained as $l = 2 \cdot 3^{-0.25}\pi^{0.5}r$ (where r is the circle radius) and eight corresponding analytical models were established according to the distribution positions and shapes of the larger rocks.

5.2. Results of the Analysis

5.2.1. Influence of Rock Shape on Slope Stability. As can be seen from the Figure 12, the slope safety factors of round rocks and triangular rocks were 1.268 and 1.273, respectively, when the larger rocks were distributed in area I. From the perspective of the plastic zone development, large round rocks had less influence on the plastic zone development of slope, while larger triangular rocks had a greater impact, which was mainly reflected in the development of the partial plastic zone affected by triangular rock edges and corners.

It could be seen from the Figure 13 that the phenomenon of "bypass" occurred in the plastic zone of the slope under the influence of circular rocks, and the plastic zone continued to develop mainly along rock edges with the slope safety factor increasing to 1.311. When triangular rocks were distributed at the top of the slope, the plastic zone was not single but developed multidirectionally along the rock gap and stress concentrations appeared at the rock edges and corners. The reason was that the circular rock surface was smoother when the plastic zone developed adjacent to the rock surface, which made it easy to guide the plastic zone development along the circular rock edge. However, the triangular rock had sharp edges and corners that did not have a streamlined shape, so the plastic zone developed freely and a more chaotic situation ensued. In addition, the existence of rock edges and corners made stress concentrations easy to occur at rock corners.

When the larger rocks were distributed at the middle part of the slope, the slope safety factors of circular and triangular rocks were 1.421 and 1.516, respectively (Figure 14). The plastic zone mainly developed along rock edges in the slope that consisted of circular rocks. In triangular rocks, a part of the plastic zone developed along the rock edges while there appeared a "flurry" development route due to the action of the rock edges and corners. In addition, stress concentrations developed at the rock edges and corners.

For the larger rocks distributed in area IV (Figure 15), the circular and triangular rocks all impeded the plastic zone development at the foot of the slope, causing the plastic zone at the foot of slope to shift upwards. When the slope consisted of circular rocks, the plastic zone developed along the rock edges, which differed considerably from the plastic zone development path taken when the slope consisted of triangular rocks. The reason was that the relatively smooth edge of the circular rocks had a certain inductive effect on the plastic zone development of the slope.

5.2.2. Discreteness Analysis of Rock Shapes. The distortion values pertaining to the circular and triangular rocks were 1.0 and 0.5, respectively. However, the distortion coefficients of the larger rocks in the actual slope were random, necessitating the statistical analysis of the slope stability under different distortion values, as shown in the Figure 16.

Through the analysis on the slope safety factor with larger rocks under different distortion coefficients in the same area, it was found that the slope safety factor went on increasing as the distortion coefficient decreased. The reason was that the larger the distortion value of rock shape, the closer the rock was to the circle, and the easier it was to induce the plastic zone of the slope to develop along rock edges and form the main plastic zone. However, rocks with relatively small distortion values could easily guide the plastic zone development along their edges and corners, resulting in the phenomenon of "diversion," which made it difficult to produce a main plastic zone. Therefore, the slope safety decreased with the increase of distortion values.

Slope safety factors with larger rocks distributed in different areas were compared, and the order was obtained as Area IV > Area III > Area II > Area I, which was same as the result from the impact analysis and influence of larger rock distribution positions on slope stability obtained in Section 3. In different areas, the differences in rock distortion values made the slope safety factor have a certain degree of dispersion, the magnitude relationship was obtained as Area IV > Area III > Area II > Area I; the difference was mainly reflected in the influence of rock shapes on the plastic zone development path of slope. Circular rocks had smoother outlines compared to polygonal rock, which were more likely to guide the development path of the plastic zone.

6. Conclusion

In this study, Adobe Photoshop software was used to complete the preprocessing of soil-rock slope images, and a custom program was developed to extract the characteristic

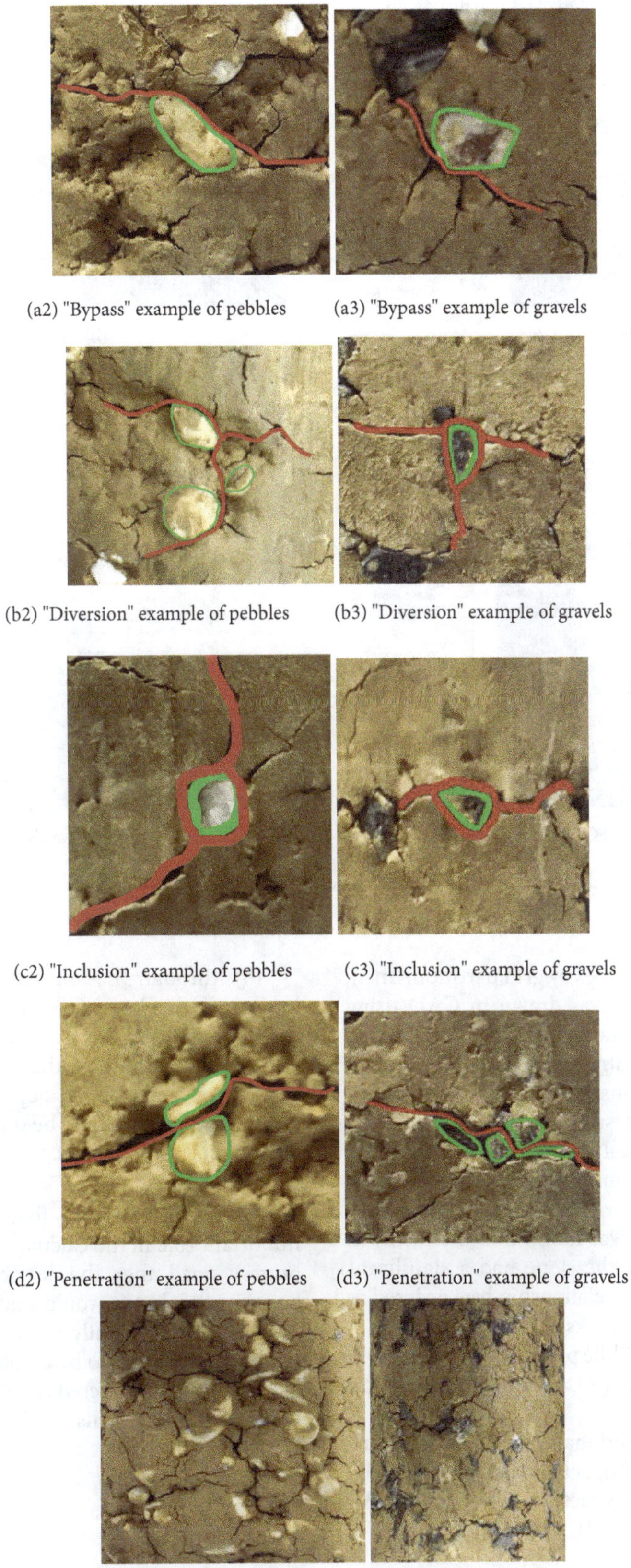

(a2) "Bypass" example of pebbles (a3) "Bypass" example of gravels

(b2) "Diversion" example of pebbles (b3) "Diversion" example of gravels

(c2) "Inclusion" example of pebbles (c3) "Inclusion" example of gravels

(d2) "Penetration" example of pebbles (d3) "Penetration" example of gravels

(e2) Cracks of pebbles (e3) Example of gravels

FIGURE 11: Cracks of pebbles and gravels.

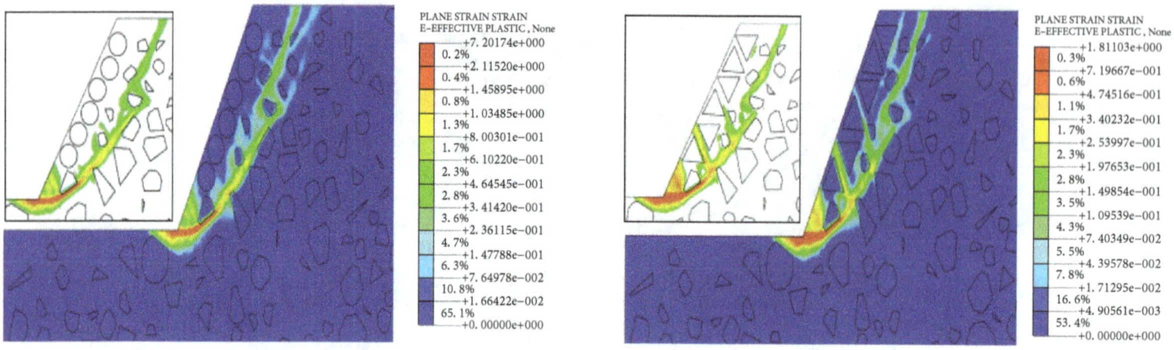

(a) Pebbles FOS=1.268 (b) Gravels FOS=1.273

FIGURE 12: Plastic zone development cloud chart of slope with accumulation of larger rocks at slope surface.

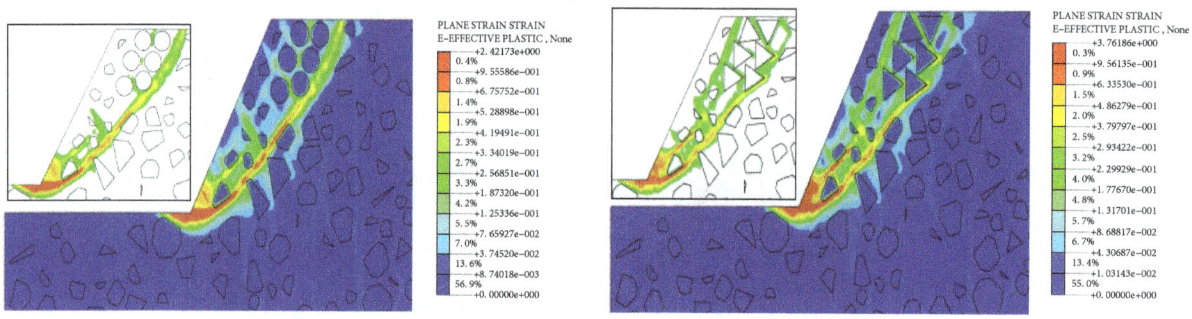

(a) Pebbles FOS=1.311 (b) Gravels FOS=1.334

FIGURE 13: Plastic zone development cloud chart of slope with accumulation of larger rocks at the top of slope.

parameters of rock in soil-rock slopes by grid partition. Then, the corresponding contour map was drawn in CAD using a VBA code. Finally, the slope stability was analyzed based on the method of finite element strength reduction. Through analysis, the following conclusions were drawn:

(1) Considering the lack of sufficient stability analyses data regarding traditional soil-soil mixture slopes, a stability analysis method for the soil-soil mixture slope based on rock characteristic parameters of the original slope was proposed.

(2) Through the analysis, it was found that the distribution position of larger rocks in the slope had a significant impact on slope stability, and the relationship was ordered as Area IV (accumulation of larger rocks at the foot of slope) > Area III (accumulation at the middle part of slope) > Areas II (accumulation at the top of the slope) > Area I (accumulation at the slope surface).

(3) In the analysis, it was found that the relative positions of the larger rocks had a certain impact on slope stability. The discrete relationship of slope safety factor was obtained as the distribution of larger rocks in Area III > Area IV > Area II > Area I.

(4) Rock shapes also had a certain effect on slope stability. The discrete relationship of slope safety factor was the distribution of larger rocks in Area IV > Area III > Area II > Area I.

(5) Through the analysis, it was found that the distribution of the larger rocks had a great influence on the development and stability of the plastic zone of the slope. When the method in this study is used to calculate the stability of the soil-rock slopes, the actual position of larger rocks in the slope should be surveyed to improve the accuracy of slope stability analysis.

It should be noted that the size effect may play an important role in the calculation result. If the slop is very big or very small, then the distribution and shape effect would be still valid, but it would make difference to the calculated absolute value. To fully capture the performance, some new methods may need to be employed [38]. As a limitation, size effect was not considered in this study, which forms the scope of future investigations.

Conflicts of Interest

The authors declare that they have no conflicts of interest.

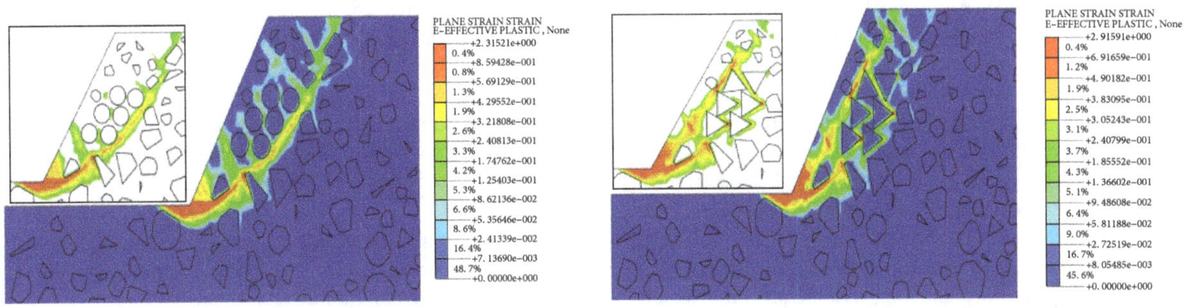

(a) Pebbles FOS=1.421 (b) Gravels FOS=1.516

FIGURE 14: Plastic zone development cloud chart of slope with accumulation of larger rocks at the middle part of slope.

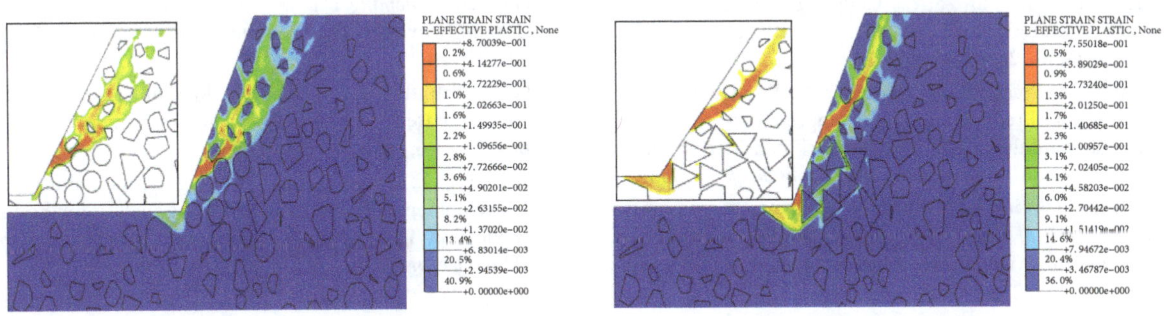

(a) Pebbles FOS=1.441 (b) Gravels FOS=1.572

FIGURE 15: Plastic zone development cloud chart of slope with accumulation of larger rocks at the foot of slope.

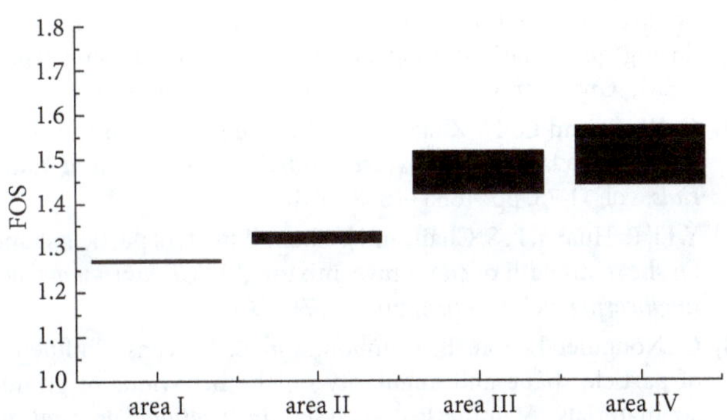

FIGURE 16: Slope safety factor under different distortion coefficients.

Acknowledgments

This work was supported by the National Natural Science Foundation of China (41772311; 51579199), the Zhejiang Provincial Natural Science Foundation of China (LY17E080016), and the Natural Science Fund for Colleges and Universities of Jiangsu Province (Grant no. 17KJB560003). China Postodoctoral Science Foundation (2018M630722) and Hainan Natural Science Foundation (518MS122, 518QN307) are also acknowledged.

References

[1] G. Vessia, J. Kozubal, and W. Puła, "High dimensional model representation for reliability analyses of complex rock–soil slope stability," *Archives of Civil and Mechanical Engineering*, vol. 17, no. 4, pp. 954–963, 2017.

[2] R. K. Goel and B. Singh, "Engineering rock mass classification," *Petroleum*, pp. 357–365, 2012.

[3] Q. Liao, X. Li, and Z. Hao, "Current status and future trends of studies on rock and soil aggregates," *Engineering Geology*, vol. 14, no. 6, pp. 801–807, 2006.

[4] D. Guo, C. He, C. Xu, and M. Hamada, "Analysis of the relations between slope failure distribution and seismic ground motion during the 2008 Wenchuan earthquake," *Soil Dynamics and Earthquake Engineering*, vol. 72, pp. 99–107, 2015.

[5] M. L. Napoli, M. Barbero, E. Ravera, and C. Scavia, "A stochastic approach to slope stability analysis in bimrocks," *International Journal of Rock Mechanics and Mining Sciences*, vol. 101, pp. 41–49, 2018.

[6] A. Zhou, C. Li, P. Jiang, K. Yao, N. Li, and W. Wang, "Slip Line Theory Based Stability Analysis on the Influence of Deep Excavation on Adjacent Slope," *Mathematical Problems in Engineering*, vol. 2018, Article ID 2041712, 2018.

[7] W. Wang, C. Zhang, N. Li, F. F. Tao, and K. Yao, "Mechanical performance of nanometer magnesia and cement reinforced seashore soft soil by direct shear test," *Marine Georesources & Geotechnology*, 2018.

[8] P. Jiang, L. Qiu, N. Li, W. Wang, A. Zhou, and J. Xiao, "Shearing Performance of Lime-Reinforced Iron Tailing Powder Based on Energy Dissipation," *Advances in Civil Engineering*, vol. 2018, pp. 1–8, 2018.

[9] W. Wang, N. Li, F. Zhang, A. Zhou, and S. Chi, "Experimental and Mathematical Investigations on Unconfined Compressive Behaviour of Costal Soft Soil under Complicated Freezing Processes," *Polish Maritime Research*, vol. 23, no. 4, pp. 112–116, 2016.

[10] Y. Pan, Y. Liu, H. Xiao, F. H. Lee, and K. K. Phoon, "Effect of spatial variability on short- and long-term behaviour of axially-loaded cement-admixed marine clay column," *Computers & Geosciences*, vol. 94, pp. 150–168, 2018.

[11] A. Hamidi, V. Yazdanjou, and N. Salimi, "Shear strength characteristics of sand-gravel mixtures," *International Journal of Geotechnical Engineering*, vol. 3, no. 1, pp. 29–38, 2009.

[12] Y. Wang, X. Li, B. Zheng, and C. F. Ma, "An Experimental Investigation of the Flow–Stress Coupling Characteristics of Soil–Rock Mixture Under Compression," *Transport in Porous Media*, vol. 112, no. 2, pp. 429–450, 2016.

[13] H. Sonmez, M. Ercanoglu, A. Kalender, G. Dagdelenler, and C. Tunusluoglu, "Predicting uniaxial compressive strength and deformation modulus of volcanic bimrock considering engineering dimension," *International Journal of Rock Mechanics and Mining Sciences*, vol. 86, pp. 91–103, 2016.

[14] J. Gong and J. Liu, "Analysis on the mechanical behaviors of soil-rock mixtures using discrete element method," *Procedia Engineering*, vol. 102, pp. 1783–1792, 2015.

[15] S. Zhang, H. Tang, H. Zhan, G. Lei, and H. Cheng, "Investigation of scale effect of numerical unconfined compression strengths of virtual colluvial-deluvial soil-rock mixture," *International Journal of Rock Mechanics and Mining Sciences*, vol. 77, pp. 208–219, 2015.

[16] Q. X. Meng, H. L. Wang, W. Y. Xu, and M. Cai, "A numerical homogenization study of the elastic property of a soil-rock mixture using random mesostructure generation," *Computers & Geosciences*, vol. 98, pp. 48–57, 2018.

[17] R. J. Fragaszy, W. Su, and F. H. Siddiqi, "Effects of oversize particles on the density of clean granular soils," *Geotechnical Testing Journal*, vol. 13, no. 2, pp. 106–114, 1990.

[18] Z.-L. Zhang, W.-J. Xu, W. Xia, and H.-Y. Zhang, "Large-scale in-situ test for mechanical characterization of soil–rock mixture used in an embankment dam," *International Journal of Rock Mechanics and Mining Sciences*, vol. 86, pp. 317–322, 2016.

[19] O. Karikari-Yeboah, "Stability of Slopes characterized by colluviums: investigation, analysis, and stabilization," *Géotechnique*, vol. 10, pp. 47–53, 2000.

[20] S. R. Hencher and D. P. McNicholl, "Engineering in weathered rock," *Quarterly Journal of Engineering Geology and Hydrogeology*, vol. 28, no. 3, pp. 253–266, 1995.

[21] Y. Liu, H. Xiao, K. Yao, J. Hu, and H. Wei, "Rock-soil slope stability analysis by two-phase random media and finite elements," *Geoscience Frontiers*, vol. 9, no. 6, pp. 1649–1655, 2018.

[22] Q. Wu, Y. An, and Q. Liu, "SPH-Based Simulations for Slope Failure Considering Soil-Rock Interaction," *Procedia Engineering*, vol. 102, pp. 1842–1849, 2015.

[23] W.-J. Xu, L.-M. Hu, and W. Gao, "Random generation of the meso-structure of a soil-rock mixture and its application in the study of the mechanical behavior in a landslide dam," *International Journal of Rock Mechanics and Mining Sciences*, vol. 86, pp. 166–178, 2016.

[24] C.-S. Li, D. Zhang, S.-S. Du, and B. Shi, "Computed tomography based numerical simulation for triaxial test of soil-rock mixture," *Computers & Geosciences*, vol. 73, pp. 179–188, 2016.

[25] Y. Wang, C. H. Li, and Y. Z. Hu, "Use of X-ray computed tomography to investigate the effect of rock blocks on meso-structural changes in soil-rock mixture under triaxial deformation," *Construction and Building Materials*, vol. 164, pp. 386–399, 2018.

[26] M. Afifipour and P. Moarefvand, "Mechanical behavior of bimrocks having high rock block proportion," *International Journal of Rock Mechanics and Mining Sciences*, vol. 65, pp. 40–48, 2014.

[27] Z. G. Gao and G. G. Liu, "Study on random aggregate model of two-dimensional concrete," *Journal of Tsinghua University (Science and Technology)*, vol. 43, no. 5, pp. 710–714, 2003.

[28] J. Zhang, N.-G. Jin, X.-Y. Jin, and J.-J. Zheng, "Numerical simulation method for polygonal aggregate distribution in concrete," *Zhejiang Daxue Xuebao (Gongxue Ban)/Journal of Zhejiang University (Engineering Science)*, vol. 38, no. 5, pp. 581–585, 2004.

[29] G. De Schutter and L. Taerwe, "Random particle model for concrete based on Delaunay triangulation," *Materials and Structures*, vol. 26, no. 2, pp. 67–73, 1993.

[30] Y. Zheng, C. Deng, S. Zhao, X. Tang, M. Liu, and L. Zhang, "Development of finite element limiting analysis method and its applications in geotechnical engineering," *International Journal of Engineering Science*, vol. 3, pp. 10–36, 2007.

[31] P. J. BARRETT, "The shape of rock particles, a critical review," *Sedimentology*, vol. 27, no. 3, pp. 291–303, 1980.

[32] L. Yan, Q. Meng, W. Xu et al., "A numerical method for analyzing the permeability of heterogeneous geomaterials based on digital image processing," *Journal of Zhejiang University-Science A*, vol. 18, no. 2, pp. 124–137, 2017.

[33] J. Kozicki, J. Tejchman, and Z. Mróz, "Effect of grain roughness on strength, volume changes, elastic and dissipated energies during quasi-static homogeneous triaxial compression using DEM," *Granular Matter*, vol. 14, no. 4, pp. 457–468, 2012.

[34] C. Wang and C.-N. Zhan, "Gravel content effect on the shear strength of clay-gravel mixtures," *Applied Mechanics and Materials*, vol. 71-78, pp. 4685–4688, 2011.

[35] Y. Li, R. Huang, L. S. Chan, and J. Chen, "Effects of particle shape on shear strength of clay-gravel mixture," *KSCE Journal of Civil Engineering*, vol. 17, no. 4, pp. 712–717, 2013.

[36] C. Nouguier-Lehon, B. Cambou, and E. Vincens, "Influence of particle shape and angularity on the behaviour of granular materials: A numerical analysis," *International Journal for Numerical and Analytical Methods in Geomechanics*, vol. 27, no. 14, pp. 1207–1226, 2003.

[37] X. Wen-Jie, X. Qiang, and H. Rui-Lin, "Study on the shear strength of soil-rock mixture by large scale direct shear test," *International Journal of Rock Mechanics and Mining Sciences*, vol. 48, no. 8, pp. 1235–1247, 2011.

[38] Y. Liu, J. Hu, H. Wei, and A.-L. Saw, "A direct simulation algorithm for a class of beta random fields in modelling material properties," *Computer Methods Applied Mechanics and Engineering*, vol. 326, pp. 642–655, 2017.

10

Pullout Behavior of GFRP Anti-Floating Anchor Based on the FBG Sensor Technology

Ming-yi Zhang,[1] Zheng Kuang,[1] Xiao-yu Bai,[1] and Xiao-yu Chen[2]

[1]*College of Civil Engineering, Qingdao University of Technology, Qingdao, Shandong 266033, China*
[2]*Faculty of Construction and Environment, The Hong Kong Polytechnic University, Hung Hom, Hong Kong*

Correspondence should be addressed to Zheng Kuang; kz9403@163.com

Academic Editor: Nicolas J. Leconte

Building anti-floating anchors have been increasingly used in recent years, but conventional steel anchors under service conditions are easily subjected to chemical erosion. Glass fiber reinforcement polymer (GFRP) is a promising solution to this problem. In this study, field pullout tests were conducted on three full instrumented GFRP anti-floating anchors in weathered granite. Specifically, the GFRP anchors during pultrusion were innovatively embedded with bare fiber Bragg grating (FBG) sensors to monitor the axial force distribution along depth. It was found that the embedded FBG could reliably monitor the axial force distribution of GFRP anchors. The ultimate pullout force of a GFRP anchor with diameter of 28 mm and anchorage length of 5 m was up to 400 kN. The GFRP anchor yielded at 0.8 m underground. Force distribution and field photos at failure indicated shear failure occurred at the anchor/bolt interface at the end of the tests. The feasibility of the GFRP anti-floating anchor was also verified in civil engineering. Finally, an elastic mechanical model and Mindlin's displacement solution are used to get distribution functions of axial force and shear stress along the depth, and the results accord with the test results.

1. Introduction

Building floating issues due to the increasingly larger and deeper underground basement at high water level have attracted wide attention recently. The anti-floating anchor, a kind of anti-floating method through connecting the anchor and soil, is widely used to solve these issues owing to its lower cost, shorter construction period, strong stratum adaptability, and less point force than traditional methods (kentledge, anti-floating pile, and drainage) [1–3]. However, due to the corrosive ions in groundwater, the ordinary steel anchor under aggressive environment inevitably suffers from chemical erosion, which seriously weakens the long-term structure strength.

Moreover, the stray current produced from the direct current power system subjects metal anchors to electrochemical erosion, especially in urban rail projects [4]. In order to solve the problems mentioned above, nonmetallic anchors appear [5].

Nowadays, glass fiber reinforcement plastic (GFRP), fabricated by extruding molten glass through an orifice, is one of the most popular alternatives of all fiber reinforcement plastics owing to its lower cost, higher tensile strength, and excellent insulation [6]. In-depth investigations of GFRP properties including pull-out performance and mechanism compared to ordinary steel anchors suggest (a) GFRP bars could alternate ordinary steel bars in most projects based on model tests and (b) the GFRP bar, owing to lower elasticity modulus, has a larger slippage at failure than steel bars [7]. What is more, a series of experiments have been conducted and revealed that higher concrete strength is beneficial for improving the pullout capacity and bond durability of GFRP anchor [8, 9]. However, GFRP bars and furthermore GFPR bars as anti-floating anchors are still unfamiliar to practicing Chinese engineers.

The force transfer mechanism and failure mode of GFRP anti-floating anchors should be comprehensively studied to encourage their use in buildings. For this purpose, the first investigation should be the measurement method of axial force along depth. Conventionally, the axial force along depth is monitored by strain gage or fiber Bragg grating (FBG) sensors directly attached on the anti-floating anchor.

TABLE 1: Mechanical parameters of test field.

Uniaxial compressive strength	Unit weight	Poisson ratio	Elastic modulus
32 MPa	24.5 kN/m^3	0.33	35 MPa

However, these sensors may fall off during anchor installation and pullout tests. Another option is to embed a sensor in the anchor by initially cutting the anchor body, but the cutting-produced defects seriously damage the integrity of fiber reinforcing anchors, reducing the pullout strength of the anchor [10].

In this study, field pullout tests were conducted on three full-instrumented GFRP anti-floating anchors installed into rock base, and the naked FBG was innovatively embedded into the anchors during pultrusion to monitor the axial force. This study aimed to (a) verify the applicability of the GFRP anti-floating anchors in rock base, (b) evaluate the feasibility of FBG installation on these anchors, (c) investigate the head-displacement response of these anchors during pullout tests, and (d) study the load transfer mechanism and failure mode based on field tests.

2. Pullout Test Program

2.1. Description of Field Geology. The test field, with soft rock, contains moderately weathered granite and other mechanical parameters are shown in Table 1.

2.2. Installation of the FBG Sensor. Wavelength γ_B of FBG, which varies with the temperature and structure strain at the granted region, could be directly measured during whole tests [11]. Due to the less time during whole tests and stable underground temperature, the temperature change during whole tests is small enough to be ignored, and the strain along the anchor is simplified as

$$\Delta\varepsilon = \frac{\Delta\gamma_B}{K_e} \qquad (1)$$

where $\Delta\varepsilon$ is the strain along the anchor and K_e is strain sensitivity coefficient of FBG.

The naked FBG was innovatively embedded into each GFRP anchor: we firstly embed the FBG sensor in the center of the glass fiber bundle along the longitudinal direction of the rod body and place into the mold and then pour the anchor base material (synthetic resin), through the procedures of pultrusion, thread winding, and solidification; GFRP anchor rod body is manufactured. Naked optional fiber of SMF28-C, diameter of 900 um, was selected here due to its low volume, high temperature resistance, and ease in installation. Bragg gratings (details in Table 2) were distributed along the naked optional fiber as illustrated in Figure 1.

To prevent FBG sensors from damage during the tests, we put the sensors into a capillary tube and encapsulated them by filling with hot epoxy resins by needles. The tip of the FBG was weld with 3-interdiameter armored cable, covered by

TABLE 2: Parameters of Bragg gratings.

Type	Length	Reflection rate	Width	SLSR
Bragg gratings	10 mm	80%	0.3 nm	15 dB

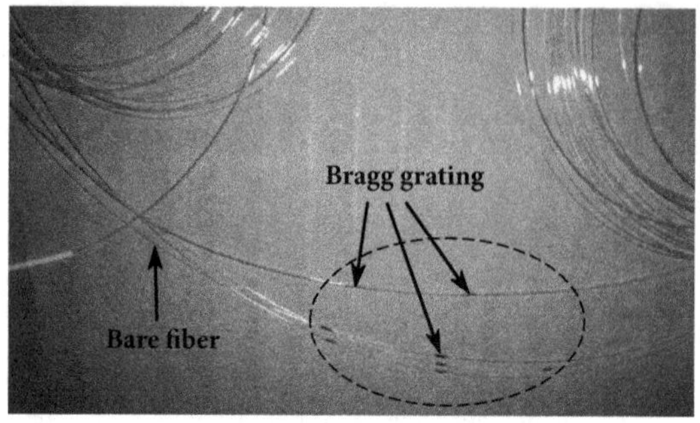

FIGURE 1: Photo of FBG sensors.

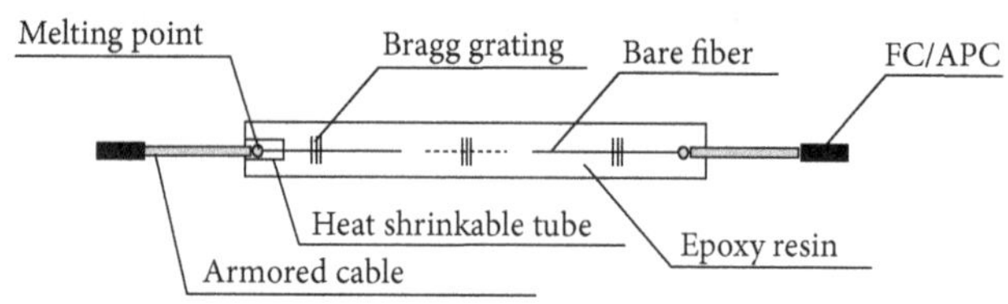

FIGURE 2: Schematic view of encapsulated FBG sensors.

polymer to blend and protect the connection, and finally the connection was put with heat shrinkable tubing (Figure 2).

The encapsulated FBG sensors were embedded in the GFRP anchors during the pultrusion. The survival rates of FBG sensors before full-scale tests were evaluated by both laboratory and field tests. Results show all FBG sensors worked well even at failure of anchors. Hence, the innovative FBG installation technique not only maintains the integrity of GFRP anchors, but also provides more high-resolution results.

2.3. Description of GFRP Anchors. All GFRP anchors (YF-H50-28) embedded with FBG sensors were manufactured by the pultrusion process in a Chinese company, and the properties were 75% glass fiber + 25% resin (unsaturated polyester resin), with section area = 590 mm^2, density = 2.1 g/cm^3, and weight = 1195 g/m (Figure 3). Accessory ingredients such as curing agent and accelerant are also applied during the anchor manufacture process. Each anchor at the hole center was supported by a centralizer (Figure 3). Three GFRP anti-floating anchors, named GFRP-01, 02, and 03 with a nominal diameter of 28 mm, total length of 6.5 m, and anchorage length of 5 m, were used in the tests. Mechanical parameters of GFRP anchors (Table 3) indicate that the tensile strength is 1.5-2.4 times and the modulus of elasticity is 20% that of ordinary steel bars. Figure 4 shows the arrangement of the FGB sensors along depth.

2.4. Pullout Test. Holes were drilled vertically by an in-the-hole-drill, in diameter of 110 mm and length 0.5 m larger than the anchor length. A GFRP anchor then was vertically

TABLE 3: Mechanical parameters of GFRP bar.

Type	Tensile strength	Shear strength	Ultimate load	Elastic modulus
GFRP bar	702 MPa	150 MPa	432 KN	51 MPa

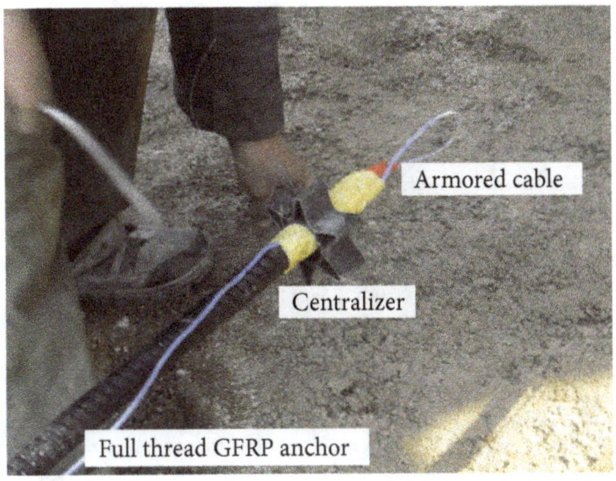

FIGURE 3: Photo of GFRP anchor.

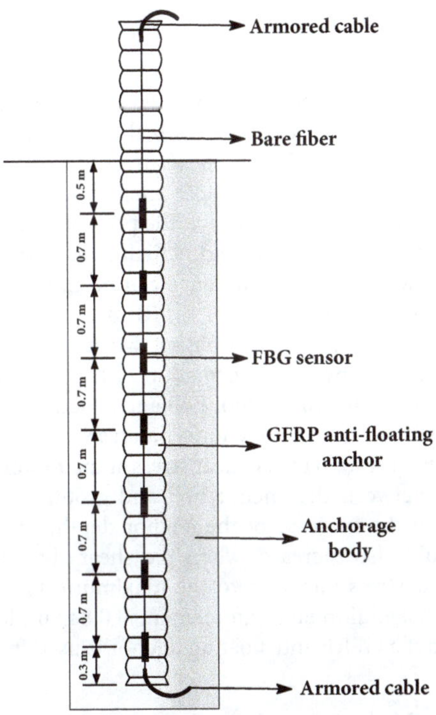

FIGURE 4: Anchor bolt after installing FBG sensors.

placed into the hole by workers and fixed at the hole center by centralizers. Finally, M32.5 cement mortar was poured into the hole.

Due to the low shear resistance of GFRP bars, stress concentration at the anchorage would damage the GFRP anchor. To protect and hold the GFRP anchor, we innovatively applied one steel tube with diameter, length, and wall thickness of 1.2 mm, 50 mm, and 5 mm, respectively, before mounting a clip anchorage on the anchor. The tube was filled with a mixture of epoxy resin and curing agent to form a strong bond on tube and GFRP anchor. In order to guarantee that the mixture fills the tube, we firstly smear the mixture on the anchor surface and then continuously rotate the tube so that the mixture fills the inner wall of the tube, until the inner wall is filled. The test devices are presented in Figure 5.

In accordance with *Chinese Technical Code*, field pullout tests were conducted at a slow applied rate of 0.2 kN/s, without loading-unloading loops. The first stage load was 50 kN, and the subsequent stage loads were kept at magnitudes 100, 150, and 200 kN to the failure state. The displacement of anchor head is read immediately after each stage.

3. Results and Discussion

3.1. Head Displacement Response. Head displacement response of each GFRP anchor during whole tests is presented in Figure 6. The head displacement is elevated linearly, without significant fluctuation, along with the increasing pullout force, which is different from the ordinary steel anchor undergoing a sudden increase at failure [12]. This difference could be attributed to the lower elastic modulus of GFRP. In addition, the GFRP anchor outperformed the steel anchor in terms of elongation at failure. The ultimate pullout force of the 28-mm-diameter GFRP anchor is defined as 400 kN in accordance with *Chinese Technical Code* [13], which is higher than that (345 kN) of the steel anchor under the same condition (Bai et al. [14]). Thus, the GFRP anti-floating anchor not only resists chemical erosion, but also provides higher pullout capacity in anti-floating structures.

3.2. Distribution of Axial Force. Figure 7 shows the distribution of axial force along depth during whole tests, which was calculated as follows, assuming the same transformation between FBG sensors and anchor:

$$N_i = E_G A_G \cdot \Delta\varepsilon_i = E_G A_G \cdot \frac{\gamma_{Bi}}{K_e} \quad (2)$$

where $\Delta\varepsilon_i$ is the strain at FBG sensor i; N_i is the axial force at FBG sensor i; and E_G and A_G are Young's modulus and cross-sectional area of the GFRP anchor, respectively.

Axial force along the GFRP anchor significantly decreased at depth of 0~3 m, while the deeper part supported relatively less pullout force, which indicates the nonuniform mobilization of axial force along the GFRP anchor.

The distribution of axial force along depth also suggests the presence of a critical load transfer depth for the GFRP anchor. Specifically, little force transfer to the anchor below 3~3.5 m was found in GFRP-01, 02, and 03, which means prolonging the anchor length, when it is up to the critical depth, does not obviously enhance the pullout force. The conclusion is consistent with the research results by Bai et al. [14]. However, adhesion between rock and anchorage bolt cannot be guaranteed if an over-short anchor length is employed.

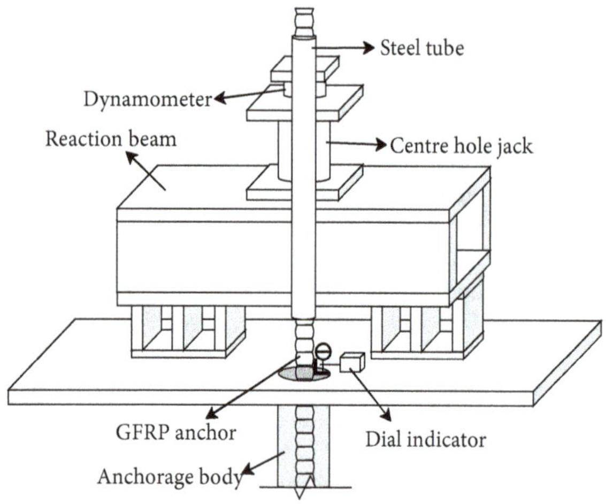

Figure 5: Photo of pull-out test devices.

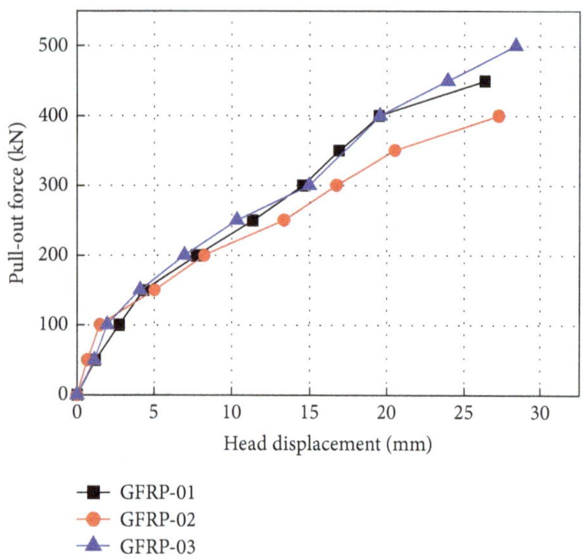

Figure 6: Head displacement response.

Nevertheless, further research is needed to determine the critical depth under different conditions.

The maximum measured axial force of GFRP anchors at failure is smaller than the ultimate tensile strength (500 kN vs 702 kN), which can be explained by the local failure along the GFRP anchor.

3.3. Distribution of Shear Stress. Figure 8 presents the shear stress distribution along depth during whole tests, which can be calculated as follows by assuming the uniform distribution of load between two contiguous FBG sensors:

$$\tau_i = \frac{(N_i - N_{i-1})}{\pi d \Delta L} \tag{3}$$

where τ_i is the shear stress along depth at point i; d is the diameter of the GFRP anchor; and ΔL is the distance between two contiguous FBG sensors.

The peak shear stress along depth was elevated and its influence area was enlarged with the increasing applied pullout force. The maximum shear stress reached 3~4 MPa at failure. Compared to the steel anchor with peak shear stress appearing at ground surface, the peak of GFRP anchors occurred at 0.8 m underground, which is consistent with a previous study [15]. This difference can be explained by the low stiffness of the GFRP anchor: GFRP anti-floating anchors produce greater elastic deformation when bearing pullout loads and cause obvious deformation of the anchor section near the hole so that the anchor bolt and grouting body show the shear slip. Hence, the anchor separates with surrounding concrete that means a lower shear stress at their interface. The shear slip between the anchor bolt and grouting body will reduce with the increase of the anchor depth, so the shear stress would also increase; when the shear slip reduces to 0, the shear stress increases to the maximum. Figure 8 also indicates that additional reinforcement at 0.8 m underground is required if a GFRP anti-floating anchor is used in projects.

3.4. Failure Mode. When the pullout force is up to 60%-70% of the ultimate loading, the GFRP anchor produces slight crack noise, and its head displacement rises constantly with 8-10 cm crack around anchorage body (Figure 9). Relative slippage between the anchor and anchorage bolt occurred at failure (all test anchors) (Figure 10(a)); surface fiber got pilled off and white-spot crack appeared on surface (only GFRP-02 anchor) (Figure 10(b)); a part of the fiber plate was detached (only GFRP-03 anchor) (Figure 10(c)). Installing additional ribs on the anchor surface could solve the problem of fiber plate detachment [16].

From the pictures of GFRP anchors at failure, it is figured out that the damage of GFRP anchors occurred at the anchorage part instead of the free part that supported the

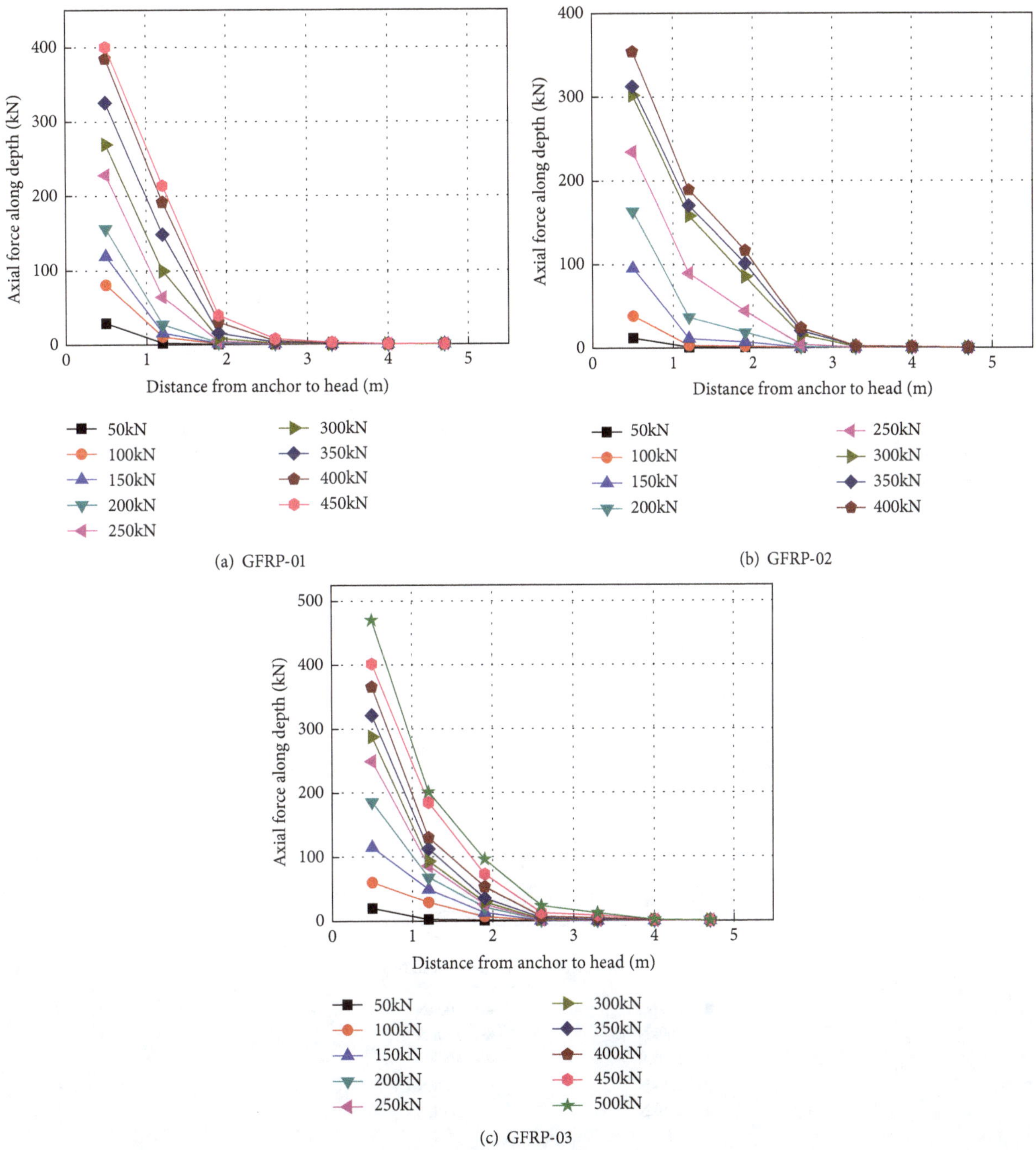

FIGURE 7: Distribution of axial force along depth.

largest pull force. In addition, the shear stress was maximized at 0.8 m underground, which is consistent with the failure location. These results indicate that the failure mode is shear failure rather than pull-out failure.

4. Theoretical Analysis

We introduced the mathematical function model studied by You [17] to predict the distribution rule of anchor's axial force and shear stress, which is beneficial for avoiding the disadvantages of the above-mentioned test, such as high cost and time waste. Assuming that the anchor is an elastomer and the properties are the same for both rock mass and anchorage body, we have an analytical model for a fully grouted anchor which is described in detail in Figure 11. The pullout force, shear stress, the length of the unit, axial force, and shear stress in a certain depth are denoted by P, τ, dx, $P(x)$, and $\tau(x)$, respectively.

Taking into account the static equilibrium condition, we obtain the following relations:

$$dp(x) = -2\pi r\tau(x)\,dx \tag{4}$$

where r is the radius of the anchor.

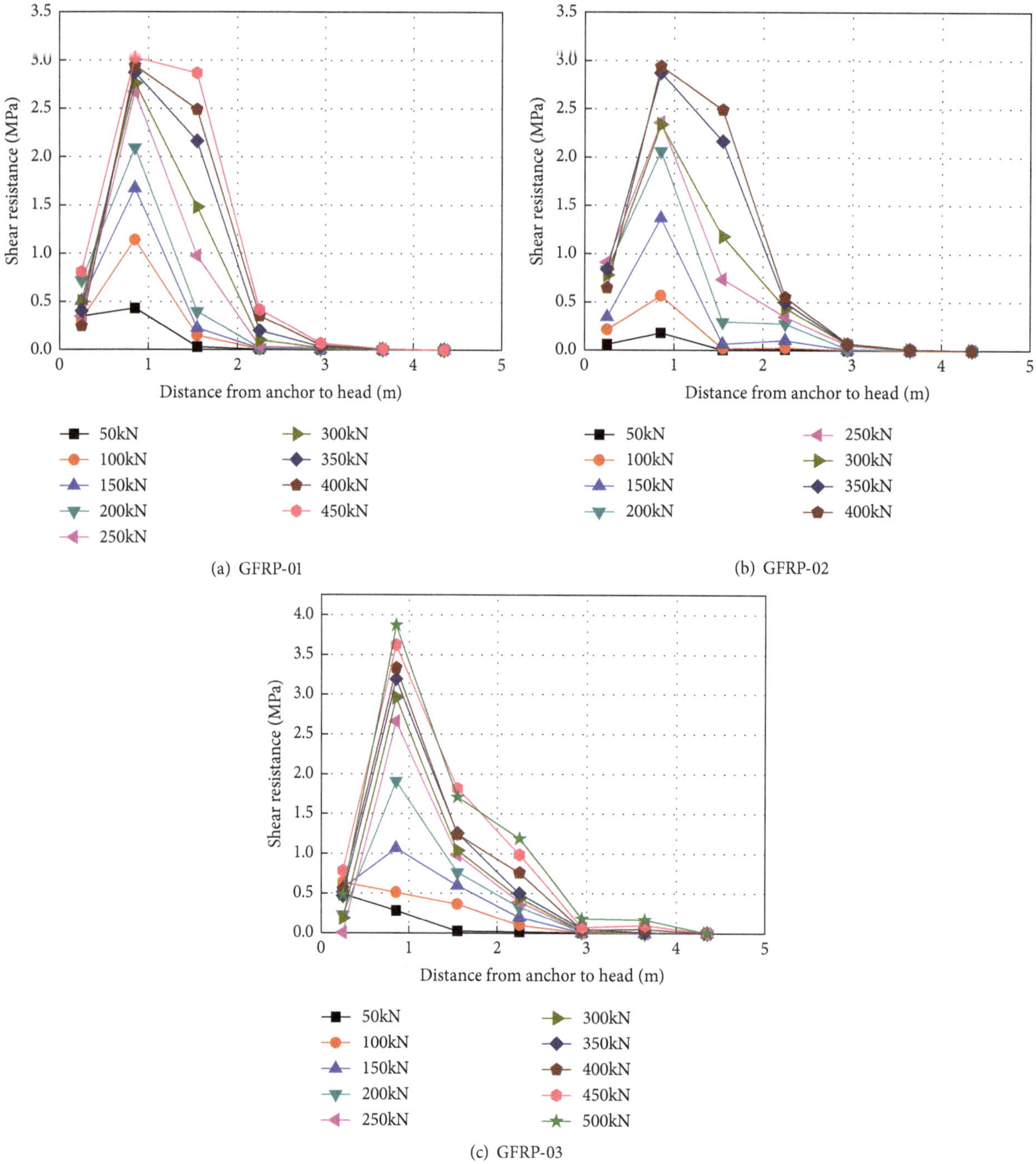

(a) GFRP-01

(b) GFRP-02

(c) GFRP-03

FIGURE 8: Distribution of shear stress along depth.

Then, integrating the above formula, we obtain the relation for $P(x)$ and $\tau(x)$:

$$P(x) = -2\pi r \int_{x}^{L_a} \tau(x)\, dx \tag{5}$$

where L_a is the anchorage length of the anchor.

As depicted in (5), $\tau(x)$ is the key issue to reach the aim of this investigation.

We will now introduce Mindlin's displacement solution to get the equation for $\tau(x)$. The assumption of the solution is that the rock mass is semi-infinite space, the origin of coordinates is located at the hole, and the anchor is along the x-axis. As shown in Figure 12, when a concentrated force, with the depth of c, is applied at the point A, the vertical displacement of the point $B\,(x, y, z)$ can be expressed as

$$s = \frac{Q(1+\mu)}{8\pi E(1-\mu)} \left[\frac{3-4\mu}{R_1} + \frac{8(1-\mu)^2 - (3-4\mu)}{R_2} \right.$$

$$\left. + \frac{(x-c)^2}{{R_1}^3} + \frac{(3-4\mu)(x+c)^2 - 2cx}{{R_2}^3} \right.$$

Figure 9: Cracks on anchorage body at failure.

(a) Relative slippage between the anchor and anchorage bolt

(b) Surface white-spot crack

(c) Fiber plate detachment

Figure 10: GFRP anchors at failure.

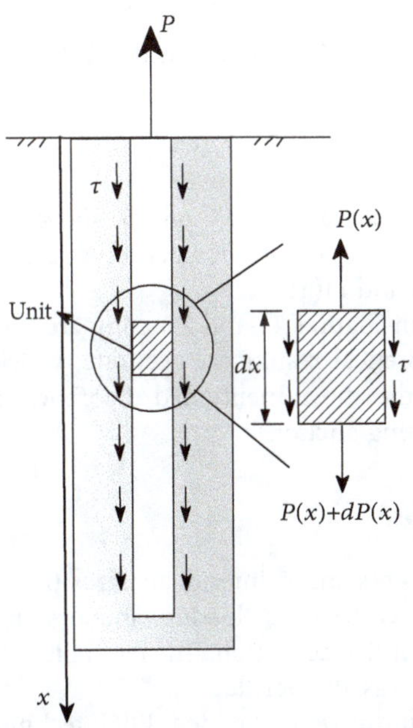

Figure 11: Mechanical model for a fully grouted anchor.

$$+ \frac{6cx\,(x+c)^2}{R_2{}^5} \Bigg] \tag{6}$$

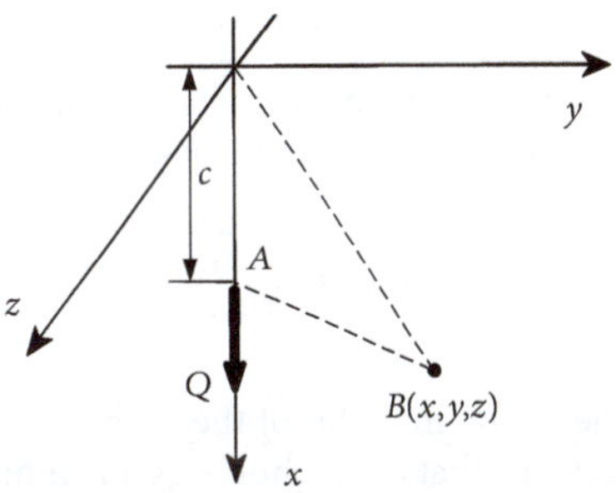

Figure 12: Schematic of Mindlin's displacement solution.

where E, μ are elastic modulus and Poisson ratio of the rock mass, respectively;

$$R_1 = \sqrt{y^2 + z^2 + (x-c)^2};$$
$$R_2 = \sqrt{y^2 + z^2 + (x+c)^2}. \tag{7}$$

At the origin of coordinates ($x = y = z = 0$), the expression of (6) becomes

$$s_o = \frac{Q(1+\mu)(3-2\mu)}{2\pi c E} \tag{8}$$

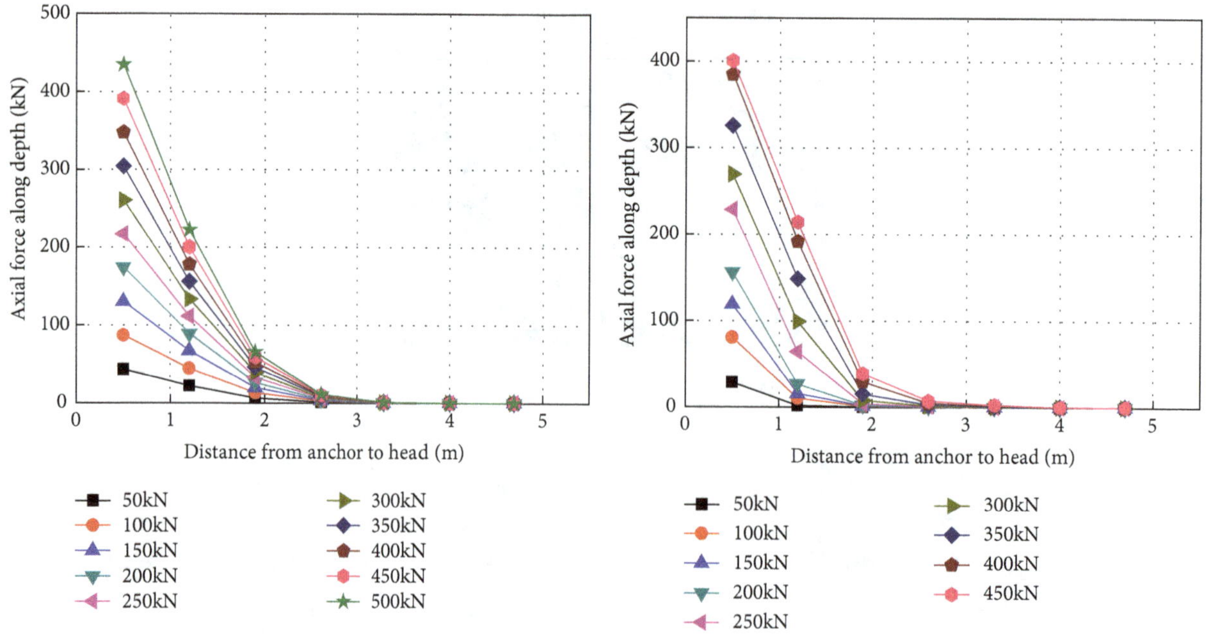

(a) Theoretical value (b) GFRP-01

FIGURE 13: Distribution of axial force along the depth.

For the rock mass, Q signifies the shear stress that the pullout force exerts on the rock mass at the point A, which is expressed as

$$Q = 2\pi r\tau \cdot dx \tag{9}$$

Additionally, for the anchor, Q denotes the pullout force transferred to the point A. So the elastic elongation of the anchor can be calculated as

$$\frac{Q - 2\pi r \int \tau dx}{E_a \cdot \pi r^2} \tag{10}$$

where E_a is the elastic modulus of the anchor.

We then assume that the anchor is a semi-infinite bolt and deforms with rock mass compatibility. The elongation of the anchor at the hole is the same as the displacement of the rock mass; the equation can be expressed as

$$\int_0^{\infty} \frac{Q - 2\pi r \int \tau(x)\,dx}{E_a \cdot \pi r^2} dx = \int_0^{\infty} \frac{2\pi r\tau(x) \cdot (1+\mu)(3-2\mu)}{2\pi x E} dx \tag{11}$$

After simplifying (11), we obtain a differential equation:

$$\tau''(x) + \left(kx - \frac{2}{x}\right)\tau'(x) + \frac{2}{x^2}\tau(x) = 0 \tag{12}$$

where $k = 2E/(1+\mu)(3-2\mu)E_a r^2$.

Transforming (12) to Weber equation [18], then substitute boundary condition, $x \longrightarrow \infty$, $\tau = 0$, into it; $\tau(x)$ can be determined as

$$\tau(x) = \frac{P}{\pi r}\left(\frac{\lambda}{2}x\right) \cdot e^{(-(\lambda/2)x^2)} \tag{13}$$

where P is the pullout force;

$$\lambda = \frac{E}{E_a(1+\mu)(3-2\mu)r^2}. \tag{14}$$

Substituting (13) into (5), we obtain the distribution of axial force:

$$P(x) = P \cdot e^{(-(\lambda/2)x^2)} \tag{15}$$

To verify the correctness of (13) and (15) above, we substitute the parameters of this test in them, and the results are shown in Figures 13(a) and 14(a).

As shown in Figures 13 and 14, compared with the test result, the theoretical results are accurate, which prove the correctness of the above-mentioned theoretical analysis for GFRP anti-floating anchor.

5. Conclusion

This paper presents and primarily analyzes the field pullout tests on three GFRP anti-floating anchors in weathered granite under service condition. The feasibility of embedded FBG technique was also verified.

(a) The innovative embedded FBG technique is able to monitor axial force accurately during whole tests, with survival rate of 100%. This technique keeps the integrity of GFRP anchors and provides reference for the monitoring of anisotropic materials.

(b) GFRP anti-floating anchors have the ultimate pullout force (400 kN) up to 2 times that of steel anchors and thus could replace conventional steel anchors in projects.

(c) Different from steel anchors, the head displacement of GFRP anchors increases linearly with the applied load and

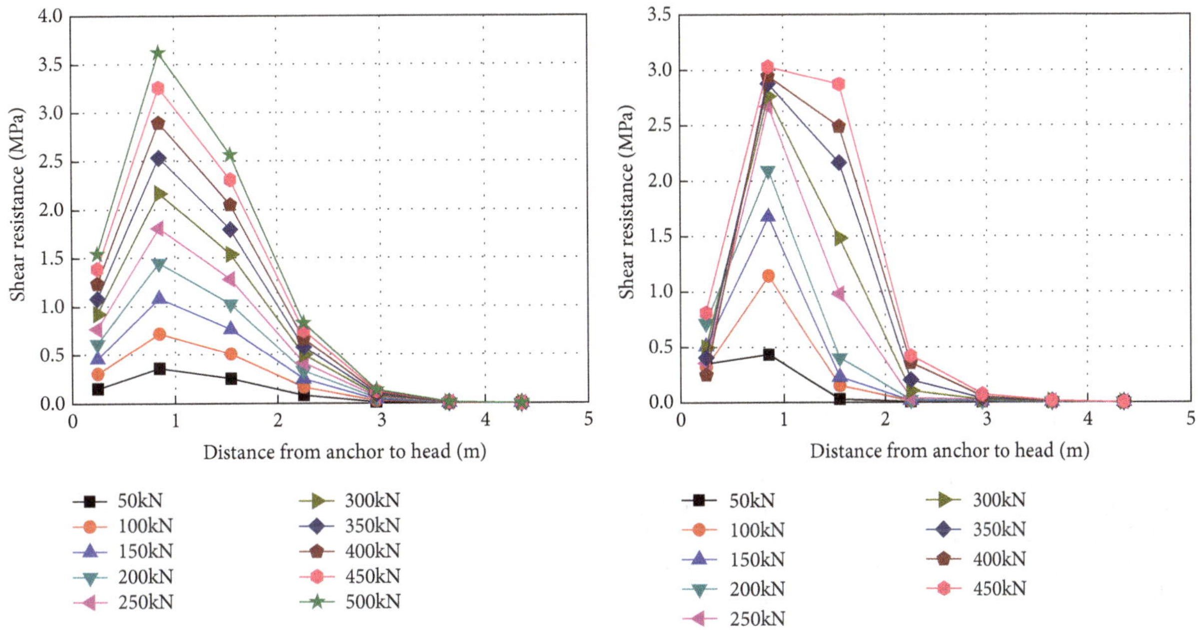

(a) Theoretical value

(b) GFRP-01

FIGURE 14: Distribution of shear stress along the depth.

has larger elongation before failure, which could be explained by their less elastic-modulus.

(d) No pullout force was transferred to 3.5 m underground, highlighting that a critical depth exists for the GFRP anchor. Hence, the ultimate capacity did not change with the anchorage length higher than critical value.

(e) Peak shear stress along depth of GFRP anchors occurred at 0.8 m underground. Therefore, additional reinforcement is required at 0.8 m underground if a GFRP anchor is used in projects.

(f) Force distribution and field photos at failure indicated that shear failure occurred at the GFRP anchor bolt interface at the end of the test.

(g) An elastic mechanical model and Mindlin's displacement solution are used to get distribution functions of axial force and shear stress along the depth, and the results accord with the test results.

Further studies are required to provide more comprehensive and conclusive observations.

Conflicts of Interest

The authors declare that they have no conflicts of interest.

Acknowledgments

The authors would like to acknowledge the support provided by Natural Science Foundation of China (No. 51078196; No. 41502304) and Natural Science Foundation of Shandong province (No. ZR2016EEQ08) and Science and Technology Project of Shandong province (No. J16LG02).

References

[1] Á. Tóth, Q. Gong, and J. Zhao, "Case studies of TBM tunneling performance in rock-soil interface mixed ground," *Tunnelling and Underground Space Technology*, vol. 38, pp. 140–150, 2013.

[2] W. G. Fu, J. G. Liu, and Z. Y. Yang, "Formulae for calculating stability of anti-floating anchor and anchor," *Chinese Journal of Geotechnical Engineering*, vol. 36, no. 11, pp. 971–1983, 2014.

[3] J.-N. Ye and G.-B. Liu, "Anti-floating safety design of structures in metro station considering friction resistance enclosure protection," *Rock and Soil Mechanics*, vol. 31, no. 1, pp. 279–283, 2010.

[4] X. Zeng, M. Lei, W. J. Zhang, and W. Z. Ji, "Discussion about "time bomb" question for bolt -an Answer to Professor Guo Ying zhong," *Chinese Journal of Rock Mechanics and Engineering*, vol. 21, no. 01, pp. 143–147, 2002.

[5] M. Pan and X. Xu, "Experimental study on bond characteristics of adhering sand and ribbed surface FRP bars with concrete," *Destech Transactions on Engineering and Technology Research*, 2017.

[6] B. Benmokrane, B. Zhang, A. Chennouf, and R. Masmoudi, "Evaluation of aramid and carbon fibre reinforced polymer composite tendons for prestressed ground anchors," *Canadian Journal of Civil Engineering*, vol. 27, no. 5, pp. 1031–1045, 2000.

[7] B. Benmokrane, H. Xu, and E. Bellavance, "Bond strength of cement grouted Glass Fibre Reinforced Plastic (GFRP) anchor bolts," *International Journal of Rock Mechanics and Mining Sciences & Geomechanics Abstracts*, vol. 33, no. 5, pp. 455–465, 1996.

[8] H. Ashrafi, M. Bazli, and A. V. Oskouei, "Enhancement of bond characteristics of ribbed-surface GFRP bars with concrete by using carbon fiber mat anchorage," *Construction and Building Materials*, vol. 134, pp. 507–519, 2017.

[9] M. Bazli, H. Ashrafi, and A. V. Oskouei, "Experiments and probabilistic models of bond strength between GFRP bar and different types of concrete under aggressive environments," *Construction and Building Materials*, vol. 148, pp. 429–443, 2017.

[10] Y. S. Kim, H. J. Sung, H. W. Kim, and J. M. Kim, "Monitoring of tension force and load transfer of ground anchor by using optical FBG sensors embedded tendon," *Smart Structures and Systems*, vol. 7, no. 4, pp. 303–317, 2011.

[11] H.-J. Sunga, T. M. Dob, J.-M. Kimc, and Y.-S. Kim, "Long-Term monitoring of ground anchor tensile forces by FBG sensors embedded tendon," *Smart Structures and Systems*, vol. 19, no. 3, pp. 269–277, 2017.

[12] M. Y. Zhang, J. Zhang, and J. W. Liu, "Experimental investigation on anti- floating anchor in moderately weathered granite," *Chinese Rock Mechanism and Engineering*, pp. 2741–2746, 2008.

[13] The Professional Standards Compilation Group of People's Republic of China, *Technical Specification for Retaining and Protection of Building Foundation Excavations*, China Architecture and Building Press, Beijing, China, 2005.

[14] X. Bai, M. Zhang, and L. Zhu, "Experimental study on shear characteristics of interface of full-bonding glass fiber reinforced polymer anti-floating anchors," *Chinese Journal of Rock Mechanics & Engineering*, 2018.

[15] E. A. Ahmed, B. Benmokrane, and M. Sansfaçon, "Case study: design, construction, and performance of the la chancelière parking garage's concrete flat slabs reinforced with GFRP bars," *Journal of Composites for Construction*, vol. 21, no. 1, pp. 6001–6013, 2017.

[16] J. Li, T. Li, and H. Zhu, "Experimental study on enhancement of additional rib to anchorage performance of frp bar," *Journal of Nanjing Tech University*, 2017.

[17] C. You, "Mechanical analysis on wholly grouted anchor," *Chinese Journal of Rock Mechanics & Engineering*, vol. 21, no. 01, pp. 143–147, 2000.

[18] E. Kamke, *Ordinary Differential Equation Manual*, China Science Publishing & Media Ltd., 1977.

11

Derivation of Universal Curves for Nonlinear Soil Consolidation with Potential Constitutive Dependences

Gonzalo García-Ros,[1] **Iván Alhama**,[1] **Manuel Cánovas**,[2] **and Francisco Alhama**[3]

[1] *Technical University of Cartagena, Civil Engineering Department, Spain*
[2] *Universidad Católica del Norte, Metallurgical and Mining Engineering Department, Chile*
[3] *Technical University of Cartagena, Applied Physics Department, Spain*

Correspondence should be addressed to Gonzalo García-Ros; gonzalo.garcia@upct.es

Academic Editor: Enrico Conte

Nonlinear consolidation scenarios, based on potential type constitutive dependences—like those proposed by Juárez-Badillo—and eliminating the more restrictive hypothesis of 1+e and *dz* constant, were characterized by the nondimensionalization process of the governing equations, providing the independent dimensionless groups that rule the main unknowns of interest. From these, universal curves have been depicted for both the characteristic time and the average degree of consolidation. The solutions were verified by numerical simulations and successfully compared in a case study, showing the simplicity of use of the curves and the high reliability of the solutions they provide.

1. Introduction

Soil consolidation has been studied by a large number of authors for almost a hundred years. Since Terzaghi announced his linear consolidation theory [1], more complex models have been proposed. These include 2D and 3D geometries [2], as well as nonlinear problems (with a nonconstant coefficient of consolidation, due to the changes in the hydraulic conductivity and void ratio during the process [3]), for whose solutions numerical techniques are required. However, most nonlinear models focus on solving a particular scenario, i.e., with a very restrictive hypothesis, with no intention of reaching solutions of a universal character which depend of the lowest number of dimensionless parameters. Geng et al. [4] and Lu et al. [5] studied nonlinear soil consolidation problems, the first under cycling loadings, and the second under time-variable loadings and vertical drains, using logarithmic type constitutive relations for the main variables of the problem (void ratio, effective stress and hydraulic conductivity), obtaining that both the initial stress state and the ratio of the coefficients of the constitutive relations are the most influential parameters that govern the consolidation process. Other references to nonlinear consolidation are the papers of Wu et al. [6] and Brandenberg [7], which also handle logarithmic dependences with the void ratio. The first author presents an analytical solution for the electroosmotic consolidation with no allusion to the dimensionless groups that rule the problem, while the second approximates the secondary consolidation.

Special attention deserve the works of Butterfield [8], Lancellotta and Preziosi [9], Zhuang et al. [10], and Conte and Troncone [11]. The first one describes the advantages of the potential dependences versus the logarithmic ones, which become evident in highly compressible soils for which, on the one hand, the slope of the e-log(σ') dependence is far from being a constant value and, on the other, the assumption of the logarithmic relation can lead to negative values for the void ratio when σ'_f/σ'_0 takes a high value. Lancellotta and Preziosi, and Conte and Troncone make references to the potential type relations, although they choose to present solutions only for logarithmic dependences. The first ones present, among others, consolidation models with impermeable or draining boundaries, as well as solid particles sedimentation models in quiescent fluids, while the latter solve a problem of step loads under the hypothesis of small strains. Finally, Zhuang et al. obtain semianalytical solutions under the small strains hypothesis and using logarithmic constitutive dependences, characterizing the problem by means of a dimensionless parameter that is the quotient between the slopes of the e-log(σ') and e-log(k) relations.

For optimal dimensionless characterization of the solution patterns (essentially the characteristic time and the average degree of consolidation, the latter in terms of settlement and pressure dissipation) of the consolidation processes based on a generalized Juárez-Badillo (J-B hereinafter) model [12], the dimensionless groups are derived by the nondimensionalization of the governing equations, a process that requires bothersome mathematical steps due to the inherent nonlinearity of the model. For the original J-B problem, which assumes the strict hypothesis of an initial void ratio that is negligible in the term of soil contraction, the number of the dimensionless groups reported by J-B is larger than the number derived in this paper, so providing a less precise solution; in addition, the author does not present universal solutions since no characteristic time is proposed. For other extended J-B models, by deleting one or more of his restrictive hypotheses (for example, $1+e_0$=constant with $e_0 \neq 0$, $1+e\neq$ constant and $dz\neq$constant), new dimensionless groups are derived, verifying the solutions by numerical simulations for the most general and complex model. The general concept of characteristic time, which may be easily defined in lineal and isotropic soils, is extended herein to nonlinear consolidation and, after its introduction into the governing equation as an unknown reference to define the dimensionless time, a new group containing the characteristic time emerges from the nondimensionalization process, allowing the universal curves to be constructed.

By nondimensionalization, the large number of isolated dimensionless parameters contained in the statement of the problem and in the constitutive relations, together with the dimensionless groups that can be formed from the relevant list of parameters and variables by applying simple rules of dimensional analysis [13], is reduced to the smallest number that best help researchers to manage the solution. As is known, the application of pi theorem—derived from the theory of homogeneous functions (Buckingham [14])—allows the unknowns of interest, expressed in their dimensionless form, to be set as a function of the mentioned dimensionless groups.

There are two techniques whose purpose is the derivation of the dimensionless groups that rule the solution of a given problem: the dimensional analysis and the method of group transformations. In the first [13], the groups are derived by simple mathematical manipulations from a list of relevant variables expressed in terms of primary quantities (length, mass and time), while in the second [15], they are obtained from the mathematical model in its dimensionless form after some mathematical steps. The technique applied in this paper, which we call nondimensionalization of governing equations [16, 17], starts from the governing equation in order to deduce the dimensionless groups. In this, after normalizing the variables, these and their changes are averaged - in fact, the equation itself is averaged - and assumed to be of the order of unity, a valid hypothesis in problems with relatively smooth nonlinearities, thus allowing the coefficients of the equations to be of the same order of magnitude and unequivocally providing the most precise solution as demonstrated in many studies [18]. Based on this methodology, Manteca et al. [19] study the nonlinear consolidation problem with constitutive dependencies of logarithmic type, providing as a solution the universal curves and the dimensionless groups that govern the process.

Classical nonlinear consolidation models (such as Davis and Raymond [20], Juárez-Badillo [12], and Cornetti and Battaglio [21]) differ from one another in the nature of the constitutive relations between the parameters, void ratio and permeability, and the effective soil pressure, when trying to reflect the behavior of real soils. In general, these dependences converge to provide the same results in problems with small changes in these parameters (quasilinear problems) but different results when the working range is wide. In addition, most nonlinear models assume restrictive hypotheses, such as a constant soil thickness in the contraction term of the governing equation, which distance the solutions from those obtained without these restrictions; these solutions are by the side of safety in consolidation and unsafety in swelling. In addition to the original J-B problem, an extended model with both void ratio and thickness of the volume element continuously changing during the consolidation process is analyzed and solved.

To verify the obtained results, a comparison is made with numerical solutions based on the network simulation method [22]. This tool, widely used in other fields, is an efficient and computationally fast numerical method that has demonstrated its reliability in many linear and nonlinear engineering problems [23]. The dependences of characteristic time and average degree of consolidation on the rest of the dimensionless groups are checked; i.e., whenever the dimensionless parameters retain the same values, against changes in the individual parameters contained within the groups, neither the characteristic time nor average degree of consolidation change. After the verification and presentation of universal curves for a wide range of values of the parameters that sufficiently cover all real cases, contributions and conclusions are summarized.

2. The Original Juárez-Badillo Model

This author [12] presented his model twenty years after the nonlinear model of Davis and Raymond [20] and applied it to the odometer test to improve understanding of the primary consolidation phenomenon. Juárez-Badillo assumes incompressibility for water and soil particles and compressibility for the soil structure for which he sets the constitutive relation $dV/V = -\gamma_v(d\sigma'/\sigma')$. Doing so, it is immediate to write $m_v = \gamma_v/\sigma'$. As regards permeability, following his deductions (Juárez-Badillo [24]), J-B assumes a constitutive k-V dependence in the form $k/k_1 = (V/V_1)^{\kappa}$, which is equivalent to a proportionality dependence between the relative deviations of these parameters $dk/k = \kappa(dV/V)$. As a result, the dependence k-σ' takes the form $k/k_1 = (\sigma'/\sigma'_1)^{-\gamma_v\kappa}$, or, in terms of unitary deviations, $dk/k = -\gamma_v\kappa(d\sigma'/\sigma')$.

The original J-B model, under odometer conditions (small volume change and negligible specific weight of water), is defined by

$$\frac{d\sigma'}{dt} = -\frac{1}{\gamma_w m_v}\frac{\partial}{\partial z}\left(k\frac{\partial u}{\partial z}\right) \quad (1)$$

Substituting the above dependences and using the parameters

$$\lambda = 1-\gamma_v\kappa,$$
$$\frac{k_1\sigma'_1}{\gamma_w\gamma_v} = \frac{k_1}{\gamma_w m_{v,1}} = c_{v,1}, \quad (2)$$
$$c_v = \frac{k}{\gamma_w m_v} = \frac{k_1\sigma'_1}{\gamma_w\gamma_v}\left(\frac{\sigma'}{\sigma'_1}\right)^{1-\gamma_v\kappa} = c_{v,1}\left(\frac{\sigma'}{\sigma'_1}\right)^{\lambda}$$

(1) reduces to

$$\frac{\partial\sigma'}{\partial t} = -\sigma' c_{v,1}\frac{1}{\sigma'_1}\frac{\partial}{\partial z}\left[\left(\frac{\sigma'}{\sigma'_1}\right)^{-\gamma_v\kappa}\frac{\partial u}{\partial z}\right] \quad (3)$$

The simplifications assumed by the odometer test, $\partial u/\partial z = -\partial\sigma'/\partial z$ and $\partial u/\partial t = -\partial\sigma'/\partial t$ allow us to write the last equation as

$$\frac{\partial\sigma'}{\partial t} = \sigma' c_{v,1}\frac{\partial}{\partial z}\left[\left(\frac{\sigma'}{\sigma'_1}\right)^{-\gamma_v\kappa}\frac{1}{\sigma'_1}\frac{\partial\sigma'}{\partial z}\right] \quad (4)$$

or, in terms of the excess pore pressure variable, as

$$\frac{\partial u}{\partial t} = \sigma' c_{v,1}\frac{\partial}{\partial z}\left[\left(\frac{\sigma'}{\sigma'_1}\right)^{-\gamma_v\kappa}\frac{1}{\sigma'_1}\frac{\partial u}{\partial z}\right] \quad (5)$$

In this equation σ'_1 can be substituted by any other pressure, for example σ'_2, providing that c_v is also substituted by $c_{v,2}$. The case λ=0 (or $\gamma_v\kappa$=1), i.e., constant c_v, is the nonlinear model of Davis and Raymond, which always provides an effective stress that is lower (or higher excess pore pressure) than that obtained with the Terzaghi model. However, the different dependences V-σ' (or e-σ') for J-B and Davis and Raymond result in a different average degree of consolidation ($\bar{U}_s$) for both models. In fact, for J-B [12], this unknown obeys the following expression:

$$\bar{U}_s = \frac{1-(1/H_1)\int_0^{H_1}(\sigma'/\sigma'_1)^{-\gamma_v}dz}{1-(\sigma'_2/\sigma'_1)^{-\gamma_v}} \quad (6)$$

or, in terms of the new normalized variable $v = (\sigma'/\sigma'_2)^{\lambda}$

$$\bar{U}_s = \frac{1-(H_2/H_1)\int_0^{1}v^{-\gamma_v/\lambda}dz'}{1-H_2/H_1} \quad (7)$$

Finally, J-B does not report analytical or numerical solutions for the excess pore pressure or for the effective pressure, except for the case λ=0, probably due to the large number of variables involved. He only reports [25] a consolidation abacus $\bar{U}_s \sim T$ for the case λ=0 ($\gamma_v\kappa$=1) and a constant c_v, using H_2/H_1 as parameter, and a set of abacuses for the case λ/γ_v=$(1/\gamma_v)$-$\kappa \neq 0$, $c_v \neq$ constant. In these abacuses the time factor is defined as $T = tc_{v,2}/H^2$. Despite these results, no physical meaning is attributed to λ, T, or v. In short, in view of (7), J-B concludes that his model depends on three groups, H_2/H_1, $(1/\gamma_v)$-κ, and the time factor T (since v is a function of T).

3. Nondimensionalization of Original and Extended Juárez-Badillo Models. Dimensionless Groups

To improve identification of the parameters involved in the constitutive dependences we redefine these as follows:

$$\frac{dV}{V} = -\gamma_{v,\sigma'}\frac{d\sigma'}{\sigma'},$$
$$\frac{dk}{k} = -\gamma_{k,\sigma'}\frac{d\sigma'}{\sigma'} \quad (8)$$
$$\frac{dk}{k} = \gamma_{k,v}\frac{dV}{V}$$

or, in their integral form

$$\frac{V}{V_o} = \left(\frac{\sigma'}{\sigma'_o}\right)^{-\gamma_{v,\sigma'}}$$
$$\frac{k}{k_o} = \left(\frac{\sigma'}{\sigma'_o}\right)^{-\gamma_{k,\sigma'}} \quad (9)$$
$$\frac{k}{k_o} = \left(\frac{V}{V_o}\right)^{\gamma_{k,v}}$$

with

$$\gamma_{k,v} = \frac{\gamma_{k,\sigma'}}{\gamma_{v,\sigma'}} \quad (10)$$

3.1. Nondimensionalization of the J-B Model in Terms of Pressure. Four cases are presented: (1) original J-B model, (2) J-B model assuming 1+e is not constant and *dz* is constant, (3) J-B model assuming 1+e is constant (a simplification of the former), and (4) the less restrictive model with both 1+e and *dz* not constant.

3.1.1. Original J-B Model. The original J-B model [12] assumes small deformations compared with the soil thickness and a negligible initial void ratio ($e_o \approx 0$). Developing the term between brackets of (3), we can write

$$\frac{1}{\sigma'}\frac{\partial\sigma'}{\partial t} = -\frac{c_{v,1}\gamma_{k,\sigma'}}{(\sigma'_1)^2}\left(\frac{\sigma'}{\sigma'_1}\right)^{-\gamma_{k,\sigma'}-1}\left(\frac{\partial\sigma'}{\partial z}\right)^2 + \frac{c_{v,1}}{\sigma'_1}\left(\frac{\sigma'}{\sigma'_1}\right)^{-\gamma_{k,\sigma'}}\frac{\partial^2\sigma'}{\partial z^2} \quad (11)$$

or, by mathematical manipulation,

$$\frac{\partial \sigma'}{\partial t} = \frac{\sigma'_o k_o}{\gamma_w \gamma_{v,\sigma'}} \left(\frac{\sigma'}{\sigma'_o}\right)^{1-\gamma_{k,\sigma'}} \left[-\frac{\gamma_{k,\sigma'}}{\sigma'}\left(\frac{\partial \sigma'}{\partial z}\right)^2 + \frac{\partial^2 \sigma'}{\partial z^2}\right] \quad (12)$$

To make this equation dimensionless, the following variables are used:

$$(\sigma')' = \frac{\sigma' - \sigma'_1}{\sigma'_2 - \sigma'_1}, \quad t' = \frac{t}{\tau_{o,\sigma'}}, \quad z' = \frac{z}{H_1} \quad (13)$$

where $\tau_{o,\sigma'}$ is a characteristic time or unknown reference, chosen as the time required for the effective pressure (on average along the whole domain) to reach a high percentage of its change, for example, 90%. With these variables and provided that (on average) $d\sigma' = d((\sigma')'(\sigma'_2 - \sigma'_1) + \sigma'_1) \approx \sigma'_2 - \sigma'_1$, (12) can be written as

$$\frac{(\sigma'_2 - \sigma'_1)}{\sigma'_m \tau_{o,\sigma'}} \frac{\partial (\sigma')'}{\partial t'} = -\frac{c_{v,1}\gamma_{k,\sigma'}}{(\sigma'_1)^2 H_1^2}\left(\frac{\sigma'_m}{\sigma'_1}\right)^{-\gamma_{k,\sigma'}-1} (\sigma'_2 - \sigma'_1)^2 \left(\frac{\partial (\sigma')'}{\partial z'}\right)^2 + \frac{c_{v,1}}{\sigma'_1 H_1^2}\left(\frac{\sigma'_m}{\sigma'_1}\right)^{-\gamma_{k,\sigma'}} (\sigma'_2 - \sigma'_1) \frac{\partial^2 (\sigma')'}{\partial z'^2} \quad (14)$$

where σ'_m is an average value of the effective pressure which we will talk about later. The dimensional coefficients of this equation, in terms of the parameters of the problem

$$\frac{1}{\tau_{o,\sigma'}}, \quad \left(\frac{k_1 \sigma'_1}{\gamma_w \gamma_{v,\sigma'}}\right) \frac{\gamma_{k,\sigma'}}{H_1^2} \left(\frac{\sigma'_2}{\sigma'_1} - 1\right)\left(\frac{\sigma'_m}{\sigma'_1}\right)^{-\gamma_{k,\sigma'}}, \quad \left(\frac{k_1 \sigma'_1}{\gamma_w \gamma_{v,\sigma'}}\right) \frac{1}{H_1^2} \left(\frac{\sigma'_m}{\sigma'_1}\right)^{1-\gamma_{k,\sigma'}} \quad (15)$$

give rise to two dimensionless independent groups:

$$\pi_I = \frac{\tau_{o,\sigma'} k_1 \sigma'_1}{\gamma_w \gamma_{v,\sigma'} H_1^2} \left(\frac{\sigma'_m}{\sigma'_1}\right)^{1-\gamma_{k,\sigma'}}, \quad \pi_{II} = \gamma_{k,\sigma'} \left(\frac{\sigma'_2 - \sigma'_1}{\sigma'_m}\right) \quad (16)$$

Based on the pi theorem, the solution of the order of magnitude of the characteristic time, derived from $\pi_I = \psi(\pi_{II})$, is given by

$$\tau_{o,\sigma'} \sim \left(\frac{\gamma_w \gamma_{v,\sigma'} H_1^2}{k_1 \sigma'_1}\right)\left(\frac{\sigma'_m}{\sigma'_1}\right)^{\gamma_{k,\sigma'}-1} \Psi_\tau \left(\gamma_{k,\sigma'} \left(\frac{\sigma'_2 - \sigma'_1}{\sigma'_m}\right)\right) \quad (17)$$

As regards $\bar{U}_{\sigma'}$, since it also depends on time, the solution (dependent on two groups) is given by

$$\bar{U}_{\sigma'} = \Psi_U \left[\frac{t}{\tau_{o,\sigma'}}, \gamma_{k,\sigma'} \left(\frac{\sigma'_2 - \sigma'_1}{\sigma'_m}\right)\right] \quad (18)$$

where Ψ_τ and Ψ_U are unknown functions of their arguments.

3.1.2. J-B Model Assuming 1+e Is Not Constant and dz Is Constant. Deleting the restrictive hypothesis $e_o \approx 0$ used by J-B for the solutions of σ' and $\bar{U}_{\sigma'}$, the latter through the dependence V-σ' for which such a hypothesis is not satisfied, and assuming that 1+e≠constant, the governing equation is

$$\frac{\partial}{\partial z}\left(\frac{k}{\gamma_w}\frac{du}{dz}\right) = \frac{\partial}{\partial t}\left(\frac{e}{1+e}\right) = \frac{1}{(1+e)^2}\left(\frac{\partial e}{\partial t}\right) \quad (19)$$

Using the dependence $V/V_1 = (\sigma'/\sigma'_1)^{-\gamma_{v,\sigma'}} = (1+e)/(1+e_o)$ or $1 + e = (1 + e_o)(\sigma'/\sigma'_o)^{-\gamma_{v,\sigma'}}$ and its derivative form $de = -((1 + e_o)\gamma_{v,\sigma'}/\sigma'_o)(\sigma'/\sigma'_o)^{-\gamma_{v,\sigma'}-1} d\sigma'$, the right and left terms of (19) can be written as

$$\frac{\partial}{\partial t}\left(\frac{e}{1+e}\right) = \frac{1}{(1+e)^2}\left(\frac{\partial e}{\partial t}\right) = -\frac{\gamma_{v,\sigma'}}{(1+e_o)\sigma'_o}\left(\frac{\sigma'}{\sigma'_o}\right)^{\gamma_{v,\sigma'}-1}\frac{\partial \sigma'}{\partial t} \quad (20)$$

On the other hand, with $k/k_o = (\sigma'/\sigma'_o)^{-\gamma_{k,\sigma'}}$, the left term of (19) writes as

$$\frac{\partial}{\partial z}\left(\frac{k}{\gamma_w}\frac{\partial u}{\partial z}\right) = \frac{k_o}{\gamma_w}\left(\frac{\sigma'}{\sigma'_o}\right)^{-\gamma_{k,\sigma'}}\left[\frac{\gamma_{k,\sigma'}}{\sigma'}\left(\frac{\partial u}{\partial z}\right)^2 + \frac{\partial^2 u}{\partial z^2}\right] \quad (21)$$

Equating both terms, (19) in terms of the effective pressure takes the form

$$\frac{\partial \sigma'}{\partial t} = \frac{(1+e_o)\sigma'_o k_o}{\gamma_w \gamma_{v,\sigma'}}\left(\frac{\sigma'}{\sigma'_o}\right)^{1-\gamma_{v,\sigma'}-\gamma_{k,\sigma'}}\left[-\frac{\gamma_{k,\sigma'}}{\sigma'}\left(\frac{\partial \sigma'}{\partial z}\right)^2 + \frac{\partial^2 \sigma'}{\partial z^2}\right] \quad (22)$$

Proceeding as in the previous case, the coefficients of the dimensionless form of the former equation

$$\frac{1}{\tau_{o,\sigma'}},$$

$$\left(\frac{(1+e_o)\,k_1\sigma'_1}{\gamma_w\gamma_{v,\sigma'}}\right)\frac{\gamma_{k,\sigma'}}{H_1^2}\left(\frac{\sigma'_2-\sigma'_1}{\sigma'_m}\right)\left(\frac{\sigma'_m}{\sigma'_1}\right)^{1-\gamma_{v,\sigma'}-\gamma_{k,\sigma'}}, \quad (23)$$

$$\left(\frac{(1+e_o)\,k_1\sigma'_1}{\gamma_w\gamma_{v,\sigma'}}\right)\frac{1}{H_1^2}\left(\frac{\sigma'_m}{\sigma'_1}\right)^{1-\gamma_{v,\sigma'}-\gamma_{k,\sigma'}}$$

give rise to the dimensionless groups

$$\pi_I = \frac{\tau_{o,\sigma'}k_1\sigma'_1(1+e_o)}{\gamma_w\gamma_{v,\sigma'}H_1^2}\left(\frac{\sigma'_m}{\sigma'_1}\right)^{1-\gamma_{k,\sigma'}-\gamma_{v,\sigma'}}, \quad (24)$$

$$\pi_{II} = \gamma_{k,\sigma'}\left(\frac{\sigma'_2-\sigma'_1}{\sigma'_m}\right)$$

Thus, the solutions for $\tau_{o,\sigma'}$ and $\bar{U}_{\sigma'}$ are given by

$$\tau_{o,\sigma'} = \left(\frac{\gamma_{v,\sigma'}\gamma_w H_1^2}{(1+e_o)\,k_1\sigma'_1}\right)\left(\frac{\sigma'_m}{\sigma'_1}\right)^{\gamma_{v,\sigma'}+\gamma_{k,\sigma'}-1} \cdot \Psi_\tau\left(\gamma_{k,\sigma'}\left(\frac{\sigma'_2-\sigma'_1}{\sigma'_m}\right)\right) \quad (25)$$

$$\bar{U}_{\sigma'} = \Psi_U\left[\frac{t}{\tau_{o,\sigma'}}, \gamma_{k,\sigma'}\left(\frac{\sigma'_2-\sigma'_1}{\sigma'_m}\right)\right] \quad (26)$$

3.1.3. J-B Model Assuming 1+e Is Constant (A Simplification of the Former). Based on the above, it is easy to simplify the hypothesis to 1+e being constant. With $(\sigma'/\sigma'_1)^{-\gamma_{v,\sigma'}} = 1$, the consolidation equation reduces to

$$\frac{\partial\sigma'}{\partial t} = \frac{(1+e_o)\,\sigma'_o k_o}{\gamma_w\gamma_{v,\sigma'}}\left(\frac{\sigma'}{\sigma'_o}\right)^{1-\gamma_{k,\sigma'}}\left[-\frac{\gamma_{k,\sigma'}}{\sigma'}\left(\frac{\partial\sigma'}{\partial z}\right)^2 + \frac{\partial^2\sigma'}{\partial z^2}\right] \quad (27)$$

an equation dependent on the initial void ratio. Using (13) and averaging, this equation provides three-dimensional coefficients

$$\frac{1}{\tau_{o,\sigma'}},$$

$$\left(\frac{(1+e_o)\,k_1\sigma'_1}{\gamma_w\gamma_{v,\sigma'}}\right)\frac{\gamma_{k,\sigma'}}{H_1^2}\left(\frac{\sigma'_2-\sigma'_1}{\sigma'_m}\right)\left(\frac{\sigma'_m}{\sigma'_1}\right)^{1-\gamma_{k,\sigma'}}, \quad (28)$$

$$\left(\frac{(1+e_o)\,k_1\sigma'_1}{\gamma_w\gamma_{v,\sigma'}}\right)\frac{1}{H_1^2}\left(\frac{\sigma'_m}{\sigma'_1}\right)^{1-\gamma_{v,\sigma'}-\gamma_{k,\sigma'}}$$

and two dimensionless independent groups

$$\pi_I = \frac{\tau_{o,\sigma'}k_1\sigma'_1(1+e_o)}{\gamma_w\gamma_{v,\sigma'}H_1^2}\left(\frac{\sigma'_m}{\sigma'_1}\right)^{1-\gamma_{k,\sigma'}}, \quad (29)$$

$$\pi_{II} = \gamma_{k,\sigma'}\left(\frac{\sigma'_2-\sigma'_1}{\sigma'_m}\right)$$

Thus, the solutions for $\tau_{o,\sigma'}$ and $\bar{U}_{\sigma'}$ are given by

$$\tau_{o,\sigma'} = \left(\frac{\gamma_{v,\sigma'}\gamma_w H_1^2}{(1+e_o)\,k_1\sigma'_1}\right)\left(\frac{\sigma'_m}{\sigma'_1}\right)^{\gamma_{k,\sigma'}-1} \cdot \Psi_\tau\left(\gamma_{k,\sigma'}\left(\frac{\sigma'_2-\sigma'_1}{\sigma'_m}\right)\right) \quad (30)$$

$$\bar{U}_{\sigma'} = \Psi_U\left[\frac{t}{\tau_{o,\sigma'}}, \gamma_{k,\sigma'}\left(\frac{\sigma'_2-\sigma'_1}{\sigma'_m}\right)\right] \quad (31)$$

3.1.4. The Less Restrictive Model with Both 1+e and dz Not Constant. For this model, 1+e and dz not constants, it is enough to consider (22) plus the condition dz not constant, or $dz = dz_o(H/H_1)$. So, after mathematical manipulation, the governing equation writes as

$$\frac{\partial\sigma'}{\partial t} = \frac{(1+e_o)\,\sigma'_o k_o}{\gamma_w\gamma_{v,\sigma'}}\left(\frac{\sigma'}{\sigma'_o}\right)^{1+\gamma_{v,\sigma'}-\gamma_{k,\sigma'}}\left[-\frac{\gamma_{k,\sigma'}}{\sigma'}\left(\frac{\partial\sigma'}{\partial z}\right)^2 + \frac{\partial^2\sigma'}{\partial z^2}\right] \quad (32)$$

The new coefficients

$$\frac{1}{\tau_{o,\sigma'}},$$

$$\left(\frac{(1+e_o)\,k_1\sigma'_1}{\gamma_w\gamma_{v,\sigma'}}\right)\frac{\gamma_{k,\sigma'}}{H_1^2}\left(\frac{\sigma'_2-\sigma'_1}{\sigma'_m}\right)\left(\frac{\sigma'_m}{\sigma'_1}\right)^{1+\gamma_{v,\sigma'}-\gamma_{k,\sigma'}}, \quad (33)$$

$$\left(\frac{(1+e_o)\,k_1\sigma'_1}{\gamma_w\gamma_{v,\sigma'}}\right)\frac{1}{H_1^2}\left(\frac{\sigma'_m}{\sigma'_1}\right)^{1+\gamma_{v,\sigma'}-\gamma_{k,\sigma'}}$$

provide the groups

$$\pi_I = \frac{\tau_{o,\sigma'}k_1\sigma'_1(1+e_o)}{\gamma_w\gamma_{v,\sigma'}H_1^2}\left(\frac{\sigma'_m}{\sigma'_1}\right)^{1-\gamma_{k,\sigma'}+\gamma_{v,\sigma'}}, \quad (34)$$

$$\pi_{II} = \gamma_{k,\sigma'}\left(\frac{\sigma'_2-\sigma'_1}{\sigma'_m}\right)$$

and the solutions

$$\tau_{o,\sigma'} = \left(\frac{\gamma_{v,\sigma'}\gamma_w H_1^2}{(1+e_o)k_1\sigma_1'}\right)\left(\frac{\sigma_m'}{\sigma_1'}\right)^{\gamma_{k,\sigma'}-\gamma_{v,\sigma'}-1} \cdot \Psi_\tau\left(\gamma_{k,\sigma'}\left(\frac{\sigma_2'-\sigma_1'}{\sigma_m'}\right)\right) \tag{35}$$

$$\bar{U}_{\sigma'} = \Psi_U\left[\frac{t}{\tau_{o,\sigma'}}, \gamma_{k,\sigma'}\left(\frac{\sigma_2'-\sigma_1'}{\sigma_m'}\right)\right] \tag{36}$$

3.2. Nondimensionalization of the J-B Model in Terms of Settlements. Three cases, defined by the hypotheses (1) 1+e≠ constant and dz constant, (2) 1+e and dz constants, and (3) 1+e and dz not being constant, are considered. For this study let us introduce a new variable directly related with the settlement, with a clear physical meaning, "$\zeta = e - e_o$", a kind of local degree of settlement or differential void index.

3.2.1. 1+e≠Constant and dz Constant. For 1+e≠constant, using the variable ζ and the dependence V-σ', it is straightforward to write

$$\zeta = e - e_o = (1+e_o)\left[\left(\frac{\sigma'}{\sigma'_1}\right)^{-\gamma_{v,\sigma'}} - 1\right] \tag{37}$$

Making use of the derivatives of (37)

$$\left(\frac{\partial\sigma'}{\partial z}\right)^2 = \frac{{\sigma'_1}^2}{(1+e_o)^2{\gamma_{v,\sigma'}}^2}\left(\frac{\sigma'}{\sigma'_1}\right)^{2\gamma_{v,\sigma'}+2}\left(\frac{\partial\zeta}{\partial z}\right)^2$$
$$\frac{\partial^2\sigma'}{\partial z^2} = -\frac{\sigma'_1}{(1+e_o)\gamma_{v,\sigma'}}\left(\frac{\sigma'}{\sigma'_1}\right)^{\gamma_{v,\sigma'}+1}\frac{\partial^2\zeta}{\partial z^2} + \frac{\sigma'_1(\gamma_{v,\sigma'}+1)}{(1+e_o)^2{\gamma_{v,\sigma'}}^2}\left(\frac{\sigma'}{\sigma'_1}\right)^{2\gamma_{v,\sigma'}+1}\left(\frac{\partial\zeta}{\partial z}\right)^2 \tag{38}$$

and the assumptions of the odometer test, after cumbersome mathematical steps, the nonlinear dimensional consolidation (19) is written in the form

$$\frac{\partial\zeta}{\partial t} = \frac{k_o{\sigma'}_o(\gamma_{k,\sigma'}-\gamma_{v,\sigma'}-1)}{\gamma_w{\gamma_{v,\sigma'}}^2}\left(\frac{\sigma'}{\sigma'_o}\right)^{1-\gamma_{k,\sigma'}}\left(\frac{\partial\zeta}{\partial z}\right)^2 + \frac{k_o{\sigma'}_o(1+e_o)}{\gamma_w\gamma_{v,\sigma'}}\left(\frac{\sigma'}{\sigma'_o}\right)^{1-\gamma_{v,\sigma'}-\gamma_{k,\sigma'}}\frac{\partial^2\zeta}{\partial z^2} \tag{39}$$

Introducing the variables

$$(\zeta)' = \frac{\zeta-\zeta_1}{\zeta_2-\zeta_1} = \frac{\zeta_1-\zeta}{\zeta_1},$$
$$z' = \frac{z}{H_1},$$
$$t' = \frac{t}{\tau_{o,s}} \tag{40}$$

into the former equation, with $\tau_{o,s}$, a reference chosen as the time required for the settlement to reach a high percentage of its change along the whole domain, its dimensionless form provides three coefficients

$$\frac{1}{\tau_{o,s}},$$
$$\frac{k_o{\sigma'}_o(\gamma_{k,\sigma'}-\gamma_{v,\sigma'}-1)}{\gamma_w{\gamma_{v,\sigma'}}^2}\left(\frac{\sigma_m'}{\sigma_1'}\right)^{1-\gamma_{k,\sigma'}}\frac{(e_f-e_o)}{H_1^2}, \tag{41}$$
$$\frac{k_o{\sigma'}_o(1+e_o)}{\gamma_w\gamma_{v,\sigma'}}\left(\frac{\sigma_m'}{\sigma_1'}\right)^{1-\gamma_{v,\sigma'}-\gamma_{k,\sigma'}}\frac{1}{H_1^2}$$

Dividing by the last, the resulting dimensionless groups are

$$\pi_I = \frac{\tau_{o,s}k_o{\sigma'}_o(1+e_o)}{\gamma_w\gamma_{v,\sigma'}H_1^2}\left(\frac{\sigma_m'}{\sigma_1'}\right)^{1-\gamma_{v,\sigma'}-\gamma_{k,\sigma'}}$$
$$\pi_{II} = \frac{(\gamma_{k,\sigma'}-\gamma_{v,\sigma'}-1)}{\gamma_{v,\sigma'}}\frac{(e_f-e_o)}{(1+e_o)}\left(\frac{\sigma_m'}{\sigma_1'}\right)^{\gamma_{v,\sigma'}} \tag{42}$$

Making use of the constitutive dependence $dV/V = -\gamma_{v,\sigma'}(d\sigma'/\sigma')$ is possible to write $\Delta H/H = -\gamma_{v,\sigma'}(\Delta\sigma'/\sigma')$, and since $(e_f - e_o)/(1 + e_o) = \Delta H/H_o$, π_{II} finally takes the form

$$\pi_{II} = (1-\gamma_{k,\sigma'}+\gamma_{v,\sigma'})\frac{\sigma_2'-\sigma_1'}{\sigma_m'} \tag{43}$$

The solution for $\tau_{o,s}$ is

$$\tau_{o,s} \approx \frac{\gamma_w\gamma_{v,\sigma'}H_1^2}{k_o{\sigma'}_o(1+e_o)}\left(\frac{\sigma_m'}{\sigma_1'}\right)^{\gamma_{v,\sigma'}+\gamma_{k,\sigma'}-1}\Psi_\tau\left[(1-\gamma_{k,\sigma'}+\gamma_{v,\sigma'})\frac{\sigma_2'-\sigma_1'}{\sigma_m'}\right] \tag{44}$$

As for the average degree of consolidation, the following dependence arises:

$$\bar{U}_s = \Psi_U\left[\frac{t}{\tau_{o,s}}, (1-\gamma_{k,\sigma'}+\gamma_{v,\sigma'})\frac{\sigma_2'-\sigma_1'}{\sigma_m'}\right] \tag{45}$$

3.2.2. 1+e and dz Constants. The hypothesis 1+e=constant is a simplification of the former case. Proceeding in the same way, the resulting coefficients

$$\frac{1}{\tau_{o,s}},$$
$$\frac{k_o{\sigma'}_o(1+\gamma_{v,\sigma'}-\gamma_{k,\sigma'})}{\gamma_w{\gamma_{v,\sigma'}}^2}\left(\frac{\sigma_m'}{\sigma_1'}\right)^{1-\gamma_{k,\sigma'}+\gamma_{v,\sigma'}}\frac{(e_f-e_o)}{H_1^2}, \tag{46}$$
$$\frac{k_o{\sigma'}_o(1+e_o)}{\gamma_w\gamma_{v,\sigma'}}\left(\frac{\sigma_m'}{\sigma_1'}\right)^{1-\gamma_{k,\sigma'}}\frac{1}{H_1^2},$$

provide the dimensionless groups

$$\pi_{\mathrm{I}} = \frac{\tau_{o,s} k_o \sigma'_o (1+e_o)}{\gamma_w \gamma_{v,\sigma'} H_1^2} \left(\frac{\sigma'_m}{\sigma'_1}\right)^{1-\gamma_{k,\sigma'}}$$
$$\pi_{\mathrm{II}} = (1-\gamma_{k,\sigma'} + \gamma_{v,\sigma'}) \frac{\sigma'_2 - \sigma'_1}{\sigma'_m} \quad (47)$$

The solutions for $\tau_{o,s}$ and $\bar{U}_s$ are

$$\tau_{o,s} \approx \frac{\gamma_w \gamma_{v,\sigma'} H_1^2}{k_o \sigma'_o (1+e_o)} \left(\frac{\sigma'_m}{\sigma'_1}\right)^{\gamma_{k,\sigma'}-1} \Psi_\tau \left[(1-\gamma_{k,\sigma'} + \gamma_{v,\sigma'}) \frac{\sigma'_2 - \sigma'_1}{\sigma'_m}\right] \quad (48)$$

$$\bar{U}_s = \Psi_U \left[\frac{t}{\tau_{o,s}}, (1-\gamma_{k,\sigma'} + \gamma_{v,\sigma'}) \frac{\sigma'_2 - \sigma'_1}{\sigma'_m}\right] \quad (49)$$

The solutions for the original J-B model are also (48) and (49) with the simplification of e_o=0. The author, however, does not talk about the characteristic time and obtains the average degree of consolidation given by (7), in which this variable depends on three groups: $(1-\gamma_{k,\sigma'})/\gamma_{v,\sigma'}$, H_2/H_1, and $tc_{v,2}/H_1^2$, undoubtedly a less precise solution.

3.2.3. 1+e and dz Not Being Constant. When 1+e and *dz* are not constants, substituting dz = $dz_o(H/H_1)$ in (39), after cumbersome mathematical manipulation, the resulting coefficients are

$$\frac{1}{\tau_{o,s}},$$
$$\frac{k_o \sigma'_o (\gamma_{k,\sigma'} - \gamma_{v,\sigma'} - 1)}{\gamma_w \gamma_{v,\sigma'}{}^2} \left(\frac{\sigma'_m}{\sigma'_1}\right)^{1+2\gamma_{v,\sigma'}-\gamma_{k,\sigma'}} \frac{(e_f - e_o)}{H_1^2}, \quad (50)$$
$$\frac{k_o \sigma'_o (1+e_o)}{\gamma_w \gamma_{v,\sigma'}} \left(\frac{\sigma'_m}{\sigma'_1}\right)^{1+\gamma_{v,\sigma'}-\gamma_{k,\sigma'}} \frac{1}{H_1^2}$$

Dividing by the last, the resulting dimensionless groups are

$$\pi_{\mathrm{I}} = \frac{\tau_{o,s} k_o \sigma'_o (1+e_o)}{\gamma_w \gamma_{v,\sigma'} H_1^2} \left(\frac{\sigma'_m}{\sigma'_1}\right)^{1+\gamma_{v,\sigma'}-\gamma_{k,\sigma'}}$$
$$\pi_{\mathrm{II}} = (1-\gamma_{k,\sigma'} + \gamma_{v,\sigma'}) \frac{\sigma'_2 - \sigma'_1}{\sigma'_m} \quad (51)$$

The solutions for $\tau_{o,s}$ and $\bar{U}_s$ are

$$\tau_{o,s} \approx \frac{\gamma_w \gamma_{v,\sigma'} H_1^2}{k_o \sigma'_o (1+e_o)} \left(\frac{\sigma'_m}{\sigma'_1}\right)^{+\gamma_{k,\sigma'}-\gamma_{v,\sigma'}-1} \Psi_\tau \left[(1-\gamma_{k,\sigma'} + \gamma_{v,\sigma'}) \frac{\sigma'_2 - \sigma'_1}{\sigma'_m}\right] \quad (52)$$

$$\bar{U}_s = \Psi_U \left[\frac{t}{\tau_{o,s}}, (1-\gamma_{k,\sigma'} + \gamma_{v,\sigma'}) \frac{\sigma'_2 - \sigma'_1}{\sigma'_m}\right] \quad (53)$$

Table 1 summarizes the results derived in this section.

4. Verification of the Results by Numerical Simulations

This section is devoted to checking the solutions given in Table 1. To shorten the exposure, we restrict the tests for the more general model (1+e and *dz* not constants) and for the unknowns $\tau_{o,s}$ and $\bar{U}_s$ since they are the ones of greater interest in civil engineering. Eight sets of simulations, arranged in three blocks, have been run, Table 2. In each one, some of the soil parameters or initial values of the problem have been changed to give the same or different values, as required, to the dimensionless group π_{II}, in the search for the same or different solutions of π_{I}. Changes in the values of the individual parameters are sufficient to cover all real scenarios.

Firstly, a reference set (set 1) is established to which all the other sets can be referred and which permits them to be compared with each other. The physical and geometrical characteristics that change are: $\gamma_{v,\sigma'}$, $\gamma_{k,\sigma'}$, e_o, H_1 (m), σ'_o (N/m^2), σ'_f (N/m^2), and k_o (m/year). The values of $\Psi = (\sigma'_f/\sigma'_o)$, H_2 (m), k_f (m/year), and π_{II} are derived from them, while π_{I} is obtained once the characteristic time $\tau_{o,s}$ is read from the simulation. The criterion for the choice of $\tau_{o,s}$ is the time required by the soil to reach 90 % of the total settlement.

It is worth mentioning here the emergence of σ'_m, an averaged value of the effective pressure, in the two dimensionless groups. Undoubtedly, the existence of σ'_m in the monomials can be explained not only by the nonlinearity of the problem but also by the kind of dependences in the constitutive relations (there are other nonlinear problems in which averaged variables do not emerge [18]). How do we choose σ'_m? We could take the mean value $\sigma'_m = (\sigma'_1 + \sigma'_2)/2$, or even $\sigma'_m = \sigma'_1$ or $\sigma'_m = \sigma'_2$, but expecting that the expression $\pi_{\mathrm{I}} = \psi(\pi_{\mathrm{II}})$ is slightly dependent on that choice (we will return to this question later). In this paper, we have taken the approximation $\sigma'_m = (\sigma'_2 + \sigma'_1)/2$.

In block I, the parameters that change in each set are: e_o, H_1, k_o and σ'_o, while maintaining the ratio $\Psi = (\sigma'_f/\sigma'_o)$; thus, both π_{II} and factor $(\sigma'_m/\sigma'_o)^{1-\gamma_{k,\sigma'}+\gamma_{v,\sigma'}}$ of π_{I} take the same value in the four sets of the block. As a result, we expect a characteristic time proportional to the ratio $\gamma_w \gamma_{v,\sigma'} H_1^2 / k_o \sigma'_o (1+e_o)$ and a same value of π_{I} in all cases, according to the solution $\pi_{\mathrm{I}} = \psi(\pi_{\mathrm{II}})$. Indeed, the results are consistent with expectations. Block II contains an only case for which π_{II} varies in comparison to block I; as a result, π_{I} also changes, as expected. Finally, block III contains three cases with the same value of π_{II} of block I, but with different values of the parameters $\gamma_{v,\sigma'}$, $\gamma_{k,\sigma'}$, and the ratio $\Psi = (\sigma'_f/\sigma'_o)$. The parameters of set 6 have been chosen in such a way that the values $\gamma_{k,\sigma'}$ and $\gamma_{v,\sigma'}$ are compensated to maintain constant $(1-\gamma_{k,\sigma'} + \gamma_{v,\sigma'})$; we expect π_{I} to be also unchanged. In effect, simulation provides $\pi_{\mathrm{I}} = 0.7139$, a value quite close to that of set 1 (0.7131). The negligible differences are due, no doubt, to the different evolution of the effective pressure along the process caused by the different values of the coefficients $\gamma_{k,\sigma'}$ and $\gamma_{v,\sigma'}$. Note that as $\gamma_{v,\sigma'}$ duplicates the denominator of the factor $\tau_{o,s} k_o \sigma'_o (1+e_o)/\gamma_w \gamma_{v,\sigma'} H_1^2$, the characteristic time has to double too, as indeed happens. The last two sets study the influence of Ψ, a parameter whose change forces the value

TABLE 1: Dimensionless groups that characterize the original and extended J-B models.

	Juárez-Badillo original	Juárez-Badillo 1+e=constant *dz* constant	Juárez-Badillo 1+e≠constant *dz* constant	Juárez-Badillo 1+e≠constant *dz* variable
π_I $\tau_{o,\sigma'}$ or $\tau_{o,s}$	$\frac{\tau_o k_o \sigma'_o}{\gamma_w \gamma_{v,\sigma'} H_1^2}\left(\frac{\sigma'_m}{\sigma'_o}\right)^{1-\gamma_{k,\sigma'}}$	$\frac{\tau_o k_o \sigma'_o (1+e_o)}{\gamma_w \gamma_{v,\sigma'} H_1^2}\left(\frac{\sigma'_m}{\sigma'_o}\right)^{1-\gamma_{k,\sigma'}}$	$\frac{\tau_o k_o \sigma'_o (1+e_o)}{\gamma_w \gamma_{v,\sigma'} H_1^2}\left(\frac{\sigma'_m}{\sigma'_o}\right)^{1-\gamma_{k,\sigma'}-\gamma_{v,\sigma'}}$	$\frac{\tau_o k_o \sigma'_o (1+e_o)}{\gamma_w \gamma_{v,\sigma'} H_1^2}\left(\frac{\sigma'_m}{\sigma'_o}\right)^{1-\gamma_{k,\sigma'}+\gamma_{v,\sigma'}}$
π_{II}	$\gamma_{k,\sigma'}\left(\frac{\sigma'_2-\sigma'_1}{\sigma'_m}\right)$ for pressure		$(1-\gamma_{k,\sigma'}+\gamma_{v,\sigma'})\left(\frac{\sigma'_2-\sigma'_1}{\sigma'_m}\right)$ for settlement	

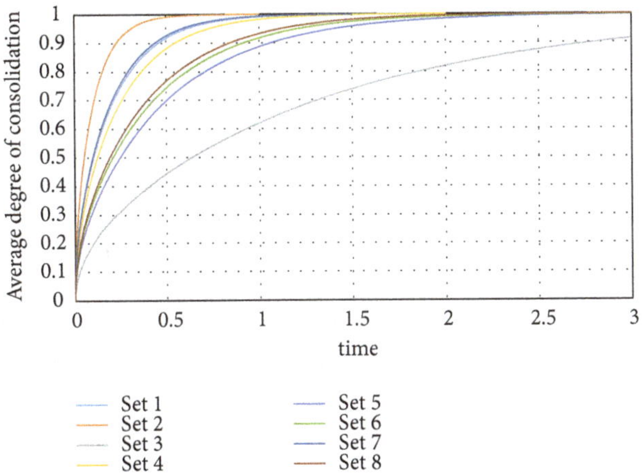

FIGURE 1: Average degree of consolidation versus time (eight sets).

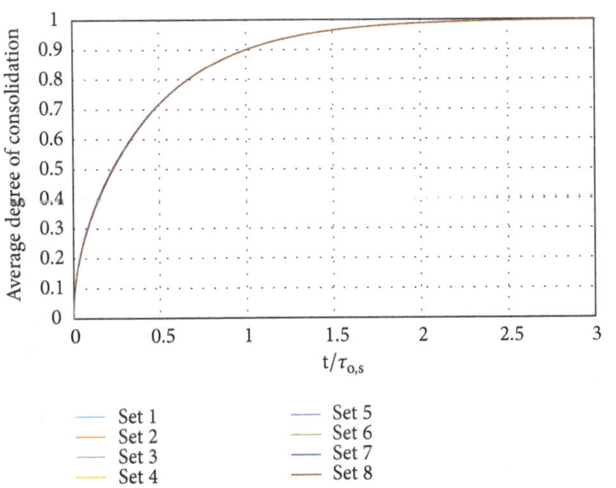

FIGURE 2: Average degree of consolidation versus dimensionless time (eight sets).

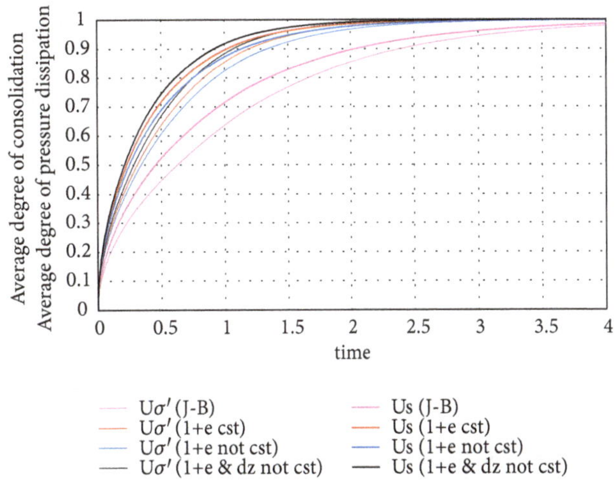

FIGURE 3: Average degree of consolidation and pressure dissipation (set 6).

$(1-\gamma_{k,\sigma'}+\gamma_{v,\sigma'})$ to be also changed in order to maintain π_{II} constant. We expect that π_I does not change as nearly occurs. The small differences are again due to the existence of σ'_m within the monomials. Note, however, that as Ψ diminishes the influence of σ'_m in the results also diminishes, since the values of σ'_1 and σ'_2 are closer.

For a better understanding of the results, Figures 1 and 2 are presented. In the first, $\bar{U}_s$ as a function of time is shown for the eight sets in Table 2. It is clear that $\tau_{o,s}$ covers a wide range of values, from 0.2283 (set 2) to 2.7968 (set 3), the last more than ten times the first. The second figure graphs $\bar{U}_s$ in front of the dimensionless time $t/\tau_{o,s}$. As shown, the eight previous curves nearly converge to a single (or universal) graph, having a common point at ($\bar{U}_s$=0.9, $t/\tau_{o,s}$=1). This means that monomial π_{II}, for the range of values assigned to the parameters, in contrast to π_I, scarcely influences the value of the average degree of consolidation. This is not in contradiction with the results derived by nondimensionalization.

Although in this section we have only checked the expressions of $\bar{U}_s$ and $\tau_{o,s}$ for the most general model, we consider it interesting to present both $\bar{U}_s$ and $\bar{U}_{\sigma'}$ for the rest of the models of Table 1 and for set 6 as a typical case, Figure 3. These results emphasize the importance of using the less restrictive (more general) model in relation to the others, in particular with the original model of J-B, in which the influence of e_o is despised. Note that the smaller of the characteristic times (related to settlement) corresponds to the more general model (black and bold line), so the use of the others, although on the safety side, provides characteristic times oversized, far from an optimal solution for the engineer. For example, for the case of Figure 3 in which e_o=1, the error of J-B original model (purple and bold line) in comparison with the extended model (black and bold line) for $\bar{U}_s$=90% is around 150%. This error increases appreciably as e_o gets larger.

5. Universal Curves

These easy to use and universal abacuses obtained by numerical simulations allow engineers to read the characteristic time ($\tau_{o,s}$) and the average degree of consolidation ($\bar{U}_s$) in terms of the dimensionless groups given in Table 1. Only the most general (less restrictive) J-B model, 1+e and *dz* not constant, is presented.

To obtain the characteristic time curves a large number of simulations have been carried out studying, separately, the influence of the parameters Ψ, $\gamma_{v,\sigma'}$, and $\gamma_{k,\sigma'}$, whose values cover most real soils: Ψ ranges from 1.01 to 8, $\gamma_{v,\sigma'}$ from 0.05 to 0.3, and $\gamma_{k,\sigma'}$ from 0.1 to 1.7. From these tests, it has been deduced, as expected from the results of the previous section, that the relatively more influential parameter is Ψ, while $\gamma_{v,\sigma'}$ and $\gamma_{k,\sigma'}$ do not produce significant changes in the values of the characteristic time. Thus, the first universal group of lines shows the dependence of π_I versus π_{II} for different values of Ψ (with $\gamma_{v,\sigma'}$=0.1), Figure 4. As the range of values of π_{II} depends on Ψ, according to (43), the range of values of each line is different. As shown in the figure, there exists a universal

TABLE 2: Values of the parameters and dimensionless groups of simulations.

Block	Set	$\gamma_{v,\sigma'}$	$\gamma_{k,\sigma'}$	e_o	H_1 (m)	k_o (m/yr)	σ'_o (N/m^2)	σ'_f (N/m^2)	Ψ σ'_f/σ'_o	H_2 (m)	k_f (m/yr)	π_{II}	$\tau_{o,s}$ (yr)	π_I
I	1	0.1	0.500	1.0	1.00	0.02	30000	60000	2	0.933	0.0141	0.40	0.4566	0.7131
	2	0.1	0.500	1.0	1.00	0.02	60000	120000	2	0.933	0.0141	0.40	0.2283	0.7131
	3	0.1	0.500	1.0	1.75	0.01	30000	60000	2	1.633	0.0071	0.40	2.7968	0.7131
	4	0.1	0.500	1.3	1.42	0.04	22000	44000	2	1.325	0.0283	0.40	0.5459	0.7131
II	5	0.1	1.500	1.0	1.00	0.02	30000	90000	3	0.896	0.0038	-0.40	1.0597	0.9834
III	6	0.2	0.600	1.0	1.00	0.02	30000	60000	2	0.871	0.0132	0.40	0.9142	0.7139
	7	0.1	0.767	1.0	1.00	0.02	30000	120000	4	0.871	0.0069	0.40	0.4360	0.7246
	8	0.2	0.920	1.0	1.00	0.02	30000	180000	6	0.699	0.0038	0.40	0.8409	0.7312

line for each value of Ψ, all converging in the central zone for values of π_{II} around 0. This set of straight lines can be fitted by $\pi_I = m\pi_{II} + n$, being n=0.846 and m given in Figure 5, as a function of Ψ. The curve that best fits the points in Figure 5 is given by

$$m = -0.235e^{(-0.868\Psi)} - 0.306e^{(-0.012\Psi)} \quad (54)$$

Choosing the longest line corresponding to Ψ=8 as the only representative line of the problem, the maximum errors in the reading of π_I (and $\tau_{o,s}$) are 3.8%.

In short, the use of these curves is as follows: from the data of the problem, Figure 4 allows us to read π_I from π_{II} and Ψ, and from (51) the characteristic time $\tau_{o,s}$ is derived.

As for the average degree of consolidation, the universal curves for 96 different scenarios of π_{II} are shown in Figure 6. It is appreciated that these curves are very close to each other, showing again that the $\bar{U}_s$ only depends on the dimensionless time. In the worst case, the errors produced by changes in value of the π_{II} group are less than 2.8% when using the curve of Figure 7.

TABLE 3: Experimental values of e, σ' and k obtained by Abbasi et al. [26].

e	σ'_o (N/m^2)
1.10	14000
1.05	28000
0.95	59000
0.85	120000
0.75	220000
0.65	450000
0.60	925000
e	**k (m/yr)**
1.75	0.683
1.60	0.315
1.20	0.084
0.85	0.021
0.65	0.011

6. Case Study

Below is a practical application of the universal curves presented in this paper. For this purpose, we take as a basis for our study the data obtained experimentally by Abbasi et al. [26] for different samples of clay whose moisture in the liquid limit is equal to 42. As can be seen (Table 3), it is a series of discrete (tabulated) values that relate the void ratio with the effective pressure and the hydraulic conductivity.

We will address a real case of consolidation, in which we have a layer of soil 2 meters thick (H_o), with an initial effective pressure (σ'_o) of 28 kN/m^2, on which a load of 31 kN/m^2 is applied. In this way, the effective pressure at the end of the consolidation process (σ'_f) will have the value of 59 kN/m^2. From the tabulated data (e-σ') it follows that e_o = 1.05 and e_f = 0.95 and, since the factor 1+e is proportional to the soil volume, from (9) we have $\gamma_{v,\sigma'} = 0.067$.

To obtain $\gamma_{k,\sigma'}$ we first calculate the value of $\gamma_{k,v}$ from the tabulated data (e-k), to then obtain $\gamma_{k,\sigma'}$ from (10). With this, we have, for the load step considered, k_o = 0.035 m/yr, $\gamma_{k,v}$ = 8.108, and $\gamma_{k,\sigma'} = 0.543$.

To solve the problem we will use the universal curves presented in the previous section, corresponding to the most general J-B model (analyzed in terms of settlement). Thus, from the values of σ'_o, σ'_f, $\gamma_{v,\sigma'}$, and $\gamma_{k,\sigma'}$, by means of (51) it is obtained that π_{II} = 0.373 (with σ'_m = 43.5 kN/m^2). Then, π_I is read from Figure 4, where for Ψ = 2.1 we have π_I = 0.72. An identical result is obtained if we use the proposed fitted equation $\pi_I = m\pi_{II} + n$, where n is a constant of value 0.846 and m is obtained from (54) or Figure 5. Finally, known the values of H_o, k_o and e_o, from (51) we get $\tau_{o,s}$ = 0.748 years. Likewise, once the characteristic time of the problem (in terms of settlement) is known, from the universal curve of Figure 7 we easily obtain the evolution of the average degree of consolidation along the whole process, Figure 8.

These solutions have been compared with those that result from the numerical simulation of the most general J-B model, for which a value of $\tau_{o,s}$ = 0.75 years has been obtained, showing clearly that the relative error committed when applying the universal solutions is negligible (Table 4). The simulation of the problem has also been carried out for a model where 1+e and *dz* are also not constant but in which the constitutive relations of potential type have been replaced by pairs of data in tabulated form, obtaining in this case $\tau_{o,s}$ = 0.817 years. The relative error, which this time increases to 8.2%, finds its explanation in the strong nonlinearity of the problem, which will always bring differences between the

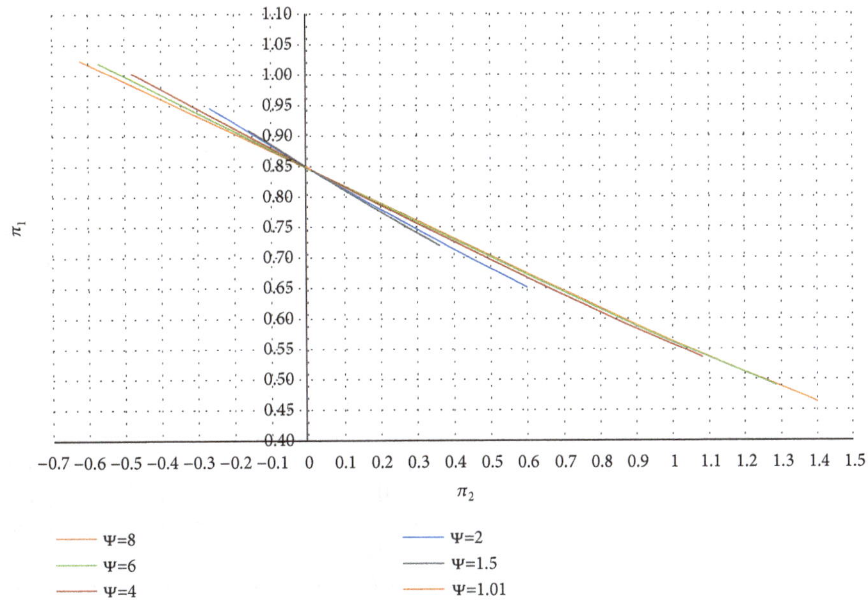

FIGURE 4: π_I as a function of π_{II} with Ψ as parameter.

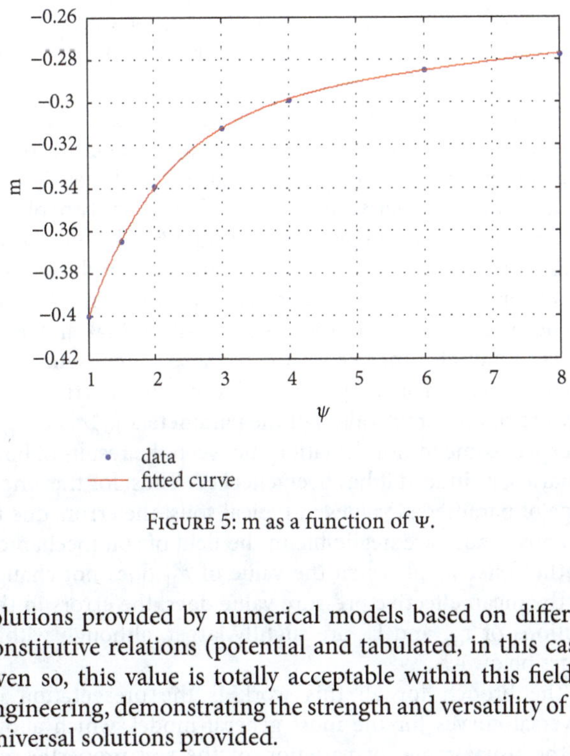

FIGURE 5: m as a function of Ψ.

solutions provided by numerical models based on different constitutive relations (potential and tabulated, in this case). Even so, this value is totally acceptable within this field of engineering, demonstrating the strength and versatility of the universal solutions provided.

7. Discussion

This work derives the dimensionless groups that rule the nonlinear soil consolidation process based on the original J-B model, following a formal nondimensionalization procedure that starts from the governing equation and does not need any other assumption or mathematical manipulation. In addition, more real and precise extended J-B models, which assume both the void ratio and the thickness of the finite volume element of the soil to be not constant, have been

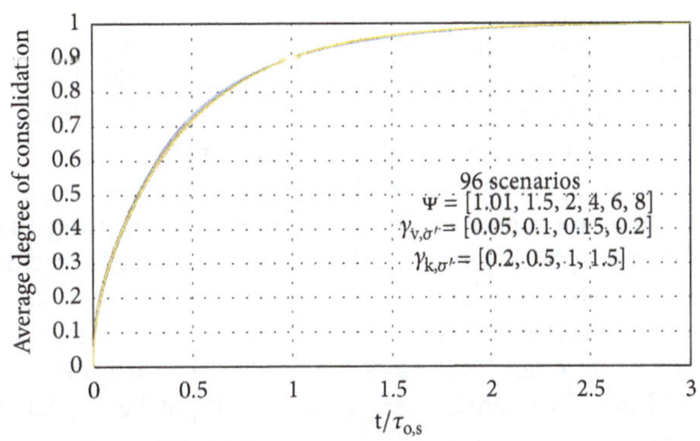

FIGURE 6: Average degree of consolidation versus dimensionless time for different scenarios.

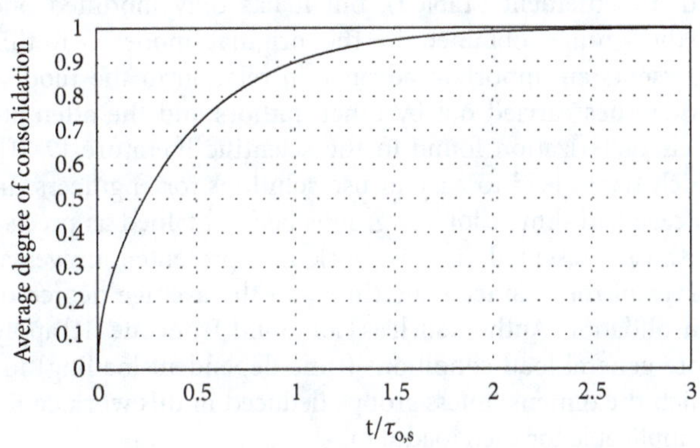

FIGURE 7: Proposed solution for the average degree of consolidation.

nondimensionalized and their dimensionless groups derived. The assumption of constitutive dependences (e-σ', e-k and k-σ') of potential type in these models is interesting since they

Table 4: Case study solution for the characteristic time $\tau_{o,s}$. Comparison with numerical models.

Universal curves	J-B most general model		Tabulated model	
$\tau_{o,s}$ (yr)	$\tau_{o,s}$ (yr)	Rel. error %	$\tau_{o,s}$ (yr)	Rel. error %
0.748	0.750	0.26	0.817	8.20

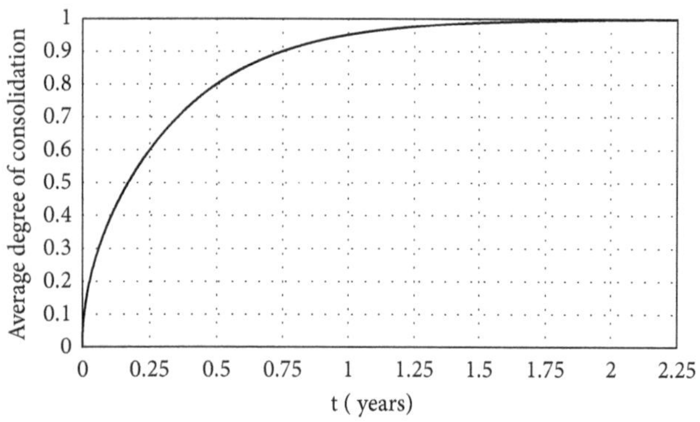

Figure 8: Case study solution for the average degree of consolidation $\bar{U}_s$ from the universal curves.

can be applied to highly compressible soils, for which the e-log(σ') relation moves away from being a straight line and in which negative values of the void ratio could be obtained under the application of high load ratios (σ'_f/σ'_o) [8].

The study has been developed both in terms of pressure dissipation and in terms of settlement, resulting that the dimensionless groups derived from each model are, in general, different. The extension to more complex models, which implies the emergence of new parameters or equations within the mathematical model, does not necessarily give rise to additional dimensionless groups, as might be expected, but to new expressions of the groups or new regrouping of the parameters involved. Thus, the elimination of the hypothesis "dz constant" in the most general consolidation model has not given rise to any new group, both in pressure and in settlement (Table 1), but it has only modified one of the groups obtained in the original model. All this represents an important advance in relation to the models and studies carried out by other authors and the attempts of characterization found in the scientific literature [9–11], which try to lead to easy-to-use solutions for engineers. In no case had dimensionless groups been obtained as precise as those presented in this work nor presented universal curves of the characteristic time and the average degree of consolidation. Authors such as Conte and Troncone [11] apply more general load conditions (time-dependent loading) for which the dimensionless groups deduced in this work could be applicable for each load step.

As regards time, to make it dimensionless, an unknown reference (τ_o) is introduced in the form mentioned by Scott [27] and used in many other problems with asymptotic solutions [28]. The precise definition of τ_o necessarily involves a criterion in relation to the percentage of average pressure decrease (or settlement) associated with the end of the consolidation process. The order of magnitude of τ_o is derived after nondimensionalization, making τ_o only appear in one of the independent dimensionless groups. In this way, the average degree of consolidation (in terms of pressure or settlement) depends on the "time/characteristic time" ratio, in addition to the dimensionless groups (without unknowns) that arise.

In short, the procedure has led to a new, very precise, solution that involves two monomials: one (π_I) that contains the characteristic time (τ_o) and another (π_{II}) that is a mathematical function of the soil parameters. In this way, τ_o only depends on π_{II} through the function $\pi_I = \psi(\pi_{II})$, while $\bar{U}$ depends on π_I and π_{II} through the expression $\bar{U} = \psi(\pi_I, \pi_{II})$. The results obtained have been verified for the most general and complex consolidation model through the numerical simulation of a large number of scenarios in which the values of the dimensionless groups and the individual parameters of the problem have been changing to convenience. The high errors produced by the original J-B model with respect to the most general justify the use of the latter in the search for more precise solutions.

In addition, the emergence of mean values of the effective pressure when averaging the governing equation, in order to assign an order of magnitude unit to the dimensionless variables and their derivatives, is an added difficulty to obtain the dimensionless groups. This is, undoubtedly, due to the nonlinearities (constitutive relations) involved in the consolidation process. This average effective pressure of reference has been assumed as the semisum of its initial and final values, although other criteria could also be chosen. Since two scenarios can have equal π_{II} with the same effective mean pressure but different values of the parameters $\gamma_{v,\sigma'}$ and $\gamma_{k,\sigma'}$, we expect some minor deviations between the results of both scenarios, as indeed it has been checked. Thus, for the whole range of parameters values in typical soils, the errors due to the above cause are negligible in the field of soil mechanics. On the other hand, when the value of π_{II} does not change but the mean effective pressure value does, the errors in the solutions of $\tau_{o,s}$ and $\bar{U}_s$ are slightly larger, although within ranges below 4%.

The brooch for all this work is the presentation of universal curves for the most general model, which allow, for the usual range of variation of the soil properties, to obtain in a simple way the value of the characteristic time of consolidation and the evolution of the average degree of settlement throughout the entire process. Thus, starting from the nonlinear parameters of the soil ($\gamma_{v,\sigma'}$ and $\gamma_{k,\sigma'}$) and the load ratio (σ'_f/σ'_o), the value of the monomial π_{II} is obtained, from which the value of π_I is directly deducted thanks to the universal curve $\pi_I = \psi(\pi_{II})$ provided. Finally, $\tau_{o,s}$ is easily obtained from the expression of π_I. On the other hand, the curve that represents the $\bar{U}_s$ evolution along the process has turned out to be (practically) unique when it is represented as a function of the dimensionless time in the form $t/\tau_{o,s}$, regardless of the soil parameters and the loads applied. All this supposes a clear advantage when solving the consolidation problem, also showing the strength of the nondimensionalization method presented.

Finally, the case study addressed has demonstrated the simplicity and reliability of the universal curves proposed. From a series of tabulated data (e-σ' and e-k) the parameters that govern the nonlinear consolidation problem for the most extended J-B model have been adjusted, with which the value of the monomial π_{II} has been derived. With this, and using the universal curves, the solutions for the characteristic time $\tau_{o,s}$ and the average degree of consolidation $\bar{U}_s$ have been obtained. These results have been satisfactorily verified with the solutions provided both by the numerical extended model of J-B presented here and by another that substitutes the constitutive dependences of potential type by data in tabulated form.

8. Conclusions

The application of the nondimensionalization technique has allowed us to deduce the dimensionless groups that rule the nonlinear consolidation problem in which the constitutive dependencies e-σ' and e-k have the form of potential functions. This type of relations approximates the soil behavior better than the logarithmic functions e-log(σ') and e-log(k), particularly in very compressible soils and under high load ratios (σ'_f/σ'_o) in which inconsistent negative values of the void ratio may appear.

Several scenarios have been addressed, from the original model of Juárez-Badillo to extended models based on the same constitutive dependences but eliminating one or more classic (and restrictive) hypotheses in this process. In all cases, only two dimensionless groups emerge, one of them consisting of a grouping of the soil parameters and the other containing the unknown characteristic time. Based on these results, universal curves have been depicted: the first to determine the characteristic time of the process and the second that represents the evolution of average degree of settlement.

The solutions have been verified by numerical simulations, changing the values of the soil parameters and the dimensionless groups conveniently. Finally, the reliability and effectiveness of the proposed solutions have been shown through a case study.

Abbreviations

c_v: Consolidation coefficient (m^2/s or m^2/yr)
$c_{v,1}$: Initial consolidation coefficient (m^2/s or m^2/yr)
$c_{v,2}$: Final consolidation coefficient (m^2/s or m^2/yr)
dz: Element of differential length in the direction of the spatial coordinate z (m)
dz_o: Element of differential length in the direction of the spatial coordinate z at the initial time (m)
e: Void ratio (dimensionless)
e_o: Initial void ratio (dimensionless)
e_f: Final void ratio (dimensionless)
H: Soil thickness up to the impervious boundary or drainage length (m)
H_1, H_o: Initial soil thickness (m)
H_2: Final soil thickness (m)
k: Hydraulic conductivity (or permeability) (m/s or m/yr)
k_1, k_o: Initial hydraulic conductivity (m/s or m/yr)
k_f: Final hydraulic conductivity (m/s or m/yr)
m_v: Coefficient of volumetric compressibility (m^2/N)
$m_{v,1}$: Initial coefficient of volumetric compressibility (m^2/N)
t: Time. Independent variable (s or yr)
T: Juarez-Badillo's time factor (dimensionless)
u: Excess pore pressure (N/m^2)
$\bar{U}$: Average degree of consolidation or pressure dissipation (dimensionless)
$\bar{U}_s$: Average degree of consolidation (dimensionless)
$\bar{U}_{\sigma'}$: Average degree of pressure dissipation (dimensionless)
V: Soil volume (m^3)
V_1, V_o: Initial soil volume (m^3)
z: Vertical spatial coordinate (m)
γ_v: Nonlinear coefficient of compressibility of Juárez-Badillo (dimensionless)
$\gamma_{k,v}$: Nonlinear coefficient of change of permeability with volume (dimensionless)
$\gamma_{k,\sigma'}$: Nonlinear coefficient of change of permeability with effective pressure (dimensionless)
$\gamma_{v,\sigma'}$: Nonlinear coefficient of compressibility or coefficient of volumetric contraction (dimensionless)
γ_w: Specific weight of water (N/m^3)
Ψ: Loading factor. Ratio between the final and initial effective pressure (dimensionless)
κ: Nonlinear coefficient of change of permeability with volume of Juárez-Badillo (dimensionless)
λ: Parameter of Juárez-Badillo. $\lambda = 1 - \gamma_v\kappa$ (dimensionless)
ν: Auxiliary variable of Juárez-Badillo (dimensionless)
π_i: Dimensionless group or number (dimensionless)
σ': Effective pressure of the soil (N/m^2)
σ'_1, σ'_o: Initial effective pressure (N/m^2)
σ'_2, σ'_f: Final effective pressure (N/m^2)
σ'_m: Mean value of the effective pressure (N/m^2)
τ_o: Characteristic time (s or yr)

$\tau_{o,s}$: Characteristic time which takes the settlement to reach approximately its final value (s or yr)
$\tau_{o,\sigma'}$: Characteristic time which takes the excess pore pressure to dissipate until approximately the value of zero (s or yr)
Ψ: Arbitrary mathematical function
ζ: Differential void ratio (dimensionless)
ζ_1, ζ_2: Initial and final differential void ratio (dimensionless)

Subscripts

I, II . . .: Denote different numbers

Superscripts

$'$: Denote dimensionless magnitude.

Conflicts of Interest

The authors declare that they have no conflicts of interest.

References

[1] K. Terzaghi, "Die Berechnung der Durchlässigkeitszifferdes tones aus dem verlauf der hydrodynamischen spanunngserscheinnungen," Technical report II a, 132 N 3/4, 125-138, Akademie der Wissenschaften in Wien, Sitzungsberichte Mathnaturwiss Klasse Abt, 1923.

[2] R. T. Walker, "Vertical drain consolidation analysis in one, two and three dimensions," *Computers & Geosciences*, vol. 38, no. 8, pp. 1069–1077, 2011.

[3] G. Y. Zheng, P. Li, and C. Y. Zhao, "Analysis of non-linear consolidation of soft clay by differential quadrature method," *Applied Clay Science*, vol. 79, pp. 2–7, 2013.

[4] X. Y. Geng, C. J. Xu, and Y. Q. Cai, "Non-linear consolidation analysis of soil with variable compressibility and permeability under cyclic loadings," *International Journal for Numerical and Analytical Methods in Geomechanics*, vol. 30, no. 8, pp. 803–821, 2006.

[5] M. Lu, S. Wang, S. W. Sloan, B. Indraratna, and K. Xie, "Nonlinear radial consolidation of vertical drains under a general time-variable loading," *International Journal for Numerical and Analytical Methods in Geomechanics*, vol. 39, no. 1, pp. 51–62, 2015.

[6] H. Wu, L. Hu, W. Qi, and Q. Wen, "Analytical Solution for Electroosmotic Consolidation Considering Nonlinear Variation of Soil Parameters," *International Journal of Geomechanics*, vol. 17, no. 5, Article ID 06016032, 2017.

[7] S. J. Brandenberg, "iConsol.js: JavaScript Implicit Finite-Difference Code for Nonlinear Consolidation and Secondary Compression," *International Journal of Geomechanics*, vol. 17, no. 6, Article ID 04016149, 2017.

[8] R. Butterfield, "A natural compression law for soils (an advance on e-log p')," *Géotechnique*, vol. 29, no. 4, pp. 469–480, 1979.

[9] R. Lancellotta and L. Preziosi, "A general nonlinear mathematical model for soil consolidation problems," *International Journal of Engineering Science*, vol. 35, no. 10-11, pp. 1045–1063, 1997.

[10] Y.-C. Zhuang, K.-H. Xie, and X.-B. Li, "Nonlinear analysis of consolidation with variable compressibility and permeability," *Journal of Zhejiang University (Engineering Science Edition)*, vol. 6, no. 3, pp. 181–187, 2005.

[11] E. Conte and A. Troncone, "Nonlinear consolidation of thin layers subjected to time-dependent loading," *Canadian Geotechnical Journal*, vol. 44, no. 6, pp. 717–725, 2007.

[12] R. Yong and F. Townsend, "General Consolidation Theory for Clays," Soil Mechanics Series, Report No. 8, Graduate School of Engineering, National University of Mexico, 1983.

[13] J. C. Gibbings, *Dimensional Analysis*, Springer-Verlag, London, UK, 2011.

[14] E. Buckingham, "On physically similar systems; Illustrations of the use of dimensional equations," *Physical Review A: Atomic, Molecular and Optical Physics*, vol. 4, no. 4, pp. 345–376, 1914.

[15] R. Seshadri and T. Y. Na, "A Survey of Methods for Determining Similarity Transformations," in *Group Invariance in Engineering Boundary Value Problems*, pp. 35–61, Springer, New York, NY, USA, 1985.

[16] V. D. Zimparov and V. M. Petkov, "Application of discriminated analysis to low Reynolds number swirl flows in circular tubes with twisted-tape inserts. Pressure drop correlations," *International Review of Chemical Engineering*, vol. 1, no. 4, pp. 346–356, 2009.

[17] M. Capobianchi and A. Aziz, "A scale analysis for natural convective flows over vertical surfaces," *International Journal of Thermal Sciences*, vol. 54, pp. 82–88, 2012.

[18] M. Conesa, J. F. Sánchez Pérez, I. Alhama, and F. Alhama, "On the nondimensionalization of coupled, nonlinear ordinary differential equations," *Nonlinear Dynamics*, vol. 84, no. 1, pp. 91–105, 2016.

[19] I. A. Manteca, G. García-Ros, and F. Alhama, "Universal solution for the characteristic time and the degree of settlement in nonlinear soil consolidation scenarios. A deduction based on nondimensionalization," *Communications in Nonlinear Science and Numerical Simulation*, vol. 57, pp. 186–201, 2018.

[20] E. H. Davis and G. P. Raymond, "A non-linear theory of consolidation," *Géotechnique*, vol. 15, no. 2, pp. 161–173, 1965.

[21] P. Cornetti and M. Battaglio, "Nonlinear consolidation of soil modeling and solution techniques," *Mathematical and Computer Modelling*, vol. 20, no. 7, pp. 1–12, 1994.

[22] C. F. González-Fernández, "Applications of the network simulation method to transport processes," in *Network Simulation Method*, J. Horno, Ed., Research Signpost, Trivandrum, India, 2002.

[23] J. Serna, F. J. S. Velasco, and A. Soto Meca, "Application of network simulation method to viscous flows: The nanofluid heated lid cavity under pulsating flow," *Computers & Fluids*, vol. 91, pp. 10–20, 2014.

[24] E. Juárez-Badillo, "General Permeability Change Equations for Soils," in *Proceedings of the International Conference on Constitutive Laws*, pp. 205–209, University of Arizona, Tueson, Ariz, USA, 1983.

[25] E. Juárez-Badillo and B. Chen, "Consolidation curves for clays," *Journal of Geotechnical Engineering*, vol. 109, no. 10, pp. 1303–1312, 1983.

[26] N. Abbasi, H. Rahimi, A. A. Javadi, and A. Fakher, "Finite difference approach for consolidation with variable compressibility and permeability," *Computers & Geosciences*, vol. 34, no. 1, pp. 41–52, 2007.

[27] R. F. Scott, *Principles of Soils Mechanics*, Addison-Wesley Publishing Company, 1963.

[28] F. Alhama and C. N. Madrid, *Análisis Dimensional Discriminado En Mecánica De Fluidos Y Transferencia De Calor*, 2012.

12

Thermomechanical Behavior of Energy Pile Embedded in Sandy Soil

Xu Huang[1] **Yajun Wu**[1,2] **Huaifeng Peng**[1] **Yaohu Hao**[1] **and Chenyang Lu**[2]

[1]*Key Laboratory of Ministry of Education for Geomechanics and Embankment Engineering, Hohai University, Nanjing 210098, China*
[2]*Department of Civil Engineering, Shanghai University, Shanghai 200072, China*

Correspondence should be addressed to Yajun Wu; wyjdldz@163.com

Academic Editor: Marco Pizzarelli

The traditional energy pile (solid energy pile) has been implemented for decades. However, the design of different kinds of energy piles is still not well understood. In this study, a series of model tests were performed on an aluminum pipe energy pile (PEP) in dry sandy soil to investigate the thermal effects on the mechanical behaviors of pipe energy pile. The thermal responses of the PEP were also analyzed. Steady temperatures of the PEP under different working conditions were also compared with that of the solid energy pile. Different loading tests were carried out on four pipe energy piles under three different temperatures of 5, 35, and 50°C, respectively. The bearing capacity change can be interpreted through the load-displacement curves. Experiment results were also compared with the solid energy pile to evaluate bearing capacities of the PEP and the solid energy pile under different temperature conditions. The mobilized shaft resistance was also calculated and compared with the solid energy pile data and the results show that the PEP has a similar load transfer mechanism with the solid energy pile. It could also be found that, for PEPs under working load, plastic displacement would appear after a whole heating cycle.

1. Introduction

Environment pollution due to the use of natural energy resources has been a crucial issue. The ground-source heat exchange system (GSHE) is one of the new techniques that has been applied by civil engineering researchers to reduce the energy demand in heating and cooling systems [1]. However, the additional cost of drilling boreholes for the GSHEs prevents their development [2].

Energy pile system is a new geotechnical technique which was developed to overcome the disadvantage of GSHEs. Energy piles typically include closed-loop tubes in concrete pile foundations, making the pile act not only as load-bearing elements, but also as heat exchange elements. The circulation of fluid (usually water) in heat exchangers allows the energy pile to exchange heat and energy with the ground since the temperature below 4 m depth underground remains constant [3]. However, the temperature change of pile foundations during heating and cooling may lead to some irreversible volumetric strain for the surrounding soils [4, 5]. It would also cause the changes in the toe resistance and shaft friction along the pile, which would affect the serviceability and stability of pile foundations [6–8].

Field tests have been conducted to investigate the thermomechanical responses of energy piles. Laloui et al. [9, 10] conducted a single pile test at the Swiss Federal Institute of Technology in Lausanne, Switzerland. The 25.8 m long single concrete pile with a diameter of 0.88 m was located at the edge of a four-storey building. Optical fiber sensors and extensometers were installed in the concrete pile, as well as a load cell on the base of the pile. A heating and recover cycle was applied to the pile before the construction and after the constructions of each floor (load increased from 0 kN to 1300 kN). The results show that heating would lead to an increment in axial load, which is larger than that on the pile head load, and the thermal effect was observed more obviously in the pile toe. Bourne-Webb et al. also conducted tests [11] at Lambeth College, London, using a 23 m long concrete pile with a diameter of 0.55 m. A sustained load-cyclic thermal test was applied on the pile at a working load of 1200 kN. The results revealed that the axial load distribution

Table 1: Experimental program of the model tests.

Number	Model pile	Loading condition	Thermal condition (°C)	Test Model	Velocity of water (L/min)
(1)	RPEP	Until failure	Room temperature	L	—
(2)	PEP1	Until failure	35	H→L	—
(3)	PEP2	Until failure	50	H→L	—
(4)	PEP3	Until failure	5	H→L	—
(5)	PEP1	—	50	H	5
(6)	PEP1	—	50	H	15
(7)	PEP2	—	35	H	5
(8)	PEP2	—	35	H	15
(9)	PEP3	—	5	C	5
(10)	RPEP	working load	35	L-H-R-C-R	5

on the pile depended on the restraint conditions of pile ends. Murphy et al. [12, 13] also conducted tests on full-scale energy foundations to study the thermal axial stress and the displacement. The seasonal thermal responses of energy foundations were also analyzed during the operation period.

Based on the field tests, Amatya et al. [14] conducted further investigation on the effect of end restraints on the mechanical behaviors. Assuming that the pile was embedded in a nonconsolidating ground and the shaft resistance was uniform along the pile, the simplified load transfer mechanisms was explicated.

McCartney and Rosenberg [15, 16] and Murphy et al. [17] performed a series of centrifuge tests to investigate the ultimate pile capacities and thermal effects on the shaft resistance of energy foundations. It is proved that the pile capacities would increase with the increasing temperature, which may be the result of the increase in side shear resistance due to the radial expansion of foundation. Stewart and McCartney [18] carried out a centrifuge test to investigate the thermomechanical behaviors of end-bearing energy pile under cyclic heating. The maximum thermal axial stress was observed near the toe of foundation since the bottom part was fully constrained by the model container. Moreover, Goode et al. [19] conducted further centrifuge tests to study the effects of head and end restraints on energy foundations.

Ng et al. [7] carried out centrifuge tests in the saturated sandy soil to investigate the thermal effects on single pile capacity as well as settlement patterns. Load tests were carried out on four aluminum model piles under different temperature conditions and loading sequences. The neutral plane was observed to shift towards the pile toe during heating. The pile capacities increase by 13% with a temperature increment of 15°C, which was mainly attributed to the increase in ultimate shaft resistance. On the other hand, the mobilized toe resistance takes most of the applied axial load with the temperature increment of 30°C.

Different types of piles are widely used nowadays; however, understanding on these different types of energy piles is still limited. Particularly, the thermal effects on the bearing capacity and effects of different operation model on the thermomechanical behavior of the energy pile have not been fully revealed. In this study, load tests were carried out on four aluminum pipe piles in typical sandy soil. This study aims to investigate the thermal effects on the bearing capacity and load-displacement pattern of the pipe energy pile. In addition, the load transfer mechanism was also investigated. Considering that the material of the model pile is different with the concrete and some boundary effects exist in this study, all the results can only be applied for the model piles and could not be used in the design of energy pile in situ directly.

2. Experimental Program

All the tests were performed in four 1/20-scaled model containers with size of 750 mm × 750 mm × 1200 mm (length × width × height). Pipe energy piles in saturated sandy soil with different temperatures, different velocities of fluid in heat exchangers, and different loading sequences were investigated. The test program is summarized in Table 1. Firstly, a load test was carried out on the reference pipe energy pile (RPEP) under the ambient temperature of 15°C. Since only the mechanical loading was applied to the RPEP, the loading sequence can be denoted as "L." In Tests 2 to 4 (as shown in Table 1), other pipe energy piles (PEP1, PEP2, and PEP3) were heated up or cooled down to temperatures of 35, 50, and 5°C, respectively (the heating period denoted as "H" and cooling period as "C"). After that, load tests were conducted (the whole process is denoted as "H→L"). Thermal responses of piles were investigated in Tests 5 to 6 (heating period denoted as "H"). And working load was applied on the RPEP in Test 7 with a whole cycle (the RPEP was first heated to 35°C, then recovering to ambient temperature; cooling to 5°C after recovering; and finally recovering to ambient temperature; the recovering period could be denoted as "R"). All these tests were conducted in separated model containers and all four piles recovered to the ambient temperature before a new test started.

3. Soil Properties and Model Preparation

All the tests were carried out in the saturated sandy soils. The natural dry density and maximum and minimum dry densities are 1.55, 1.74, and 1.47 × 10^3 kg/m^3, respectively. The relative density is 0.8 for the tests. The uniformity coefficient and curvature coefficient are 3.0 and 0.9, respectively. The sandy soil has a mean diameter of D_{50} = 0.34 mm, and the ratio of

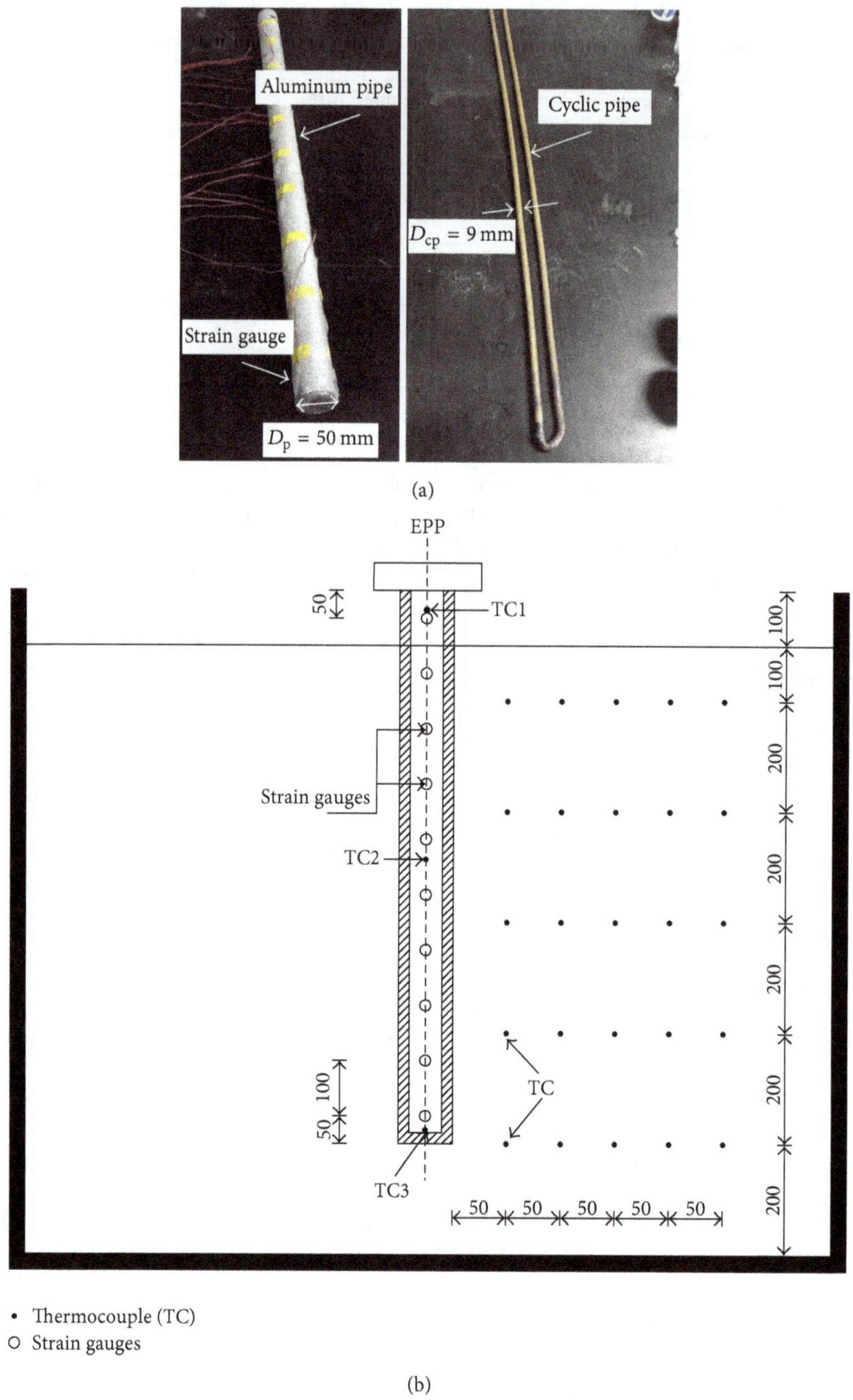

FIGURE 1: Details of model pile and pile instrumentation.

pile diameter to mean particle size (D/D_{50} = 147) is larger than the limiting value recommended by Fioravante [20], so the particle-size scaling effect could be negligible. Direct shear tests were conducted for the pile-soil interface. In the saturated sandy soil, the pile-soil interface angles range from 30.6° to 31.9° within the temperature range from 20 to 35°C.

Figure 1 shows the diagram of containers for the tests. The sandy soil was placed into the containers by tamping. The soil needed was calculated based on the size of model containers and divided into 20 layers to be rained and tamped to ensure the relative compaction in each of the containers to be the same.

4. Pile Properties and Instrumentation

The model energy piles were prefabricated from aluminum alloy tubes with Young's modulus of 70 GPa. The tube has an inner diameter and outer diameter of 40 mm and 50 mm, respectively. The aluminum pipe energy pile has a thermal expansion coefficient and thermal conductivity of 23×10^{-6} m/(m·K) and 237 W/(m·K), respectively. The thermal expansion of the model energy pile is almost three times than that of a concrete pile (8.6×10^{-6} m/(m·K)). As a result, the thermal expansion of the model pile and the thermally induced changes in the pile-soil interaction are overestimated in the model tests. The length of model energy pile is 1000 mm, and only 900 mm is embedded in the sand as shown in Figure 1(b). The distance between the pile toe and bottom of the containers is 20 cm which equals 4 times of the pile diameter and less than 6 times of pile diameter. Thus some boundary effects would influence the results.

A copper pipe with a diameter of 9 mm was put into the aluminum pile as heat exchangers. The toe of pile was sealed and the cavity of pipe pile was also filled with water to help the heat transfer from the copper pipe to the aluminum pipe (the model energy pile).

Ten sets of strain gauges were installed on the outer surface of the aluminum pile at a distance of 100 mm (2 m in prototype) and an epoxy resin with a thickness of 2 mm was applied on the surface to protect the gauges from damage and water. It is demonstrated that the effects of the epoxy resin on the side friction could be negligible. Since only a small part of the side surface was covered by the strain gauges, the effects on the shaft friction could be negligible. The wires connected with the strain gauges were also placed just next to the side wall of pile and kept in loose condition so that the effects of the wires could be reduced to the minimum and be negligible. To reduce the temperature effects on the strain gauges, two strain gauges, connected in the form of a half-full-bridge circuit, were bonded to a single polyimide carrier. Prior to installing the pile in the model container, the strain gauges were calibrated under an axial load of 1 kN on the top of the aluminum pile under temperatures of 15, 30, and 50°C.

Thermocouples were also installed on the inner side of the pile to monitor the temperature changes. Prior to installation, all the thermocouples were calibrated under temperatures ranging from 15 to 50°C. The epoxy resin with a thickness of 2 mm was also applied on the surface of thermocouples to protect them from damage and water. Figure 1 show the photograph and schematic plan of the model energy pile.

5. Experiment Results

5.1. Temperature Profiles along the Pipe Energy Pile. The temperature distribution has great influences on the mechanical behaviors of energy piles, and the temperature along the pile body would give more details to explain the thermal induced redistributions of stress and strain.

After heating energy piles continuously for 5 hours (4.17 days in prototype) at a constant temperature of 15°C without any load on the top of pile, the measured temperature profiles along the pipe energy piles are shown in Figure 2. The temperature differences of readings have been obtained from the three thermocouples along the pile (TC1, TC2, and TC3). At the end of heating, for the PEP1 in Test 5 and Test 6, the maximum temperature along the pile rose up to 47.75°C (target temperature of 50°C) from the initial temperature of 22.88°C at the top part of the pile, while the temperature at the bottom of PEP2 went up from 23.44 to 38.25°C. There is a 9.5°C difference between the top and bottom of the PEP1. At the same time, for the PEP2, in Test 7 and Test 8, the maximum temperature along the pipe energy pile rose up to 33.63°C (target temperature of 35°C) at the top of the pile, from the initial temperature of 21.44°C. And a relatively lower temperature was recorded by the TC3. The temperature near bottom part of the PEP2 went up from 21.56 to 29.3°C. There is a 4.25°C difference between the two readings (TC1 and TC3) of the PEP2. Normally, the temperature of energy pile under operation conditions does not show obvious change along the depth because of the heat conduction in the working period. However, the duration of tests in this paper was not long enough for that heat conduction.

The temperature changes of the PEP1 were shown in Figure 2(c). Comparing the temperature changes in Test 5 and Test 6, it could be found that the value of change in Test 6 is relatively higher than change in Test 5. Normally, the water has a lower heat conduction coefficient than the solid materials. However, the fluid has a different heat conduction types with the solid medium. And the velocity of fluid has effects on its heat conduction coefficient. Both of the velocities in these tests are relatively slow and the type of fluid is laminar flow. Since the velocity of water in Test 6 is higher than that in Test 5, it could be found that the energy pile with higher velocity of water in the cyclic pipe could have higher heat conduction coefficient between water and the pile under the conditions of this study.

Since the temperature of pile still increased a little bit after 5 hours of heating, models were built in software to simulate an extended heating period for 24 hours to make sure that the temperatures of the pipe energy piles reached a thermal equilibrium. And the thermomechanical characters of energy piles would be accurate. Numerical simulation model is shown in Figure 3.

Nonisothermal laminar flow conjugated with heat transfer is included in the analysis, based on main formulations of continuity, momentum, and energy equations [19, 20].

For calculation of fluid convection inside the pipes, the following equations are used:

$$\begin{aligned} &\frac{\partial}{\partial t}u + \nabla \cdot (\rho u) = 0, \\ &\rho \frac{\partial}{\partial t}u + \rho u \cdot \nabla u = -\nabla p + \nabla \\ &\qquad \cdot \left\{\mu\left(\nabla u + (\nabla u)^T\right) - \frac{2}{3}\mu(\nabla \cdot u) I\right\} \\ &\qquad + F, \end{aligned} \tag{1}$$

where ρ is density of heat-carrier-fluid, u is the heat-carrier-fluid velocity vector, and F is the body force vector.

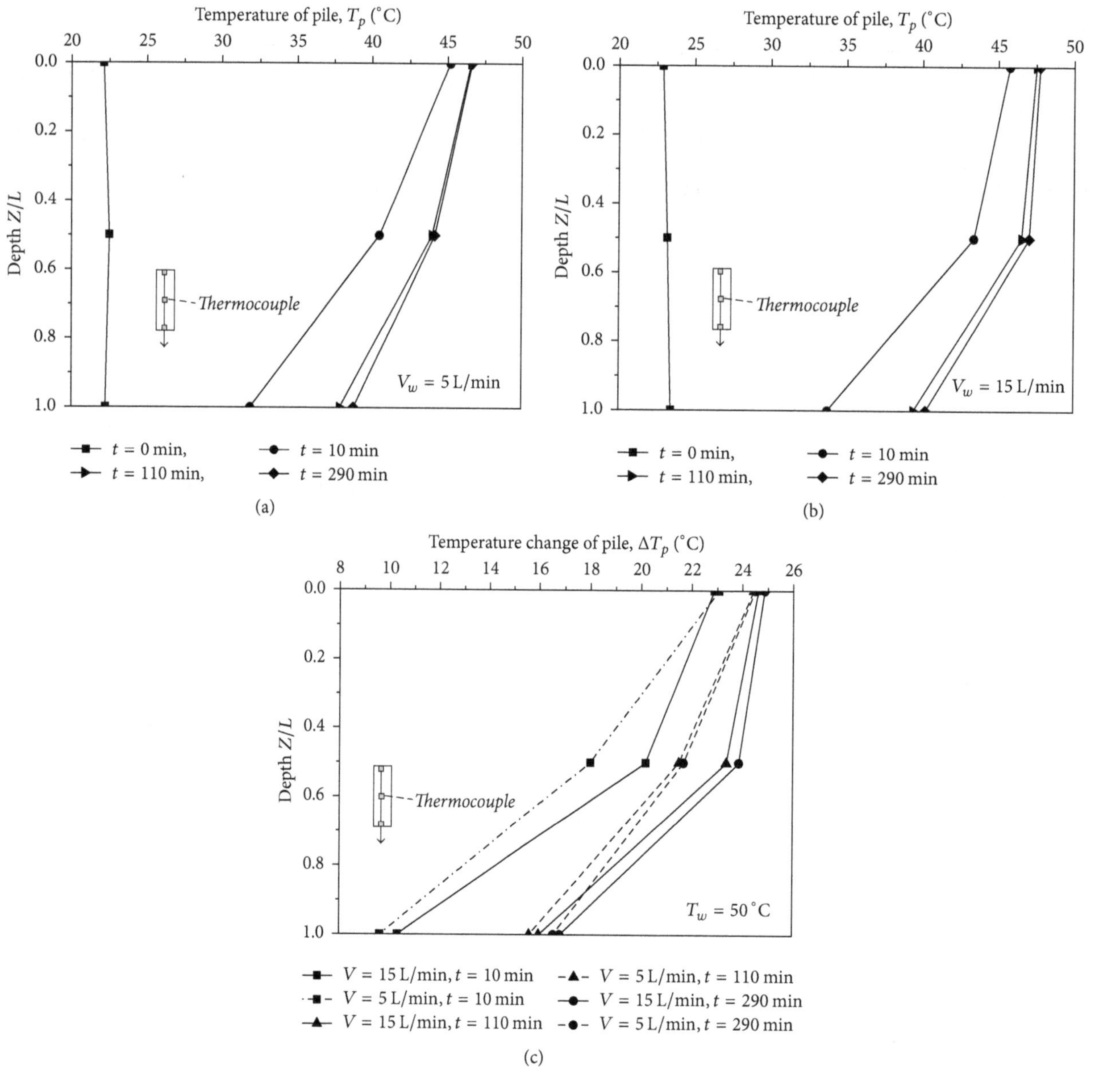

FIGURE 2: Temperature profiles of the PCC energy pile without load: (a) Test 5; (b) Test 6; (c) temperature variation in Test 5 and Test 6.

The fluid heat transfer equation is solved as

$$\rho C_p\left(\frac{\partial T}{\partial t}+u\cdot\nabla T\right)+\nabla\cdot q = \alpha_p T\left(\frac{\partial p}{\partial t}+u\cdot\nabla p\right)+\tau:\nabla u+Q, \tag{2}$$

$$\alpha_p=-\frac{1}{\rho}\frac{\partial\rho}{\partial T},$$

where C_p is the fluid heat capacity at constant pressure, T is the absolute temperature, α_p is the coefficient of thermal expansion, τ is viscous stress tensor, and Q is heat sources other than viscous heating.

The interface supports heat transfer in solid media, for example, pipe walls:

$$\rho C_p\frac{\partial}{\partial t}T=-(\nabla\cdot q)-T\frac{\partial E}{\partial t}+Q, \tag{3}$$

while heat transfer in porous media uses the following (Darcy porous medium):

$$\left(\rho C_p\right)_{\text{eq}}\frac{\partial T}{\partial t}+\rho Cp\cdot u\cdot\nabla T=\nabla\cdot\left(k_{\text{eq}}\nabla T\right)+Q, \tag{4}$$

$$K_{\text{eq}}=\theta_p\cdot k_p+\theta_L\cdot k, \tag{5}$$

$$\left(\rho C_p\right)_{\text{eq}}=\theta_p\cdot\rho_p\cdot C_{p,p}+\theta_L\cdot\rho\cdot C_p, \tag{6}$$

$$\theta_p+\theta_L=1, \tag{7}$$

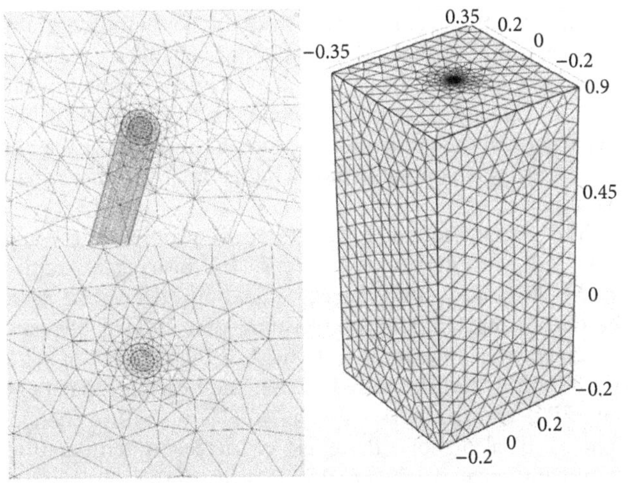

FIGURE 3: Numerical simulation model of energy pile.

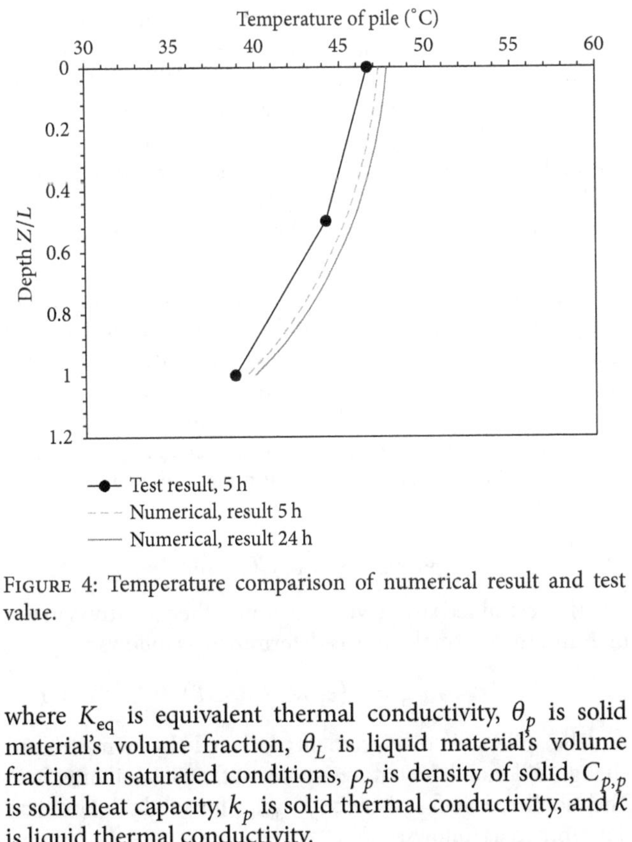

FIGURE 4: Temperature comparison of numerical result and test value.

where K_{eq} is equivalent thermal conductivity, θ_p is solid material's volume fraction, θ_L is liquid material's volume fraction in saturated conditions, ρ_p is density of solid, $C_{p,p}$ is solid heat capacity, k_p is solid thermal conductivity, and k is liquid thermal conductivity.

The numerical results were compared with experimental values in Figure 4. And good agreement was found. And the temperature of pile after heating for 5 hours and 24 hours was also compared. The results show that the average temperature of pile after heating for 5 hours would reach more than 95% of the average temperature after heating for 24 hours; thus the energy pile has already reached a condition of temperature equilibrium.

Temperature variations of the PEP after heating and cooling were compared with the solid energy pile in Figure 5. The temperature variations after cooling along the PEP were −6, −7.812, and −2.812°C, respectively. During the same period, the temperature variations along solid energy pile were −5, −5.34, and −6.2°C, respectively. Maximum temperature variations of both piles occurred near the middle part; however, the temperature variation of PEP near the toe dropped rapidly. On the same time, the temperature variations after heating along PEP were 24.875, 23.875, and 16.813 with 19, 19.43, and 18.31 along solid pile, respectively. It can be concluded that the thermal responses of the PEP were slight obviously compared to the solid energy pile. And it can also have some effect on the results of mechanical characters.

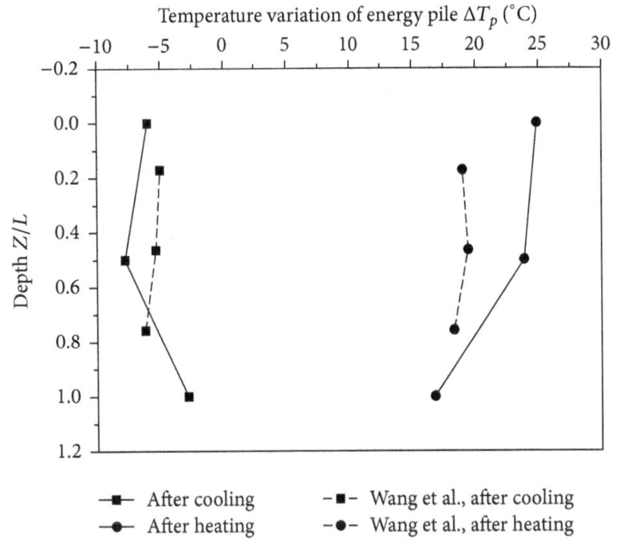

FIGURE 5: Comparison on temperature profiles of the pipe energy pile and solid energy pile.

5.2. Comparison of Load-Settlement Relationships with Solid Energy Pile. Conventionally, the ultimate shaft resistance along a pile can be estimated by the following:

$$\tau_s = K_o \sigma'_{vo} \tan\delta = \left(1 - \sin \emptyset'\right) \sigma'_{vo} \tan\delta, \quad (8)$$

where K_o is the earth pressure coefficient at rest; σ'_{vo} denotes the effective vertical stress; $\emptyset'$ is the internal friction angle; and δ is the friction angle at the pile-soil interface.

The ultimate shaft resistance can also be estimated by the following:

$$\tau_s = \sigma'_{ho} \tan\delta, \quad (9)$$

where σ'_{ho} is the horizontal stress.

The increase in pile shear stresses during heating can be estimated by the following:

$$\tau_s = \left(\sigma'_{ho} + \Delta\sigma'_{ho}\right) \tan\delta, \quad (10)$$

where $\Delta\sigma'_{ho}$ is the increment of effective horizontal stress caused by heating.

The heating caused the pile to be expanded horizontally, which increases the normal stress on the pile-soil interface. The increment of normal stress leads to the increase of shaft

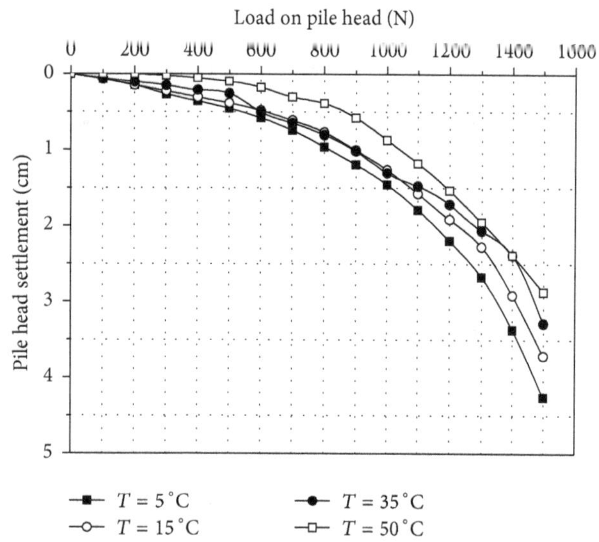

FIGURE 6: Curves on load-displacement of pile head over time under different heating conditions.

resistance at the pile-soil interface. The horizontal expansion of pile can be estimated using the following:

$$\Delta a = a_0 \alpha \Delta T, \tag{11}$$

where a_0 is the pile radius; α is the expansion coefficient of the material of pile (aluminum in the model test and concrete in the prototype); ΔT is the temperature increase of energy pile.

According to Yu [21], the limiting horizontal stress is

$$\Delta\sigma'_{\text{ho}} = \Delta\sigma'_{\text{lim}} = \frac{\left(b^2 - a_0{}^2\right)\left(\alpha' - 1\right)}{\left(1 + \alpha'\right) b^2 + \left(\alpha' - 1\right) a_0{}^2}\sigma'_{\text{ho}}, \tag{12}$$

where b ($\approx 15\, a_0$ in the model test) is the distance between the pile center and model boundary; and $\alpha' = (1 + \sin\emptyset')/(1 - \sin\emptyset')$.

The load-settlement curves in the model tests are shown in Figure 6. When the displacement of pile head reached 1 time of pile diameter (5 cm), the piles were thought to get into the failure condition. The results show that the bearing capacity increases a little with increasing temperature. When the pile settlement is less than 10% D (5 mm in the model test), all the model piles behaved linearly and the difference of the pile settlements is not significant. After that, higher bearing capacities were measured in the pile under higher temperatures. As mentioned above, when the temperature was raised, the pile would have a radius expansion Δa which increases the radial stress between pile-soil interface and thus the increase of shaft resistance. When the temperature of cyclic water in pipe is lower than the ambient temperature (the line "T = 15°C"), the radius expansion would convert to radius contraction and Δa will be negative. The dropping down of the shaft resistance follows. It can also be observed that, in the final stage, all the settlements have a similar gradient. That is because, with further increment in the axial load, the toe resistance started to take up more applied load; however, the toe resistance would not have a significant increment with the temperature since the sandy soil generally shows a limited affection by temperature changes. As shown in the figure, the results of 50°C show evident changes in the bearing capacity. This phenomenon may be caused by the preparation of the sandy soil. The surrounding soil for the test under 50°C was thought to have a higher relative density than the other tests.

The comparison of load-settlement profile with Wang et al. [22] was shown in Figure 7. The bearing capacities of the PEP and solid energy pile both decreased with the decreasing temperature. Assume that 20% of ultimate settlement was inflexion point. The bearing capacity of solid pile in Wang's test decreased by 6%, and bearing capacity of the PEP decreased by 11% after a −10°C temperature change. Figure 7(b) also shows the effect of increasing temperature on bearing capacity. The bearing capacity of the solid energy pile increased by 13%; during the same period, the bearing capacity of the PEP increased by 17% with a temperature increment of 35°C.

As results have shown, the PEP has a similar trend of change in bearing capacity with temperature variation. It is also indicated that bearing capacity of the PEP is likely more sensitive to the temperature variation compared with the traditional solid energy pile. It is partly due to the fact that the mobilized shaft resistance takes most part of the PEP's applied load and the end capacity has a little influence on the total bearing capacity, compared with solid energy pile.

5.3. Shaft Resistance on Pile-Soil Interface and Axial Strain. The pile placed in the ground could not expand (or contract) freely, owing to the shaft resistance between pile-soil interface and any end restraint. The measured axial strain $\varepsilon_{T\text{-Obs}}$ caused by temperature changes will be different from the pile which could expand freely. The restrained axial strain $\varepsilon_{T\text{-Rstr}}$ will be created for the pile in the ground, and it could be determined as follows [23]:

$$\varepsilon_{T\text{-Rstr}} = \varepsilon_{T\text{-Free}} - \varepsilon_{T\text{-Obs}}. \tag{13}$$

The restrained strain will also induce thermal stress in the pile, and the thermal stress is determined as follows:

$$\sigma_{T\text{-Rstr}} = E\left(\varepsilon_{T\text{-Obs}} - \alpha_C \Delta T\right). \tag{14}$$

The pile-soil displacement induced by temperature change will mobilize shaft resistance at the pile-soil interface. The mobilized shaft resistance $f_{s\text{-}s,\text{mob},j}$ for solid pile could be determined as follows:

$$f_{s\text{-}s,\text{mob},j} = \frac{\left(\sigma_{T\text{-Rstr},j} - \sigma_{T\text{-Rstr},j-1}\right) D}{4\Delta l}. \tag{15}$$

The toe of energy pipe pile in this test was sealed, and the mobilized shear stress $f_{s\text{-}p,\text{mob},j}$ for the energy pipe pile could be determined as follows:

$$f_{s\text{-}p,\text{mob},j} = \frac{\left(\sigma_{T\text{-Rstr},j} - \sigma_{T\text{-Rstr},j-1}\right)\left(D_{\text{out}}^2 - D_{in}^2\right)}{4D_{\text{out}}\Delta l}. \tag{16}$$

Figure 8 shows the mobilized shaft resistance of the PEP. When only mechanical load (working load 900 N)

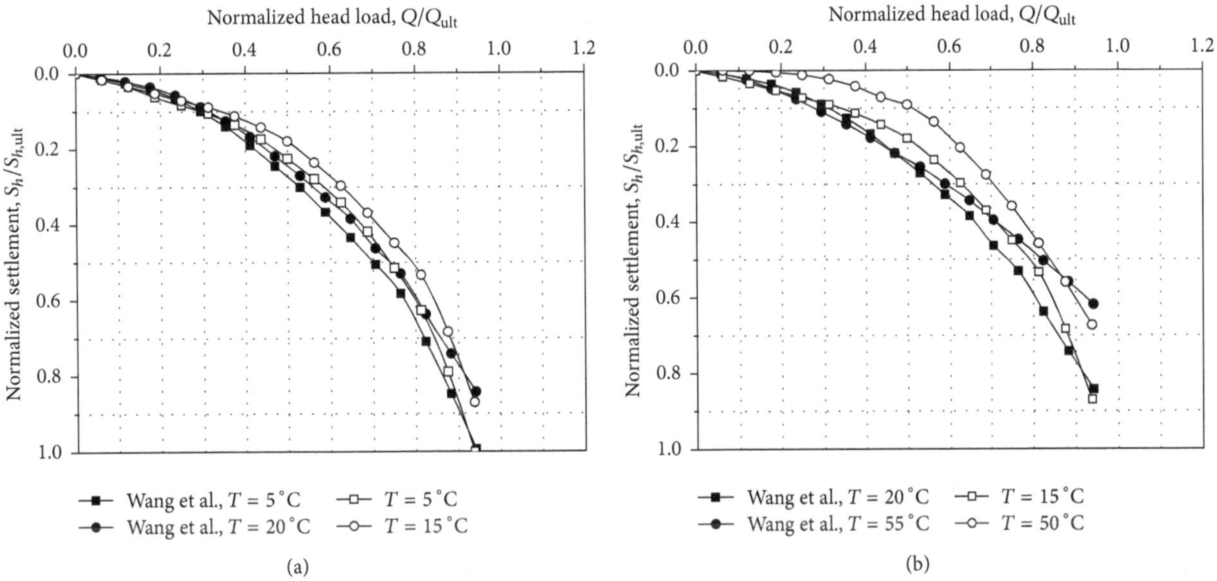

FIGURE 7: Comparison on normalized pile head settlement with the solid energy pile: (a) cooling; (b) heating.

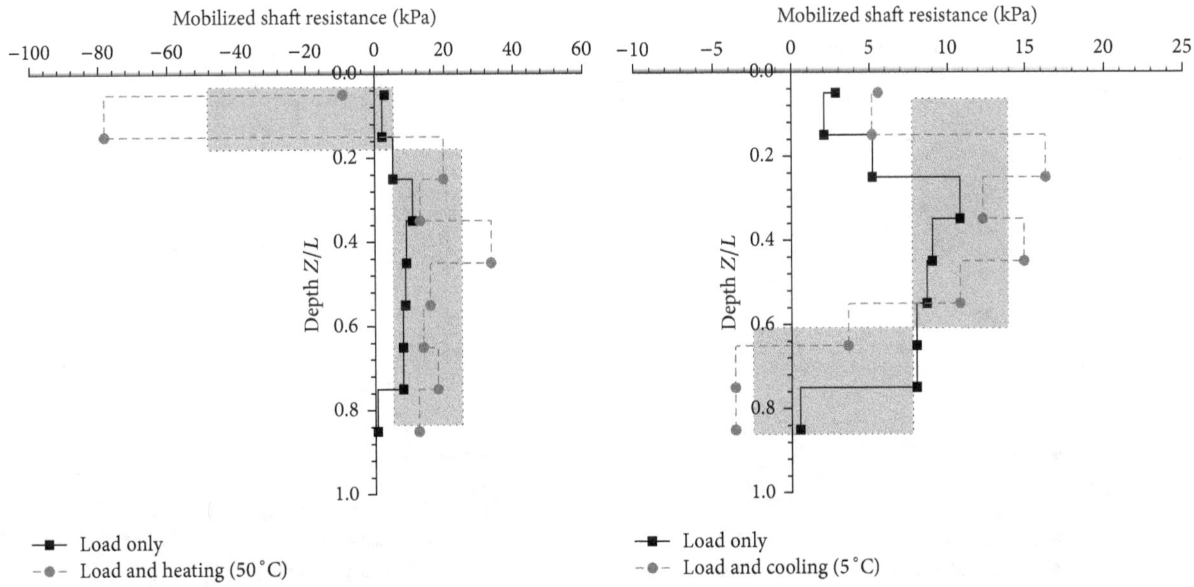

FIGURE 8: Mobilized shaft resistance distribution along the PEP under different temperatures: (a) heating; (b) cooling.

was applied, the maximum shaft resistance was 10.8 kPa, which occurred in depth of 0.35 m, and the minimum shaft resistance was 0.54 kPa near the pile toe. The shaft resistance along the pile did not change evidently; meanwhile, it was also not a constant as shown in Figure 8(a) (Bourne-Webb et al., 2013). The soil near pile toe and the loading on the pile top both have a restraint to the displacement between pile-soil interfaces, which is different with the freely expanding (contract) solid pile.

After heating, the shaft resistance near pile top changed to negative which means it was downwards. The maximum value of negative shaft resistance was −78.28 kPa. And the shaft stress under depth of 0.25 m was positive and larger than that with mechanical load only. The maximum and minimum positive shaft stress were 19.75 kPa and 12.54 kPa, respectively. The shaded area in Figure 6 represents change due to heating. As shown in Figures 8(a) and 9(a), the shaft stress of the PEP under heating and loading has a similar trend with the solid pile. Since the model PEP was made by aluminum, its thermal expansion was large compared to that made by concrete and the negative shaft stress near the top part was more obvious than the concrete solid pile in Bourne-Webb et al.'s test.

The comparison of shear stress under mechanical load only and cooling load was also shown in Figure 8(b). The

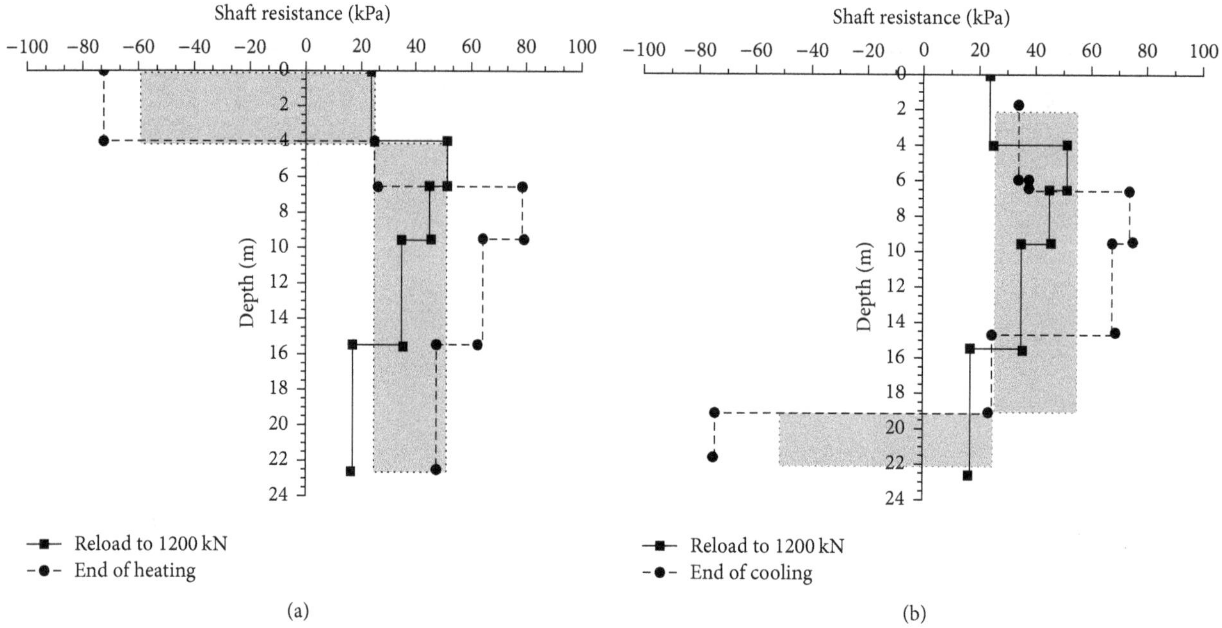

FIGURE 9: Shaft resistance distribution of the solid energy pile (Bourne-Webb et al., 2013): (a) heating; (b) cooling.

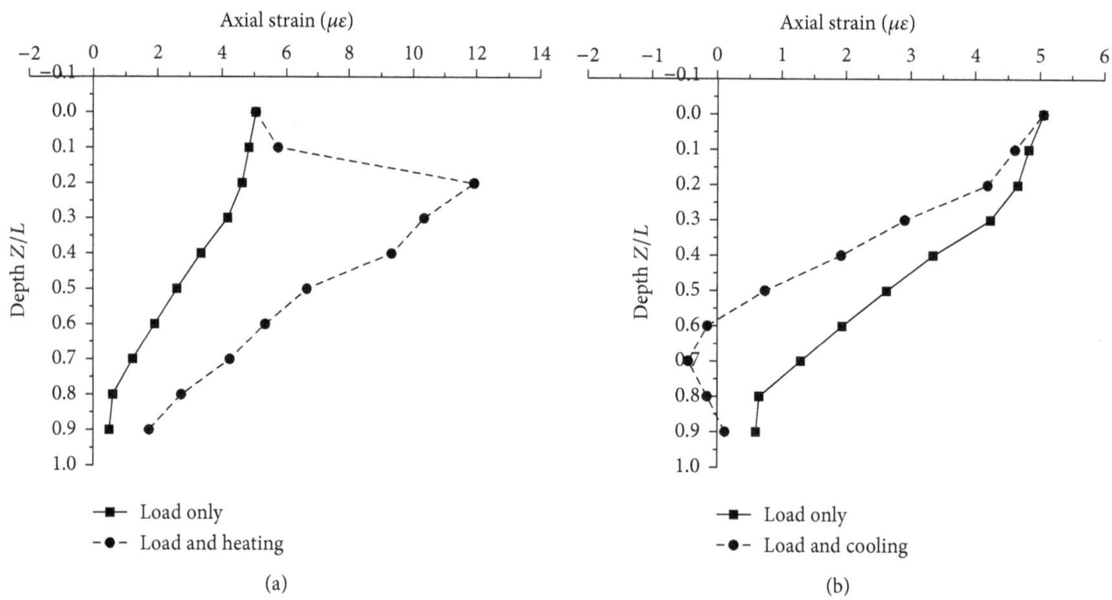

FIGURE 10: Axial strain distribution of the PEP: (a) heating; (b) cooling.

maximum shear stress and minimum shear stress were 16.3 kPa and −3.6 kPa, respectively. After cooling, the mobilized shear stress in the upper part of pile shows an increase relative to that under mechanical load only and reverses in the lower part of pile (from upwards to downwards). When the pile was cooled and contracted, the soil under pile and load part on the head of pile would not have an evident restraint to the pile. The shaded area shows the cooling effect on shaft stress, and it also shows a similar trend compared with the solid pile as shown in Figure 9(b).

The axial strain distribution is also shown in Figure 10. When only mechanical load was applied to the pile, the axial strain decreased with depth because the shaft resistance led to the decreasing of axial load. After heating, the axial strain increased along the pile depth, while after cooling, the axial strain decreased along the pile and the direction of axial strain even changed at the toe part of pile.

5.4. Displacement of Pile Head under Long-Term Thermal Effects. Figure 11 shows the displacement of pile head under a heating-cooling cycle. The pile head would heave when the pile was heated. However, after heating, the pile head would settle until the temperature of pile decreased to the room temperature. At the end of the process, the settlement of pile

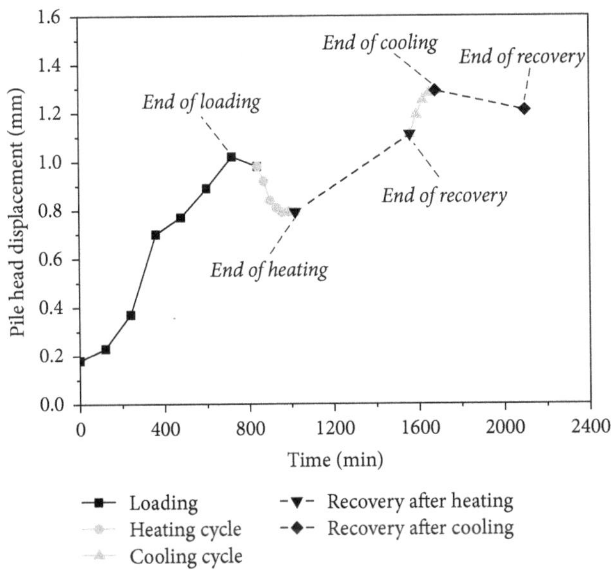

FIGURE 11: The pile head displacement under a heating-cooling cycle.

head is more than the heave displacement, which means that uncovered plastic displacement was created along the pile. The pile head would settle for the constriction of pile under decreasing temperature. When the temperature recovered, the displacements of the pile head would not return to the initial value. Plastic deformation also occurred along the pile in this period. The plastic deformation is likely caused by the contraction of soil may during the thermal cycles with temperature variation, which need to be studied further in the future.

6. Conclusions

In this study, load tests on four instrumented aluminum model piles were carried out to investigate thermal effects on the settlement patterns and capacities of single piles in saturated sand. Based on this study, the following conclusions can be drawn:

(1) With the increasing velocity of cyclic water, the static temperature of pile would also be higher. On the same time, after cooling or heating, the temperature variations of the PEP are larger than the solid energy pile.

(2) The bearing capacity of pipe energy pile showed similar trend with the solid energy pile under different temperature variation. It increased with the increasing temperature of cyclic water and decreased when the PEP was cooling by cyclic water. The PEPs behaved softer (bearing capacity decreased by 11%) after being cooled than the solid energy pile (bearing capacity decreased by 6%). On the same time, the bearing capacity of the PEP increased 17% with temperature increment of 35°C, which is more than the solid energy pile with 13% during the same period. It is indicated that PEPs seem to be more sensitive to the temperature variation than the solid energy pile.

(3) The shaft resistance along pile-soil interface was calculated and compared with the solid energy pile. The direction of the shaft resistance near the top part of pile converted after heating because of the upward displacement of pile head, and the shaft resistance under 0.25 m was positive and larger than the shaft resistance under load only. After cooling, the shaft resistance near top part increased with the contraction of the pile and then converted to negative near the toe of the pile. The mobilized shaft resistance profile is similar to the solid pile without restraints at both ends.

(4) During the full heating-cooling cycle, pile head heaved with increasing temperature and the upward displacements would decrease after heating. However, when the temperature recovers to the room temperature, there are still some unrecoverable displacements because the thermal expansion of pile toe inside the sand led to plastic deformation of the sand. On the following cooling period, the pile head displacements were downward because of the contraction under cooling. After recovering to the room temperature, plastic displacements were also observed due to the contraction of sand.

Conflicts of Interest

The authors declare that they have no conflicts of interest.

Acknowledgments

The author would like to acknowledge the National Natural Science Foundation of China (Grant no. 51378178) and the Research Innovation Program for College Graduates of Jiangsu Province (Grant no. KYZZ15_0145), which collectively funded this project.

References

[1] S. Kavanaugh, k. Rafferty, and M. Geshwiler, "Ground-source heat pumps-design of geothermal systems for commercial and industrial buildings," *ASHRAE*, article 167, 1997.

[2] P. Hughes, "Geothermal (Ground-Source)Heat Pumps: Market Status, Barriers to Adoption, and Actions to Overcome Barriers," Oak Ridge Nat. Lab. Report ORNL-2008/232, 2008.

[3] B. Saner, "Shallow geothermal energy," *GHC Bulletin*, vol. 22, no. 4, pp. 19–25, 2001.

[4] J. G. Agar, N. R. Morgenstern, and J. D. Scott, "Thermal expansion and pore pressure generation in oil sands.," *Canadian Geotechnical Journal*, vol. 23, no. 3, pp. 327–333, 1986.

[5] J. G. Agar, N. R. Morgenstern, and J. D. Scott, "Shear strength and stress-strain behaviour of Athabasca oil sand at elevated temperatures and pressures," *Canadian Geotechnical Journal*, vol. 24, no. 1, pp. 1–10, 1987.

[6] C. G. Olgun, T. Y. Ozudogru, and C. F. Arson, "Thermo-mechanical radial expansion of heat exchanger piles and possible effects on contact pressures at pile–soil interface," *Geotechnique Letters*, vol. 4, pp. 170–178, 2014.

[7] C. W. W. Ng, C. Shi, A. Gunawan, L. Laloui, and H. L. Liu, "Centrifuge modelling of heating effects on energy pile performance in saturated sand," *Canadian Geotechnical Journal*, vol. 52, no. 8, pp. 1-13, 2015.

[8] J. C. Goode and J. S. McCartney, "Centrifuge modeling of end-restraint effects in energy foundations," *Journal of Geotechnical and Geoenvironmental Engineering*, vol. 141, no. 8, article 04015034, 2015.

[9] L. Laloui, M. Moreni, and L. Vulliet, "Comportement d'un pieu bi-fonction, fondation et échangeur de chaleur," *Canadian Geotechnical Journal*, vol. 40, no. 2, pp. 388–402, 2003.

[10] L. Laloui, M. Nuth, and L. Vulliet, "Experimental and numerical investigations of the behaviour of a heat exchanger pile," *International Journal for Numerical and Analytical Methods in Geomechanics*, vol. 30, no. 8, pp. 763–781, 2006.

[11] P. J. Bourne-Webb, B. Amatya, K. Soga, T. Amis, C. Davidson, and P. Payne, "Energy pile test at lambeth college, London: geotechnical and thermodynamic aspects of pile response to heat cycles," *Geotechnique*, vol. 59, no. 3, pp. 237–248, 2009.

[12] K. D. Murphy, J. S. McCartney, and K. S. Henry, "Evaluation of thermo-mechanical and thermal behavior of full-scale energy foundations," *Acta Geotechnica*, vol. 10, no. 2, pp. 179–195, 2015.

[13] K. D. Murphy and J. S. McCartney, "Seasonal response of energy foundations during building operation," *Geotechnical and Geological Engineering*, vol. 33, no. 2, pp. 1–14, 2015.

[14] B. L. Amatya, K. Soga, P. J. Bourne-Webb, T. Amis, and L. Laloui, "Thermo-mechanical behaviour of energy piles," *Geotechnique*, vol. 62, no. 6, pp. 503–519, 2012.

[15] J. S. McCartney, "Engineering performance of energy foundations," in *Proceedings of the 14th Pan-American Conference on Soil Mechanics and Geotechnical Engineering*, Canadian Geotechnical Society, Ottawa, Canada, 2011.

[16] J. S. McCartney and J. E. Rosenberg, "Impact of heat exchange on side shear in thermo-active foundations," in *Proceedings of the Geo-Frontiers 2011: Advances in Geotechnical Engineering*, J. Han and D. E. Alzamora, Eds., pp. 488–498, Reston, VA, USA, March 2011.

[17] K. D. Murphy, J. S. McCartney, and H. K. Henry, "Thermomechanical characterization of a full-scale energy foundation," in *From Soil Behavior Fundamentals to Innovations in Geotechnical Engineering: Honoring Roy E. Olson*, M. Iskander, J. E. Garlanger, and M. H. Hussein, Eds., pp. 617–628, American Society of Civil Engineers, Reston, VA, USA, 2014.

[18] M. A. Stewart and J. S. McCartney, "Centrifuge modeling of soil-structure interaction in energy foundations," *Journal of Geotechnical and Geoenvironmental Engineering*, vol. 140, no. 4, article 04013044, 2014.

[19] J. C. Goode, M. Zhang, and J. S. McCartney, "Centrifuge modelling of energy foundations in sand," in *Proceedings of the 8th International Conference on Physical Modelling in Geotechnics 2014 (ICPMG2014)*, pp. 729–736, CRC Press, Perth, Australia, 2014.

[20] V. Fioravante, "On the shaft friction modelling of non-displacement piles in sand," *Soils and Foundations*, vol. 42, no. 2, pp. 23–33, 2002.

[21] H. Yu, *Cavity Expansion Methods in Geomechanics*, Springer, Dordrecht, Netherlands, 2000.

[22] C. L. Wang, H. L. Liu, G. Q. Kong, C. W. W. Ng, and Wu. Di, "Model tests of energy piles with and without vertical load," *Geotechnical Engineering*, vol. 3, no. 4, pp. 203–213, 2016.

[23] P. J. Bourne-Webb, B. Amatya, and K. Soga, "A framework for understanding energy pile behaviour," *Geotechnical Engineering*, vol. 166, no. 2, pp. 170–177, 2013.

13

An Exact Dynamic Stiffness Formulation for Predicting Natural Frequencies of Moderately Thick Shells of Revolution

Xudong Chen[1] **and Kangsheng Ye**[2]

[1]*School of Civil Engineering, Suzhou University of Science and Technology, Suzhou 215011, China*
[2]*Department of Civil Engineering, Tsinghua University, Beijing 100084, China*

Correspondence should be addressed to Xudong Chen; chenxd06@mails.tsinghua.edu.cn

Academic Editor: M. I. Herreros

An exact dynamic stiffness formulation is proposed for calculating the natural frequencies of shells of revolution based on the first-order Reissner-Mindlin theory. Equations of motion are reduced to be one-dimensional, should the circumferential wave number is specified, and then are rewritten in Hamilton form. Therefore, a shell of revolution with a moderate thickness is analysed as a special skeletal structure with five degrees of freedom at element ends. Dynamic stiffnesses are constructed directly from the one-dimensional governing vibration equations using classic skeletal theory. Number of natural frequencies below a specific value is counted by applying the Wittrick-Williams (W-W) algorithm, and exact eigenvalues can be simply obtained using bisection method. A solution to the number of clamped-end frequencies J_0 in the W-W algorithm is also proposed and proven to be reliable. Numerical examples on a variety of shells with different boundary conditions are investigated, and results are well compared and validated. Influences of a variety of shell parameters on natural frequencies of both cylindrical and spherical shells are discussed in detail, demonstrating that the proposed dynamic stiffness formulation is applicable to analyse natural frequencies of moderately thick shells of revolution with high accuracy.

1. Introduction

Among all the structure types, shells of revolution have been widely used in tubes, containment vessels, aircraft fuselages, etc. Their natural frequencies are important parameters when performing further dynamic structural analysis. Therefore, exact free vibration analysis of shells of revolution is of considerable scientific and engineering significance.

As the simplest shell of revolution, circular cylindrical shells are the most widely employed in engineering applications as they are easy to manufacture. Since Love [1] established the basis of thin elastic shell theory, loads of effort have been devoted to the free vibration analysis of circular cylindrical shells over decades [2, 3]. Besides circular cylindrical shells, other types of shells like conical, spherical, elliptical, parabolic, and hyperbolic shells are also frequently used in civil, mechanical, aeronautical, and marine engineering. The earliest study on free vibration of elliptical shells can be traced back to Lamb [4] in 19th century. Following researchers [5, 6] investigated the vibration behaviour of both prolate and oblate spheroidal shells. Lin and Lee [7] and Hoppmann et al. [8] studied the free vibration of paraboloidal shells. Carter et al. [9] presented a finite element approach to calculate the natural frequencies of hyperbolic shells of revolution. Li et al. [10] solved the natural frequencies of conical shells with Hamilton's principle and Rayleigh-Ritz method. In recent decades, general solutions to free vibration of shells of revolution were provided; e.g., Luah and Fan [11] analysed the free vibration of shells of revolution by introducing the spline finite element method; Tan [12] predicted the natural frequencies of general shells of revolution using a substructuring approach.

However, all the studies above are based on thin shell theories. For shells with moderate thickness, the employment of Reissner-Mindlin assumptions with the consideration of transverse shear deformation and rotary inertia is necessary. There are some researches conducted with moderately thick shell theories. Naghdi and Cooper [13] analysed the free vibration of circular cylindrical shells including shear deformation and rotary inertia using an elastic wave propagation approach. Mirsky and Herrmann [14] investigated nonaxially symmetric vibration motions of moderately thick circular cylindrical shells. Relevant research was collected by Leissa in his monograph [15].

Recently, Hosseini-Hashemi and Fadaee [16] obtained the closed-form solutions to the free vibration of moderately thick spherical shell panels using new auxiliary and potential functions in conjunction with the separation method of variables. Su et al. [17] analysed the free vibration of moderately thick functionally graded open shells with general boundary conditions using the Rayleigh-Ritz method and Fourier series. Tornabene and Viola [18] conducted free vibration examination on parabolic shells with Generalized Differential Quadrature (GDQ) method based on the first-order shear deformation theory. For even thicker shells, Kang and Leissa [19] employed three-dimensional theory to analyse the free vibration of thick hyperbolic shells with Ritz method. Al-Khatib and Buchanan [20] investigated the free vibration of a parabolic shell of revolution using the three-dimensional finite element method.

Unlike thin shells, free vibration of moderately thick shells of revolution with general meridian shapes and arbitrary boundary conditions is not that extensively investigated due to its complexity in governing equations. In this paper, a dynamic stiffness approach for the free vibration of moderately thick shells of revolution is formulated in conjunction with the Wittrick-Williams (W-W) algorithm [21, 22]. It treats structures as continuous systems with infinite degrees of freedom and dynamic stiffnesses are calculated directly from vibration governing equations. Natural frequencies are solved computationally by employing the W-W algorithm together with the bisection method. Since the governing partial differential equations (PDE) are degraded analytically into ordinary differential equations (ODE) when calculating the dynamic stiffnesses, the proposed formulation is more accurate over traditional finite element approaches where the differential equations are discretized into linear algebraic equations.

The dynamic stiffness method and the W-W algorithm have been used in free vibration analysis of skeletal structures such as nonuniform Timoshenko beams [23], three-layered sandwich beams [24] and functionally graded beams [25]. However, limited attention is paid to the plates and shells from the computational community. Pagani et al. [26] analysed the free vibration of thin-walled structures with exact dynamic stiffness elements. El-Kaabazi and Kennedy [27] applied the dynamic stiffness method and the W-W algorithm to calculate the frequencies and modes of thin circular cylindrical shells with variable thickness.

The present authors proposed the idea of dynamic stiffness method in investigating the free vibration of shells of revolution [28], which was summarised in [29] based on the classic thin shell theory. Meanwhile, endeavours on predicting the free vibration characteristics of moderately thick circular cylindrical shells [30] and elliptical shells [31] were made. A further comparison study [32] on the dynamic stiffness approach between thin and moderately thick circular cylindrical shells was conducted, showing that frequencies from moderately thick circular cylindrical shells can be used as alternatives to those from thin shells with no observation of shear locking when the thickness to radius ratio is small.

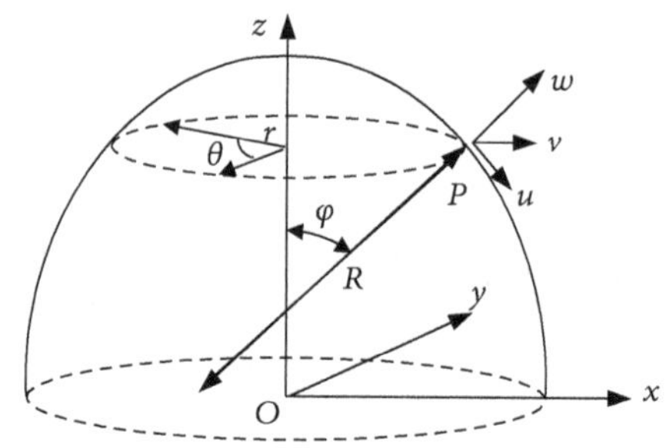

FIGURE 1: A general shell of revolution and the coordination.

In this paper, a concise but complete formulation for free vibration of moderately thick shells of revolution with general shaped meridians and arbitrary boundary conditions is presented. Since the number of circumferential half-waves is an integer, the original vibration partial differential equations are degraded into series of ordinary differential equations, and hence W-W algorithm is employed. A solution of J_0 in the W-W algorithm is provided. Numerical results on natural frequencies of moderately thick circular cylindrical shells, elliptical shells, and parabolic and hyperbolic shells are given, showing that the proposed dynamic stiffness formulation is applicable and accurate.

2. Equations of Motion

Consider a homogeneous, isotropic, and circumferentially closed shell of revolution with moderate thickness (see Figure 1). Referring to Figure 1, u, v, and w are the translational displacements along the meridian, circumferential, and radial directions on the middle surface at point P. φ is the angle between the normal at point P and z-axis, and θ is the angle along the circumferential direction. r and R are the radii of the parallel circle and meridian curvature, respectively. Young's modulus is E, Poisson's ratio is ν, density is ρ, and the thickness is h.

Based on the Reissner-Mindlin hypothesis, vibration of a moderately thick shell is represented by the middle surface displacement vector $\{u, v, w, \psi_\varphi, \psi_\theta\}^T$, where ψ_φ and ψ_θ are the angular displacements of the middle surface along φ and θ directions, respectively.

The strain-displacement relationship of a shell of revolution is given in

$$\{\boldsymbol{\varepsilon}\} = [\mathbf{L}]\{\Delta\} \quad (1)$$

where the strain vector $\{\boldsymbol{\varepsilon}\} = \{\varepsilon_\varphi, \varepsilon_\theta, \varepsilon_{\varphi\theta}, \chi_\varphi, \chi_\theta, \chi_{\varphi\theta}, \gamma_\varphi, \gamma_\theta\}^T$, and the displacement vector $\{\Delta\} = \{u, v, w, \psi_\varphi, \psi_\theta\}^T$. The differential matrix $[\mathbf{L}]$ can be referred to in [15].

Similarly, the force-strain relationship is

$$\{\mathbf{N}\} = [\mathbf{D}]\{\boldsymbol{\varepsilon}\} \quad (2)$$

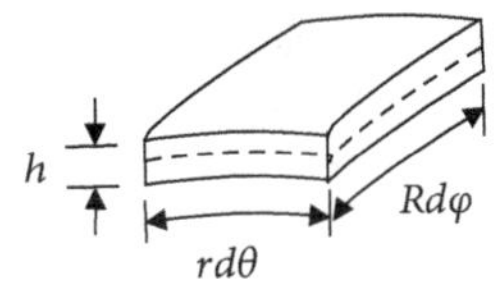

FIGURE 2: An infinitesimal portion in a moderately thick shell of revolution.

where the internal force vector $\{\mathbf{N}\} = \{N_\varphi, N_\theta, N_{\varphi\theta}, M_\varphi, M_\theta, M_{\varphi\theta}, Q_\varphi, Q_\theta\}^{\mathrm{T}}$ and $[\mathbf{D}]$ is the stiffness matrix

$$[\mathbf{D}] = \begin{bmatrix} \mathbf{D}_1 & 0 \\ 0 & \mathbf{D}_2 \end{bmatrix} \tag{3}$$

where

$$[\mathbf{D}_1] = \frac{Eh}{1-\nu^2}\begin{bmatrix} 1 & \nu & 0 & 0 & 0 & 0 \\ & 1 & 0 & 0 & 0 & 0 \\ & & \dfrac{1-\nu}{2} & 0 & 0 & 0 \\ & & & \dfrac{h^2}{12} & 0 & 0 \\ & sym. & & & \dfrac{h^2}{12} & 0 \\ & & & & & \dfrac{h^2(1-\nu)}{24} \end{bmatrix} \tag{4}$$

$$[\mathbf{D}_2] = \begin{bmatrix} \overline{\kappa}Gh & 0 \\ 0 & \overline{\kappa}Gh \end{bmatrix}. \tag{5}$$

In (5), $\overline{\kappa} = \pi^2/12$, G is the shear modulus, and $G = E/2(1+\nu)$.

Referring to Figure 2, dA denotes the area of an infinitesimal portion in the middle surface of a moderately thick shell of revolution, and $dA = rRd\varphi d\theta$. The total strain energy U for free vibration of a moderately thick shell can be obtained by integrating the middle surface S as

$$U = \iint_S \frac{1}{2}\{\boldsymbol{\varepsilon}\}^{\mathrm{T}}[\mathbf{D}]\{\boldsymbol{\varepsilon}\}\,dA \tag{6}$$

and the corresponding total kinetic energy is

$$T = \frac{\rho}{2}\iint_\Omega \left\{ h\left[\left(\frac{\partial u}{\partial t}\right)^2 + \left(\frac{\partial v}{\partial t}\right)^2 + \left(\frac{\partial w}{\partial t}\right)^2\right] + \frac{h^3}{12}\left[\left(\frac{\partial \psi_x}{\partial t}\right)^2 + \left(\frac{\partial \psi_\theta}{\partial t}\right)^2\right]\right\} dA. \tag{7}$$

Hamilton's principle suggests the total strain energy and kinetic energy satisfy

$$\delta \int_{t_1}^{t_2} (U - T)\,dt = 0 \tag{8}$$

where time interval (t_1, t_2) is arbitrary. Displacement functions u, v, w, ψ_φ, and ψ_θ can be assumed as

$$\begin{aligned} u &= u_n(\varphi)\cos n\theta \sin \omega t \\ v &= v_n(\varphi)\sin n\theta \sin \omega t \\ w &= w_n(\varphi)\cos n\theta \sin \omega t \\ \psi_\varphi &= \psi_{\varphi n}(\varphi)\cos n\theta \sin \omega t \\ \psi_\theta &= \psi_{\theta n}(\varphi)\sin n\theta \sin \omega t \\ & \qquad n = 0, 1, 2\ldots \end{aligned} \tag{9}$$

where n is the circumferential wave number and ω is a circular frequency. Equation (9) is widely accepted and suitable for general vibration analysis of moderately thick shells of revolution. The Symplectic components (generalized displacements and forces) can be obtained through displacement substitution and integration manipulation. Similar to [29], when integrating along the meridian of a shell with a positive Gaussian curvature, generalized displacements and forces obtained from (8) are given by

$$\{u\} = \begin{cases} u_1 = u_n \\ u_2 = v_n \\ u_3 = w_n \\ u_4 = \psi_{\varphi n} \\ u_5 = \psi_{\theta n} \end{cases} \tag{10}$$

$\{v\}$

$$= \begin{cases} v_1 = \dfrac{Eh\pi}{1-\nu^2}\left[\nu\cos\varphi u_n + \dfrac{r}{R}u_n' + \nu n v_n + \left(\nu\sin\varphi + \dfrac{r}{R}\right)w_n\right] \\ v_2 = \dfrac{Eh\pi}{1+\nu}\left[-\left(\dfrac{n}{2} + \dfrac{h^2}{24}\dfrac{n\sin\varphi}{rR}\right)u_n - \left(\dfrac{\cos\varphi}{2} + \dfrac{h^2}{24}\dfrac{\sin\varphi\cos\varphi}{rR}\right)v_n + \left(\dfrac{r}{2R} + \dfrac{h^2}{24}\dfrac{\sin^2\varphi}{rR}\right)v_n' - \dfrac{h^2}{24}\dfrac{n\sin\varphi}{r}\psi_{\varphi n} - \dfrac{h^2}{24}\dfrac{\sin\varphi\cos\varphi}{r}\psi_{\theta n} + \dfrac{h^2}{24}\dfrac{\sin\varphi}{R}\psi_{\theta n}'\right] \\ v_3 = \overline{\kappa}Gh\pi\left[-\dfrac{r}{R}u_n + \dfrac{r}{R}w_n' + r\psi_{\varphi n}\right] \\ v_4 = \dfrac{Eh\pi}{1-\nu^2}\dfrac{h^2}{12}\left[\nu\cos\varphi\psi_{\varphi n} + \dfrac{r}{R}\psi_{\varphi n}' + \nu n\psi_{\theta n}\right] \\ v_5 = \dfrac{Eh\pi}{1+\nu}\dfrac{h^2}{24}\left[-\dfrac{n}{R}u_n - \dfrac{\cos\varphi}{R}v_n + \dfrac{\sin\varphi}{R}v_n' - n\psi_{\varphi n} - \cos\varphi\psi_{\theta n} + \dfrac{r}{R}\psi_{\theta n}'\right] \end{cases} \tag{11}$$

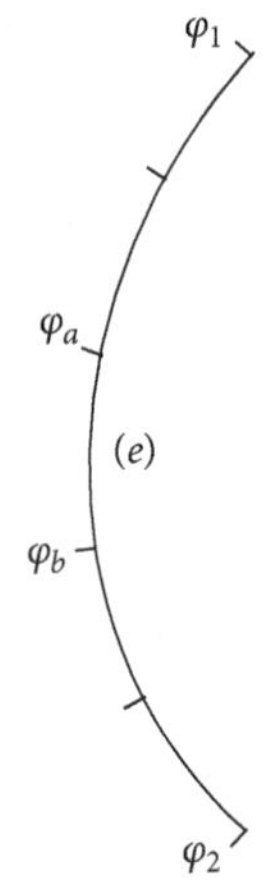

FIGURE 3: Mesh divisions along the meridian.

where $()'$ denotes the differentiation with respect to φ. The free vibration governing equations can be obtained in the Hamilton form as

$$[\mathbf{J}]\{\mathbf{z}'\} = [\mathbf{S}]\{\mathbf{z}\} \tag{12}$$

where $\{\mathbf{z}\} = \{\{u\}^{\mathrm{T}}\ \{v\}^{\mathrm{T}}\}^{\mathrm{T}}$; $[\mathbf{J}] = [\begin{smallmatrix} \mathbf{0} & -\mathbf{I} \\ \mathbf{I} & \mathbf{0} \end{smallmatrix}]$, $\mathbf{I}$ is a fifth-order identity matrix; $[\mathbf{S}]$ is a symmetric tenth-order matrix. For shells of revolution with a positive Gaussian curvature, $[\mathbf{S}] = [\mathbf{S_P}]$. For those with a negative Gaussian curvature, $[\mathbf{S}] = [\mathbf{S_N}]$. Details on $[\mathbf{S_P}]$ and $[\mathbf{S_N}]$ are given in the Appendix.

3. Dynamic Stiffness

Governing vibration equations in (12) is one-dimensional with respect to φ when circumferential wave number n is known. Consequently, generalized one-dimensional skeletal theory can be seamlessly extended to the two-dimensional moderately thick shells of revolution. Therefore, a moderately thick shell of revolution is treated as a special 'beam' with five degrees of freedom $\{u_1\ u_2\ u_3\ u_4\ u_5\}^{\mathrm{T}}$ and five corresponding internal force components $\{v_1\ v_2\ v_3\ v_4\ v_5\}^{\mathrm{T}}$.

Figure 3 illustrates a possible mesh division of a moderately thick shell of revolution, which is divided into couple of shell elements along its meridian. Taking the instance of element (e), the element-end displacement vector $\{\mathbf{d}\}^e$ and force vector $\{\mathrm{F}\}^e$ are

$$\begin{aligned}\{\mathbf{d}\}^e = \{&u_1(\varphi_a), u_2(\varphi_a), u_3(\varphi_a), u_4(\varphi_a), u_5(\varphi_a),\\ &u_1(\varphi_b), u_2(\varphi_b), u_3(\varphi_b), u_4(\varphi_b), u_5(\varphi_b)\}^{\mathrm{T}}\end{aligned} \tag{13}$$

and

$$\begin{aligned}\{\mathrm{F}\}^e = \{&-v_1(\varphi_a), -v_2(\varphi_a), -v_3(\varphi_a), -v_4(\varphi_a),\\ &-v_5(\varphi_a), v_1(\varphi_b), v_2(\varphi_b), v_3(\varphi_b), v_4(\varphi_b), v_5(\varphi_b)\}^{\mathrm{T}}\end{aligned} \tag{14}$$

where φ_a and φ_b are angle coordinates at the element ends shown in Figure 3, respectively.

Should boundary conditions in (15) be applied to (12) in turn

$$\{\mathbf{d}\}^e = \{e_j\}, \quad j = 1, 2 \ldots, 10 \tag{15}$$

($\{e_j\}$ is a unit vector with the j-th element equals to one), the corresponding element dynamic stiffness matrix $[\mathbf{k_i}]^e$ is obtained as

$$[\mathbf{k_i}]^e = [\{\mathrm{F}\}^e_{j=1}\ \{\mathrm{F}\}^e_{j=2}\ \{\mathrm{F}\}^e_{j=3}\ \{\mathrm{F}\}^e_{j=4}\ \{\mathrm{F}\}^e_{j=5}\ \{\mathrm{F}\}^e_{j=6}\ \{\mathrm{F}\}^e_{j=7}\ \{\mathrm{F}\}^e_{j=8}\ \{\mathrm{F}\}^e_{j=9}\ \{\mathrm{F}\}^e_{j=10}]. \tag{16}$$

Equation (12) is too complicated to be solved analytically and hence an adaptive ODE solver COLSYS is employed. Details on COLSYS can be referred to in [37, 38]. After obtaining element dynamic stiffness matrix $[\mathbf{k_i}]^e$, global dynamic stiffness matrix $\mathbf{K}$ is assembled in a direct way along the meridian with regular element location vectors.

4. Wittrick-Williams Algorithm

The Wittrick-Williams algorithm is a counting algorithm which gives the number of frequencies below a trial value $\omega*$ as

$$J = J_0 + J_K \tag{17}$$

where J is the total number of frequencies below $\omega*$; $J_K = s\{\mathbf{K}\}$ is the sign count of the global dynamic stiffness matrix $\mathbf{K}$ and equals to the number of negative elements on the diagonal of an upper triangular matrix obtained from $\mathbf{K}$ by applying standard Gaussian elimination without row exchange; J_0 is the number of clamped-end frequencies exceeded by $\omega*$ and can be obtained by accumulating J_0^e of each element over the whole meridian as

$$J_0 = \sum J_0^e \tag{18}$$

Taking advantage of the self-adaptability of COLSYS, J_0^e can be calculated using a substructuring method. When computing $[\mathbf{k_i}]^e$ of element (e), submesh (termed as $(\hat{e})$) is defined by COLSYS as is schematically shown in Figure 4.

Applying the Wittrick-Williams algorithm again on submesh $(\hat{e})$,

$$J_0^e = \sum J_0^{(\hat{e})} + s\{\mathbf{K}^{(\hat{e})}\} \tag{19}$$

since COLSYS is capable of controlling error tolerance due to its self-adaptability, the number of clamped-end frequencies

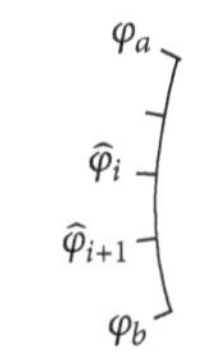

FIGURE 4: Submesh by COLSYS on element (e).

below $\omega*$ on submesh $(\widehat{e})$ has to be zero, indicating $J_0^{(\widehat{e})} = 0$. Otherwise, the clamped-end modes can be enlarged infinitely, resulting in unsatisfaction of any given error tolerance.

For a submesh element $(\widehat{\varphi}_i, \widehat{\varphi}_{i+1})$, its dynamic stiffness matrix $[\widehat{\mathbf{k}}_{\mathbf{i}}]^{(\widehat{e})}$ can be obtained by linearly combining ten solutions of (12) and (15). The element-end displacements and force vectors for submesh are denoted as $\{\mathbf{d}\}^{(\widehat{e})}$ and $\{\mathbf{F}\}^{(\widehat{e})}$, respectively, the submesh element stiffness matrix $[\widehat{\mathbf{k}}_{\mathbf{i}}]^{(\widehat{e})}$ satisfies

$$\left[\widehat{\mathbf{k}}_{\mathbf{i}}\right]^{(\widehat{e})} \{\mathbf{d}\}^{(\widehat{e})} = \{\mathbf{F}\}^{(\widehat{e})}. \tag{20}$$

Applying the boundary conditions in (15) in turn again, (20) is expanded as

$$\left[\widehat{\mathbf{k}}_{\mathbf{i}}\right]^{(\widehat{e})} [\mathbf{B}]^{(\widehat{e})} = [\mathbf{C}]^{(\widehat{e})} \tag{21}$$

where

$$[\mathbf{B}]^{(\widehat{e})} = \left[\{\mathbf{d}\}_{j=1}^{(\widehat{e})}\ \{\mathbf{d}\}_{j=2}^{(\widehat{e})}\ \{\mathbf{d}\}_{j=3}^{(\widehat{e})}\ \{\mathbf{d}\}_{j=4}^{(\widehat{e})}\ \{\mathbf{d}\}_{j=5}^{(\widehat{e})}\ \{\mathbf{d}\}_{j=6}^{(\widehat{e})}\ \{\mathbf{d}\}_{j=7}^{(\widehat{e})}\ \{\mathbf{d}\}_{j=8}^{(\widehat{e})}\ \{\mathbf{d}\}_{j=9}^{(\widehat{e})}\ \{\mathbf{d}\}_{j=10}^{(\widehat{e})}\right] \tag{22}$$

and

$$[\mathbf{C}]^{(\widehat{e})} = \left[\{\mathbf{F}\}_{j=1}^{(\widehat{e})}\ \{\mathbf{F}\}_{j=2}^{(\widehat{e})}\ \{\mathbf{F}\}_{j=3}^{(\widehat{e})}\ \{\mathbf{F}\}_{j=4}^{(\widehat{e})}\ \{\mathbf{F}\}_{j=5}^{(\widehat{e})}\ \{\mathbf{F}\}_{j=6}^{(\widehat{e})}\ \{\mathbf{F}\}_{j=7}^{(\widehat{e})}\ \{\mathbf{F}\}_{j=8}^{(\widehat{e})}\ \{\mathbf{F}\}_{j=9}^{(\widehat{e})}\ \{\mathbf{F}\}_{j=10}^{(\widehat{e})}\right] \tag{23}$$

Since $J_0^{(\widehat{e})} = 0$, $[\mathbf{B}]^{(\widehat{e})}$ is nonsingular (otherwise $\omega*$ is one of a clamped-end frequency, which is conflicted with $J_0^{(\widehat{e})} = 0$), the submesh dynamic stiffness matrix $[\widehat{\mathbf{k}}_{\mathbf{i}}]^{(\widehat{e})}$ is formulated as

$$\left[\widehat{\mathbf{k}}_{\mathbf{i}}\right]^{(\widehat{e})} = [\mathbf{C}]^{(\widehat{e})} \left([\mathbf{B}]^{(\widehat{e})}\right)^{-1}. \tag{24}$$

Similarly, $\mathbf{K}^{(\widehat{e})}$ can be assembled regularly from $[\widehat{\mathbf{k}}_{\mathbf{i}}]^{(\widehat{e})}$, and $J_0^e = s\{\mathbf{K}^{(\widehat{e})}\}$ according to (19).

By employing the Wittrick-Williams algorithm repeatedly, an approximate upper and lower bound of each frequency can be predicted. Simple bisection method can then be employed to obtain exact natural frequencies. Based on the proposed theory, a FORTRAN code was written by the authors to determine natural frequencies of moderately thick shells of revolution with various meridian shapes and different boundary conditions.

5. Numerical Examples

Numerical examples on natural frequencies of moderately thick circular cylindrical shells, elliptical shells, and parabolic and hyperbolic shells are presented in this section using the proposed approach. Through comparison with results from existing literature, the exact dynamic stiffness formulation is proven to be reliable and accurate.

5.1. Circular Cylindrical Shells. For circular cylindrical shells, r is constant, $\varphi = 90^\circ$ and $R = \infty$, the meridian coordinate is changed to the distance coordinate s along the meridian. The equation of motion in (12) is still applicable with the form of

$$[\mathbf{J}] \frac{\mathrm{d}\{\mathbf{z}\}}{\mathrm{d}s} = [\mathbf{S}]\{\mathbf{z}\}. \tag{25}$$

5.1.1. Convergence of J_0. A moderately thick circular cylindrical shell with length L = 20m, radius r = 2m, thickness h = 0.2m, Young's modulus E=200GPa, Poisson's ratio ν=0.3, and density ρ=7800kg/m^3 is examined. The shell is divided equally along the meridian. J_0, J_K, and J in the W-W algorithm with different number of shell elements ne at n=1, $\omega*$=3000 rad/s are tabulated in Table 1. It can be observed that the

TABLE 1: Frequency counting of a moderately thick circular cylindrical shell with different ne under n=1, $\omega*$=3000 rad/s.

ne	J_0	J_K	$J = J_0 + J_K$
2	16	4	20
3	15	5	20
4	12	8	20
5	10	10	20
8	8	12	20
10	10	10	20
11	0	20	20
12	0	20	20
15	0	20	20

TABLE 2: Natural frequencies of a circular cylindrical shell ($\times\omega_0$).

h/r	m		n = 1	2	3	4	5
0.1	1	Present	1.06348	0.88147	0.80659	0.89439	1.11575
		Ref. [33]	1.06142	0.88158	0.80860	0.89819	1.12156
		η	0.19%	-0.01%	-0.25%	-0.42%	-0.52%
	2	Present	2.37850	2.72107	3.15851	3.66689	4.22897
		Ref. [33]	2.37417	2.71556	3.15287	3.66166	4.22432
		η	0.18%	0.20%	0.18%	0.14%	0.11%
	3	Present	3.96689	4.49462	5.24988	6.14424	7.12431
		Ref. [33]	3.96347	4.48736	5.23619	6.12162	7.08999
		η	0.09%	0.16%	0.26%	0.37%	0.48%
0.2	1	Present	1.19118	1.09378	1.18201	1.46592	1.88124
		Ref. [33]	1.18622	1.09885	1.19644	1.48941	1.91475
		η	0.42%	-0.46%	-1.21%	-1.58%	-1.75%
	2	Present	2.39136	2.73821	3.17697	3.68507	4.24608
		Ref. [33]	2.37498	2.71703	3.15526	3.66486	4.22776
		η	0.69%	0.78%	0.69%	0.55%	0.43%
	3	Present	3.96726	4.49576	5.25153	6.14593	6.93506
		Ref. [33]	3.95328	4.46562	5.19327	6.04694	6.97100
		η	0.35%	0.67%	1.12%	1.64%	-0.52%

solution to J_0 proposed in this paper converges to zero with the increase of ne, and the proposed submesh-based solution to J_0 presented in Section 4 is reliable.

5.1.2. A Simply Supported Circular Cylindrical Shell. Khalili et al. [33] calculated natural frequencies of a thick simply supported circular cylindrical shell using a three-dimensional refined higher-order theory. The geometry of the shell is L/r=1.0 and Poisson's ratio ν=0.3. Table 2 lists the lowest three natural frequencies under different h/r and n from both the present dynamic stiffness formulation and [33], where ω_0 is a frequency parameter as $\omega_0 = (1/L)\sqrt{E/\rho(1+\nu)}$. m denotes the order under the corresponding n. Define the relative error η as

$$\eta = \frac{\Omega_{\text{current}} - \Omega_{\text{literature}}}{\Omega_{\text{literature}}} \times 100\% \qquad (26)$$

where Ω_{current} and $\Omega_{\text{literature}}$ are the frequencies obtained from the current formulation and existing literature, respectively. The relative error η between the present formulation and [33] are also tabulated in Table 2. Sign '-' denotes the obtained value from the dynamic stiffness formulation is smaller than that from literature.

Referring to Table 2, frequencies from the present dynamic stiffness formulation agree excellently well with those in [33] when h/r=0.1, where relative errors are well less than 1.0%. For h/r=0.2, good agreement is still achieved except some individual data points whose errors are between 1.0% and 2.0%, which is still acceptable. In general, very good agreement is achieved between the present method and [33], demonstrating the proposed dynamic stiffness formulation is exact.

5.2. Elliptical Shells. The meridian equation of an elliptical shell in OXZ plane is given by

$$\frac{x^2}{a^2} + \frac{z^2}{b^2} = 1 \qquad (27)$$

where a and b are constants. Radius of meridian curvature R and radius of parallel circles r can be written with respect to φ as

$$R = a^2b^2\left(\frac{1+\cot^2\varphi}{a^2+b^2\cot^2\varphi}\right)^{3/2} \qquad (28)$$

$$r = \frac{a}{\sqrt{1+(b^2/a^2)\cot^2\varphi}}. \qquad (29)$$

In (28) and (29), φ is the only independent variable, R and r can be defined continuously and smoothly within each shell element along the meridian, guaranteeing an accurate solution to free vibration of elliptical shells should the proposed dynamic stiffness formulation is used.

5.2.1. Clamped-Free Spherical Shell Caps. Should a=b in (27), an elliptical shell becomes a standard spherical shell where R is constant and r=$R\sin\varphi$. Gautham and Ganesan [34] analysed the free vibration of moderately thick hemispherical shell caps with 30° and 60° cutout shown in Figure 5. The boundary condition is clamped-free.

The present authors calculated the lowest three nondimensional frequency parameters Ω ($\Omega = \omega a\sqrt{\rho/E}$) under different circumferential wave number n when h/a=0.2 using the proposed formulation. In this example, Poisson's ratio is 0.3. Results from both the dynamic stiffness approach and

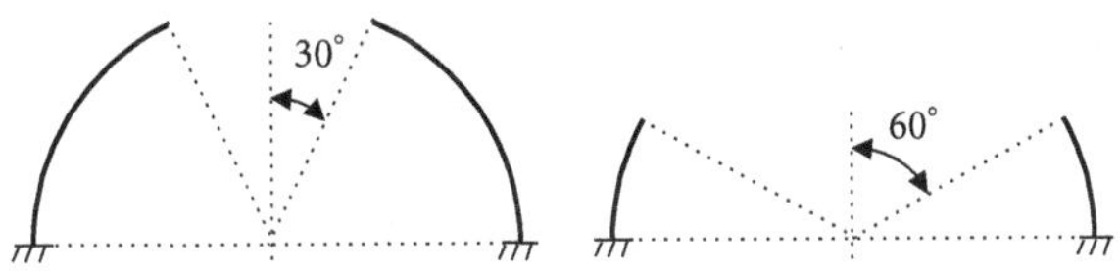

FIGURE 5: Diagram of hemispherical caps with cutout.

TABLE 3: Ω for hemispherical caps with cutout.

n	m	30° Present	30° Ref. [34]	30° η	60° Present	60° Ref. [34]	60° η
0	1	1.138	1.133	0.44%	1.276	1.289	-1.01%
	2	1.309	1.297	0.93%	2.033	2.013	0.99%
	3	1.420	1.427	-0.49%	2.729	2.747	-0.66%
1	1	0.751	0.753	-0.27%	1.143	1.152	-0.78%
	2	1.291	1.295	-0.31%	2.348	2.337	0.47%
	3	1.980	1.982	-0.10%	2.796	2.810	-0.50%
2	1	0.648	0.653	-0.77%	1.043	1.053	-0.95%
	2	1.668	1.688	-1.18%	2.794	2.822	-0.99%
	3	2.020	2.022	-0.10%	3.167	3.150	0.54%
3	1	1.137	1.157	-1.73%	1.141	1.154	-1.13%
	2	2.062	2.087	-1.20%	3.055	3.099	-1.42%
	3	2.437	2.435	0.08%	3.635	3.634	0.03%
4	1	1.682	1.719	-2.15%	1.425	1.441	-1.11%
	2	2.525	2.557	-1.25%	3.356	3.407	-1.50%
	3	3.115	3.107	0.26%	3.784	3.782	0.05%
5	1	2.187	2.235	-2.15%	1.832	1.853	-1.13%
	2	3.103	3.151	-1.52%	3.721	3.772	-1.35%
	3	3.857	3.860	-0.08%	4.020	4.025	-0.12%

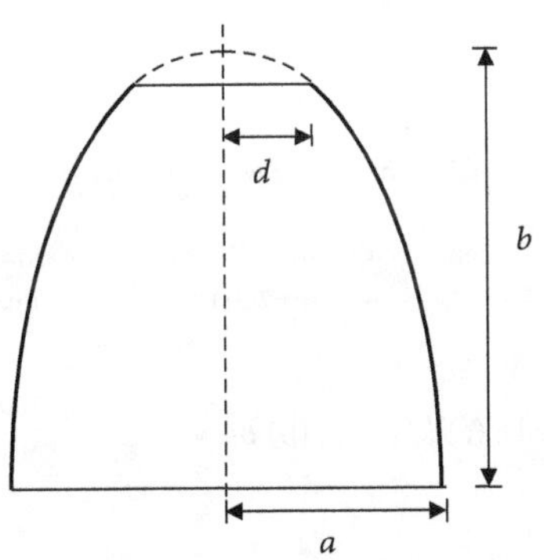

FIGURE 6: The diagram of a free-free elliptical shell.

[34] are given opposite in Table 3 for comparison. It can be observed that both data agree very well and most relative errors η are less than 1.0%. In general, most frequencies are close to results in existing literature at a 10^{-2} magnitude.

5.2.2. Free-Free Elliptical Shells. Shim and Kang [35] studied free-free elliptical shells (see Figure 6) with three-dimensional elastic theory and Ritz method. In this case, d=0.2a and Poisson's ratio ν=0.3. Table 4 gives the lowest three nonzero frequency parameters Ω ($\Omega = \omega a\sqrt{\rho/G}$ and G is the shear modulus) of both prolate and oblate elliptical shells with n from 0 to 6 when h/a=0.1.

Referring to Table 4, results from the dynamic stiffness formulation agree very well with those in [35]. This verifies that the dynamic stiffness formulation proposed in this paper is effective and accurate.

Further error analysis on frequency spectrum with different value of b/a is presented in Figure 7. It is understood that the differences between frequencies from the present method and [35] are extremely minor regardless of circumferential wave number n. Frequency curves from the dynamic stiffness formulation and [35] are almost identical, no matter the shape of the elliptical shell is prolate or oblate.

5.3. Parabolic Shells. For parabolic shells, the meridian equation in OXZ plane is

$$z = c\left[1 - \left(\frac{x}{a}\right)^2\right] \tag{30}$$

where a and c are nonzero constants. Similar to the elliptical shells, radius of meridian curvature R and the parallel circle r corresponding to coordinate φ can also be defined and can be referred to mathematical handbooks.

TABLE 4: Frequencies of free-free ellipsoidal shells in Ω with different b/a.

n	m	b/a=1/2 Present	b/a=1/2 Ref. [35]	b/a=2 Present	b/a=2 Ref. [35]
0	1	1.069	1.060	1.557	1.559
	2	1.658	1.620	1.652	1.652
	3	2.921	2.876	1.891	1.890
1	1	1.240	1.230	1.356	1.358
	2	1.977	1.933	1.620	1.616
	3	2.555	2.539	1.942	1.931
2	1	0.2321	0.2223	0.1521	0.1505
	2	1.445	1.419	0.9069	0.8924
	3	1.767	1.771	1.443	1.423
3	1	0.5702	0.5456	0.4085	0.4066
	2	2.049	2.020	1.070	1.061
	3	3.121	3.108	1.654	1.639
4	1	0.9882	0.9476	0.7527	0.7509
	2	2.651	2.611	1.343	1.328
	3	4.080	4.049	1.961	1.936
5	1	1.473	1.416	1.175	1.175
	2	3.306	3.254	1.764	1.746
	3	4.948	4.912	2.417	2.385
6	1	2.017	1.944	1.670	1.672
	2	4.017	3.953	2.290	2.270
	3	5.819	5.787	2.980	2.946

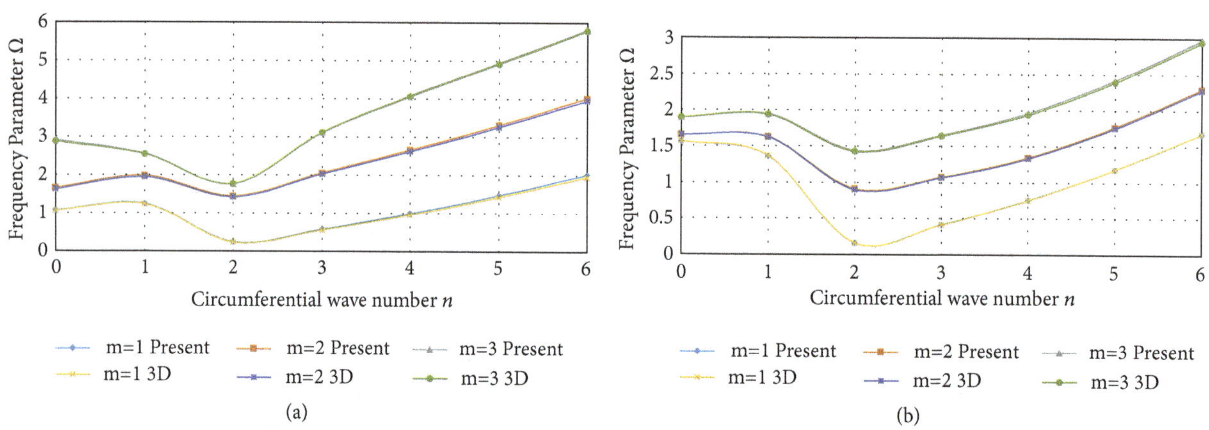

FIGURE 7: Comparison on frequency spectrum of elliptical shells: (a) b/a=1/2; (b) b/a=2.

5.3.1. A Clamped-Free Parabolic Dome. Tornabene and Viola [18] calculated natural frequencies of a clamped-free parabolic dome using the Generalized Differential Quadrature (GDQ) method. The geometry and material properties of the examined parabolic dome are a=4m, c=2.133m, height H=2m, thickness h=0.1m, density ρ=7800kg/m^3, Young's modulus E=2.1×10^{11}N/m^2, and Poisson's ratio ν=0.3. The lowest five natural frequencies calculated from the present dynamic stiffness formulation and [18] are tabulated in Table 5. It is observed that frequencies obtained from the present dynamic stiffness formulation agree very well with data from [18], showing good reliability and accuracy of the proposed formulation.

5.3.2. Apex-Closed Parabolic Caps. Al-Khatib and Buchanan [20] examined the free vibration of clamped parabolic shell caps with ν=0.3, h=1/20c. Apexes of parabolic caps are closed. A tiny circular hole with the size of $\varphi_0 = 0.005°$ is employed at the apexes so that singularity of differential matrix [**L**] in (1) can be avoided whilst not reducing accuracy. Table 6 lists the first natural frequencies of each n when a=c. Table 7 shows the first three frequencies of n=0 and 1 when a=2c. Frequencies are presented in the form of nondimensional parameter Ω, where $\Omega = \omega c\sqrt{\rho/G}$.

According to Tables 6 and 7, good agreement is achieved between the present method and [20], demonstrating that the proposed dynamic stiffness formulation is capable of

TABLE 5: Comparison on frequencies of a clamped-free parabolic dome (Hz).

Total order	n	Present	Ref. [18]	η
1	2	91.68	91.24	0.48%
2	3	132.33	131.29	0.79%
3	4	154.79	154.91	-0.08%
4	1	160.63	157.99	1.67%
5	5	168.19	169.46	-0.75%

TABLE 6: Frequencies in Ω for an apex-closed parabolic cap with a=c.

n	Present	Ref [20]	η
0	1.798	1.774	1.35%
1	1.162	1.159	0.26%
2	1.180	1.166	1.20%
3	1.157	1.133	2.12%
4	1.283	1.240	3.47%

TABLE 7: Frequencies in Ω for an apex-closed parabolic cap with a=2c.

n	m	Present	Ref [20]	η
	1	0.6998	0.6955	0.62%
0	2	0.8004	0.7896	1.37%
	3	0.9263	0.9177	0.94%
	1	0.6058	0.6043	0.25%
1	2	0.7325	0.7252	1.01%
	3	0.8793	0.8612	2.10%

analysing the free vibration of apex-closed parabolic shells of revolution.

5.4. A Hyperbolic Cooling Tower Shell. Cooling towers are commonly found in civil engineering infrastructures, e.g., power plants. Their meridian can be approximated as a hyperbolic curve. Based on the illustration of Figure 8, the meridian equation of a cooling tower shell in OXZ plane is given by

$$\frac{x^2}{a^2} - \frac{z^2}{b^2} = 1. \quad (31)$$

Accordingly, the radius of meridian curvature R and parallel circle r corresponding to coordinate φ are defined as

$$R = a^2 b^2 \left(\frac{1 + \cot^2\varphi}{a^2 - b^2\cot^2\varphi} \right)^{3/2} \quad (32)$$

$$r = \frac{a}{\sqrt{1 - (b^2/a^2)\cot^2\varphi}}. \quad (33)$$

In this section, natural frequencies of a cooling tower at Didcot Power Station [36] are calculated. Actually the cooling tower is assumed to be a general shell of revolution; however, its basic shape can be approximated as a hyperboloid. Parameters of the hyperboloid are a=25.1205m,

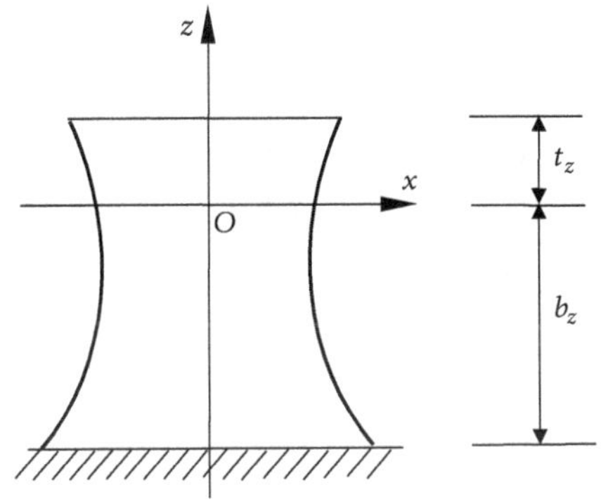

FIGURE 8: A hyperbolic shell and its coordinate system.

b=60.2019m, t_z=20.7264m, and b_z=93.5736m. Thickness of the hyperbolic shell is h=0.1777m. The properties of material are E=2.7579×10^{10}N/m^2, ρ=2323kg/m^3, ν=0.2. The boundary condition is clamped-free.

Strictly speaking, the cooling tower should be considered as a thin shell since h/a or h/b is well smaller than 1/20, which is the boundary between thin and thick shells. However, since the proposed dynamic stiffness formulation is exact, it is capable of producing accurate natural frequencies even if the shell is very thin. For comparison, results obtained from the dynamic stiffness formulation using current moderately thick shell theory, thin shell theory (previously proposed by present authors in [29]), as well as those in [36] by Deb Nath using ring shell elements are tabulated in Table 8.

Referring to Table 8, conclusion can be reached that frequencies predicted both from the moderately thick and from thin dynamic stiffness formulae agree very well with data in [36]. A further comparison on the error of $\eta_{\text{thick-[38]}}$ and $\eta_{\text{thin-[38]}}$ shows that results obtained with moderately thick dynamic stiffness formulation enjoys a much better precision over their thin counterpart, especially when circumferential wave number n increases. This observation suggests that although the proposed dynamic stiffness formulation is based on moderately thick shell theory, the present method can be applied to predict natural frequencies of thin shells, without observation of any shear locking. As a result, the exactness of the proposed method is validated.

The actual Didcot tower is slightly different from the approximation of a hyperboloid. Figure 9 plots the frequency curves calculated both as a moderately thick hyperbolic shell with present formulation and a general shell of revolution (with slightly different meridian shape based on the actual construction drawings as is explained in [36])

Table 8: Frequencies of Didcot cooling tower (Hz).

n	m	Thick	Thin	Deb Nath [36]	$\eta_{thick\text{-}[38]}$	$\eta_{thin\text{-}[38]}$
0	1	6.7463	6.7463	6.7493	-0.04%	-0.04%
	2	8.1912	8.1913	8.1956	-0.05%	-0.05%
	3	12.110	12.111	12.129	-0.16%	-0.15%
1	1	3.4558	3.4560	3.4575	-0.05%	-0.04%
	2	7.0329	7.0331	7.0364	-0.05%	-0.05%
	3	11.004	11.005	11.011	-0.06%	-0.05%
2	1	1.9364	1.9394	1.9373	-0.05%	0.11%
	2	3.6550	3.6567	3.6571	-0.06%	-0.01%
	3	6.9282	6.9292	6.9320	-0.05%	-0.04%
3	1	1.5935	1.6027	1.5942	-0.04%	0.53%
	2	2.0114	2.0192	2.0125	-0.05%	0.33%
	3	4.1627	4.1661	4.1653	-0.06%	0.02%
4	1	1.2721	1.2909	1.2730	-0.07%	1.41%
	2	1.8300	1.8488	1.8308	-0.04%	0.98%
	3	2.6961	2.7042	2.6979	-0.07%	0.23%
5	1	1.3206	1.3468	1.3218	-0.09%	1.89%
	2	1.6820	1.7104	1.6837	-0.10%	1.59%
	3	2.4524	2.4710	2.4536	-0.05%	0.71%
6	1	1.5014	1.5253	1.5030	-0.11%	1.48%
	2	1.8584	1.8942	1.8604	-0.11%	1.82%
	3	2.3832	2.4112	2.3861	-0.12%	1.05%
7	1	1.6851	1.7048	1.6868	-0.10%	1.07%
	2	2.2235	2.2546	2.2263	-0.13%	1.27%
	3	2.5470	2.5861	2.5501	-0.12%	1.41%
8	1	1.8771	1.8953	1.8792	-0.11%	0.86%
	2	2.5006	2.5263	2.5038	-0.13%	0.90%
	3	3.1195	3.1548	3.1236	-0.13%	1.00%

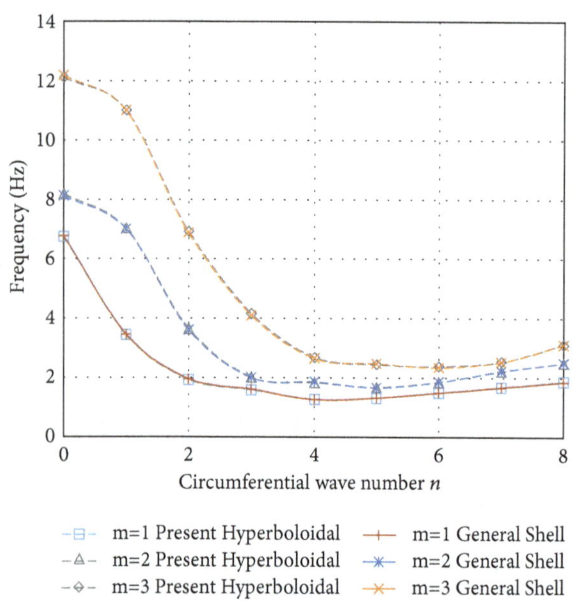

Figure 9: Comparison on frequencies obtained as a hyperboloid and a general shell.

using FEM in [36]. It can be observed that results agree very well for different n and m combinations, and plots for strict hyperboloid and general shell are almost identical. This further demonstrates that the assumption of a cooling tower as a hyperboloid is acceptable, and results from the proposed dynamic stiffness formulation are highly reliable.

6. Further Discussions

In this section, a further study is conducted to investigate the influences of parameters (e.g., radius, thickness, material properties, etc.) on natural frequencies of cylindrical and spherical shells.

6.1. Cylindrical Shells. Consider a clamped-clamped (C-C) circular cylindrical shell with length L=20m. Baseline parameters are radius r=2m, thickness h=0.2m, Poisson's ratio ν=0.3, Young's modulus E=200GPa, and density ρ=7800kg/m^3. When one of these parameters changes, the rest of them are kept unaltered. To better illustrate the inherent relations between frequencies and shell

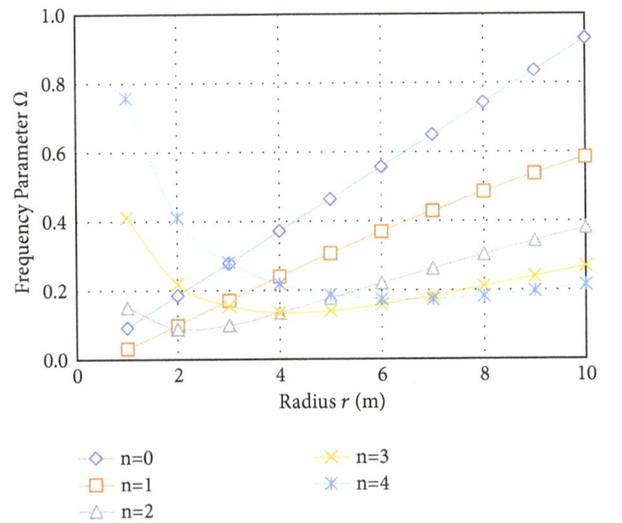

FIGURE 10: Trends of the lowest natural frequency parameter Ω against radius r.

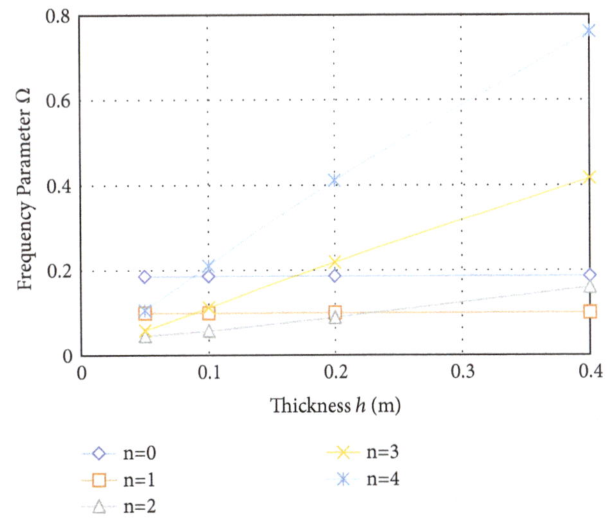

FIGURE 11: Trends of the lowest natural frequency parameter Ω against thickness h.

parameters, a nondimensional frequency parameter Ω is defined as

$$\Omega = \omega r\sqrt{\frac{\rho\left(1-\nu^2\right)}{E}}. \tag{34}$$

Therefore, irrelevant parameters such as Young's modulus E and density ρ to Ω can be ignored.

Figure 10 shows the trends of the lowest natural frequency parameter Ω with circumferential wave number n from 0 to 4 against radius r. Referring to Figure 10, frequencies with different n exhibit differently with the increase of radius r. When n is small, e.g., n=0 and 1, the lowest frequency increases almost linearly with the increase of r. However, this trend is not true any longer for n=2, 3, and 4, as the lowest natural frequency for a smaller r is much higher than that when r increases.

Figure 11 plots the trends of Ω against thickness h. It can be observed that, for a small n (n=0 and 1), increase of thickness h has little effect on the value of Ω, and both curves are almost horizontal. However, a monotonic linear increase trend is obtained for n=2, 3, and 4. A possible conclusion can be reached that vibration modes with n=0 and 1 is not sensitive to thickness h.

Influence of Poisson's ratio ν on frequency parameter Ω is given in Figure 12. Referring to Figure 12, a minor effect from Poisson's ratio ν on the frequency parameter Ω is obtained when it varies.

6.2. Spherical Shells. Here a clamped-free (C-F) spherical shell is considered (Figure 13). Same as the cylindrical shell studied in the previous section, baseline geometry and material parameters for this spherical shell are defined as follows: radius a=1.0m, thickness h=0.15m (Figure 15), density ρ=2700kg/m^3, Poisson's ratio ν=0.3, and Young's modulus E=70GPa.

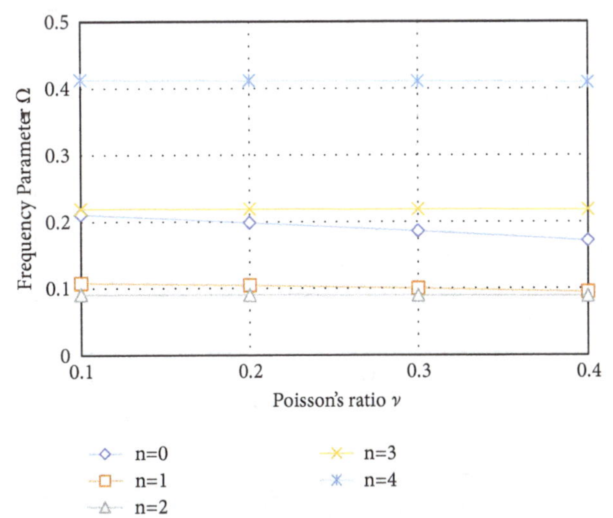

FIGURE 12: Trends of the lowest natural frequency parameter Ω against Poisson's ratio ν.

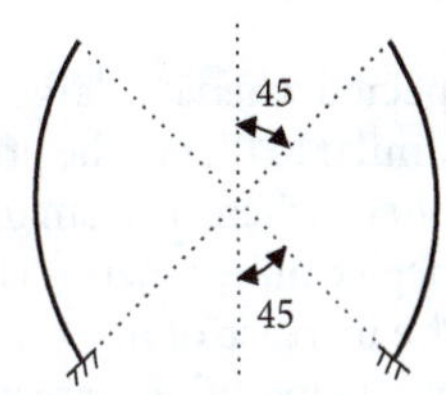

FIGURE 13: Illustration of a clamped-free spherical shell.

Engineering frequencies f (in Hz) are computed in this section and thus both Young's modulus E and density ρ play a role in the obtained natural frequencies. Parameters a, h, ρ, and E are investigated. Figure 14 shows the change of natural frequency f against radius a. The frequency decreases dramatically when a increases, and a converged value with

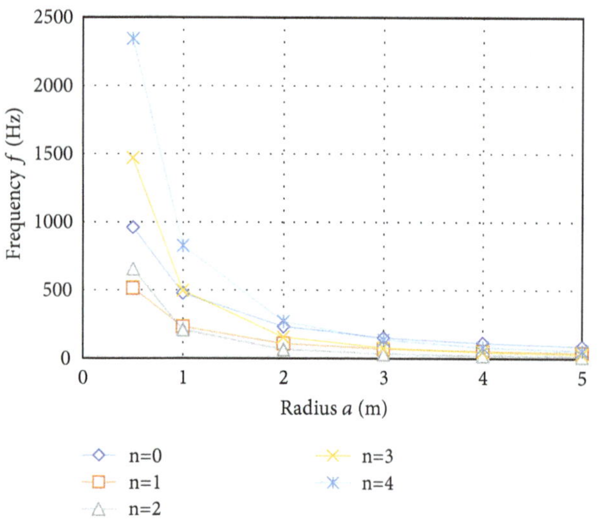

FIGURE 14: Trends of natural frequencies of a spherical shell against radius a.

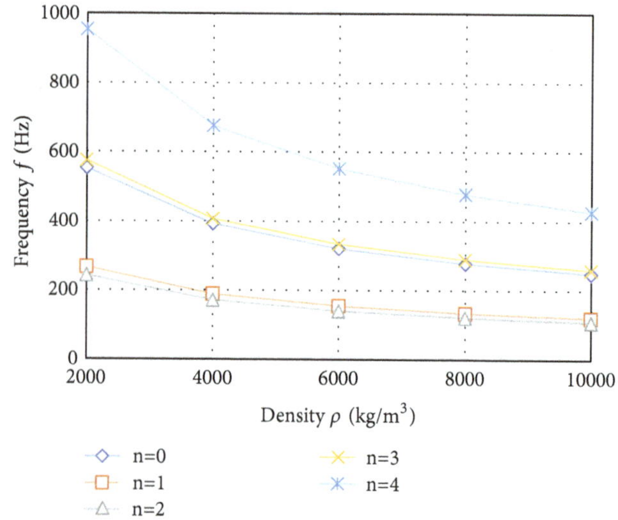

FIGURE 16: Trends of natural frequencies of a spherical shell against density ρ.

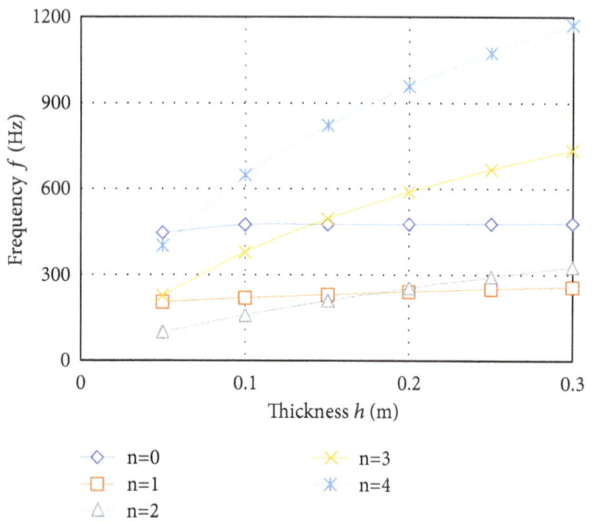

FIGURE 15: Trends of natural frequencies of a spherical shell against thickness h.

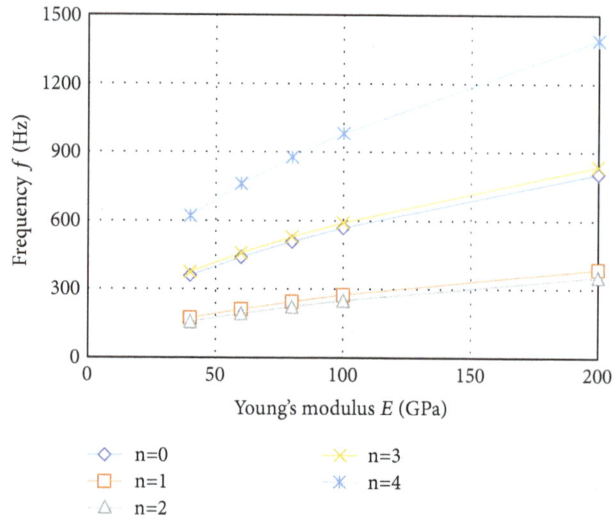

FIGURE 17: Trends of natural frequencies of a spherical shell against Young's modulus E.

different n can be expected when a is large enough. For the f-h relation, it is quite similar to that in the previously examined circular cylindrical shell: when n is small (n=0 and 1), the influence is almost neglectable; when n is large (n=2, 3 and 4), f increases with the increase of h.

Density ρ and Young's modulus E are vital to engineering frequency f. Figure 16 shows a decrease trend of f with the increase of density ρ. This can be simply interpreted. With the increase of density ρ, the mass matrix increases accordingly. Since the stiffness matrix is not changed, frequency decreases naturally according to the general eigenvalue definitions. Regarding f-E relation, plotting is given in Figure 17 and similar conclusion applies to the spherical shell. Furthermore, it is worth mentioning that the smallest frequency of a shell is not necessarily obtained with a smallest n, i.e., n=0. On the contrary, it is determined by the vibration mode that is the easiest to occur, which is usually n=2 in this example and other shell analysis with normal dimensions. It can also be summarised that the inherent characteristics of free vibration of moderately thick shells of revolution is revealed by the proposed dynamic stiffness formulation.

7. Conclusions

Free vibration behaviour and natural frequencies of shells of revolution are investigated using an exact dynamic stiffness formulation with the consideration of transverse shear deformation and rotary inertia. The W-W algorithm is employed onto the dynamic stiffnesses and natural frequencies are obtained accurately. Since the W-W algorithm is mathematically strict, no eigenvalue will be omitted. The number of clamped-end frequencies J_0 in the W-W algorithm is given, proving the proposed formulation is complete and converged. When computing dynamic stiffnesses, governing vibration PDEs are degraded into a set of ODEs analytically, and a spline ODE solver COLSYS is used to calculate the dynamic stiffness matrices numerically. Therefore, the accuracy of obtained natural frequencies can be guaranteed and the proposed method is exact.

Numerical examples on natural frequencies of a variety moderately thick shells of revolution including circular cylindrical, spherical, elliptical, and parabolic and hyperbolic shells are presented. Shells with different boundary conditions such as simply supported, clamped, and free ends are modelled. Results are compared with data in existing literature, and very good agreement is achieved. In general, the exact dynamic stiffness formulation for free vibration of moderately thick shells of revolution presented in this paper is reliable and accurate. Though equations are derived from Reissner-Mindlin assumptions, the formulation well applies to very thin shells. Furthermore, it outperforms its counterpart based on thin shell theory when comparing with data in open literature, without observation of shear locking.

Further discussions on the influences of cylindrical and spherical shell parameters such as radius, thickness, Young's modulus, Poisson's ratio, and density are presented. It is found that the smallest natural frequency is determined by the vibration mode which is the easiest to occur (usually obtained with n=2 for shells with a normal dimension). The proposed dynamic stiffness formulation is proven to be a useful tool for analysing free vibration characteristics of moderately thick shells of revolution.

Appendix

$$[\mathbf{S}_\mathrm{P}] = \begin{bmatrix} s_{11} & s_{12} & s_{13} & 0 & 0 & s_{16} & s_{17} & s_{18} & 0 & s_{1*10} \\ & s_{22} & s_{23} & 0 & s_{25} & s_{26} & s_{27} & 0 & 0 & s_{2*10} \\ & & s_{33} & 0 & s_{35} & s_{36} & 0 & 0 & 0 & 0 \\ & & & s_{44} & s_{45} & 0 & 0 & s_{48} & s_{49} & s_{4*10} \\ & & & & s_{55} & 0 & 0 & 0 & s_{59} & s_{5*10} \\ & & & & & s_{66} & 0 & 0 & 0 & 0 \\ & sym. & & & & & s_{77} & 0 & 0 & s_{7*10} \\ & & & & & & & s_{88} & 0 & 0 \\ & & & & & & & & s_{99} & 0 \\ & & & & & & & & & s_{10*10} \end{bmatrix} \tag{A.1}$$

$$[\mathbf{S}_\mathrm{N}] = \begin{bmatrix} -s_{11} & -s_{12} & -s_{13} & 0 & 0 & s_{16} & s_{17} & s_{18} & 0 & s_{1*10} \\ & -s_{22} & -s_{23} & 0 & -s_{25} & s_{26} & s_{27} & 0 & 0 & s_{2*10} \\ & & -s_{33} & 0 & -s_{35} & s_{36} & 0 & 0 & 0 & 0 \\ & & & -s_{44} & -s_{45} & 0 & 0 & s_{48} & s_{49} & s_{4*10} \\ & & & & -s_{55} & 0 & 0 & 0 & s_{59} & s_{5*10} \\ & & & & & -s_{66} & 0 & 0 & 0 & 0 \\ & sym. & & & & & -s_{77} & 0 & 0 & -s_{7*10} \\ & & & & & & & -s_{88} & 0 & 0 \\ & & & & & & & & -s_{99} & 0 \\ & & & & & & & & & -s_{10*10} \end{bmatrix} \tag{A.2}$$

where nonzero elements in the upper triangular domain are

$$s_{11} = \rho\pi\omega^2 rRh - Eh\pi\frac{R}{r}\cos^2\varphi,$$

$$s_{12} = -Eh\pi n\frac{R}{r}\cos\varphi,$$

$$s_{13} = -Eh\pi\frac{R}{r}\sin\varphi\cos\varphi,$$

$$s_{16} = -\nu\frac{R}{r}\cos\varphi,$$

$$s_{17} = n\frac{R}{r},$$

$$s_{18} = 1,$$

$$s_{1*10} = \frac{n}{r}\left(1 - \frac{R}{r}\sin\varphi\right),$$

$$
\begin{aligned}
s_{22} &= \rho\pi\omega^2 rRh - Eh\pi n^2\frac{R}{r} - \overline{\kappa}Gh\pi\frac{R}{r}\sin^2\varphi,\\
s_{23} &= -\frac{R}{r}nh\pi\sin\varphi\,[E+\overline{\kappa}G],\\
s_{25} &= \overline{\kappa}Gh\pi R\sin\varphi,\\
s_{26} &= -\nu n\frac{R}{r},\\
s_{27} &= \frac{R}{r}\cos\varphi,\\
s_{2*10} &= \frac{\cos\varphi}{r}\left(1-\frac{R}{r}\sin\varphi\right),\\
s_{33} &= \rho\pi\omega^2 rRh - Eh\pi\frac{R}{r}\sin^2\varphi - \overline{\kappa}Ghn^2\pi\frac{R}{r},\\
s_{35} &= \overline{\kappa}Gh\pi nR,\\
s_{36} &= -\left(1+\nu\frac{R}{r}\sin\varphi\right),\\
s_{44} &= \rho\pi\omega^2 rR\frac{h^3}{12} - \pi\frac{Eh^3}{12}\frac{R}{r}\cos^2\varphi,\\
s_{45} &= -Eh\pi n\frac{h^2}{12}\frac{R}{r}\cos\varphi,\\
s_{48} &= -R,\\
s_{49} &= -\nu\frac{R}{r}\cos\varphi,\\
s_{4*10} &= n\frac{R}{r},\\
s_{55} &= \rho\pi\omega^2 rR\frac{h^3}{12} - \overline{\kappa}Gh\pi rR - \pi n^2\frac{Eh^3}{12}\frac{R}{r},\\
s_{59} &= -\nu n\frac{R}{r},\\
s_{5*10} &= \frac{R}{r}\cos\varphi,\\
s_{66} &= \frac{R}{r}\frac{1-\nu^2}{Eh\pi},\\
s_{77} &= \frac{2R}{r}\frac{1+\nu}{Eh\pi},\\
s_{7*10} &= -\frac{2R\sin\varphi}{r^2}\frac{1+\nu}{Eh\pi},\\
s_{88} &= \frac{R}{r}\frac{1}{\overline{\kappa}Gh\pi},\\
s_{99} &= \frac{R}{r}\frac{12}{h^2}\frac{1-\nu^2}{Eh\pi},\\
s_{10*10} &= \frac{1+\nu}{Eh\pi}\frac{R}{r}\left(\frac{24}{h^2}+\frac{2\sin^2\varphi}{r^2}\right).
\end{aligned}
\tag{A.3}
$$

Conflicts of Interest

The authors declare that they have no conflicts of interest.

Acknowledgments

The authors gratefully acknowledge the receipt of financial support from National Natural Science Foundation of China (51078198) and Natural Science Research Fund of Suzhou University of Science and Technology for this project.

References

[1] A. E. H. Love, *A Treatise on the Mathematical Theory of Elasticity*, Dover Publications, New York, NY, USA, 4th edition, 1944.

[2] R. N. Arnold and G. B. Warburton, "The flexural vibrations of thin cylinders," *Proceedings of the Institution of Mechanical Engineers*, vol. 167, no. 1, pp. 62–80, 1953.

[3] C. Wang and J. C. S. Lai, "Prediction of natural frequencies of finite length circular cylindrical shells," *Applied Acoustics*, vol. 59, no. 4, pp. 385–400, 2000.

[4] H. Lamb, "On the Vibrations of a Spherical SHEll," *Proceedings of the London Mathematical Society*, vol. 14, pp. 50–56, 1882/83.

[5] F. L. DiMaggio and A. Silbiger, "Free extensional torsional vibrations of a prolate spheroidal shell," *The Journal of the Acoustical Society of America*, vol. 33, pp. 56–58, 1961.

[6] A. M. Al-Jumaily and F. M. Najim, "An approximation to the vibrations of oblate spheroidal shells," *Journal of Sound and Vibration*, vol. 204, no. 4, pp. 561–572, 1997.

[7] Y. K. Lin and F. A. Lee, "Vibrations of thin paraboloidal shells of revolution," *Journal of Applied Mechanics*, vol. 27, no. 4, pp. 743-744, 1960.

[8] I. Hoppmann, M. I. Cohen, and V. X. Kunukkasseril, "Elastic vibrations of paraboloidal shells of revolution," *The Journal of the Acoustical Society of America*, vol. 36, pp. 349–353, 1964.

[9] R. Carter L, R. A. Robinson, and W. C. Schnobrich, "Free vibration of hyperboloidal shells of revolution," *ASCE-Engineering Mechanics*, vol. EM5, pp. 1033–1052, 1969.

[10] F.-M. Li, K. Kishimoto, and W.-H. Huang, "The calculations of natural frequencies and forced vibration responses of conical shell using the Rayleigh-Ritz method," *Mechanics Research Communications*, vol. 36, no. 5, pp. 595–602, 2009.

[11] M. H. Luah and S. C. Fan, "General free vibration analysis of shells of revolution using the spline finite element method," *Computers & Structures*, vol. 33, no. 5, pp. 1153–1162, 1989.

[12] D.-Y. Tan, "Free vibration analysis of shells of revolution," *Journal of Sound and Vibration*, vol. 213, no. 1, pp. 15–33, 1998.

[13] P. M. Naghdi and R. M. Cooper, "Propagation of Elastic Waves in Cylindrical Shells, Including the Effects of Transverse Shear and Rotatory Inertia," *The Journal of the Acoustical Society of America*, vol. 28, no. 1, pp. 56–63, 1956.

[14] I. Mirsky and G. Herrmann, "Nonaxially symmetric motions of cylindrical shells," *The Journal of the Acoustical Society of America*, vol. 29, pp. 1116–1123, 1957.

[15] A. W. Leissa, "Vibration of shells," Report No. 288, National Aeronautics and Space Administration (NASA), Washington, Wash, USA, 1973.

[16] S. Hosseini-Hashemi and M. Fadaee, "On the free vibration of moderately thick spherical shell panel - A new exact closed-form procedure," *Journal of Sound and Vibration*, vol. 330, no. 17, pp. 4352–4367, 2011.

[17] Z. Su, G. Jin, and T. Ye, "Free vibration analysis of moderately thick functionally graded open shells with general boundary conditions," *Composite Structures*, vol. 117, no. 1, pp. 169–186, 2014.

[18] F. Tornabene and E. Viola, "2-D solution for free vibrations of parabolic shells using generalized differential quadrature method," *European Journal of Mechanics - A/Solids*, vol. 27, no. 6, pp. 1001–1025, 2008.

[19] J.-H. Kang and A. W. Leissa, "Three-dimensional vibration analysis of thick hyperboloidal shells of revolution," *Journal of Sound and Vibration*, vol. 282, no. 1-2, pp. 277–296, 2005.

[20] O. J. Al-Khatib and G. R. Buchanan, "Free vibration of a paraboloidal shell of revolution including shear deformation and rotary inertia," *Thin-Walled Structures*, vol. 48, no. 3, pp. 223–232, 2010.

[21] F. W. Williams and W. H. Wittrick, "An automatic computational procedure for calculating natural frequencies of skeletal structures," *International Journal of Mechanical Sciences*, vol. 12, no. 9, pp. 781–791, 1970.

[22] W. H. Wittrick and F. W. Williams, "A general algorithm for computing natural frequencies of elastic structures," *The Quarterly Journal of Mechanics and Applied Mathematics*, vol. 24, pp. 263–284, 1971.

[23] S. Yuan, K. Ye, C. Xiao, F. W. Williams, and D. Kennedy, "Exact dynamic stiffness method for non-uniform Timoshenko beam vibrations and Bernoulli-Euler column buckling," *Journal of Sound and Vibration*, vol. 303, no. 3–5, pp. 526–537, 2007.

[24] J. R. Banerjee, C. W. Cheung, R. Morishima, M. Perera, and J. Njuguna, "Free vibration of a three-layered sandwich beam using the dynamic stiffness method and experiment," *International Journal of Solids and Structures*, vol. 44, no. 22-23, pp. 7543–7563, 2007.

[25] H. Su, J. R. Banerjee, and C. W. Cheung, "Dynamic stiffness formulation and free vibration analysis of functionally graded beams," *Composite Structures*, vol. 106, pp. 854–862, 2013.

[26] A. Pagani, M. Boscolo, J. R. Banerjee, and E. Carrera, "Exact dynamic stiffness elements based on one-dimensional higher-order theories for free vibration analysis of solid and thin-walled structures," *Journal of Sound and Vibration*, vol. 332, no. 23, pp. 6104–6127, 2013.

[27] N. El-Kaabazi and D. Kennedy, "Calculation of natural frequencies and vibration modes of variable thickness cylindrical shells using the Wittrick-Williams algorithm," *Computers & Structures*, vol. 104-105, pp. 4–12, 2012.

[28] X. Chen, *Research on dynamic stiffness method for free vibration of shells of revolution. Master Thesis, Tsinghua University [Master, thesis]*, Chinese, Tsinghua University, 2009.

[29] X. Chen and K. Ye, "Free vibration analysis for shells of revolution using an exact dynamic stiffness method," *Mathematical Problems in Engineering*, Article ID 4513520, Art. ID 4513520, 12 pages, 2016.

[30] X. Chen and K. Ye, "Analysis of free vibration of moderately thick circular cylindrical shells using the dynamic stiffness method. Engineering Mechanics," in *Analysis of free vibration of moderately thick circular cylindrical shells using the dynamic stiffness method*, pp. 40–48, Chinese, Engineering Mechanics, 2016.

[31] X.-D. Chen and K.-S. Ye, "Free vibration analysis of moderately thick elliptical shells using the dynamic stiffness method," *Zhendong yu Chongji*, vol. 35, no. 6, pp. 85–90, 2016.

[32] X. Chen and K. Ye, "Comparison study on the exact dynamic stiffness method for free vibration of thin and moderately thick circular cylindrical shells," *Shock and Vibration*, vol. 2016, Article ID 9748135, 14 pages, 2016.

[33] S. M. R. Khalili, A. Davar, and K. Malekzadeh Fard, "Free vibration analysis of homogeneous isotropic circular cylindrical shells based on a new three-dimensional refined higher-order theory," *International Journal of Mechanical Sciences*, vol. 56, no. 1, pp. 1–25, 2012.

[34] B. P. Gautham and N. Ganesan, "Free vibration analysis of thick spherical shells," *Computers & Structures*, vol. 45, no. 2, pp. 307–313, 1992.

[35] H.-J. Shim and J.-H. Kang, "Free vibrations of solid and hollow hemi-ellipsoids of revolution from a three-dimensional theory," *International Journal of Engineering Science*, vol. 42, no. 17-18, pp. 1793–1815, 2004.

[36] J. M. Deb Nath, "Free vibration, stability and 'non-classical modes' of cooling tower shells," *Journal of Sound and Vibration*, vol. 33, no. 1, pp. 79–101, 1974.

[37] U. Ascher, J. Christiansen, and R. D. Russell, "Collocation software for boundary value ODEs," *ACM Transactions on Mathematical Software*, vol. 7, no. 2, pp. 209–222, 1981.

[38] U. Ascher, J. Christiansen, and R. D. Russell, "Algorithm 569, COLSYS: collocation software for boundary value ODEs [D2]," *ACM Transactions on Mathematical Software*, vol. 7, no. 2, pp. 223–229, 1981.

14

Numerical Analysis of Stress Fields and Crack Growths in the Floor Strata of Coal Seam for Longwall Mining

Lingzhi Sun,[1,2] Yunyue Xie,[1,2] and Hongtian Xiao[1,2]

[1]College of Civil Engineering and Architecture, Shandong University of Science and Technology, China
[2]Shandong Key Lab of Civil Engineering Disaster Prevention & Mitigation, Qingdao 266510, China

Correspondence should be addressed to Hongtian Xiao; xiaohongtian@tsinghua.org.cn

Academic Editor: Andras Szekrenyes

This paper predicts the possibility of water inrush from a confined aquifer under the action of mining activities and water pressure. The study uses numerical analyses to evaluate stress redistribution and crack growth which result from coal extraction operations. Two models are presented in this study. By simplifying the distribution of the disturbed vertical stress on the coal seam and floor around a working face, a model is established to analyze the additional stresses in the floor strata induced by mining activities. And some distribution features of all the additional stress components are described. By using the superposition principle in fracture mechanics, another model is developed to analyze the crack growth in the floor strata under the action of disturbed stresses and water pressure. And the stress intensity factors at the crack tip are presented and the process of crack growth is obtained in the advancement of a working face. Because of discretizing only loading areas and crack surfaces, the present methods can obtain the accurate numerical results. Finally, some suggestions are made for preventing the water inrush from a confined aquifer.

1. Introduction

In the coalfields of North China, coal seams are underlain by Ordovician limestone, which is strongly karstified and highly permeable. In most cases, the limestone aquifer contains an abundant supply of water and is confined. The activities of mining these coal seams are threatened with the confined water from the Ordovician limestone. Water inrush accidents result in heavy losses and casualties [1, 2]. With coal mining being executed in deeper sites, the pressure of confined water becomes larger and this type of water inrushes may occur more frequently.

In order to avoid water inrushes and keep the mines safe, large-scale dewatering or depressurizing of the Ordovician aquifer was chosen. In practice, dewatering causes serious environmental problems such as surface subsidence and underground water pollution and reduces the amount of water available for supply in the area. To solve this dilemma, some researches were devoted to make full use of the uncontaminated karst water obtained from dewatering for water supply [3]. Another measure to prevent water inrushes is to use the floor strata between the coal seam and the aquifer as a protective rock pillar. However, once the strength of the floor strata is difficult to resist the invasion of high-pressure water, the confined water floods into working faces. In some cases, grouting into the rock layers is used to improve the strength of floor strata to resist the invasion of confined water. By taking these measures, the water in the Ordovician limestone can be preserved to some extent and the coal seams can be extracted in a relatively safe environment.

As early as in 1979, the field investigations of the floor strata were executed in Fengfeng coalfield, China, where mining activities were threatened by confined aquifers [4]. After that, more field investigations of floor strata were performed in different types of mines. In Jingxing coalfield, China, it was found that the joints in floor strata were impermeable before mining, and confined water passed through the joints and flooded into the mines after mining [5]. This phenomenon may be explained by the joint growth induced by mining activities and water pressure from a confined aquifer. Additionally, the test of water injection in boreholes was used for understanding the permeability variations in floor strata induced by mining activities [6, 7]. These measured data were further used to determine the failure zones directly below the coal seam and above the confined aquifer.

In the past several decades, other types of analytical methods have been employed to study mining with water pressure for preventing water inrushes. Wang et al. [8] established both physical and numerical models to gain a better understanding into fracture initiation and propagation in floor strata during a mining process. Yang et al. [9] and Li et al. [10] used finite element methods to understand the mechanisms of water inrushes. Because different coal seams have different vulnerabilities for water inrushes, Wu et al. [11] assessed the differences in vulnerability between multiple coal seams. To evaluate a major water inflow hazard in a mine, Rafiqul et al. [12] used boundary element methods to model the mining-induced stress field. Zuo et al. [13] explored the spatial morphology of karst collapse and employed the discontinuous deformation analysis method to investigate karst collapse.

Although a great deal of field investigations and numerical analyses have been performed and some research results of water inrushes have been obtained, all the mechanisms of water inrushes are not known and the water inrushes are often reported. It is estimated that more than 80 percent of water inrushes may occur in the case of the floor rocks having faults or joints. This means that faults and joints in floor strata may induce water inrushes under the action of water pressure from aquifers and mining activities. Therefore, the water inrushes induced by faults or joints in longwall mining need to be further studied.

The principal objectives of this paper are to understand the stress fields in floor strata induced by coal mining and to predict the joint growths in the floor strata of the coal seam threatened by confined water in the advancement of a working face. From now on, the term crack will be used to describe a joint. In this paper, firstly, two numerical methods to be used are briefly introduced. Secondly, a mechanical model for simulating coal mining is established and the additional stress fields induced by coal mining are presented. Thirdly, another mechanical model is established for analyzing the crack problems under the action of water pressure and mining activities, and the stress intensity factors (SIFs) are obtained for the cracks at different positions of floor strata. Finally, the crack growths are predicted using SIF values and fracture criteria, and some suggestions are made for preventing the water inrushes into working faces from confined aquifers in floor strata.

2. Analytical Methods for Elastic Fields and Crack Problems in a Layered Medium

2.1. Yue's Solution Based Numerical Method for Elastic Fields in a Layered Medium. Xiao et al. [14] developed a numerical method for assessing the elastic fields in an arbitrarily depth-heterogeneous medium subject to complex loads using Yue's solution. Yue's solution [15] is a fundamental singular solution for the generalized Kelvin problems of a multilayered elastic solid of infinite extent subject to concentrated point body vectors. The stresses and displacements at any points of a layered medium are described as

$$\sigma_{ij}(Q) = \int_S \sigma^*_{ijk}(Q,P)\,t_k(P)\,dS(P), \quad i,j,k = x,y,z \tag{1a}$$

$$u_i(Q) = \int_S u^*_{ik}(Q,P)\,t_k(P)\,dS(P), \quad i,k = x,y,z \tag{1b}$$

where $\sigma^*_{ijk}(Q,P)$ and $u^*_{ik}(Q,P)$ are, respectively, stresses and displacements of Yue's solution for the field point Q due to the unit force along the k direction at the source point P; $t_k(P)$ is the traction at the source point P; the integral domain S is the loading area. It should be noted that the subscript k is a dummy index.

The corresponding computer program LayerSmart3D was written in FORTRAN. The techniques adopted in LayerSmart3D primarily involve the discretization of the loading area into a finite number of quadrilateral elements. Values of the loads are inputted at the node points of the discretized area. Numerical verification of LayerSmart3D indicates that numerical solutions of high accuracy can be efficiently calculated for elastic fields induced by the distributed loads in a layered medium.

2.2. Yue's Solution Based DDM for Analyzing Crack Problems in a Layered Medium. Xiao and Yue [16, 17] developed a new displacement discontinuity method (DDM) for the analysis of crack problems in a layered medium of infinite extent. This approach is also based on Yue's solution [15] and the corresponding computer program LayerDDM3D was written in FORTRAN.

Assume that the crack surfaces consist of Γ^+ and Γ^-. In the global coordinates (x,y,z), the discontinuous displacements of the crack surfaces are defined as $\Delta u_j(Q_{\Gamma^+}) = u_j(Q_{\Gamma^+}) - u_j(Q_{\Gamma^-})$. u_j is a displacement along the direction $j = x,y,z$. The traction on the crack surface is defined as $t_j(P_{\Gamma^+})$. The basic equations of Yue's solution based DDM are as follows:

$$t_j(P_{\Gamma^+}) + n_i(P_{\Gamma^+}) \int_{\Gamma^+} T^*_{ijk}(P_{\Gamma^+},Q)\,\Delta u_k(Q)\,d\Gamma(Q) = 0, \quad i,j,k = x,y,z \tag{2}$$

where $n_i(P_{\Gamma^+})$ is the unit outward normal of the positive side of the crack surface at the source point P_{Γ^+} and $T^*_{ijk}(P_{\Gamma^+},Q)$ is a new kernel function which can be calculated by using the traction t^*_{ij} of Yue's solution.

In developing the DDM for (2), the nine-node continuous and discontinuous elements are used to discretize a crack surface. Kutt's numerical method is used to calculate the hypersingular integral in the integral equation. By solving the discretized equations of (2), the discontinuous displacements on the crack surface can be obtained. The numerical verification shows that the present results are in very good agreement with the existing ones. However, the proposed DDM is only suitable for the analysis of crack problems in layered media of infinite extent under the action of loadings on crack surfaces. Other types of crack problems in infinite media can be analyzed by using the two numerical methods together.

Table 1: The roof and floor strata of the extracted coal seam in No. 2710 working face.

Strata	Thickness(m)	Elastic modulus (GPa)	Poisson's ratio
Sandstone	14	25	0.28
Extracted coal seam	1.5	4	0.33
Sandstone	17	25	0.28
Limestone	5	40	0.28
Shale	4.2	15	0.29
Coal seam	1.8	4	0.33
Shale	20	20	0.30
Ordovician Limestone	Larger than 650	45	0.30

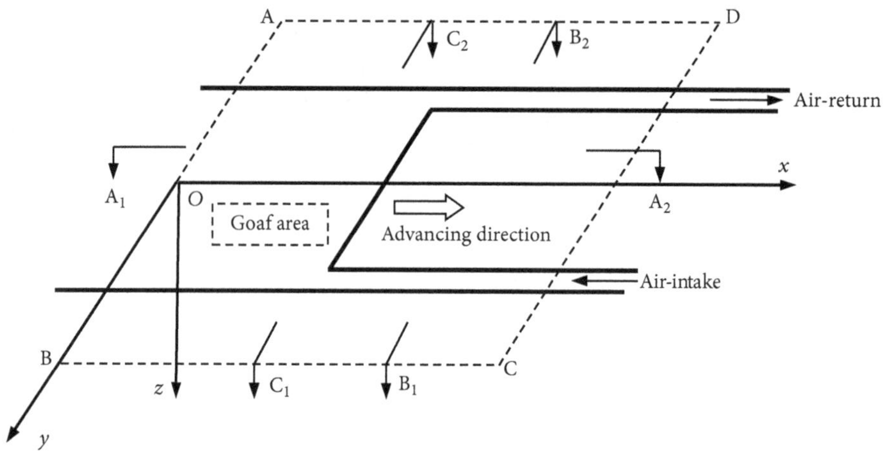

Figure 1: The disturbed area ABCD induced by coal mining.

2.3. The Analytical Methods of Crack Problems in Layered Media. In the following, the two numerical methods mentioned above, i.e., LayerSmart3D and LayerDDM3D, and the superposition principle in fracture mechanics are utilized to analyze crack problems in a layered medium. LayerSmart3D is firstly employed to obtain the stress fields of a layered medium without a crack under the action of distributed loadings within layered media. Using the superposition principle [18], the tractions, whose values are equal to the mining-induced stress and which have opposite directions, are then applied on the crack surfaces in the layered medium without the above distributed loadings. LayerDDM3D is finally employed to obtain the discontinuous displacements of the crack surfaces under the action of the above tractions. The SIF values at the crack tip can be calculated by using the discontinuous displacements of the crack surfaces and the relationship between SIFs and discontinuous displacements.

3. The Additional Stresses of Floor Strata without Cracks Induced by Mining

3.1. Geological and Mining Conditions of the Working Face. Jing et al. [4] investigated the deformation and failure of floor strata in mining the coal seam threatened by confined water in No. 2 Coal Mine of Fengfeng coalfield. No. 2710 working face in this mine was chosen for field measurements. Table 1 describes the roof and floor strata of the extracted coal seam. The coal seam is about 48 m away from the Ordovician limestone in which the water pressure ranges from 1.5 to 2 MPa. The working face is located at a depth of about 145 m and the longwall panel is 90 m long and 350 m wide. The coal seam is 1.5 m thick and the dip angle is about 1-3°. For the sake of simplicity, the rock strata are treated as horizontal media in developing a mechanical model. The working face was advanced along the strike direction of the coal seam. From August 1979 to February 1980, comprehensive measurements of floor deformation and failure were performed in the advancing process of the working face. In the field measurements, large amounts of data were obtained and will be further employed to analyze the water inrush from confined aquifers here.

3.2. A Model of Analyzing Additional Stresses of Floor Strata Induced by Mining

3.2.1. The Mining-Disturbed Area on the Interface of Coal Seam and Floor. Figure 1 illustrates the disturbed area ABCD on the interface between the mined coal seam and the floor. It is assumed that the domain out of the area ABCD is not disturbed by coal mining. According to the in situ investigations, the disturbed area ABCD is 150 m long and

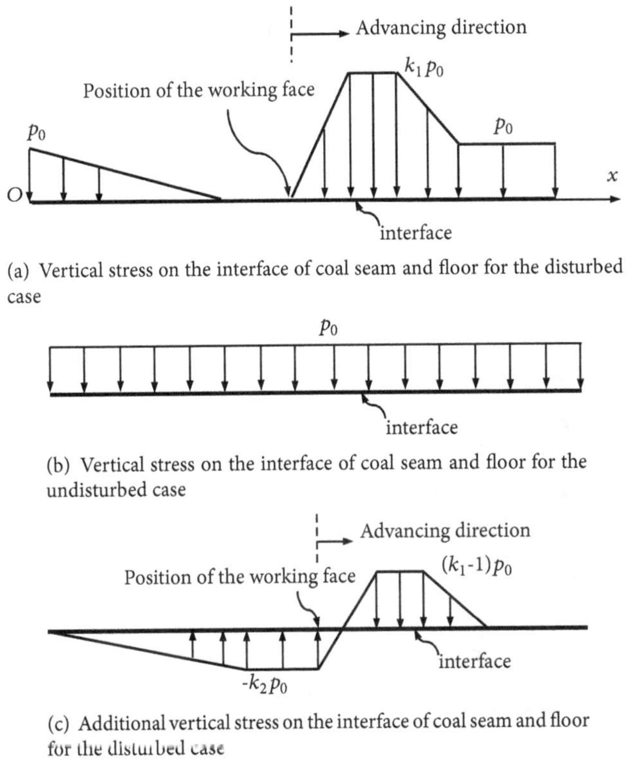

(a) Vertical stress on the interface of coal seam and floor for the disturbed case

(b) Vertical stress on the interface of coal seam and floor for the undisturbed case

(c) Additional vertical stress on the interface of coal seam and floor for the disturbed case

FIGURE 2: Analytical method of additional vertical stress on the interface of coal seam and floor along the cross section A_1-A_2.

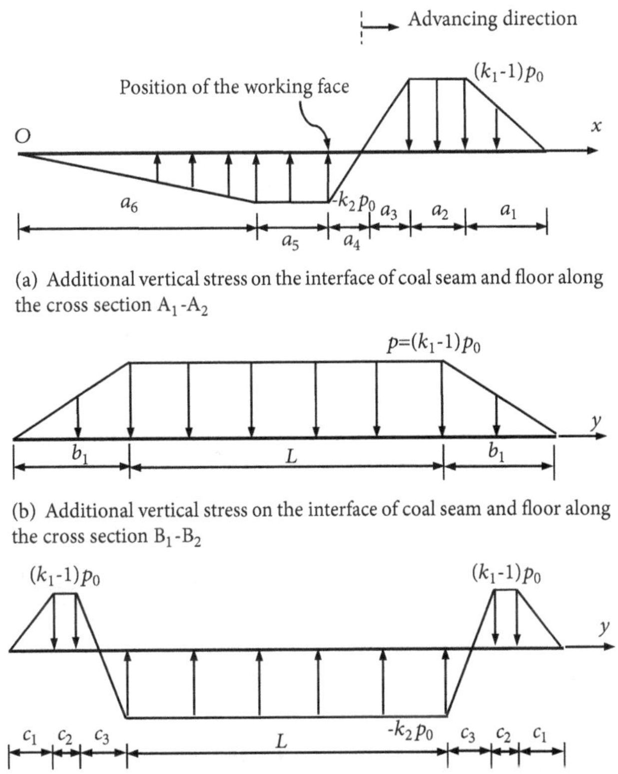

(a) Additional vertical stress on the interface of coal seam and floor along the cross section A_1-A_2

(b) Additional vertical stress on the interface of coal seam and floor along the cross section B_1-B_2

(c) Additional vertical stress on the interface of coal seam and floor along the cross section C_1-C_2

FIGURE 3: Additional vertical stress on the interface of coal seam and floor along different cross sections.

150 m wide. The vertical stress on the interface of the coal seam and the floor strata, which are not disturbed by mining, can be calculated using the following formula:

$$p_0 = \gamma h = 25 \times 145 = 3625 kN/m^2 \quad (3)$$

where γ is the average unit weight of the overlying strata and h is the depth of the extracted coal seam below the ground surface.

Mining activities induce the stress concentration on the coal pillars and floor around the working face. In the field investigations at No. 2710 working face, the disturbed area and stress concentration on the coal seams were obtained. Based on these data, we develop a model for the analysis of the influence of coal mining on floor strata.

The vertical stresses along different cross sections on the disturbed area are assumed by using the measured data. The concentration factors of the vertical stress ahead of and behind the working face along the cross section A_1-A_2 are taken to be $k_1 = 3$ and $k_2 = 1$, respectively. And the stress concentration factors along the cross sections B_1-B_2 and C_1-C_2 are taken to be $k_1 = 3$. On the floor strata near the roadways and the working face, the stresses are highly concentrated and the stress concentration factor is taken to be k_3 = 3.5. The stress concentration factors along other cross sections can be obtained by using the methods of linear or quadratic interpolation.

Figure 2 illustrates the process of obtaining the additional vertical stress on the interface of coal seam and floor along the cross section A1-A2 shown in Figure 1. It can be found that the additional vertical stress in Figure 2(c) is obtained by subtracting out the weight p_0 of the overburden above the mined coal seam in Figure 2(b) from the vertical pressure in Figure 2(a). Figure 3 presents the additional vertical stresses on the interface between coal seam and floor along three different cross sections. Based on the data from in situ measurements, the parameters of lengths shown in Figure 3(a) are presented as follows: a_1 = 40m, a_2 = 10m, a_3 = 5m, a_4 = 5m, a_5 = 30m, and a_6 = 60m. The parameters of lengths shown in Figure 3(b) are presented as follows: L = 90m and b_1 = 30m. The parameters of lengths shown in Figure 3(c) are presented as follows: L = 90m, c_1 = 30m, c_2 = 30m, and c_3 = 30m.

3.2.2. Element Mesh of the Disturbed Area and Comparison of Numerical and Field Results. In order to analyze the elastic fields using the software LayerSmart3D, the disturbed area ABCD is discretized into the mesh with 2610 nodes and 8037 elements, shown in Figure 4. The additional vertical stresses on the disturbed area ABCD (see Figure 3) are applied on the nodes of the element mesh. Figure 5 illustrates the distribution of the additional vertical stress on the interface of coal seam and floor. It can be found that the distribution of the additional vertical stress shown in Figure 5 is in good agreement with the one shown in Figure 3.

After presenting the mechanical parameters of floor strata in Table 1 and the additional vertical stress in Figure 5, the

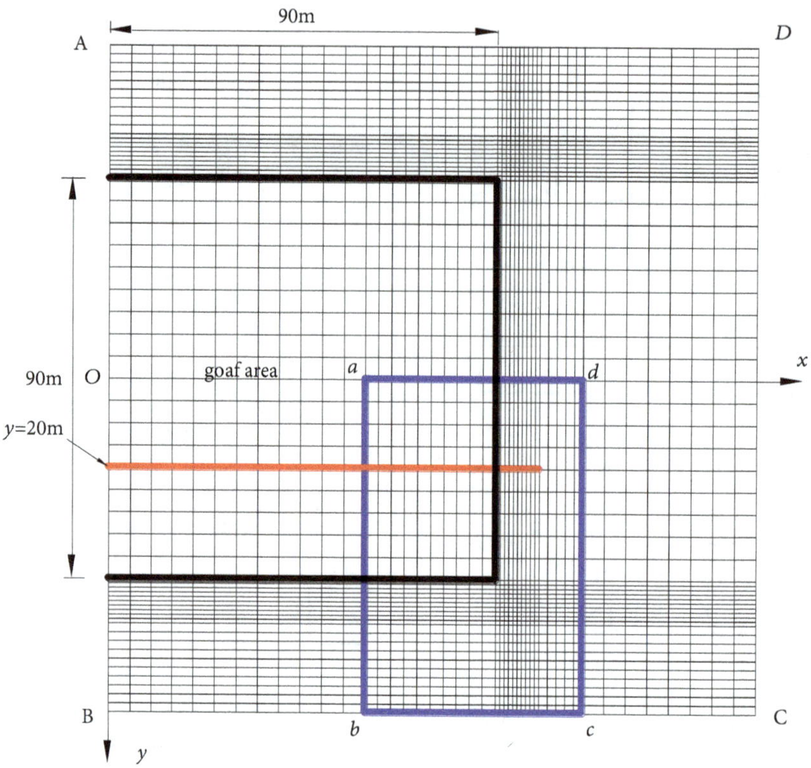

FIGURE 4: Element mesh of the disturbed area ABCD, the projection *abcd* of the calculating area at $z = 15$m, and the plane at $y = 20$m for calculating the stresses.

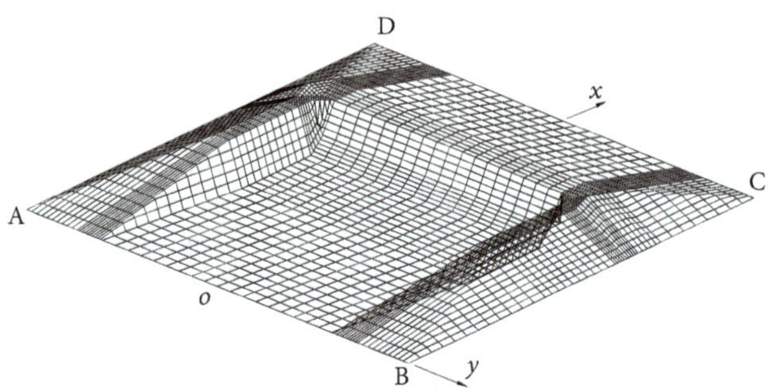

FIGURE 5: Additional vertical stress on the disturbed area ABCD.

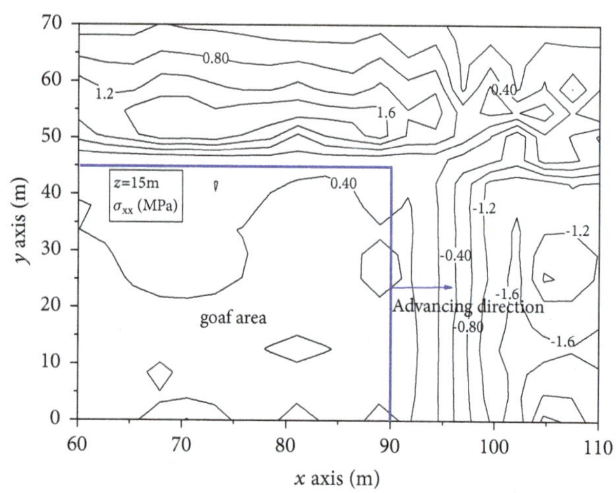

FIGURE 6: Contours of the additional stress σ_{xx} at the depth $z = 15$m.

software LayerSmart3D is used to obtain the stress and displacement fields of floor strata. Table 2 lists the vertical displacements at some points in the floor strata from numerical analyses and in situ measurements. It can be found that the displacements from the two methods are much close to each other if the differences of the geological and mining conditions for two methods are considered.

3.3. *Variations of Additional Stresses in Floor Strata Induced by Mining*

3.3.1. Distribution of Additional Stresses at the Depth $z = 15$m. As shown in Figure 4, the model is symmetric with respect to the coordinate plane *Oxz*. Thus, the rectangular area, i.e., $60\,\text{m} \le x \le 110\text{m}$, $0\,\text{m} \le y \le 45\text{m}$, and $z = 15\text{m}$, is chosen to calculate the additional stresses. This area has a projection *abcd* on the coordinate plane *Oxy* shown in Figure 4. Figures 6–11 illustrate the distribution of all the stress components on the calculating area.

Figures 6–8 show the variations of three normal stresses in the floor strata. In Figure 6, the rectangular area can be divided into three areas by considering the distribution features of σ_{xx}. Below the goaf area, σ_{xx} is tensile and has small values. Below the coal seam in the front of the working face,

TABLE 2: Comparison of maximum vertical displacements from in situ measurements and numerical analyses along the advancing direction (unit: mm).

Positions	ahead of the face wall		behind the face wall	
(y, z)	In situ measurements	This study	In situ measurements	This study
$(0, 4.9)$	8	9.9995	5	-2.7071
$(0, 6.2)$	8	9.7000	-2	-2.5608
$(0, 10.2)$	7	8.9472	1	-2.1707
$(40, 2.4)$	13	10.3224	6	8.9056
$(40, 4.7)$	14	9.3394	13	7.0610
$(40, 7.2)$	14	9.6763	16	8.2905

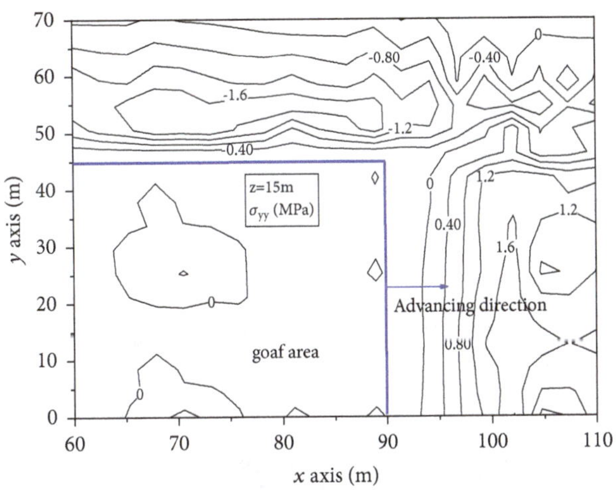

FIGURE 7: Contours of the additional stress σ_{yy} at the depth $z = 15$m.

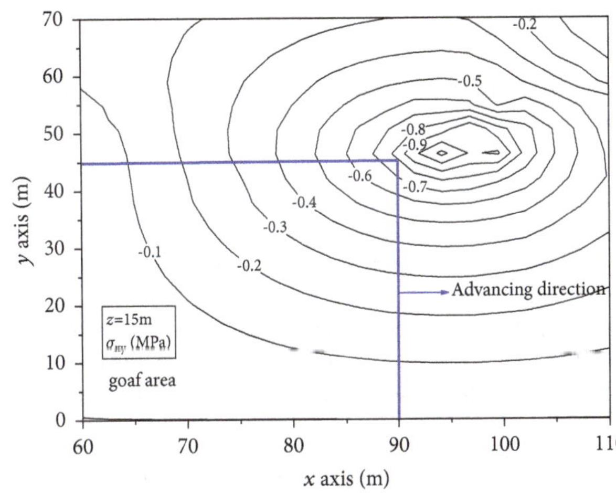

FIGURE 9: Contours of the additional stress σ_{xy} at the depth $z = 15$m.

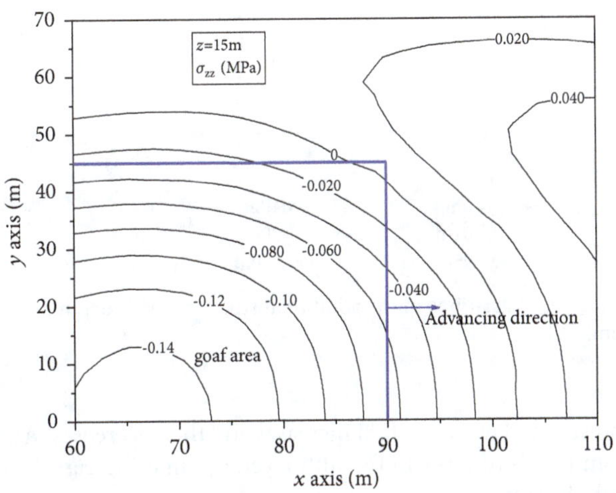

FIGURE 8: Contours of the additional stress σ_{zz} at the depth $z = 15$m.

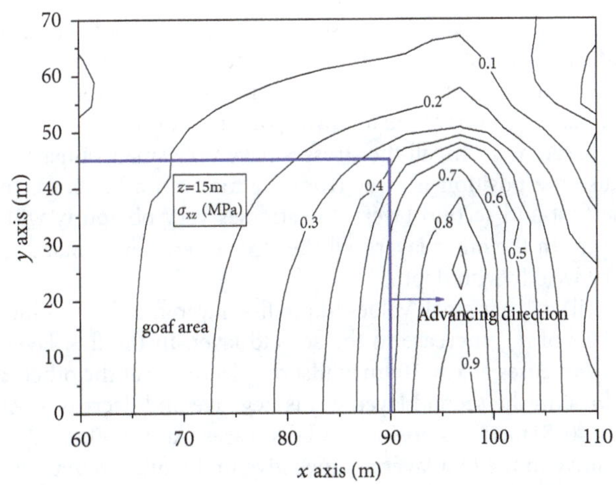

FIGURE 10: Contours of the additional stress σ_{xz} at the depth $z = 15$m.

σ_{xx} is compressive and its minimum value is about -1.6 MPa. And below the coal pillar outside the goaf area, σ_{xx} is tensile and its maximum value is about 1.6 MPa. In Figure 7, the rectangular area can also be divided into three areas. Below the goaf area, σ_{yy} approaches zero. Below the coal seam in the front of the working face, σ_{yy} is tensile and its minimum value is about 1.6 MPa. And below the coal pillar outside the goaf area, σ_{yy} is compressive and its minimum value is about 1.6 MPa. In Figure 8, the rectangular area can be divided into two areas. σ_{zz} is compressive below the goaf area and is tensile away from the goaf area.

Figures 9–11 show the variations of three shear stresses in the floor strata. In Figure 9, σ_{xy} has a maximum value of

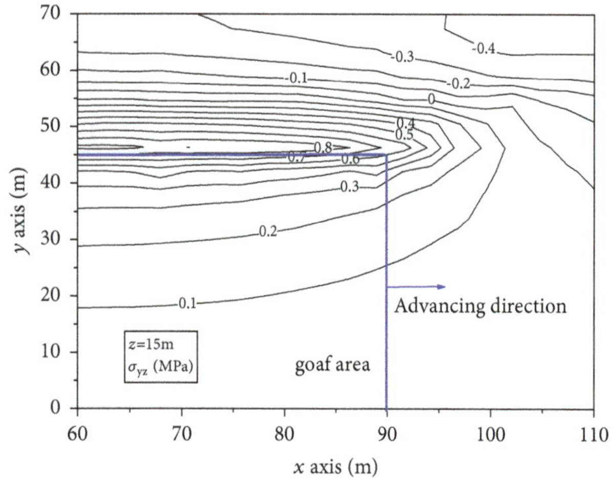

FIGURE 11: Contours of the additional stress σ_{yz} at the depth z = 15m.

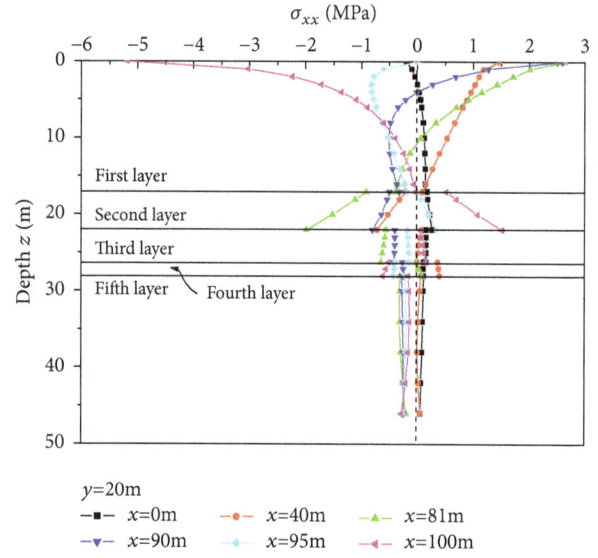

FIGURE 12: Variations of the additional stress σ_{xx} with depth at y = 20m.

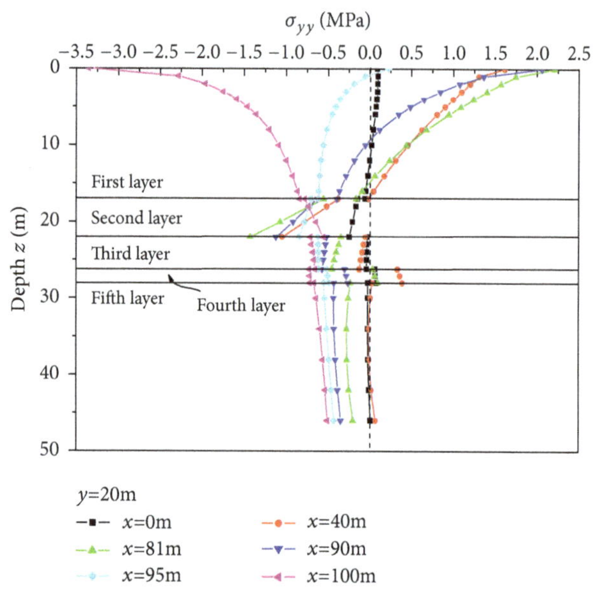

FIGURE 13: Variations of the additional stress σ_{yy} with depth at y = 20m.

about -1.0 MPa at the point $(x, y, z) = (94\text{m}, 46\text{m}, 15\text{m})$ and σ_{xy} becomes small away from this point. In Figure 10, σ_{xz} is concentrated in the front of the working face and becomes small away from this zone and its maximum value is about 1.0 MPa. In Figure 11, σ_{yz} is concentrated below the coal pillar outside the goaf area and becomes small away from this zone.

3.3.2. Distribution of the Additional Stresses in the Plane y = 20m. Figures 12–17 illustrate the distribution of all the stress components in the plane y = 20m for different x and z values. As shown in Figure 4, this plane is along the advancing direction of the working face and has a horizontal distance of 25 m from the coal pillar outside the goaf. For the sake of simplification, the stress variations are referred to as the ones with the depth increasing in the following analyses. From these figures, we can have the following observations:

(i) At x = 0m, all the stresses have very small values because the position is away from the maximum loadings. In the first and second layers, the stresses vary obviously with depth. In the other layers, all the stresses are very small and vary weakly with depth.

(ii) σ_{xx} has small values in the first layer and the absolute values of σ_{xx} increase in the second layer. In the first layer, σ_{yy} increases at x = 100m whilst σ_{yy} decreases at the other x values. In the second layer, σ_{yy} is negative and decreases. At x = 40, 81m, σ_{zz} is positive and decreases. At x = 90m, σ_{zz} is positive in the first layer and negative in the other remaining layers and decreases. σ_{zz} is negative and decreases at x = 95m, and σ_{zz} is negative and increases at x = 100m.

(iii) At x = 0, 40m, σ_{xy} decreases in the first, second, and third layers and increases in the remaining two layers. The variation of σ_{xy} at x = 81, 90, 95, 100m is different from the one at x = 0, 40m. σ_{xz} is negative at x = 0, 40m whilst σ_{xz} is positive at x = 81, 90, 95, 100m. In the different layers, σ_{xz} first increases and then decreases for a given z value from x = 81 to 100 m. The absolute values of σ_{xz} first increase and then decrease. σ_{yz} is always positive for all the x values. For x = 0m, σ_{yz} increases. Except for x = 0m, in the first and second layers, σ_{yz} first increases and then decreases. And from the third layer to the fifth layer, σ_{yz} first increases and then decreases.

4. A Model of Analyzing Crack Problems in Floor Strata

As shown in Figure 18, a rectangular crack is located in floor strata and is subject to mining-disturbed stresses and water pressure. A global coordinate system $Oxyz$, which is completely the same as the one shown in Figure 1, is attached to the model to be analyzed and a local coordinate system $O'x'y'z'$ is attached on the crack surface. The rectangular crack is 2 m

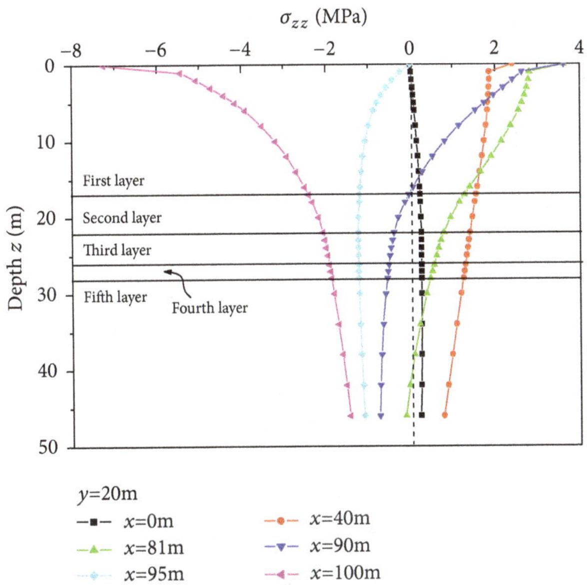

FIGURE 14: Variations of the additional stress σ_{zz} with depth at y = 20m.

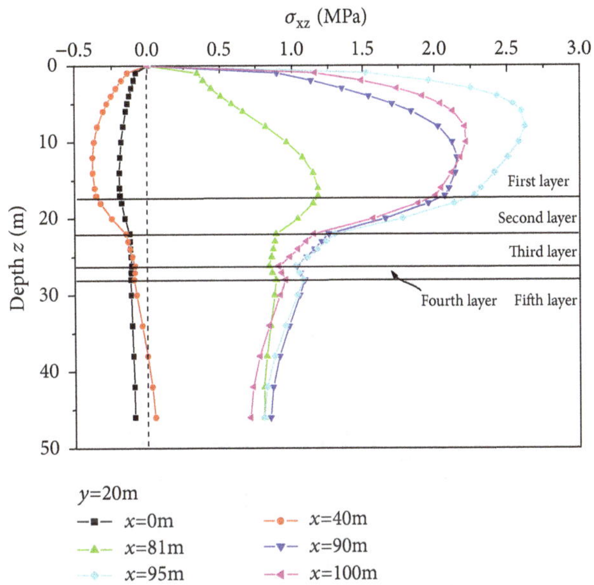

FIGURE 16: Variations of the additional stress σ_{xz} with depth at y = 20m.

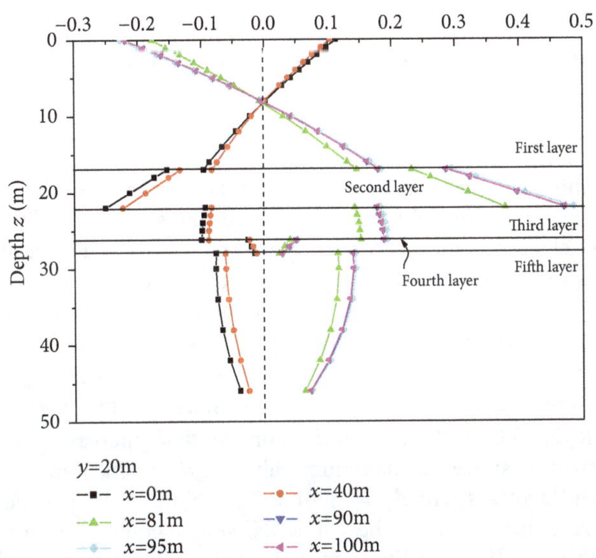

FIGURE 15: Variations of the additional stress σ_{xy} with depth at y = 20m.

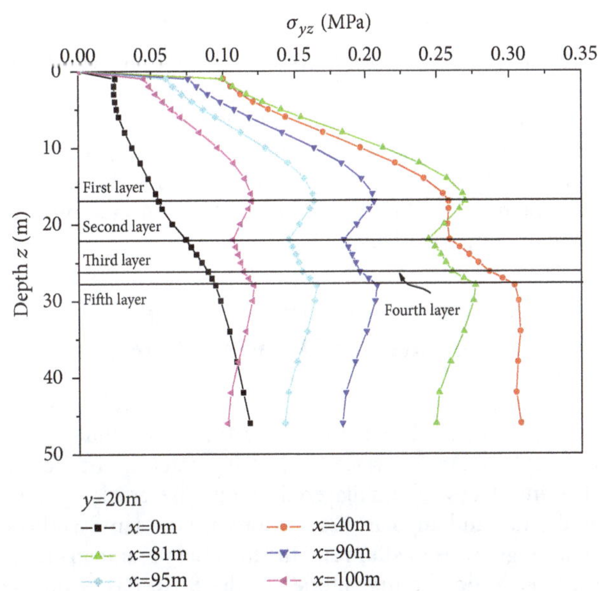

FIGURE 17: Variations of the additional stress σ_{yz} with depth at y = 20m.

wide (AB and CD sides) and 4 m long (AD and BC sides). Assume that the crack surfaces are parallel to the advancing direction of the working face and the crack center is located at the point $(x, y, z) = (d_1, d_2, 46\text{m})$.

Using the superposition principle in fracture mechanics [18], the disturbed stresses induced by mining activities can be transformed to the tractions on the crack surfaces. The disturbed stresses should be the summation of the horizontal normal stress σ_h induced by rock weights and the additional stresses $(\sigma_{yy}, \sigma_{xy}, \sigma_{zy})$ induced by mining activities. Additionally, the crack surfaces are subject to the water pressure p_w from confined aquifer. Thus, the crack surfaces are subjected to the following tractions:

$$f_x = -\sigma_{xy} \tag{4a}$$

$$f_y = -(\sigma_{yy} + \sigma_h) + p_w \tag{4b}$$

$$f_z = -\sigma_{yz} \tag{4c}$$

where p_w is taken to be 2 MPa.

In (4b), σ_h is calculated by the following formula:

$$\sigma_h = \frac{\nu}{1-\nu}\gamma h \tag{5}$$

where ν and γ are, respectively, the average values of Poisson's ratio and unit weight of rock strata and h is the depth of the

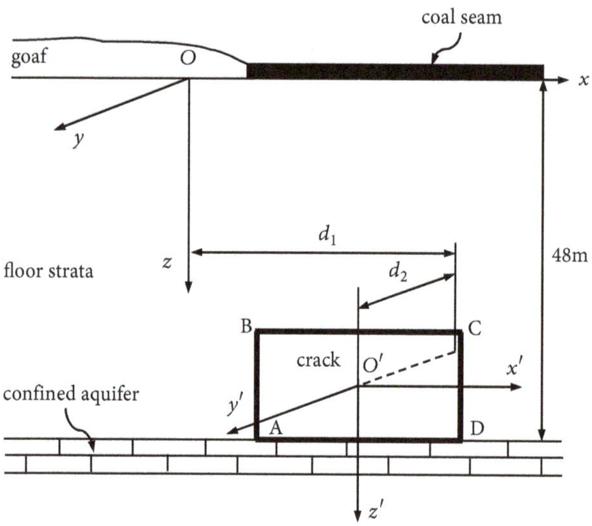

FIGURE 18: Model of analyzing crack problems in floor strata under the action of distributed stresses and water pressure.

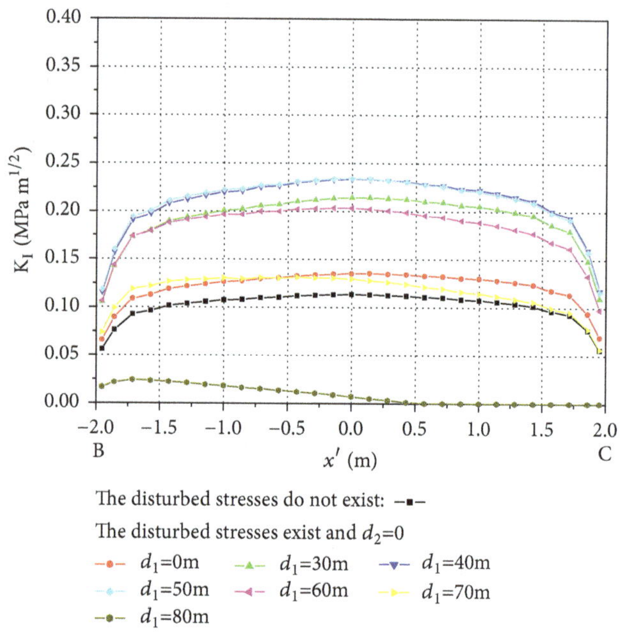

FIGURE 19: Variations of K_I values at the crack side (BC) for Case 1.

crack center below the ground surface. Take $\gamma = 25\text{kN/m}^2$, $\nu = 0.28$, and $h = 192\text{m}$. Thus, $\sigma_h \approx 1.8667\text{MPa}$.

The crack surface is discretized into 435 nodes and 98 nine-node elements with 60 nine-node isoparametric elements, 34 discontinuous elements of type I, and four discontinuous elements of type II. The discontinuous displacements of the crack surfaces can be obtained using the element mesh, the tractions (f_x, f_y, f_z) on the crack surface, and the computer code LayerDDM3D.

5. SIFs of Cracks under the Action of Water Pressure and Disturbed Stresses

5.1. General. The stress intensity factors (SIF) values at the crack tip can be calculated by using the discontinuous displacements of crack surfaces. When the crack is not located at the interfaces of multilayered solids, the crack tip field singularities and angular distributions are the same as those in a homogeneous medium and the material parameters exert an influence on the magnitudes of the SIFs. Based on the relationship of displacements and the SIFs, the formulae of the SIFs along the crack side BC are as follows:

$$K_I = \frac{E}{4(1-\nu^2)}\sqrt{\frac{\pi}{2r}}\Delta u_{y'} \quad (r, \theta = \pm\pi, \varphi = 0), \tag{6a}$$

$$K_{II} = \frac{E}{4(1-\nu^2)}\sqrt{\frac{\pi}{2r}}\Delta u_{z'} \quad (r, \theta = \pm\pi, \varphi = 0), \tag{6b}$$

$$K_{III} = \frac{E}{4(1+\nu)}\sqrt{\frac{\pi}{2r}}\Delta u_{x'} \quad (r, \theta = \pm\pi, \varphi = 0), \tag{6c}$$

where (r, θ, φ) is the spherical coordinates located at the crack front and θ is an angle in the coordinate plane normal to the crack line of crack front.

The following two cases are analyzed: Case 1: a rectangular crack is located in the coordinate plane Oxz, i.e., $d_2 = 0\text{m}$; Case 2: a rectangular crack is located in the plane $y = 40\text{m}$, i.e., $d_2 = 40\text{m}$.

5.2. Case 1: The Crack Is Located in the Coordinate Plane Oxz. In this case, the shear stresses σ_{xy} and σ_{zy} in the coordinate plane Oxz are equal to zero because of the symmetry of the model. As a result, K_{II} and K_{III} are equal to zero. Figure 19 illustrates the variations of K_I for the crack at different positions along the advancing direction of the working face. For the crack only subject to water pressure, the maximum value of K_I is $0.1134\,\text{MPa}\cdot\text{m}^{1/2}$. However, for the crack subject to water pressure and disturbed stresses, the maximum value of K_I is $0.1351\,\text{MPa}\cdot\text{m}^{1/2}$ at $d_1 = 0\text{m}$. With d_1 increasing, K_I increases, arrives at maximum values at $d_1 = 40, 50\text{m}$, and then decreases. For $d_1 = 80\text{m}$ and $x' \geq 0.5\text{m}$, K_I is equal to zero and K_I along the whole BC side is equal to zero at $d_1 \geq 90\text{m}$. These findings mean that the crack below the coal seam is closed and the crack below the goaf is opened under the action of disturbed stresses and water pressure. When the crack is located at some horizontal distance of 40-50m after the coal wall of the working face, there are the maximum opening displacements. In this case, the crack becomes easy to open and induces the uplifting of the confined water.

5.3. Case 2: The Crack Is Located in the Plane $y = 40\text{m}$, i.e., $d_2 = 40\text{m}$. In this case, the crack is close to the coal pillars outside the working face and has a horizontal distance of 5 m from the coal pillar. As discussed in Section 3, the shear stresses induced by mining exist at this position and as a result the crack has three types of fracture modes. Figures 20–22 show the variations of K_I, K_{II}, and K_{III} along the crack front BC, respectively.

In Figure 20, the K_I values for $d_1 = 0, 40\text{m}$ are very close to the ones of the crack only subject to water pressure whilst

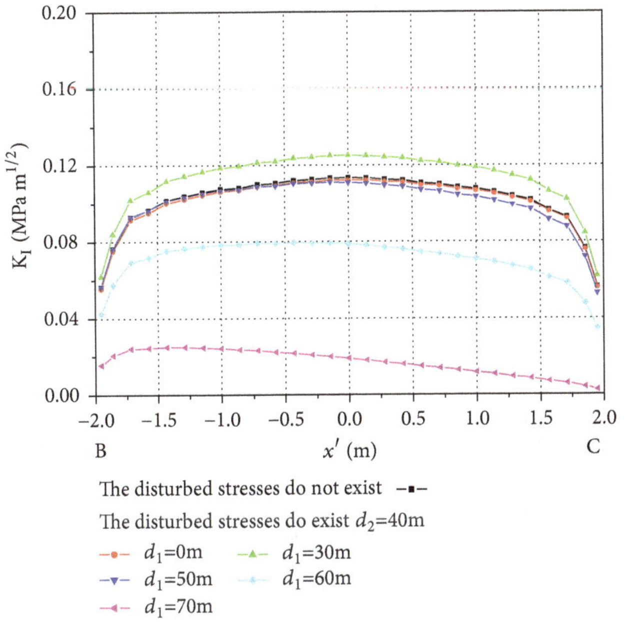

FIGURE 20: Variations of K_I values at the crack side (BC) for Case 2.

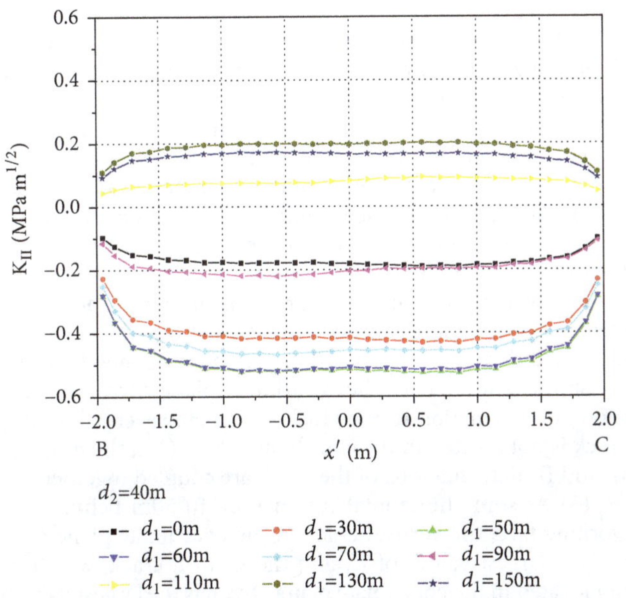

FIGURE 21: Variations of K_{II} values at the crack side (BC) for Case 2.

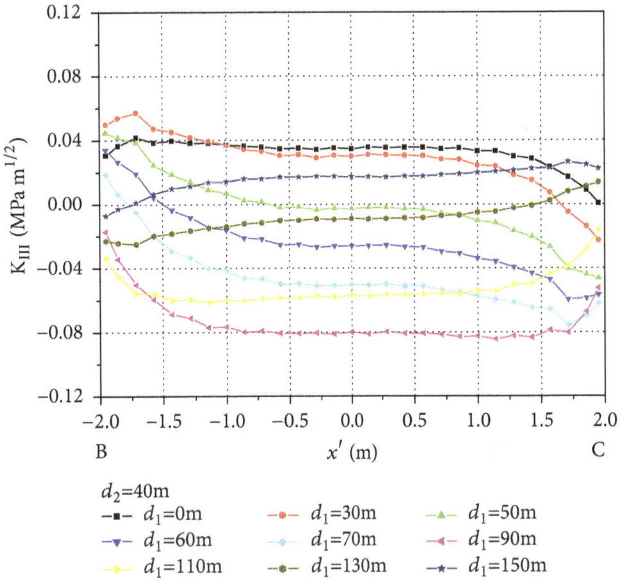

FIGURE 22: Variations of K_{III} values at the crack side (BC) for Case 2.

the K_I values for d_1 = 30m are larger than the ones of the crack only subject to water pressure. As the d_1 values increase ($d_1 \geq 30$m), the K_I values decrease. When $d_1 \geq 90$m, the K_I values are equal to zero. This means that the crack directly below the goaf is opened and the crack directly below the coal seam is closed under the action of water pressure and disturbed stresses. By comparing the values shown in Figures 19 and 20, it can be found that the K_I values for d_2 = 40m are smaller than the ones for d_2 = 0m at a given d_1 value.

In Figure 21, the K_{II} values are, respectively, negative and positive for d_1 = 0 − 90m and d_1 = 110 − 150m. For d_1 = 0−90m, as the d_1 values increase, the K_{II} values first decrease, arrive at the minimum values at d_1 = 50m, and then increase. For d_1 = 110 − 150m, as the d_1 values increase, the K_{II} values first increase, arrive at the maximum values at d_1 = 130m, and then decrease. It is believed that the K_{II} values tend to zero when the absolute values of d_1 tend to an infinite value.

In Figure 22, the K_{III} values for d_1 = 0m are positive. As the d_1 values increase, the K_{III} values decrease along the most part of the crack side BC, become negative values, and arrive at the minimum values at d_1 = 90m. Then, the K_{III} values increase along the most part of the crack side BC. It is believed that the K_{III} values tend to zero when the absolute values of d_1 tend to an infinite value.

6. Analysis of Crack Growths Induced by Mining

6.1. Case 1: The Crack Is Located in the Coordinate Plane Oxz. As discussed in Section 5.2, the crack has only mode I deformation type in this case. The crack growth can be analyzed by using the fracture criterion $K_I = K_{IC}$. Chong et al. [19] investigated the fracture toughness K_{IC} of layered rocks by using three-point-bend specimens and presented the K_{IC} values of oil shale. Using an analogical method, the K_{IC} value of the shale in Table 1 is estimated to be 0.74 MPa $\cdot$ m$^{1/2}$. As shown in Figure 19, the maximum K_{Imax} values are about 0.2335 MPa $\cdot$ m$^{1/2}$ and $K_{Imax} < K_{IC}$. This means that this type of crack does not grow and the confined water does not uplift under the action of water pressure and disturbed stresses.

6.2. Case 2: The Crack Is Located in the Plane y = 40m, i.e., d_2 = 40m. As discussed in Section 5.3, the mode I, II, and III deformations of the crack are coupled together. Growth of a crack in elastic solids subject to complex stress states can be assessed using the so-called S-criterion proposed by

Sih [20]. The S-criterion was developed on the basis of the strain energy density concept. It states that local instability is assumed to occur when the local minimum energy factor S_{min} reaches a critical value S_{cr}. Using the fracture criterion and the SIF values in Section 5.3, the crack growth for Case 2 will be discussed in the following.

The strain energy density factor S of a three-dimensional crack can be defined as [21]

$$S(\theta) = a_{11}(\theta) K_I^2 + 2a_{12} K_I K_{II} + a_{22}(\theta) K_{II}^2 + a_{33}(\theta) K_{III}^2 \quad (7)$$

where

$$a_{11}(\theta) = \frac{1}{16\pi\pi\mu}(3 - 4\mu - \cos\theta)(1 + \cos\theta)$$

$$a_{12}(\theta) = \frac{1}{8\pi\mu}\sin\theta(\cos\theta - 1 + 2\nu)$$

$$a_{22}(\theta) = \frac{1}{16\pi\mu}[4(1-\nu)(1-\cos\theta) + (3\cos\theta - 1)(1 + \cos\theta)] \quad (8)$$

$$a_{33} = \frac{1}{4\pi\mu}$$

where μ is the shear modulus of elasticity, ν is Poisson's ratio, and θ is an angle in the coordinate plane normal to the crack line of crack front.

Using the SIF values presented in Section 5.3 and (7), the local minimum energy factor S_{min} can be calculated. Figure 23 illustrates the variations of S_{min} values with the distance d_1. At $d_1 = 0, 90, 110, 130, 150$m, the S_{min} values are smaller than the ones at $d_1 = 30, 50, 60, 70$m. From $d_1 = 0$ to 90 m, S_{min} first increases, arrives at the maximum value at $d_1 = 50$m, and then decreases. From $d_1 = 90$ to 150 m, S_{min} first decreases, arrives at the minimum value at $d_1 = 110$m, and then increases. It can be estimated that S_{min} tends to zero as the floor strata are far from the coal wall of the working face.

The following conclusions can be drawn:

(i) For $d_1 \leq 90$m, S_{min} is the largest at $d_1 = 50$m whilst for $d_1 > 90$m, S_{min} is the largest at $d_1 = 110$m. This means that S_{min} has two largest values before and after the coal pillar of the working face. Whether the crack grows or not depends on the S_{cr} magnitude of the shale.

(ii) The maximum values of S_{min} at $d_1 = 50$m are larger than the ones at $d_1 = 110$m. The crack growth becomes easy when the crack is located below the goaf. It has been found that the K_I values for Case 1 are the largest at $d_1 = 40, 50$m. This means that the growth of any cracks parallel to the coordinate plane Oxz is easy at some horizontal distance of 40-50m behind the coal wall of the working face.

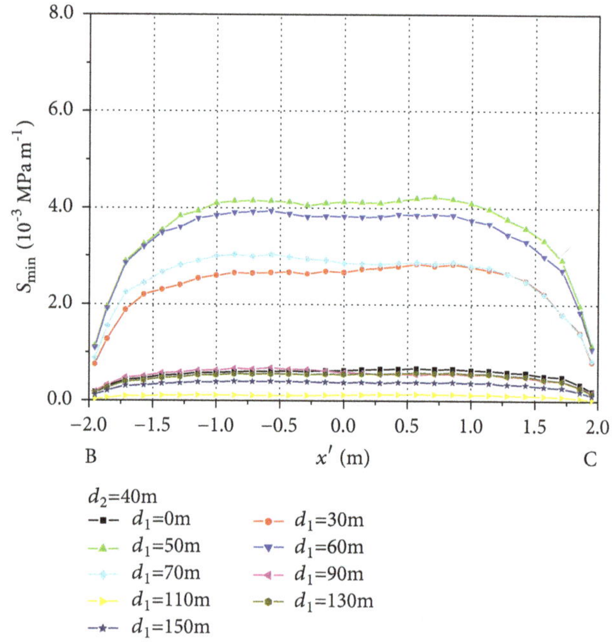

FIGURE 23: Variations of S_{min} values at the crack side (BC) for Case 2.

7. Conclusions

Numerical methods and mechanical models have been proposed for analyzing the stress fields and crack growths in floor strata for longwall mining. The additional stresses in the floor strata induced by mining are calculated and the crack problems subject to mining-induced stresses and water pressure are investigated. The research results presented in this paper can be summarized as follows:

(1) The additional stresses of the floor strata for longwall mining are described in detail and the distribution features of all the stress components with the depth below the floor and the distance from the working face are obtained.

(2) By analyzing the crack problems under the action of water pressure and disturbed stresses, the SIF values of the crack at different positions of floor strata are obtained. When the crack is located in the coordinate plane Oxz, only the mode I deformation of the crack exists. However, when the crack is not located in the coordinate plane Oxz, the mode I, II, and III deformations of the crack are coupled together.

(3) At some horizontal distance of 40-50m behind the working face, the vertical crack in the coordinate plane Oxz has the largest values of K_I and the vertical crack, which is not located in the coordinate plane Oxz, has the largest values of S_{min}. At these positions, the crack growths become easier. Therefore, the field measures should be taken in the advancement of the working face to prevent the water inrushes into the goaf behind working faces.

It is the authors' belief that the analytical methods presented in this paper can be a powerful numerical tool, which can apply to other types of cracks in the floor strata of the coal seam threatened by confined aquifers. Some of the related problems are currently under investigation by the authors for further understanding the complex mechanisms of water inrushes.

Conflicts of Interest

The authors declare that they have no conflicts of interest.

Acknowledgments

The work presented in this paper was supported by the National Natural Science Foundation of China (Grant No. 41672291) and SDUST Research Fund (2015TD-JH104).

References

[1] W. F. Zhou, "Karst water control and management in the Hanxing mining are of North China," *Environmental Geology*, vol. 3-4, no. 30, pp. 280–284, 1997.

[2] G. Li and W. Zhou, "Impact of karst water on coal mining in North China," *Environmental Geology*, vol. 49, no. 3, pp. 449–457, 2006.

[3] Q. Wu, W. Zhou, D. Li, Z. Di, and Y. Miao, "Management of karst water resources in mining area: Dewatering in mines and demand for water supply in the Dongshan Mine of Taiyuan, Shanxi Province, North China," *Environmental Geology*, vol. 50, no. 8, pp. 1107–1117, 2006.

[4] Z. G. Jing, B. Y. Li, and Z. P. Sun, "In situ measurement of the floor strata failure in the No. 2 coal mine of Fengfeng coalfield," SDUST Project 1979-1, 1979 (Chinese).

[5] B. Y. Li, Z. P. Sun, and Z. C. Liu, "In situ measurement of the floor strata failure in the No. 3 coal mine of Jingxing coalfield," SDUST Project 1986-1, 1987.

[6] H. T. Xiao and H. Y. Xu, "In situ permeability measurements to establish the influence of slice mining on floor rocks," *International Journal of Rock Mechanics and Mining Sciences*, vol. 37, no. 5, pp. 855–860, 2000.

[7] J. Zhang, "Investigations of water inrushes from aquifers under coal seams," *International Journal of Rock Mechanics and Mining Sciences*, vol. 42, no. 3, pp. 350–360, 2005.

[8] J. A. Wang and H. D. Park, "Coal mining above a confined aquifer," *International Journal of Rock Mechanics and Mining Sciences*, vol. 40, no. 4, pp. 537–551, 2003.

[9] T. H. Yang, J. Liu, W. C. Zhu, D. Elsworth, L. G. Tham, and C. A. Tang, "A coupled flow-stress-damage model for groundwater outbursts from an underlying aquifer into mining excavations," *International Journal of Rock Mechanics and Mining Sciences*, vol. 44, no. 1, pp. 87–97, 2007.

[10] L. C. Li, T. H. Yang, Z. Z. Liang et al., "Numerical investigation of groundwater outbursts near faults in underground coal mines," *International Journal of Coal Geology*, vol. 85, no. 3-4, pp. 276–288, 2011.

[11] Q. Wu., Y. Z. Liu, and Y. Liu, "Using the Vulnerable Index Method to Assess the Likelihood of a Water Inrush through the Floor of a Multi-seam Coal Mine in China," *Mine Water and the Environment*, vol. 30, no. 1, pp. 54–60, 2011.

[12] M. R. Islam, D. Hayashi, and A. B. M. Kamruzzaman, "Finite element modeling of stress distributions and problems for multi-slice longwall mining in Bangladesh, with special reference to the Barapukuria coal mine," *International Journal of Coal Geology*, vol. 78, no. 2, pp. 91–109, 2009.

[13] J. P. Zuo, S. P. Peng, Y. J. Li, Z. H. Chen, and H. P. Xie, "Investigation of karst collapse based on 3-D seismic technique and DDA method at Xieqiao coal mine, China," *International Journal of Coal Geology*, vol. 78, no. 4, pp. 276–287, 2009.

[14] H. T. Xiao, Z. Q. Yue, and X. M. Zhao, "A generalized Kelvin solution based method for analyzing elastic fields in heterogeneous rocks due to reservior water impoundment," *Computers & Geosciences*, vol. 43, no. 6, pp. 126–136, 2012.

[15] Z. Q. Yue, "On generalized Kelvin solutions in a multilayered elastic medium," *Journal of Elasticity*, vol. 40, no. 1, pp. 1–43, 1995.

[16] H. T. Xiao and Z. Q. Yue, "A three-dimensional displacement discontinuity method for crack problems in layered rocks," *International Journal of Rock Mechanics and Mining Sciences*, vol. 48, no. 3, pp. 412–420, 2011.

[17] H. T. Xiao and Z. Q. Yue, *Fracture Mechanics in Layered And Graded Materials: Analysis Using Boundary Element Methods*, De Gruyter, Berlin, Germany, 2014.

[18] R. J. Sanford, *Principles of Fracture Mechanics*, Prentice Hall: Pearson Education, Inc., 2003.

[19] K. P. Chong, M. D. Kuruppu, and J. S. Kuszmaul, "Fracture toughness determination of layered materials," *Engineering Fracture Mechanics*, vol. 28, no. 1, pp. 43–54, 1987.

[20] G. C. Sih and B. C. K. Cha, "A fracture criterion for three-dimensional crack problems," *Engineering Fracture Mechanics*, vol. 6, no. 4, pp. 699–723, 1974.

[21] R. J. Hartranft and G. C. Sih, "Stress singularity for a crack with arbitrary curved front," *Engineering Fracture Mechanics*, vol. 9, no. 3, pp. 705–718, 1976.

15

Slip Line Theory Based Stability Analysis on the Influence of Deep Excavation on Adjacent Slope

Aizhao Zhou,[1] Chengchao Li,[1] Pengming Jiang,[1] Kai Yao,[2] Na Li,[3] and Wei Wang[3]

[1]*Department of Civil and Architecture Engineering, Jiangsu University of Science and Technology, Zhenjiang, Jiangsu 212003, China*
[2]*Department of Civil & Environmental Engineering, National University of Singapore, Singapore 117576*
[3]*School of Civil Engineering, Shaoxing University, Shaoxing, Zhejiang 312000, China*

Correspondence should be addressed to Kai Yao; yaokai@u.nus.edu and Wei Wang; wellswang@usx.edu.cn

Academic Editor: Edoardo Artioli

The impact of deep excavation to the stability of adjacent slope is evaluated based on the slip line theory. Stress field of slope under various excavation conditions is simulated by finite element method, while slip line field is determined by non-associated flow rule. Factor of safety is obtained by integrating the skid-resistance and the shear stress on each slip line, and the slip line with minimum factor of safety corresponds to the critical slip surface. Two typical displacement constraint boundaries are considered. The results indicate that the critical slip surface moves towards to the slope surface and develops downwards. The factor of safety decreases with the excavation process. For flexible displacement constraint boundary, large deformation of supporting pile causes obvious variation of critical slip surface and factor of safety. In terms of the stiff displacement constraint boundary with internal supports, deep excavation only has limited effect on the slope stability.

1. Introduction

The buildings are extensively crowded with the rapid development of city, leading to complex construction sites for deep excavation. The situation is more serious in hilly areas like Chongqing in China, as there are numerous slopes adjacent to the excavation sites. Deep excavation adjacent to a slope brings about a lot of challenges for the engineers, as the sliding mechanism of slope tends to be more complicated. The excavation disrupts the original earth pressure balance in the field. Moreover, the supporting structures for deep excavation impose restrictions for the displacement of slope toe and the soils behind the structures. Thus, the commonly used limit equilibrium method (LEM) is no more available [1, 2]. Luo [3] investigated the slope stability whereby deep excavation is around the slope toe by strength reduction method (SRM), considering the effect of excavation depth and soil parameters. The stress and displacement fields can be obtained through SRM under the failure state of slope, in order to achieve the factor of safety based on the constitutive model [4, 5]. However, SRM could not analyze the influence of supporting structure stiffness on the slope stability. Therefore, the variation of factor of safety and critical slip surface is unavailable under different working conditions.

According to the theory of plastic mechanics, slip line method is to take the soils as a rigid-plastic material and generate a slip line field for the ultimate volume force which satisfies the boundary conditions [6–8]. Zhu et al. [9] proposed a more practical "potential slip line theory" by treating soil as elastoplasticity material. This method is able to calculate the factor of safety and identify the critical slip surface of slope based on the stress field generated by finite element method (FEM). In this paper, the stress fields of slope under various excavation conditions are simulated by FEM, considering the stiffness of supporting structure. Based on the above analysis, the distribution of slip lines in the slope are then obtained according to stress characteristic equation. Finally the corresponding factor of safety on each slip line is figured out by integrating the stress in the slip line field. The slip line with the minimum factor of safety is the critical slip surface of slope for various excavation conditions. Two typical displacement constraint boundaries are considered,

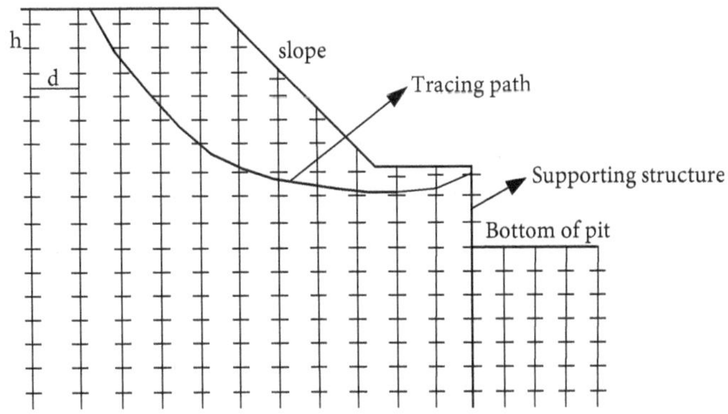

FIGURE 1: Tracing paths and control points (h and d refer to the element dimension).

namely, flexible displacement constraint boundary and stiff displacement constraint boundary.

2. Stress Field during Excavation

To simulate the stress filed during excavation by FEM, the calculation process can be divided into several steps:

(1) The initial stress field $\{\sigma_0\}$ and displacement field $\{\delta_0\}$ in the site caused by deadweight and external load are calculated before deep excavation. Taking into account the fact that the soil is completely consolidated before excavation, the initial displacement field $\{\delta_0\} = 0$.

(2) Calculate the subsidiary stress field $\{\Delta\sigma_i\}$ and displacement field$\{\Delta\delta_i\}$ caused by the first unloading due to excavation and the construction of supporting structures. Then the total stress field $\{\sigma_i\}$ and displacement field $\{\delta_i\}$ in the site are as follows:

$$\begin{aligned} \{\sigma_i\} &= \{\sigma_{i-1}\} + \{\Delta\sigma_i\} \\ \{\delta_i\} &= \{\delta_{i-1}\} + \{\Delta\delta_i\} \end{aligned} \tag{1}$$

(3) The second step is repeated until the final stress field $\{\sigma\}$ and displacement field $\{\delta\}$ are obtained:

$$\begin{aligned} \{\sigma\} &= \{\sigma_0\} + \sum \{\Delta\sigma_i\} \\ \{\delta\} &= \{\delta_0\} + \sum \{\Delta\delta_i\} \end{aligned} \tag{2}$$

In order to generate the slip line field and ensure the uniformity of slip line variation in the tracing path, the control points and grids are applied, as shown in Figure 1.

3. Slip Line Field

For the plane strain condition, two mutually perpendicular principal stresses are there at any point on the plane [10]. By connecting the lines representing the principal stress direction at each point, two clusters of curves orthogonal to each other can be obtained. They are called principal stress traces, which can be shown as lines 1-1 and 2-2 in Figure 2.

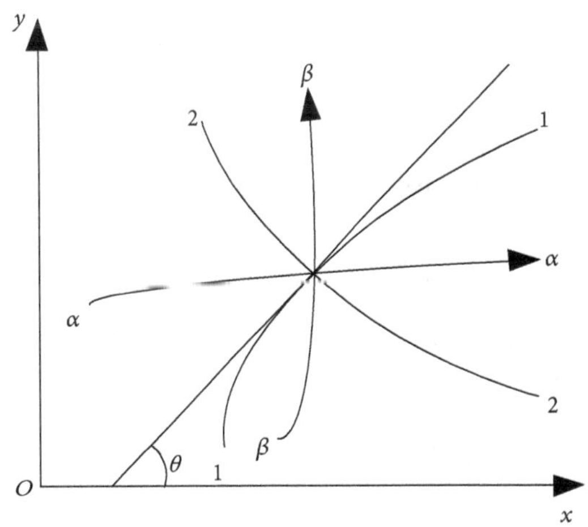

FIGURE 2: Principal stress traces and slip lines.

For soils in plastic state, there are two shear failure surfaces at each point [10]. By connecting the shear failure surface (or slip surface) of each point, two clusters of curves named slip line are generated shown as lines $\alpha - \alpha$ and $\beta - \beta$ in Figure 2. The tangential direction of one point on the slip line is the direction of the sliding surface of the corresponding point. θ is the angle between the tangent direction of major principal stress and x axis.

For Mohr-Coulomb model based on the associated flow rule, the angle μ between the slip line and the principal stress is calculated by the following [11]:

$$\mu = \frac{\pi}{2} - \frac{\varphi}{2} \tag{3}$$

in which, φ is the internal friction angle.

However, it is reported that soil does not follow the associated flow rule (e.g., [12, 13]). Moreover, the associated flow rule overestimates the dilatancy of soils, resulting in nonconservative calculation of slope stability [13–15]. Up to now, Davis [16], Drescher and Detournay [14], and Yang and Huang [17] generated the slip line field by non-associated flow rule with adjusted soil strength parameters. It was indicated that the relationship between normal stress and shear stress on the slip line is similar to Mohr-Coulomb criterion:

$$F = \tau' - c' - \sigma_n' \tan\varphi' \quad (4)$$

where the strength parameters c' and φ' have similar mechanical function with parameters c and φ of Mohr-Coulomb criterion. Their relationships can be represented as follows:

$$\frac{c'}{c} = \frac{\tan\varphi'}{\tan\varphi} = \frac{\cos\psi\cos\varphi}{1 - \sin\psi\sin\varphi} \quad (5)$$

where ψ is the dilatancy angle of soil.

The main difference between non-associated flow rule and associated flow rule lies on the selection of dilatancy angle. When the dilatancy angle ψ is taken to be equal to the angle of shearing resistance φ, it is associated flow rule. When $0 \le \psi < \varphi$, the soil obeys non-associated flow rule. Kong et al. [18] investigated the dilatancy angle for geomaterials under the non-associated flow rule and found that ψ can be selected as $\varphi/2$ instead of just treating ψ to be zero. Therefore, in this paper, ψ is set as $\varphi/2$.

In terms of non-associated flow rule, the angle between slip line and the major principal stress (μ') in the plastic region should be adjusted to the following:

$$\mu' = \frac{\pi}{4} - \frac{\psi}{2} \quad (6)$$

while the characteristic equations for slip line α and line β are described as

$$\frac{dy}{dx} = \tan\left(\theta - \mu'\right) \quad (\textit{slip line } \alpha)$$
$$\frac{dy}{dx} = \tan\left(\theta + \mu'\right) \quad (\textit{slip line } \beta) \quad (7)$$

According to the "potential slip line theory" proposed by Zhu et al. [9], there are two sliding surfaces orthogonal to each other with the smallest vertical shear capacity for each point in the elastic region. Corresponding to angle μ' in plastic region, the angle in elastic region μ_e is adjusted to be

$$\mu_e = \frac{\pi}{4} - \frac{\theta_{\max}}{2} \quad (8)$$

where $\theta_{\max} = \arcsin(((\sigma_1 - \sigma_3)/2)/(-(\sigma_1 - \sigma_3)/2 + c\cot\varphi))$. σ_1 and σ_3 are the maximum principal stress and minimum principal stress, respectively.

As shown in Figure 1, mesh size is controlled by the point spaces h and d. Small space will lead to high accuracy of the determination of slip line field. The stress of unit node is obtained by FEM, while the direction of slip line at each point is calculated by the above method. The first point can be anyone except those on the top of the slope. The direction of slip line at next nearby point follows that of the current point. In general, the intersection of the slip line and the control path does not fall on the control points. It is necessary to determine the sliding direction of the intersection by linear interpolation of the adjacent upper and lower control points. Then the above steps are repeated until a complete slip line is formed to the boundary supporting structure. Various slip lines will be generated for different starting points, by which the slip line field is determined.

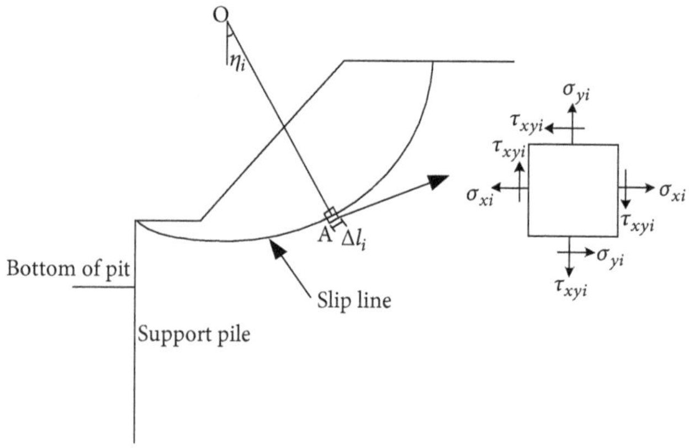

FIGURE 3: Stresses on the slip line.

4. Factor of Safety

Conventionally, factor of safety is defined as the ratio of the slip resistance to the downward force, while, for each slip line, it is the ratio of the total shear strength to the shear stress on the arc segment. During the path tracing of slip line, the stresses σ_{xi}, σ_{yi}, and τ_{xyi} at each point on slip line and the maximum principal stress angle θ can be calculated by FEM. In Figure 3, the slip line can be treated as a combination of many continuous small segments Δl_i. The equivalent stress on each segment can be represented by the average stress of the two nearby points. For example of the equivalent center point A on the segment Δl_i, its stresses are as follows:

$$\overline{\sigma}_{xi} = \frac{\sigma_{xi} + \sigma_{x(i+1)}}{2}$$
$$\overline{\sigma}_{yi} = \frac{\sigma_{yi} + \sigma_{y(i+1)}}{2} \quad (9)$$
$$\overline{\tau}_{xyi} = \frac{\sigma_{xyi} + \sigma_{xy(i+1)}}{2}$$

while its angle between the tangent direction of major principal stress and x axis is

$$\overline{\theta}_i = \frac{(\theta_i + \theta_{i+1})}{2} \quad (10)$$

Assuming that the angle between the segment Δl_i and horizon is η_i, the equivalent normal stress $\overline{\sigma}_{ni}$ and shear stress $\overline{\tau}_i$ are

$$\overline{\sigma}_{ni} = \frac{1}{2}\left(\overline{\sigma}_{xi} + \overline{\sigma}_{yi}\right) - \frac{1}{2}\left(\overline{\sigma}_{xi} - \overline{\sigma}_{yi}\right)\cos 2\eta_i + \overline{\tau}_{xyi}\sin 2\eta_i \quad (11)$$

$$\overline{\tau}_i = -\frac{1}{2}\left(\overline{\sigma}_{xi} - \overline{\sigma}_{yi}\right)\sin 2\eta_i + \overline{\tau}_{xyi}\cos 2\eta_i \quad (12)$$

where $\eta_i = \theta_i - \mu$ and $\mu = \pi/4 - \varphi/4$, according to the non-associated flow rule.

After getting the equivalent normal stress and shear stress of all the segments on the slip line, the total shear force and

TABLE 1: Deep excavation process.

Process	Finite element analysis
Step 1	Initial stress calculation (dead-weight plus a uniform load of 20kPa)
Step 2	Construction of bored piles and first support
Step 3	Excavation to 3m depth and construction of second support
Step 4	Excavation to the bottom of the pit
Step 5	Construction of baseboard

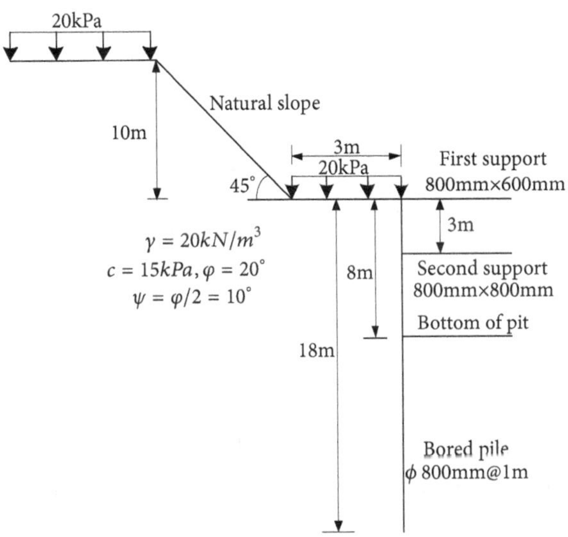

FIGURE 4: Modelling and soil parameters.

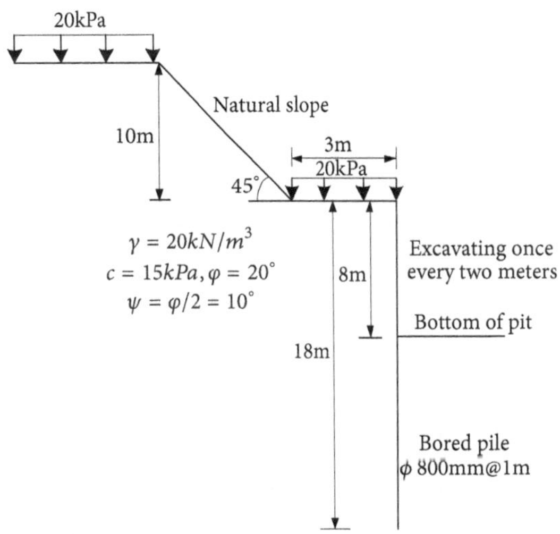

FIGURE 5: Calculation model with weak displacement constraint boundary.

shear strength along the entire slip line are achieved. Thus, the factor of safety corresponding to each slip line is

$$F_s = \frac{\sum_{i=1}^{n} \left(c + \overline{\sigma}_{ni} \tan\varphi\right) \Delta l_i}{\sum_{i=1}^{n} \overline{\tau}_i \Delta l_i} \tag{13}$$

where n represents the number of elements on the slip line. Since the factor of safety can be obtained by integration of shear stress and shear strength on each slip line, the critical slip surface is the slip line with the minimum factor of safety.

5. Case Study

5.1. Modelling and Soil Parameters. Figure 4 shows the modelling and soil parameters for the slope adjacent to deep excavation. The soil is simplified as a single homogeneous layer. The supporting structures are 3m adjacent to the slope toe, while the excavation depth is 8m. The process for deep excavation is summarized in Table 1.

5.2. Flexible Displacement Constraint Boundary. According to Shen [19], the problems of earth pressure can be divided into three categories according to different boundary conditions. Bearing capacity of foundation and earth pressure of retaining wall belong to the first category, while earth pressure of tunnel belongs to the third category. Slope stability and deep excavation support are classified under the second category, which is to calculate the ultimate load resulting in the rigid body sliding with knowledge of the external load. The boundary is unconstraint for natural slope. Oppositely, the boundary of deep excavation support is displacement constraint. For deep excavation adjacent to a steep slope like the case shown in Figure 4, the bored pile is to take the earth pressure. Due to the unbalanced earth pressure around the retaining pile, there may be some amount of deformation on the pile resulting in further sliding of the slope to the excavation side. Thus, the boundary is flexible displacement constraint. Figure 5 shows the calculation model, considering the bored piles as support structure. Excavation is conducted once every two meters, by then the position of critical slip surface and factor of safety under different excavation conditions are also calculated. Specific analysis of excavation process is shown in Table 2.

Figure 6 illustrates that variation of critical slip surface with the process of excavation. It can be identified that the critical slip surface gradually moves to the slope surface and develops downward. In Figure 7, it can be seen that the factor of safety almost reduces linearly with the step of excavation. The supporting bored pile restrains the displacement of slope toe, preventing the slope from sliding towards the foundation pit. Therefore from step 1 to step 2, the end of slip surface moves towards the slope toe, resulting in smaller factor of safety (increased downward force with reduced slip resistance). Due to the excavation process from step 3, the deformation of bored pile gradually increases leading to

Table 2: Excavation process under weak displacement constraint boundary.

Process	Finite element analysis
Step 1	Initial stress calculation (dead-weight plus a uniform load of 20kPa)
Step 2	Construction of bored piles
Step 3~6	Excavating once every two meters

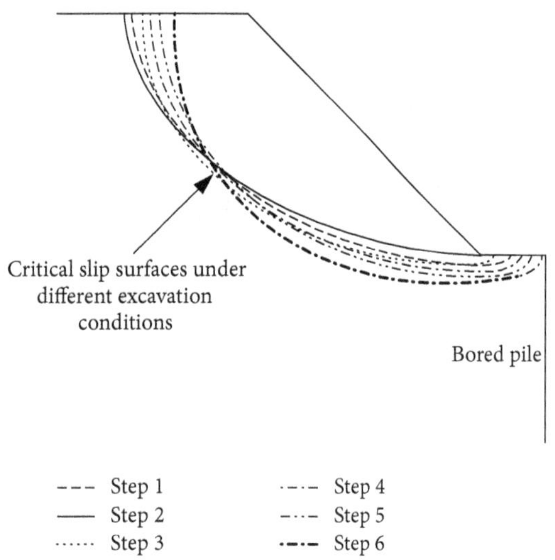

Figure 6: Distribution of critical slip surfaces under different excavation process.

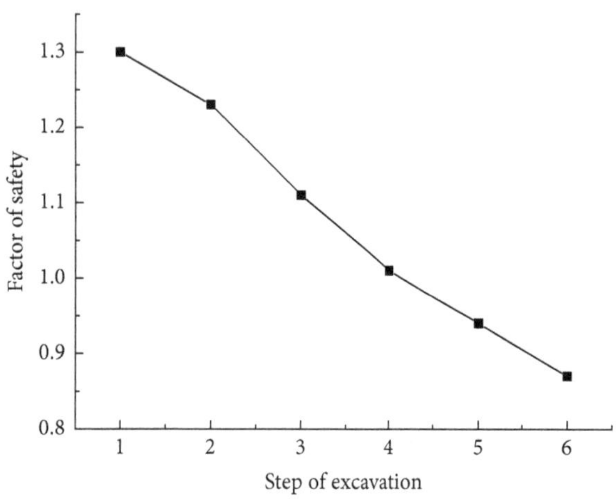

Figure 7: Variation of factor of safety under different excavation process.

gradually weakened displacement constraint boundary. Then the soil behind the supporting pile slides towards the pit, and the shape of slip surface develops from the original arc to an approximately logarithmic helix. For the whole excavation process, the calculated factor of safety gradually reduces from the initial 1.30 (step 1) to 0.87 (step 6). So it reveals that deep excavation has a significant impact on the stability of adjacent slope. The construction of supporting structure leads to the variation of critical slip surface. If only replying on bored piles

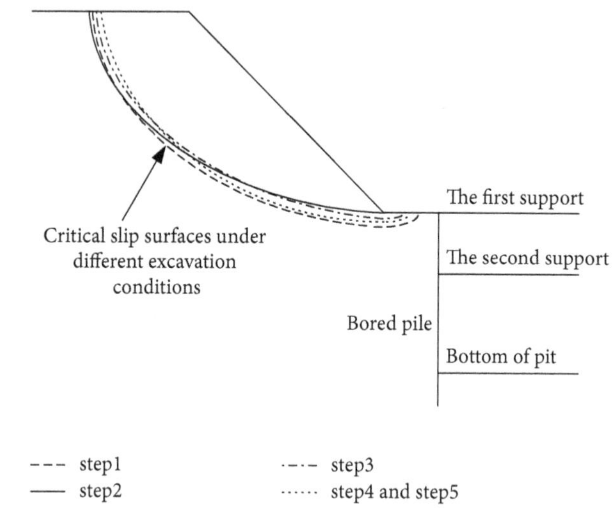

Figure 8: Distribution of critical slip surfaces under different excavation process.

as supporting structure, its deformation is large. However, the pile does not fail as the sliding force is less than the shear strength of pile. Bending of the pile will push the soil to slide to the pile head, so the displacement of pile head is large. Thus, restriction of the pile head movement is an effective way for improving the slope stability.

5.3. Stiff Displacement Constraint Boundary. With the internal support for the bored pile (like first and second support in Figure 8), the deformation of supporting pile is rigidly limited. So the support structure with both bored pile and internal supports is considered as a stiff displacement constraint boundary. For this case, the calculation model and excavation process can refer to Figure 4 and Table 1, respectively.

It can be identified from Figure 8 that the excavation has a limited effect on the critical slip surface because of the stiff displacement constraint boundary. The trend of slip surfaces is similar to that of the flexible displacement constraint boundary, as the critical slip surface also gradually moves to the slope surface and develops downward. However, the variation of slip surfaces is smaller. Compared with the critical slip surface of slope under natural state, the size of sliding block is slightly reduced due to the excavation. As shown in Figure 9, factor of safety reduces from 1.30 to 1.02 during the excavation. However, the construction of baseboard (step 5 in Table 1) does not make any difference to slope stability. The above findings are due to that the

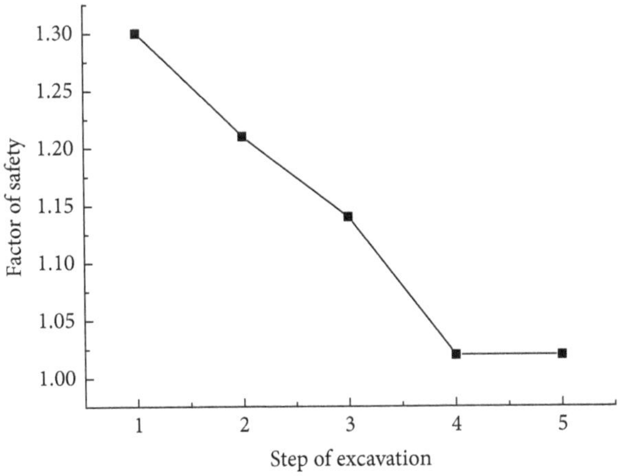

FIGURE 9: Variation of factor of safety under different excavation process.

two internal supports have greatly limited the deformation of the pile, weakening the influence of excavation on the slope stability. But some impact still exists due to the deep excavation.

6. Conclusion

Based on the "potential slip line theory" proposed by Zhu et al. [9], corresponding slip line field is built by considering the stress fields under different excavation conditions with FEM. Factor of safety is obtained with numerical integration of shear stress and shear strength on each slip line, and the critical slip surface is the one with the least factor of safety. Compared with traditional computing methods of slope stability, the method adopted in this study can effectively simulate the variation of slip surface and factor of safety under different excavation conditions. Two typical boundary conditions (i.e., flexible displacement constraint boundary and stiff displacement constraint boundary) are taken into account for the slope stability. For each boundary condition, the variation of critical slip surfaces and factor of safety are investigated. For the case with flexible displacement constraint boundary, due to excavation, the critical slip surface gradually moves to slope surface and develops downward, while the factor of safety gradually reduces. As the function of internal support of stiff displacement constraint boundary, excavation has smaller effect on slope stability compared with the case of flexible displacement constraint boundary, although the variations of critical slip surface and factor of safety show similar trend. It implies that the stiff supporting structure is beneficial for the safety of adjacent slope during deep excavation.

It should be noted that the limitation of this study is that uniform soil profile is adopted for the analysis. However, the soil condition is always complex and randomly distributed in the construction site. Since spatial variations of soil properties have been reported recently (e.g., [20]), further research considering nonhomogeneous soils will be conducted in the future.

Conflicts of Interest

The authors declare that they have no conflicts of interest.

Acknowledgments

This study was financially cosupported by the National Natural Science Foundation of China (Grants nos. 51579119 and NO.41772311) and the Natural Science Foundation of Zhejiang Province (no. LY17E080016).

References

[1] A. W. Bishop, "The use of the slip circle in the stability analysis of slopes," *Géotechnique*, vol. 5, no. 1, pp. 7–17, 1955.

[2] D. G. Fredlund and J. Krahn, "Comparison of slope stability methods of analysis," *Canadian Geotechnical Journal*, vol. 14, no. 3, pp. 429–439, 1977.

[3] L. Peng, *The soil stability analysis of soft soil slope toe excavation. Master Thesis [Master, thesis]*, Hefei University of Technology, Hefei, China, 2013 (Chinese).

[4] T. Matsui and K. C. San, "Finite element slope stability analysis by shear strength reduction technique," *Soils and Foundations*, vol. 32, no. 1, pp. 59–70, 1992.

[5] H. Zheng, G. H. Sun, and D. F. Liu, "A practical procedure for searching critical slip surfaces of slopes based on the strength reduction technique," *Computers & Geosciences*, vol. 36, no. 1-2, pp. 1–5, 2009.

[6] W. Chen, "Preface," in *Limit Analysis and Soil Plasticity*, vol. 7 of *Developments in Geotechnical Engineering*, pp. IX–X, Elsevier, 1975.

[7] D. C. Drucker and W. Prager, "Soil mechanics and plastic analysis or limit design," *Quarterly of Applied Mathematics*, vol. 10, pp. 157–165, 1952.

[8] W. F. Chen and X. L. Liu, *Limit analysis in soil mechanics*, Elsevier, 2012.

[9] Y.-W. Zhu, C.-Q. Wu, and Y.-Q. Cai, "Determination of slip surface in slope based on theory of slip line field," *Yanshilixue Yu Gongcheng Xuebao/Chinese Journal of Rock Mechanics and Engineering*, vol. 24, no. 15, pp. 2609–2616, 2005.

[10] Y. R. Zheng, C. J. Deng, and J. L. Wang, "The study of slip line field and upper bound method based on the non-associated flow rule," *Engineering Science*, vol. 8, no. 12, pp. 56–69, 2010 (Chinese).

[11] O. C. Zienkiewicz, C. Humpheson, and R. W. Lewis, "Associated and non-associated visco-plasticity and plasticity in soil mechanics," *Géotechnique*, vol. 25, no. 4, pp. 671–689, 1975.

[12] P. A. Vermeer and R. de Borst, "Non-associated plasticity for soils, concrete and rock," *Heron*, vol. 29, no. 3, 1984.

[13] Y.-J. Wang, J.-H. Yin, and C. F. Lee, "The influence of a non-associated flow rule on the calculation of the factor of safety of soil slopes," *International Journal for Numerical and Analytical Methods in Geomechanics*, vol. 25, no. 13, pp. 1351–1359, 2001.

[14] A. Drescher and E. Detournay, "Limit load in translational failure mechanisms for associative and non-associative materials," *Géotechnique*, vol. 43, no. 3, pp. 443–456, 1993.

[15] S. Melentijevic, A. Serrano, C. Olalla, and R. A. Galindo, "Incorporation of non-associative flow rules into rock slope stability analysis," *International Journal of Rock Mechanics and Mining Sciences*, vol. 96, pp. 47–57, 2017.

[16] E. H. Davis, "Theories of plasticity and the failure of soil masses," in *Soil Mechanics Selected Topics*, K. Lee, Ed., 1968.

[17] X.-L. Yang and F. Huang, "Slope stability analysis considering joined influences of nonlinearity and dilation," *Journal of Central South University of Technology (English Edition)*, vol. 16, no. 2, pp. 292–296, 2009.

[18] W.-X. Kong, Y.-Q. Rui, and B.-D. Dong, "Determination of dilatancy angle for geomaterials under non-associated flow rule," *Yantu Lixue/Rock and Soil Mechanics*, vol. 30, no. 11, pp. 3278–3282, 2009.

[19] Z. J. Shen, *Discontinuous solution in limit earth pressure computation application*, Tsinghua University Press, Beijing, China, 2005.

[20] Y. Liu, W. Zhang, L. Zhang, Z. Zhu, J. Hu, and H. Wei, "Probabilistic stability analyses of undrained slopes by 3D random fields and finite element methods," *Geoscience Frontiers*, 2017.

16

Determination and Application of Comprehensive Specific Frictional Resistance in Heating Engineering

Yanan Tian,[1] Haosen Wang,[2] Yongjiang Shi,[1] Jiayu Rong,[1] Zhenhua Ma,[2] and Cong Xie[2]

[1]*College of Energy and Environmental Engineering, Hebei Institute of Architecture and Civil Engineering, Zhangjiakou, Hebei 075000, China*

[2]*College of Mathematics and Physics, Hebei Institute of Architecture and Civil Engineering, Zhangjiakou, Hebei 075000, China*

Correspondence should be addressed to Yongjiang Shi; yy1114916122@163.com

Academic Editor: Andrea L. Facci

In this study, we analyze the deficiencies of specific frictional resistance in heating engineering. Based on economic specific frictional resistance, we put forward the concept of comprehensive specific frictional resistance, which considers the multiple factors of technology, economy, regulation modes, pipe segment differences, and medium pressure. Then, we establish a mathematical model of a heating network across its lifespan in order to develop a method for determining the comprehensive specific frictional resistance. Relevant conclusions can be drawn from the results. As an application, we have planned the heating engineering for Yangyuan County in China, which demonstrates the feasibility and superiority of the method.

1. Introduction

With the development of modern society, improvements in heating engineering have attracted attention. In the field of heating engineering, the specific frictional resistance refers to the resistance loss of pipe per meter [1, 2], which is very important in China. In this paper, we discuss specific frictional resistance during the designing or planning stage, when specific frictional resistance acts as an intermediary to select a suitable pipe diameter.

At present, studies mainly involve two aspects of specific frictional resistance: one is the weighted average specific frictional resistance; the other is economic specific frictional resistance. For a heating network with regulated change in flow rate at different stages, the specific frictional resistance undergoes a step change due to the change in flow rate. In order to measure and reflect this characteristic of the specific frictional resistance, weighted average specific frictional resistance is recommended for selecting pipe diameters. Hong [3] discussed the differences between weighted average frictional resistance and traditional specific frictional resistance and analyzed the power-saving effect by taking a heating network with flow rate regulation as an example. The economic specific frictional resistance is an ideal value relative to the specific frictional resistance from the economic perspective and is widely used in China at present. Ge et al. [4] established a mathematical model, which took the minimum cost of the heating network as the objective function, and discussed the relationship between the economic specific frictional resistance and the economic heating radius. Liu et al. [5] offered the concept of the new economic specific frictional resistance by establishing the transportation distance limit model and the transport energy consumption ratio model. Wang et al. [6] took the indoor heating horizontal single tube and vertical single tube system as examples to establish a mathematical model with minimum cost as the objective function, by which they obtained the economic specific frictional resistance under different conditions. Li et al. [7] studied the hydraulic calculation methods and the existing problems on the condensate pipe net, and they used the differential method to calculate the economic specific frictional resistance of the condensate pipe net.

Through analysis of the various research [3–11] on specific frictional resistance, we identified problems related to the value of specific frictional resistance, as follows:

(i) In China, the value of specific frictional resistance used in engineering still follows the traditional economic specific frictional resistance value of 30~70 Pa/m, which was introduced by the former Soviet Union in the last century. With the rise in prices and changes in materials, this value is not applicable to the current state of practice in heating engineering.

(ii) The value of specific frictional resistance should be related to the regulation modes, which can bring about different changes in flow rate and medium temperature. However, the traditional model of economic specific frictional resistance treats the different regulation modes in the same way [3–11].

(iii) With the development of the central heating network, pipe networks have become more massive and complex, which has led to increasing differences among pipe segments. Thus, it is not suitable to use the same specific frictional resistance standard of 30~70 Pa/m to determine the pipe diameter in a network system. Instead, each pipe segment should be differentiated. That is to say, each pipe segment should have a specific frictional resistance of its own [3–11].

(iv) The distribution of pressure in the pipes is now more complicated because of the mass and complexity of the pipe networks. It has a direct relationship to the value of specific frictional resistance whether the pressure meets the requirements. However, the value of traditional economic specific frictional resistance of 30~70 Pa/m does not take into account the constraints of the medium pressure [3–11].

In order to solve the above problems, we propose the concept of comprehensive specific frictional resistance. It refers to a measure of resistance loss of pipe per meter that incorporates a number of related factors. By means of it, heating engineering applications can achieve better economic effects during their lifespan [12–14]. The related factors considered in this paper include technology, economy [15, 16], operation regulation modes [17, 18], pipe segment differences, and the medium pressure [19–22].

2. The Mathematical Model

In this section, we give the method for obtaining the value of comprehensive specific friction resistance as follows.

2.1. Factors Related to Comprehensive Specific Frictional Resistance Theory. Through the analysis of technology, economy, operation regulation modes, pipe segment differences, and the medium pressure in a heating system, we have drawn several conclusions about the comprehensive specific friction resistance model theory:

(i) Pipe Segment Differences. With pipe networks becoming massive and complex, the difference among pipe segments becomes more apparent. For example, it is common for the pipe diameter close to the heat source to be DN1400, while the pipe diameter can be reduced to DN200 at the user end point. This difference should not be ignored. In order to account for the differences among pipe segments, a single pipe segment is taken as an object in establishing the mathematical model in this paper. In this way, we get different values corresponding to different pipe segments.

(ii) Operation Regulation Modes. We take into account the influence of operation regulation modes on comprehensive specific frictional resistance due to the fact that different operation regulation modes incur different electricity consumption and different heat loss.

(iii) Medium Pressure. The value of comprehensive specific frictional resistance takes into account the medium pressure constraint, which can prevent heating accidents on the operational stage, such as pipe cracking or medium vaporization. In the following study, the medium pressure constraint is mainly embodied in the constraint conditions of the mathematical model.

(iv) Technology and Economy. The Present Value Cost is chosen as the objective function [23, 24] by analyzing the advantages and disadvantages of each economic effect evaluation index.

(v) Technology and Economy. The value of specific frictional resistance is sensitive to the factors of primary network construction fee (including civil engineering cost and pipe installation fee), electricity fee, and fuel cost across the lifespan of heating engineering application. These factors must be embodied in the mathematical model. Heat source construction fee, heat exchange station construction fee, thermal user fuel cost, water fee, maintenance cost, and labor wages are irrelevant factors that can be ignored in the mathematical model.

2.2. Establishment of the Objective Function. In this section, we establish the objective function taking the single segment i as the object. The objective function is the Present Value Cost of pipe segment i across the lifespan of n years, and the cost includes civil engineering cost, pipe installation fee, electricity fee, and fuel cost.

For the different regulations, the civil engineering cost and the pipe installation fee are the same, but the electricity fee and fuel cost are different, because the heat load varies continuously during heating season. The following is the establishment process of the objective function.

2.2.1. The Civil Engineering Cost. Taking the directly buried laying method as an example, a set of data about pipe diameter d and the corresponding civil engineering cost m_1 from Chinese markets is shown in Figure 1.

The fitted curve is $m_1 = 255.673d^2 + 498.819d + 113.133$, whose R-squared value is as high as 99.95%. Hence, the relationship between the civil engineering cost and the pipe diameter can be expressed by a quadratic polynomial, as follows:

$$m_1 = a_1 d^2 + b_1 d + c_1, \tag{1}$$

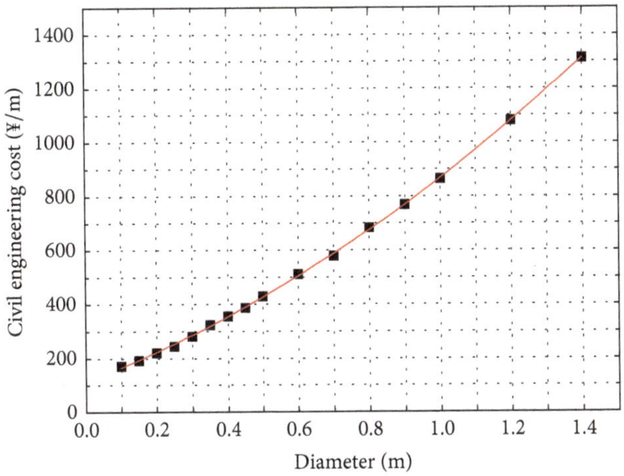

FIGURE 1: Fitting curve of heating pipe diameter and civil engineering cost.

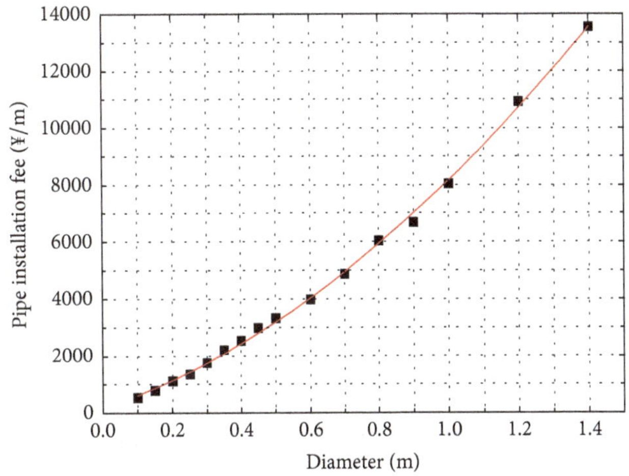

FIGURE 2: Fitting curve of heating pipe diameter and pipe installation fee.

where m_1 is the civil engineering cost of pipe with a diameter d (¥/m), d is the pipe diameter (m), and a_1, b_1, and c_1 are the fitting polynomial coefficients of civil engineering cost, which are determined by the specific situation.

As we all know, the Darcy-Weisbach Formula is

$$h_f = \lambda \frac{l}{d} \cdot \frac{v^2}{2g}, \tag{2}$$

where h_f is the head loss (m), λ is the coefficient of frictional resistance [25–27], l is the length of pipe (m), v is the velocity of medium in the pipe (m/s), and g is the gravitational acceleration (m/s^2).

The specific frictional resistance R refers to the resistance loss of pipe per meter, so we get formula (3) from formula (2):

$$R = \frac{\lambda}{d} \cdot \frac{\rho v^2}{2}, \tag{3}$$

where ρ is the density of medium (kg/m^3) and R is the specific frictional resistance (Pa/m).

Therefore, the specific frictional resistance, pipe diameter, and flow rate have the following relationship:

$$R = 0.81 \times \frac{\lambda G^2}{\rho d^5}, \tag{4}$$

where G is the flow rate (kg/s).

For a heating network, the determination of the comprehensive specific frictional resistance needs to account for the change in flow rate during operation. However, the selection of the diameter through the flow rate and the specific frictional resistance still relates to the most unfavorable situation. Therefore, the flow rate G in formula (4) is the design flow rate G'. That is,

$$d_i = 0.959 \frac{\lambda^{1/5} G_i'^{(2/5)}}{(\rho R_i)^{1/5}}, \tag{5}$$

where d_i is the pipe diameter of pipe segment i (m), G_i' is the design flow rate of pipe segment i (kg/s), and R_i is the specific frictional resistance of pipe segment i (Pa/m).

Applying the basic formula of specific heat capacity $Q = cm\Delta t$ (c is the specific heat capacity, which for water is 4.2 $\times 10^3$ J/(kg·°C), m is the quality (kg), Δt is the temperature difference (°C), and Q is the heat quantity (J)) to heating engineering, the following formula can be obtained:

$$G = 2.39 \times 10^{-4} \frac{Q}{\Delta t}, \tag{6}$$

where Q is the thermal load (W), Δt is the temperature difference (°C), and G is the flow rate of medium (kg/s).

Substituting formula (6) into formula (5), we get

$$d_i = 3.4133 \times 10^{-2} \Gamma_i, \tag{7}$$

where $\Gamma_i = \lambda^{1/5} Q_i'^{(2/5)} / (\rho R_i)^{1/5} \Delta t_{ghi}'^{\,2/5}$, $\Delta t'_{ghi}$ is the design temperature difference between supply and return water of pipe segment i (°C), and Q_i' is the design thermal load of pipe segment i (w).

Substituting formula (7) into formula (1), the civil engineering cost of pipe segment i can be given by

$$M_{1i} = \left(1.1655 \times 10^{-3} a_1 \Gamma_i^{\,2} + 3.4133 \times 10^{-2} b_1 \Gamma_i + c_1\right) l_i, \tag{8}$$

where M_{1i} is the civil engineering cost of pipe segment i (¥) and l_i is the length of pipe segment i (m).

2.2.2. The Pipe Installation Fee. The data from Chinese markets about pipe diameter d and the corresponding pipe installation fee m_2 is shown in Figure 2.

The fitted curve is $m_2 = 3881.425d^2 + 4134.162d + 155.267$, whose R-squared value is as high as 99.73%. Hence, the relationship between the pipe installation fee and the pipe diameter can also be expressed by a quadratic polynomial, as follows:

$$m_2 = a_2 d^2 + b_2 d + c_2, \quad (9)$$

where m_2 is the installation fee of pipe with a diameter d (¥/m) and a_2, b_2, and c_2 are the fitting polynomial coefficients of the installation fee, which are determined by the specific situation.

Substituting formula (7) into formula (9), the installation fee of the pipe segment i can be given by

$$M_{2i} = \left(1.1655 \times 10^{-3} a_2 \Gamma_i^2 + 3.4133 \times 10^{-2} b_2 \Gamma_i + c_2\right) l_i, \quad (10)$$

where M_{2i} is the installation fee of the pipe segment i (¥).

2.2.3. The Electricity Fee

(i) For a Heating System with Quality Regulation. A heating system with quality regulation is regulated simply by changing the water-supply temperature, so the flow rate of pipe segment i is always the design flow rate G_i'. Because the specific frictional resistance R refers to the resistance loss of pipe per meter, the annual electricity consumption of pipe segment i is

$$W_3 = R_i l_i \frac{G_i'}{\rho} \tau_i, \quad (11)$$

where W_3 is the annual electricity consumption of pipe segment i (J) and τ_i is the run time of pipe segment i per year (s).

So, the annual electricity fee for the pipe segment i can be given by

$$M_{3i} = \frac{c_3}{\eta} W_3 = \frac{c_3}{\eta \rho} R_i l_i G_i' \tau_i, \quad (12)$$

where M_{3i} is the annual electricity fee for pipe segment i (¥), c_3 is the electrovalence (¥/J), and η is the pump efficiency.

Substituting formula (6) into formula (12), we get

$$M_{3i} = 2.39 \times 10^{-4} \frac{c_3}{\eta \rho \Delta t'_{ghi}} R_i l_i Q_i' \tau_i. \quad (13)$$

(ii) For a Heating System with Flow Rate Regulation. A heating system with quality regulation is regulated by changing the flow rate. The large variation on the thermal load during the entire heating period results in a wide range of flow rate changes, so calculating the electricity fee according to the design flow rate G_i' does not adequately describe the actual situation. In order to estimate the electricity consumption accurately, we need to know the local thermal load changes and the corresponding hours to draw a thermal load duration diagram [28]. Under common conditions, the thermal load duration of the pipe segment i in heating engineering can be described as Figure 3.

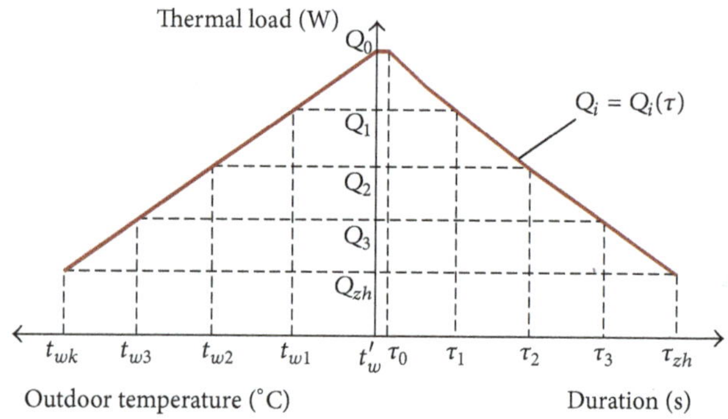

FIGURE 3: Thermal load duration diagram of the pipe segment i.

According to the definition of specific frictional resistance R, the annual electricity consumption of pipe segment i is

$$W_3 = R_i l_i \frac{1}{\rho} \int_{\tau=0}^{\tau=\tau_{zh}} G_i(\tau)\, d\tau, \quad (14)$$

where G_i is the actual flow rate of pipe segment i (kg/s), $G_i(\tau)$ is the flow rate (kg/s)-duration (s) function, and τ_{zh} is the duration corresponding to the minimum thermal load during a heating season (s).

The annual electricity fee for the pipe segment i is

$$M_{3i} = \frac{c_3}{\eta} W_3 = \frac{c_3}{\eta \rho} R_i l_i \int_{\tau=0}^{\tau=\tau_{zh}} G_i(\tau)\, d\tau. \quad (15)$$

Substituting formula (6) into formula (15), the annual electricity fee for the pipe segment i can be given by

$$M_{3i} = 2.39 \times 10^{-4} \frac{c_3}{\eta \rho \Delta t'_{ghi}} R_i l_i \Lambda_i, \quad (16)$$

where $\Lambda_i = \int_{\tau=\tau_0}^{\tau=\tau_{zh}} Q_i(\tau) d\tau$, Q_i is the actual thermal load of pipe segment i (w) and $Q_i(\tau)$ is the thermal load (w) – duration (s) function, as shown in Figure 3.

2.2.4. The Fuel Cost. The fuel cost includes two parts: one is heat loss fuel cost; the other is thermal user fuel cost. Due to the irrelevance of the specific frictional resistance to thermal user fuel cost, there is no need to embody it in the mathematical model. However, massive, complicated pipeline networks increase heat loss along the route, so heat loss fuel cost should be included in the total cost rather than ignored. The heat loss fuel cost is different for different regulation modes, because it is associated with the medium temperature.

(i) For a Heating System with Flow Rate Regulation. The heat transfer coefficient h refers to the amount of heat transfer per unit time per unit surface area for a 1°C temperature difference, so for a heating network adopting the flow rate regulation mode, the consumption of heat loss fuel is

$$W_4 = h\left(t'_{ki} - t_j\right) \pi d_i l_i \tau_i, \quad (17)$$

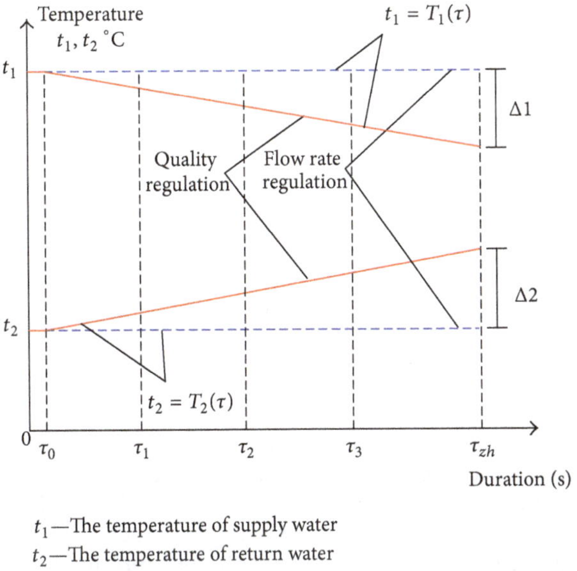

FIGURE 4: Duration of actual water-supply or water-return temperature.

where W_4 is the consumption of heat loss fuel (J), h is the heat transfer coefficient of the pipe surface (w/m^2 ·°C), t_j is the ambient temperature of pipe segment i (°C), and t'_{ki} is the design temperature of medium in pipe segment i (°C). For the water-supply pipe, $k = 1$, $t'_{ki} = t'_{gi} = t'_{hi} + \Delta t'_{ghi}$ and for the water-return pipe, $k = 2$, $t'_{ki} = t'_{hi}$, where t'_{gi} is the design water-supply temperature of pipe segment i (°C), t'_{hi} is the design water-return temperature of pipe segment i (°C).

So the annual heat loss fuel cost is

$$M_{4i} = W_4 \cdot c_4 = c_4 h \left(t'_{ki} - t_j\right) \pi d_i l_i \tau_i, \tag{18}$$

where M_{4i} is the annual heat loss fuel cost of the pipe segment i (¥) and c_4 is the fuel price (¥/J).

Substituting formula (7) into formula (18), the annual heat loss fuel cost of the pipe segment i can be given by

$$M_{4i} = 3.4133 \times 10^{-2} c_4 h \left(t'_{ki} - t_j\right) \pi \tau_i l_i \Gamma_i. \tag{19}$$

(ii) For a Heating System with Quality Regulation. For heat-supply networks adopting the quality regulation mode, calculating the heat loss fuel cost by the design temperature is unreasonable due to the variance of the medium temperature in the pipe. As we know, compared with the design water-supply temperature and design water-return temperature, a reduction in actual water-supply temperature of Δ1 is equal to an increase of Δ2 of the actual water-return temperature in the quality regulation networks [8]. This is described in Figure 4.

So, the following formula can be given:

$$t_{ki} = t'_{hi} + \frac{\Delta t'_{ghi}}{2} - (-1)^k \frac{\Delta t_{ghi}}{2}, \tag{20}$$

where t_{ki} is the actual temperature of medium in pipe segment i (°C) and Δt_{ghi} is the actual temperature difference between supply and return water of pipe segment i (°C), as shown in Figure 4.

According to the definition of heat transfer coefficient h, the consumption of heat loss fuel is

$$W_4 = h\pi d_i l_i \int_{\tau=0}^{\tau=\tau_{zh}} \left(t_{ki} - t_j\right) d\tau. \tag{21}$$

Substituting formula (20) into formula (21), we get the consumption of heat loss fuel:

$$W_4 = h\pi d_i l_i \left[\left(t'_{hi} + \frac{\Delta t'_{ghi}}{2} - t_j \right) \cdot \tau_{zh} - \frac{(-1)^k}{2} \int_{\tau=0}^{\tau=\tau_{zh}} \Delta t_{ghi} d\tau \right]. \tag{22}$$

So, the annual heat loss fuel cost of the pipe segment i is

$$M_{4i} = c_4 W_4 = c_4 h\pi d_i l_i \left[\left(t'_{hi} + \frac{\Delta t'_{ghi}}{2} - t_j \right) \cdot \tau_{zh} - \frac{(-1)^k}{2} \int_{\tau=0}^{\tau=\tau_{zh}} \Delta t_{ghi} d\tau \right]. \tag{23}$$

Substituting formula (7) into formula (23), the annual heat loss fuel cost of the pipe segment i can be given by

$$M_{4i} = 3.4133 \times 10^{-2} c_4 h\pi l_i \Gamma_i \left[\left(t'_{hi} + \frac{\Delta t'_{ghi}}{2} - t_j \right) \tau_{zh} - \frac{(-1)^k}{2} \int_{\tau=0}^{\tau=\tau_{zh}} \Delta t_{ghi} d\tau \right]. \tag{24}$$

For a heating system with quality regulation, the actual flow rate G_i is always the design flow rate G'_i, so we know

$$\frac{Q_i}{\Delta t_{ghi}} = \frac{Q'_i}{\Delta t'_{ghi}}. \tag{25}$$

Substituting formula (25) into formula (24), we get

$$M_{4i} = 3.4133 \times 10^{-2} c_4 h\pi l_i \Gamma_i \left[\left(t'_{hi} + \frac{\Delta t'_{ghi}}{2} - t_j \right) \tau_{zh} - \frac{(-1)^k \Delta t'_{ghi}}{2Q'_i} \Lambda_i \right]. \tag{26}$$

2.2.5. The Objective Function. For pipe segment i, the objective function is the Present Value Cost in its lifespan of n years. Supposing the discount rate is ε, it can be described as follows.

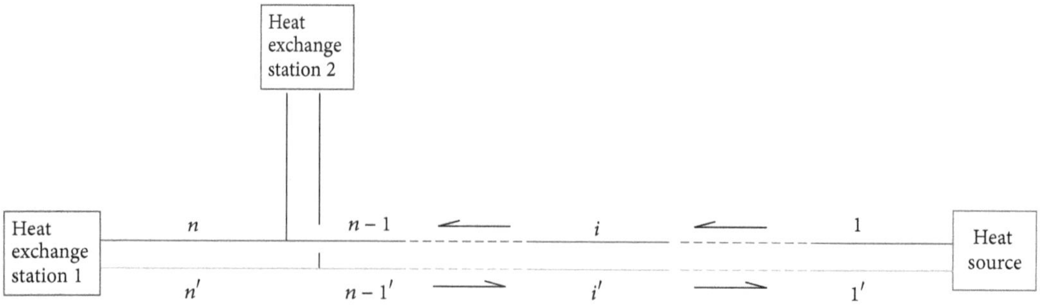

FIGURE 5: A branched network diagram.

(i) For a Heating System with Quality Regulation

$$\begin{aligned} \mathrm{PC} &= M_{1i} + M_{2i} + (M_{3i} + M_{4i}) \cdot \left(\frac{P}{A} \cdot \varepsilon \cdot n\right) \\ &= \left(1.1655 \times 10^{-3} a \Gamma_i^{\ 2} + 3.4133 \times 10^{-2} b \Gamma_i + c\right) l_i \\ &\quad + \left[2.39 \times 10^{-4} \frac{c_3}{\eta \rho \Delta t'_{ghi}} R_i l_i Q'_i \tau_i + 3.4133 \right. \\ &\quad \times 10^{-2} c_4 h \pi l_i \Gamma_i \left(\left(t'_{hi} + \frac{\Delta t'_{ghi}}{2} - t_j \right) \tau_{zh} \right. \\ &\quad \left. \left. - \frac{(-1)^k \Delta t'_{ghi}}{2 Q'_i} \Lambda_i \right) \right] \times \left(\frac{P}{A} \cdot \varepsilon \cdot n\right) \end{aligned} \tag{27}$$

(ii) For a Heating System with Flow Rate Regulation

$$\begin{aligned} \mathrm{PC} &= M_{1i} + M_{2i} + (M_{3i} + M_{4i}) \cdot \left(\frac{P}{A} \cdot \varepsilon \cdot n\right) \\ &= \left(1.1655 \times 10^{-3} a \Gamma_i^{\ 2} + 3.4133 \times 10^{-2} b \Gamma_i + c\right) l_i \\ &\quad + \left(2.39 \times 10^{-4} \times \frac{c_3}{\eta \rho \Delta t'_{ghi}} R_i l_i \Lambda_i + 3.4133 \right. \\ &\quad \left. \times 10^{-2} c_4 h \left(t'_{ki} - t_j\right) \pi \tau_i l_i \Gamma_i \right) \times \left(\frac{P}{A} \cdot \varepsilon \cdot n\right), \end{aligned} \tag{28}$$

where PC is the Present Value Cost (¥), ε is the discount rate, and n is the life period (years), $a = a_1 + a_2, b = b_1 + b_2, c = c_1 + c_2$, $(P/A \cdot \varepsilon \cdot n) = ((1+\varepsilon)^n - 1)/\varepsilon(1+\varepsilon)^n$.

2.3. Establishment of the Constraint Conditions

2.3.1. The Constraint Conditions of Medium Pressure. The pressure in the pipe must meet certain requirements in order to ensure normal operation of the system and prevent pipe burst or medium vaporization. As the research object is the primary network system, the lower pressure limit must guarantee that high-temperature water does not vaporize in the pipeline, while the upper limit is the pressure that the pipe and pipe fittings can bear. Additionally, the water-return temperature is generally less than 100°C, and for any water-supply pipe segment i, there must be a corresponding water-return pipe segment i' whose pressure is lower than the pressure in the pipe segment i. Therefore, only the water-supply pipes need to obey the pressure constraints. In addition, in order to ensure that the pressure in the pipe segment meets the requirements, the starting and ending points of the pipe must also meet the requirements.

A branched network diagram without distributed pressure pumps is shown in Figure 5. Pipe segments $1', 2', 3', \ldots, n', n, n-1, n-2, \ldots, 3, 2, 1$ constitute the trunk line. Under normal circumstances, the constant pressure point is located at the entrance of the circulating pumps, that is, the end of pipe segment $1'$. Assuming that the pressure at the constant pressure point is $P_{1'mo}$, the pressure at any section of any pipe segment can be calculated.

According to the Bernoulli equation (Bernoulli's principle), the pressure at the start of pipe segment i can be given by

$$\begin{aligned} P_{ishi} &= P_{1'mo} + \left(Z_{1'mo} - Z_{ishi}\right)\gamma + \frac{\rho v_{1'}^2}{2} - \frac{\rho v_i^2}{2} \\ &\quad + (1+\alpha)\left(\sum_{f=1'}^{i+1} R_f l_f + R_i l_i\right) + \Delta P_{\text{zhan}}. \end{aligned} \tag{29}$$

So the pressure constraint condition of pipe segment i can be given by

$$\begin{aligned} P_{tgi'\text{qihua}} &\le P_{1'mo} + \left(Z_{1'mo} - Z_{ishi}\right)\gamma + \frac{\rho v_{1'}^2}{2} - \frac{\rho v_i^2}{2} \\ &\quad + (1+\alpha)\left(\sum_{f=1'}^{i+1} R_f l_f + R_i l_i\right) + \Delta P_{\text{zhan}} \\ &\le P_{\text{cheng}}, \end{aligned} \tag{30}$$

where P_{ishi} is the pressure at the start of pipe segment i (Pa), v_i is the flow velocity of the medium in pipe segment i (m/s), Z_{ishi} is the topographic height at the start of pipe segment i (m), $P_{1'mo}$ is the pressure at the end of pipe segment $1'$ (the constant pressure point) (Pa), $v_{1'}$ is the flow velocity of the medium in pipe segment $1'$ (m/s), $Z_{1'mo}$ is the topographic height at the end of pipe segment $1'$ (m), ΔP_{zhan} is the resistance loss at heat exchange station (Pa), α is the ratio of the local resistance to the resistance along the path, $P_{tgi'\text{qihua}}$ is the saturation steam pressure corresponding to the design water-supply temperature t'_{gi} (Pa), P_{cheng} is the maximum pressure that the pipe and pipe fittings can bear (Pa), and γ is the volumetric weight (N/m^3), $\gamma = \rho g$.

In order to simplify formula (30), the item $(\rho v_{1'}^2/2-\rho v_i^2/2)$ can be moved as follows:

$$P_{tgi'\text{qihua}} - \left(\frac{\rho v_{1'}^2}{2} - \frac{\rho v_i^2}{2}\right) \le P_{1'mo} + (Z_{1'mo} - Z_{ishi})\gamma + (1+\alpha)\left(\sum_{f=1'}^{i+1} R_f l_f + R_i l_i\right) + \Delta P_{\text{zhan}} \le P_{\text{cheng}} - \left(\frac{\rho v_{1'}^2}{2} - \frac{\rho v_i^2}{2}\right). \quad (31)$$

There is little difference among different pipelines in terms of flow velocity, for which the range is 0~4 m/s [29, 30]; that is, the value of $(\rho v_{1'}^2/2 - \rho v_i^2/2)$ is −8000~8000 Pa. Compared with $P_{tgi'\text{qihua}}$ and P_{cheng}, its order of magnitude is low. Therefore, formula (31) can be simplified to the following form:

$$P_{tgi'\text{qihua}} + 8000 \le P_{1'mo} + (Z_{1'mo} - Z_{ishi})\gamma + (1+\alpha)\left(\sum_{f=1'}^{i+1} R_f l_f + R_i l_i\right) + \Delta P_{\text{zhan}} \le P_{\text{cheng}} - 8000. \quad (32)$$

In this way, not only has the formula been simplified, but also a certain margin has been set for the upper and lower limits.

From the Antoine equation [31], we know that $P_{tgi'\text{qihua}} = e^{9.3876-3826.36/(tgi'+227.68)} \times 10^6$ Pa, so the pressure constraint condition at the start of pipe segment i can be given by

$$e^{\Theta_i} \times 10^6 + 8000 \le P_{1'mo} + (Z_{1'mo} - Z_{ishi})\gamma + (1+\alpha)\left(\sum_{f=1'}^{i+1} R_f l_f + R_i l_i\right) + \Delta P_{\text{zhan}} \le P_{\text{cheng}} - 8000, \quad (33)$$

where $\Theta_i = 9.3876 - 3826.36/(t'_{hi} + \Delta t'_{ghi} + 227.68)$.

As described above, the pressure at the end of pipe segment i is equal to the pressure at the start of pipe segment $i+1$ because the velocity difference has been set as a margin for the upper and lower limits, which means the pressure constraint at the end of pipe segment i has been validated in the previous step by the pressure constraint of pipe segment $i+1$. Therefore, the pressure constraint at the end of pipe segment i can be ignored.

2.3.2. The Constraint Condition of Design Temperature Difference between Supply and Return Water. The design temperature difference between supply and return water $\Delta t'_{ghi}$ should be limited to a certain range. Too small a temperature difference can result in reduced and inadequate heat exchange, while too large a temperature difference may need a very high water-supply temperature, which may be difficult to achieve or may incur a very high cost in practice. Therefore, the design temperature difference between supply and return water should be limited by

$$\Delta t'_{gh\min} \le \Delta t'_{ghi} \le \Delta t'_{gh\max}. \quad (34)$$

2.3.3. The Constraint Condition of the Flow Velocity. The flow velocity in the pipes should be restricted to a certain range, as follows:

$$v_{\min} \le v_i = \frac{4G'_i}{\rho \pi d_i^2} \le v_{\max}. \quad (35)$$

Substituting formulas (6) and (7) into formula (35), the constraint conditions of the flow velocity can be given by

$$v_{\min} \le 0.821 \frac{Q'_i}{\rho \pi \Delta t'_{ghi} \Gamma_i^2} \le v_{\max}. \quad (36)$$

2.4. Establishment of a Mathematical Model. Based on the above analysis, a mathematical model of Present Value Cost to determine the comprehensive specific frictional resistance of pipe segment i can be given as follows.

(i) For a Heating System with Quality Regulation

$$\begin{aligned}\min \quad \text{PC} &= M_{1i} + M_{2i} + (M_{3i} + M_{4i}) \cdot \left(\frac{P}{A} \cdot \varepsilon \cdot n\right) \\ &= \left(1.1655 \times 10^{-3} a \Gamma_i^2 + 3.4133 \times 10^{-2} b \Gamma_i + c\right) l_i \\ &+ \left[2.39 \times 10^{-4} \frac{c_3}{\eta \rho \Delta t'_{ghi}} R_i l_i Q'_i \tau_i + 3.4133 \times 10^{-2} c_4 h \pi l_i \Gamma_i \left(\left(t'_{hi} + \frac{\Delta t'_{ghi}}{2} - t_j\right)\tau_{zh} - \frac{(-1)^k \Delta t'_{ghi}}{2Q'_i} \Lambda_i\right)\right] \\ &\times \left(\frac{P}{A} \cdot \varepsilon \cdot n\right).\end{aligned} \quad (37)$$

(ii) For a Heating System with Flow Rate Regulation

$$\min \quad \mathrm{PC} = M_{1i} + M_{2i} + (M_{3i} + M_{4i}) \cdot \left(\frac{P}{A} \cdot \varepsilon \cdot n\right)$$
$$= \left(1.1655 \times 10^{-3} a\Gamma_i^2 + 3.4133 \times 10^{-2} b\Gamma_i + c\right) l_i$$
$$+ \left(2.39 \times 10^{-4} \times \frac{c_3}{\eta\rho\Delta t'_{ghi}} R_i l_i \Lambda_i + 3.4133 \times 10^{-2} c_4 h \left(t'_{ki} - t_j\right) \pi\tau_i l_i \Gamma_i\right) \times \left(\frac{P}{A} \cdot \varepsilon \cdot n\right) \tag{38}$$

$$\mathrm{St} \quad \Delta t'_{gh\min} \le \Delta t'_{ghi} \le \Delta t'_{gh\max} \tag{39}$$

$$\upsilon_{\min} \le 0.821 \frac{Q'_i}{\rho\pi\Delta t'_{ghi}\Gamma_i^2} \le \upsilon_{\max} \tag{40}$$

$$e^{\Theta_i} \times 10^6 + 8000 \le P_{1'mo} + \left(Z_{1'mo} - Z_{ishi}\right)\gamma + (1+\alpha)\left(\sum_{f=1'}^{i+1} R_f l_f + R_i l_i\right) + \Delta P_{\mathrm{zhan}} \le P_{\mathrm{cheng}} - 8000. \tag{41}$$

For the water-supply pipes, the constraint conditions are the inequalities (39), (40), and (41), while, for the water-return pipes, the constraint conditions are the inequalities (39) and (40).

2.5. The Solvability Analysis of the Mathematical Model. The objective function $\mathrm{PC}(R_i, \Delta t'_{ghi})$ we established is a continuous real valued function map, and the value ranges of the decision variable R_i and $\Delta t'_{ghi}$ are compact sets. Therefore, we can know by the Weierstrass Theorem [32] that there must be a solution vector $(R_i, \Delta t'_{ghi})$ that makes the objective function a minimum.

2.6. The Analysis of Algorithm. The mathematical model established above belongs to the two-dimensional extreme value problem under inequality constraints. The Complex method is selected as the optimization algorithm and is used to solve the models, whose function is to solve the n-dimensional extremum problem under the equality and inequality constraints.

2.7. The Value of Comprehensive Specific Frictional Resistance. According to the actual engineering, the basic parameters in the objective function and the constraint conditions can be determined, while the total heat load of the heat consumers that the pipe segment i provides heat for during the heating period $\Lambda_i = \int_{\tau=\tau_0}^{\tau=\tau_{zh}} Q_i(\tau)d\tau$ can be calculated. Substitute all of them into the mathematical models (37)~(41) and solve the mathematical model. We can obtain the comprehensive specific frictional resistance of the pipe segment i R_i and the design temperature difference between supply and return water $\Delta t'_{ghi}$.

3. Application

3.1. Case. We have planned the heating engineering for Yangyuan County in China. The heating system is based on flow rate regulation as the main regulation mode, considering its energy-saving advantage. In the planning process of the network, the two schemes, comprehensive specific frictional resistance and economic specific frictional resistance, are used to optimize the pipe diameters.

According to market research, the calculation parameters have been selected as follows.

Coefficient of frictional resistance $\lambda = 0.0196$, discount rate $\varepsilon = 7\%$, electrovalence $c_3 = 2.0278 \times 10^{-7}$ ¥/J, fuel price $c_4 = 1.7088175 \times 10^{-8}$ ¥/J, the lifespan $n = 20$ years, run time per year $\tau_i = 14515200$ s (equivalent to 168 days), pump efficiency $\eta = 70\%$, ambient temperature of pipe segments $t_j = 0°$C, heat transfer coefficient of pipe surface $h = 1.22$ w/(m^2 ·°C), design water-return temperature $t'_{hi} = 70°$C, fitting polynomial coefficients of civil engineering cost $a_1 = 255.67292$, $b_1 = 498.81903$, and $c_1 = 113.13293$, fitting polynomial coefficients of pipe installation fee $a_2 = 3881.42532$, $b_2 = 4134.16167$, and $c_2 = 155.26657$, the ratio of local resistance to resistance along the path $\alpha = 0.3$, the resistance loss at heat exchange station $\Delta P_{\mathrm{zhan}} = 0.1$ MPa, $P_{1'mo} = 0.1$ MPa, upper bound of flow velocity $\upsilon_{\max} = 3.5$ m/s, lower bound of flow velocity $\upsilon_{\min} = 0.5$ m/s, upper bound of design temperature difference between supply and return water $\Delta t'_{gh\min} = 30°$C, lower bound of design temperature difference between supply and return water $\Delta t'_{gh\max} = 60°$C, upper bound of pressure $P_{\mathrm{cheng}} = 1.6$ MPa, $\rho = 1000$ kg/m^3, and $\gamma = 9807$ N/m^3, and the duration of outdoor temperature in Zhangjiakou are shown in Table 1.

According to the site investigations, and to follow the trends of the existing pipeline network, the whole heating area was divided into three independent heating areas as shown in Figure 6: the planning floor area of heating area A is 9.61×10^6 m^2, the planning floor area of heating area B is 6.57 $\times 10^6$ m^2, and the planning floor area of heating area C is 4.75 $\times 10^6$ m^2.

FIGURE 6: The map of heating areas A, B, and C.

TABLE 1: Duration of outdoor temperature in Zhangjiakou.

Number	Outdoor temperature (°C)	Duration of outdoor temperature (hours)
(1)	>5	4728
(2)	5	4032
(3)	4	3850
(4)	3	3667
(5)	2	3483
(6)	1	3299
(7)	0	3114
(8)	−1	2927
(9)	−2	2740
(10)	−3	2552
(11)	−4	2363
(12)	−5	2173
(13)	−6	1981
(14)	−7	1788
(15)	−8	1593
(16)	−9	1397
(17)	−10	1197
(18)	−11	996
(19)	−12	790
(20)	−13	579
(21)	−14	361
(22)	−15	120

In this study, we take heating area C as an example. In Figure 7, only water-supply pipes are marked, while the position and trend of water-return pipes are in agreement with the corresponding water-supply pipes. In the following description, the numbers with "′" represent the water-return pipes, while the numbers without "′" represent the water-supply pipes.

The entire calculation process is as follows:

① Choose the most distant loop from the heat source (the trunk line), that is, $100 \rightarrow 101 \rightarrow 102 \rightarrow 103 \rightarrow 104 \rightarrow 105 \rightarrow 106 \rightarrow 107 \rightarrow 107' \rightarrow 106' \rightarrow 105' \rightarrow 104' \rightarrow 103' \rightarrow 102' \rightarrow 101' \rightarrow 100'$;

② Calculate the R_i and $\Delta t'_{ghi}$ of the water-return pipes in the most distant loop with the models (38), (39), and (40);

③ Follow the order $107 \rightarrow 106 \rightarrow 105 \rightarrow 104 \rightarrow 103 \rightarrow 102 \rightarrow 101 \rightarrow 100$ to calculate the R_i and $\Delta t'_{ghi}$ of the water-supply pipes in the most distant loop with the models (38), (39), (40), and (41);

④ For the other branch loops, calculate the R_i and $\Delta t'_{ghi}$ of the water-return pipes that have not been calculated with the models (38), (39), and (40), and then calculate the R_i and $\Delta t'_{ghi}$ of the water-supply pipes that have not been calculated, following the order from far to near with the models (38), (39), (40), and (41).

The calculation results are shown in Table 2. As expected, the range of values for traditional economic specific frictional resistance is 30~70 Pa/m.

The optimization results in Table 2 show that the design water-supply temperature for pipe segment 107 is 120°C, which is different from all the others. The reasons for this

Table 2: Diameter optimization result.

Number	Thermal load (MW)	Design water-supply temperature (°C)	Design temperature difference (°C)	Design flow rate (kg/s)	Economic specific frictional resistance			Comprehensive specific frictional resistance			
					Pipe diameter (m)	Specific frictional resistance (Pa/m)	Flow rate (m/s)	Specific frictional resistance (Pa/m)	Pipe diameter (m)	Corrected pipe diameter (m)	Flow rate (m/s)
100	184.00	130	60	732.59	0.7	51	1.90	65	0.666	0.7	1.9
101	168.61	130	60	671.32	0.7	43	1.74	67	0.39	0.6	2.4
102	160.05	130	60	637.24	0.7	38	1.66	68	0.624	0.6	2.3
103	108.22	130	60	430.88	0.6	38	1.52	77	0.521	0.5	2.2
104	45.50	130	60	181.16	0.4	51	1.44	105	0.346	0.35	1.9
105	30.61	130	60	121.87	0.35	45	1.27	122	0.287	0.3	1.7
106	20.74	130	60	82.58	0.3	45	1.17	142	0.238	0.25	1.7
107	10.62	120	50	50.74	0.25	42	1.03	197	0.183	0.20	1.6
108	15.39	130	60	61.28	0.25	61	1.25	161	0.206	0.2	2.0
109	9.18	130	60	36.55	0.2	66	1.16	199	0.161	0.15	2.1
110	6.21	130	60	24.73	0.2	30	0.79	235	0.133	0.15	1.4
111	8.56	130	60	34.08	0.2	58	1.08	206	0.155	0.15	1.9
112	35.79	130	60	142.50	0.4	31	1.13	115	0.309	0.3	2.0
113	8.66	130	60	34.48	0.2	59	1.10	206	0.156	0.15	2.0
114	9.05	130	60	36.03	0.2	64	1.15	201	0.159	0.15	2.0
115	18.08	130	60	71.99	0.3	34	1.02	151	0.223	0.2	2.3
116	9.66	130	60	38.46	0.2	73	1.22	197	0.164	0.15	2.2
117	8.42	130	60	33.52	0.2	56	1.07	206	0.154	0.15	1.9
118	16.04	130	60	63.86	0.25	66	1.30	158	0.210	0.2	2.0
119	7.82	130	60	31.14	0.2	48	0.99	215	0.148	0.15	1.8
120	8.22	130	60	32.73	0.2	53	1.04	209	0.152	0.15	1.9
121	62.72	130	60	249.72	0.5	32	1.27	94	0.402	0.4	2.0
122	17.95	130	60	71.47	0.3	33	1.01	152	0.222	0.2	2.3
123	10.53	130	60	41.93	0.25	29	0.85	188	0.172	0.15	2.4
124	34.24	130	60	136.33	0.35	56	1.42	117	0.302	0.3	1.9
125	16.86	130	60	67.13	0.25	73	1.37	155	0.215	0.2	2.1
126	8.33	130	60	33.17	0.2	55	1.06	209	0.153	0.15	1.9
127	9.05	130	60	36.03	0.2	64	1.15	201	0.159	0.15	2.0
128	7.68	130	60	30.58	0.2	46	0.97	215	0.147	0.15	1.7
129	7.21	130	60	28.71	0.2	41	0.91	221	0.143	0.15	1.6
130	9.87	130	60	39.30	0.25	25	0.80	194	0.166	0.15	2.2
131	10.12	130	60	40.29	0.25	26	0.82	192	0.168	0.15	2.3
100′	184.00	130	60	732.59	0.7	51	1.90	64	0.645	0.7	2.37
101′	168.61	130	60	671.32	0.7	43	1.74	65	0.630	0.6	2.25
102′	160.05	130	60	637.24	0.7	38	1.66	74	0.525	0.6	2.19
103′	108.22	130	60	430.88	0.6	38	1.52	100	0.350	0.5	1.88
104′	45.50	130	60	181.16	0.4	51	1.44	115	0.290	0.35	1.72
105′	30.61	130	60	121.87	0.35	45	1.27	134	0.241	0.3	1.68
106′	20.74	130	60	82.58	0.3	45	1.17	176	0.174	0.25	2.39
107′	10.62	130	60	42.28	0.25	29	0.86	151	0.209	0.15	1.95
108′	15.39	130	60	61.28	0.25	61	1.25	185	0.163	0.2	2.07
109′	9.18	130	60	36.55	0.2	66	1.16	219	0.135	0.15	1.40
110′	6.21	130	60	24.73	0.2	30	0.79	192	0.157	0.15	1.93

Table 2: Continued.

Number	Thermal load (MW)	Design water-supply temperature (°C)	Design temperature difference (°C)	Design flow rate (kg/s)	Economic specific frictional resistance			Comprehensive specific frictional resistance			
					Pipe diameter (m)	Specific frictional resistance (Pa/m)	Flow rate (m/s)	Specific frictional resistance (Pa/m)	Pipe diameter (m)	Corrected pipe diameter (m)	Flow rate (m/s)
111′	8.56	130	60	34.08	0.2	58	1.08	109	0.312	0.15	2.02
112′	35.79	130	60	142.50	0.4	31	1.13	192	0.158	0.3	1.95
113′	8.66	130	60	34.48	0.2	59	1.10	188	0.161	0.15	2.04
114′	9.05	130	60	36.03	0.2	64	1.15	141	0.226	0.15	1.47
115′	18.08	130	60	71.99	0.3	34	1.02	183	0.167	0.25	2.18
116′	9.66	130	60	38.46	0.2	73	1.22	192	0.156	0.15	1.90
117′	8.42	130	60	33.52	0.2	56	1.07	148	0.213	0.15	2.03
118′	16.04	130	60	63.86	0.25	66	1.30	200	0.150	0.2	1.76
119′	7.82	130	60	31.14	0.2	48	0.99	195	0.154	0.15	1.85
120′	8.22	130	60	32.73	0.2	53	1.04	89	0.407	0.15	1.99
121′	62.72	130	60	249.72	0.5	32	1.27	142	0.225	0.4	1.46
122′	17.95	130	60	71.47	0.3	33	1.01	176	0.174	0.25	2.37
123′	10.53	130	60	41.93	0.25	29	0.85	111	0.306	0.15	1.93
124′	34.24	130	60	136.33	0.35	56	1.42	145	0.218	0.3	2.14
125′	16.86	130	60	67.13	0.25	73	1.37	195	0.155	0.2	1.88
126′	8.33	130	60	33.17	0.2	55	1.06	188	0.161	0.15	2.04
127′	9.05	130	60	36.03	0.2	64	1.15	200	0.149	0.15	1.73
128′	7.68	130	60	30.58	0.2	46	0.97	206	0.145	0.15	1.62
129′	7.21	130	60	28.71	0.2	41	0.91	181	0.168	0.15	2.22
130′	9.87	130	60	39.30	0.25	25	0.80	180	0.170	0.15	2.28
131′	10.12	130	60	40.29	0.25	51	1.90	64	0.645	0.15	2.37

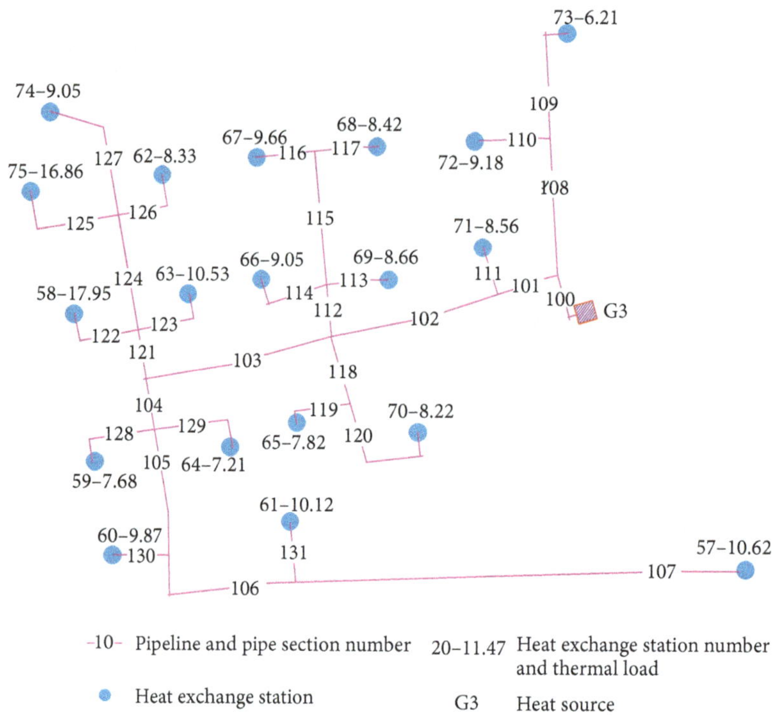

FIGURE 7: Heating network in heating area C.

are the long distance from the heat source to the pipe and the high elevation of the pipe, which lead to a lot of pressure reduction in the pipe segment. Therefore, the only way to ensure the pipe pressure is higher than the vaporization pressure is to decrease the water-supply temperature. If the pressure constraint formula (41) is removed, the result for pipe segment 107 will be $R_{107} = 188$, $\Delta t'_{g107} = 130^{\circ}\text{C}$, at which temperature the medium will vaporize. Because there must be one water-supply temperature for the entire engineering, so this optimization result is not desirable. There are two solutions to this problem: ① the entire heating system is recalculated according to a lower water-supply temperature in all the results until the optimization results $\Delta t'_{gi}$ of all the segments are same; ② the heat consumers after the pipe segments whose optimization results of water-supply temperature are different from the others take the indirect connection form. In this case, taking into account the fact that only one segment pipe is different from the others, it is obvious that the second solution is more reasonable. That is, pipe segment 107 and the heat user 57 should connect with the trunk line indirectly.

From the perspective of the entire engineering application, based on the calculation of the various costs and incomes, including heat source construction fee, heat exchange station construction fee, primary network construction fee, electricity fee, heat loss fuel cost, thermal user fuel cost, water fee, maintenance cost, labor wage, heating charges, and residual value of fixed assets, the economic effect evaluation indexes of the two schemes are shown in Table 3.

By comparison, we can see that under the premise of ensuring the same heating quality, the economic effect evaluation indexes of scheme one are better. Compared to scheme two, the Payback Period is lowered, the Initial Investment is reduced by 5%, the Present Value Cost is decreased by ¥23.6039 million, the Net Present Value has an increase of ¥34.1176 million, the Rate of Return on Investment is increased by 4.98%, and the Internal Rate of Return is increased by 0.6%, which demonstrate the feasibility and superiority of the comprehensive specific frictional resistance.

3.2. Discussion. Through the analysis of the case and the mathematical models, we show there are many factors that influence the value of comprehensive specific frictional resistance, including the coefficient of frictional resistance λ, heat transfer coefficient of pipe surface h, electrovalence c_3, fuel price c_4, discount rate ε, lifespan n, pump efficiency η, and primary network construction fee. The various values of the above parameters in different engineering applications will lead to the different values of comprehensive specific frictional resistance. In addition, the comprehensive specific frictional resistance has different sensitivities for different parameters.

Using the calculation parameters in Section 3.1, and setting the design water-supply temperature $\Delta t'_g = 130^{\circ}\text{C}$, we get the sensitivities as shown in Figure 8.

Similarly, using the calculation parameters in Section 3.1, and setting the design water-supply temperature $\Delta t'_g = 130^{\circ}\text{C}$, the comprehensive specific frictional resistance corresponding to each flow rate can be solved, by which we can draw the curves of flow rate-Comprehensive specific frictional resistance as in Figure 9.

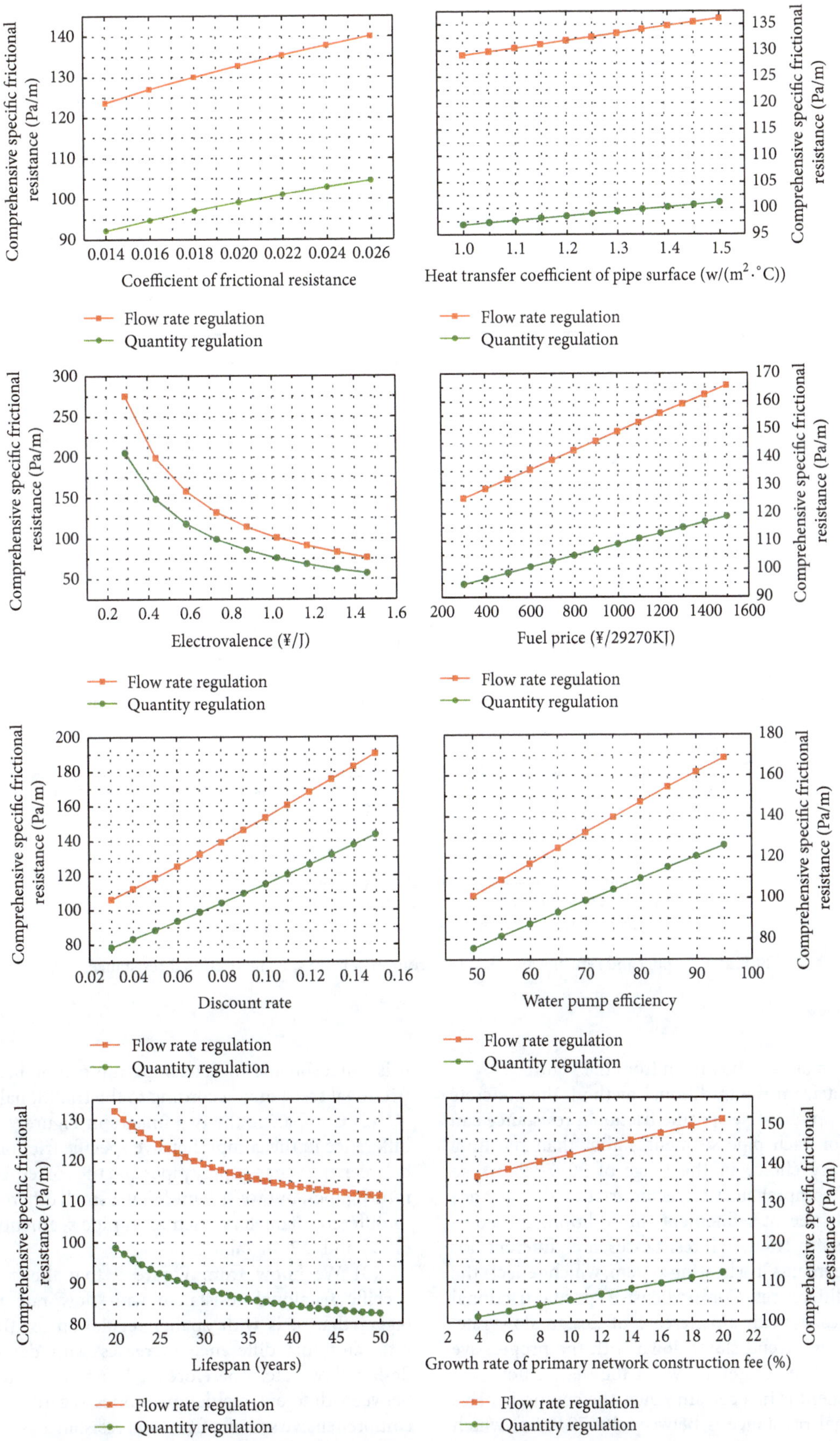

FIGURE 8: Effects of various factors on comprehensive specific frictional resistance.

TABLE 3: Comparison table of economic effect evaluation indexes.

Schemes Index	Payback Period (years)	Net Present Value (¥10000)	Rate of Return on Investment (%)	Internal Rate of Return (%)	Present Value Cost (¥10000)	Initial Investment (¥10000)
Scheme one (Comprehensive specific frictional resistance)	11.39	40872.73	39.13	11.7	32116.94	104463.1
Scheme two (Economic specific frictional resistance)	12.04	37460.97	34.15	11.1	34477.33	109679.8

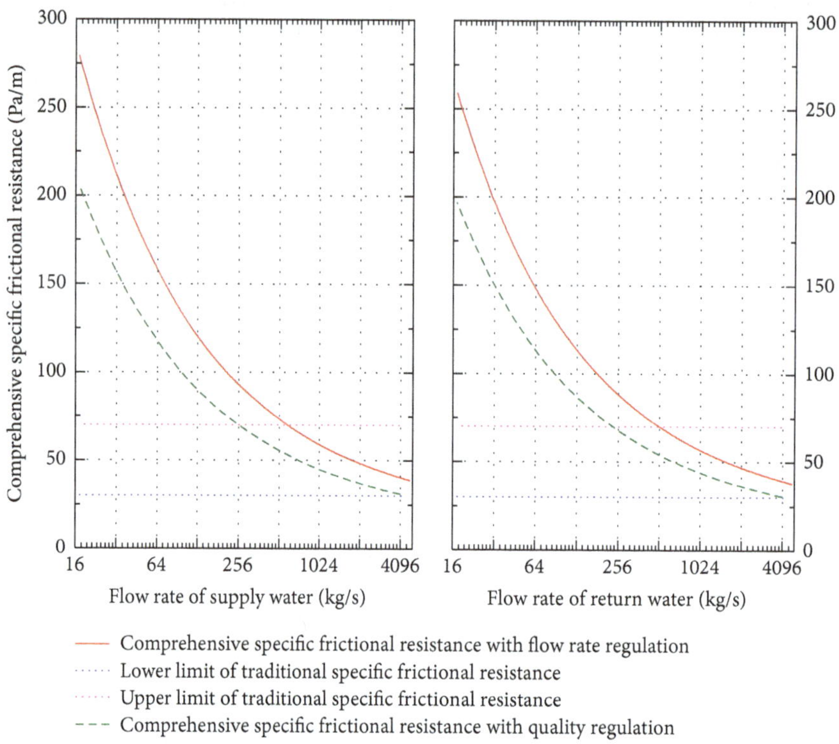

FIGURE 9: Supply and return water flow rate-comprehensive specific frictional resistance.

Some conclusions can be drawn from the Figure 9:

(1) Different from the traditional method, the concrete value of the comprehensive specific frictional resistance can be calculated for each pipe segment, rather than resulting from an arbitrary choice in the range of 30~70 Pa/m, so the method of comprehensive specific frictional resistance has a higher degree of refinement. In addition, the comprehensive specific frictional resistances are different for different pipe segments in a heating system, which reflects the pipe segment differences. Comprehensive specific frictional resistance decreases with an increase of design flow rate, and this decreasing trend slows down with the progressive increase in flow rate. Therefore, when the design flow rate of the pipe segment is in a certain range, the comprehensive specific frictional resistance is between 30~70 Pa/m, which is consistent with the traditional economic specific frictional resistance, but for the pipe segments beyond this range, it is unreasonable to determine the comprehensive specific frictional resistance according to the traditional method.

(2) Comparing the two images in Figure 9, there is little difference in the comprehensive specific frictional resistance between the water-supply pipes and the corresponding water-return pipes, but we still find that a small difference may lead to different diameters, such as in pipe segments 115 and $115'$ or 122 and $122'$ in Table 2.

(3) We know from Figure 9 that the comprehensive specific frictional resistance with flow rate regulation is bigger than it is with quality regulation for the same flow rate, and this difference increases with decreases in the design flow rate. Therefore, it is necessary to distinguish between different regulation modes regarding the issue of comprehensive specific frictional resistance.

(4) For the pipe segments with universal flow rate, comparing with the range of traditional economic specific

frictional resistance of 30~70 Pa/m, the value of comprehensive specific frictional resistance is higher and can reach up to 200 Pa/m. There are three reasons for this: ① the value of 30~70 Pa/m was obtained from experience with directly connected heating networks in the past [8]. For indirectly connected heating networks, according to the experience of Nordic countries, the value of specific frictional resistance will be higher than the above specified value; ② in recent years, the growth trend of electrovalence has been relatively flat, but the growth in pipe materials and pipe fittings has been large, resulting in an increase in specific frictional resistance; ③ for flow rate regulation, the calculation of comprehensive specific frictional resistance considers the change of the actual flow rate during operation. By this calculation, the computational electricity consumption is less than the electricity consumption calculated by design flow rate, so the value of comprehensive specific frictional resistance is greater than the traditional value.

4. Conclusions

In this paper, we put forward the concept of comprehensive specific frictional resistance, which aims to improve the precision degree of specific frictional resistance so as to promote the development of heating engineering. We provide a method to obtain the value of comprehensive specific frictional resistance by establishing a mathematical model, which is applicable to the optimization of pipe diameter in primary heating networks. Although the comprehensive specific frictional resistance has the disadvantage of a complex calculation process, it reflects the pipe segment differences, differentiates the different operation regulation modes, and takes into account the safety problems caused by pipe pressure. Therefore, it is of high practical value.

The ideal value of comprehensive specific frictional resistance is affected by the factors mentioned in this paper and also needs to be enriched and developed through further research.

Conflicts of Interest

The authors declare that there are no conflicts of interest regarding the publication of this article.

References

[1] H. V. Larsen, B. Bøhm, and M. Wigbels, "A comparison of aggregated models for simulation and operational optimisation of district heating networks," *Energy Conversion and Management*, vol. 45, no. 7-8, pp. 1119–1139, 2004.

[2] H. Shi, *The theory research of economic friction of heating pipe network [M.S. thesis]*, Xi'an University of Architecture and Technology, Xi'an, China, 2013.

[3] C. Hong, "A brief analysis of the weighted mean specific frictional resistance," *District Heating*, vol. 41, pp. 49–53, 1992.

[4] A. Ge, C. Qiao, and Y. Zhao, "The relationship between range of heat-supply service and optimal specific frictional head loss," *Energy Conservation*, vol. 35, no. 12, pp. 18–21, 2015.

[5] M. Liu, R. Wu, and Q. Yang, "Transportation energy consumption and economical specific frictional resistance design of sewage heat pump system," *Energy Conservation Technology*, vol. 32, no. 188, pp. 487–491, 2014.

[6] Z. Wang, Z. Dong, and R. Wang, "Study of economic specific frictional resistance on hot water heating system for civil buildings," *Architectural Science*, vol. 26, no. 10, pp. 31–34, 2010.

[7] X. Li, P. Zhou, and M. Duan, "Calculation of optimal specific frictional head loss in condensate network," *Gas and Heat*, vol. 26, no. 3, pp. 59–62, 2006.

[8] P. He and G. Sun, *Heating Project*, China Architecture and Building Press, Beijing, China, 2000.

[9] Ministry of Housing and Urban-Rural Construction of the People's Republic of China, *State Technical Method for Civil Building Project Design*, China Planning Press, China.

[10] X. Sun, *Research of the economic specific frictional resistance of primary network of hot water heating[M.S. thesis]*, Taiyuan University of Technology, Shanxi, China, 2010.

[11] J. Liu, "Analysis and calculation of economic specific frictional resistance in hot water heating network," *Journal of Nanjing Institute of Architectural Engineering*, vol. 15, no. 4, pp. 38–43, 1998.

[12] M. Manfren, P. Caputo, and G. Costa, "Paradigm shift in urban energy systems through distributed generation: Methods and models," *Applied Energy*, vol. 88, no. 4, pp. 1032–1048, 2011.

[13] C. Y. Xu and Q. Chen, "An entransy dissipation-based method for global optimization of district heating networks," *Energy & Buildings*, vol. 48, no. 1, pp. 50–60, 2011.

[14] M. Bojic and N. Trifunovic, "Linear programming optimization of heat distribution in a district-heating system by valve adjustments and substation retrofit," *Building and Environment*, vol. 35, no. 2, pp. 151–159, 2000.

[15] J. Fu, *Industrial Technology Economics*, Tsinghua University Press, Beijing, China, 1996.

[16] F. Wu and F. Ye, *Engineering Economics*, China Machine Press, Beijing, China, 2007.

[17] S. Xu and Y. Jia, "The analysis of the operational regulation mode in the system of district heating," *Journal of Hebei University of Engineering*, vol. 25, no. 2, pp. 36–39, 2008.

[18] J. Chen, *Research on energy saving of the operation adjustment mode of heating system [M.S. thesis]*, North China Electric Power University, Beijing, China, 2014.

[19] B. Babiarz, "An introduction to the assessment of reliability of the heat supply systems," *International Journal of Pressure Vessels and Piping*, vol. 83, no. 4, pp. 230–235, 2006.

[20] H. Boudali and J. B. Dugan, "Erratum: "A continuous-time Bayesian network reliability modeling and analysis framework" (IEEE Transactions on Realibility)," *IEEE Transactions on Reliability*, vol. 57, no. 3, p. 532, 2008.

[21] O. Fujiwara and A. U. De Silva, "Algorithm for reliability-based optimal design of water networks," *Journal of Environmental Engineering (United States)*, vol. 116, no. 3, pp. 575–587, 1990.

[22] I. C. Goulter and A. V. Coals, "Quantitative approaches to reliability assessment in pipe networks," *Journal of Transportation Engineering*, vol. 112, no. 3, pp. 287–301, 1986.

[23] D. A. Savic and G. A. Walters, "Genetic algorithms for least-cost design of water distribution networks," *Journal of Water Resources Planning and Management*, vol. 123, no. 2, pp. 67–77, 1997.

[24] K. Vairavamoorthy and M. Ali, "Optimal design of water distribution systems using genetic algorithms," *Computer-Aided*

Civil and Infrastructure Engineering, vol. 15, no. 5, pp. 374–382, 2000.

[25] X. Xia and L. Feng, "Analysis of the influence of absolute equivalent roughness," *Shanghai Gas*, vol. 54, no. 2, pp. 10–12, 2010.

[26] F. Chen, H. Yu, Z. Sun, and Y. Guo, "Determination of inner wall roughness of pipeline," *Nature Gas and Oil*, vol. 25, no. 6, pp. 8–10, 2007.

[27] Y. Dou and X. He, "Calculation method for hydraulic resistance of pipe," *Journal of Hydraulic Engineering*, vol. 40, no. 7, pp. 28–31, 1995.

[28] M. Xu and P. He, "Determination method of heating load continuity diagram," *District Heating*, vol. 4, no. 1, pp. 3–10, 1986.

[29] L. Yang and H. Tao, "Engineering experimental research on energy-saving of economical flow velocity and pipe diameter optimization in water supply system," *Water & Wastewater Engineering*, vol. 35, pp. 334–336, 2009.

[30] H.-B. Jiang, "Cost function, optimal pipe diameter and economical flow velocity of water supply system," *Harbin Gongye Daxue Xuebao/Journal of Harbin Institute of Technology*, vol. 36, no. 1, pp. 84–90, 2004.

[31] Z. Chen, Z. Cai, and C. Qian, *Chemical Engineering Thermodynamics*, Chemical Industry Press, China, 2015.

[32] J. Chen, *Real Function Theory*, Science Press, China, 1978.

17

Evaluation of Rock Quality of Tunnel Wall Rock Based on Rough Set Theory and Unascertained Measurement Theory

Changfu Huang,[1,2] Shuguang Tian,[1] Qun Li,[2] and Jun Huang[1]

[1]*China Railway 16th Bureau Group Co., Ltd., Beijing 100018, China*
[2]*School of Civil and Resources Engineering, University of Science and Technology Beijing, Beijing 100083, China*

Correspondence should be addressed to Shuguang Tian; tsg880917@gmail.com

Academic Editor: Anna Vila

A method for classification of the tunnel wall rock is established based on the rough set theory and unascertained measurement theory. The saturated uniaxial compressive strength, rock mass integrity index, structural surface condition, seepage measurement of groundwater, and the angle between the hole axis and main structural surface are selected as the evaluation indexes. The problem of weight coefficients for these evaluation indexes is converted into that of significance estimating on the attributes in the rough set theory. The proposed method is verified by the wall rock data of Yuanyanghui tunnel. The results show that the proposed method has an excellent performance in good agreement with the practical situation of wall rock and is feasible to classify the wall rock. Finally, the proposed method is applied to Xihualing tunnel of Zhu-Yong Expressway. The results are basically the same as those from Delphi-ideal point method, set pair analysis method, and the actual situation, which proves that the rough set theory and unascertained measurement theory are effective for classification of the wall rock.

1. Introduction

Reasonable classification of the wall rock plays an important role in guiding the construction of the tunnel and is an also important basis for evaluating the stability of the wall rock and selecting suitable support. However, due to the complexity of the engineering geological conditions in the tunnel, the classification of the wall rock is always not accurate enough, leading to a certain difference with the actual rock grade. These problems result in the changes in the design during the tunnel construction and seriously affect the construction progress. Therefore, it is of great significance to develop an accurate and simple classification method for wall rock.

In the tunnel construction, the wall rock of the construction area should be classified and its quality also be evaluated. At present, there are many kinds of rock quality evaluation methods including traditional methods (RQD method [1], RMR method [2], RMi method [3, 4], etc.) and new methods (fuzzy theory [5], artificial neural networks [6, 7], extension theory [8], fractal theory [9], etc.). The above methods and theory have been introduced into the study of tunnel rock classification and some achievements have been achieved. However, each theory or method has its own characteristics and shortcomings; it cannot well serve the actual tunnel construction. Therefore, it is necessary to explore the new method of classification of wall rock.

In 1990, Wang [10] proposed the unascertained information theory, which is different from the fuzzy information [11], random information [12], and grey information [13]. On this basis, Liu and Lin et al. [14, 15] established the unascertained measurement evaluation model and obtained satisfactory results by using the model to evaluate the reliability of an electricity supply system. Since then, the theory and evaluation model have been widely applied in many diverse fields. Parsons and Wen [16] used unascertained measurement theory to create an evaluation method to determine the condition of cement concrete pavement. Yao [17] applied unascertained theory and blind number theory in describing the uncertainty of the mechanical system failure process to help develop appropriate design validation and verification program for reliability assessment. Zhou [18] established an UM theory of blastability of rock mass predictions on the basis of unascertained measurement theory, and the model can provide an effective and credible method for estimation the blastability of rock mass. An unascertained measurement model of water and mud inrush risk evaluation is established founded on unascertained measurement theory, and the results show that the evaluation results agree well with practical construction situation [19].

In 1982, Pawlak [20] proposed rough set theory, which is a mathematical tool for dealing with fuzzy and uncertain knowledge. The main idea is based on indiscernibility relations that the obtained information is used to describe indistinguishability of objects. On the basis, Pawlak and Grzymala-Busse [21, 22] et al. discussed and analyzed in greater detail in their papers. Beaubouef [23] used the model based on intuitionistic and rough set developed here is applied to databases, which provided greater management of uncertainty for databases applications in a less than certain world. Yasdi [24] utilized the method combining rough set theory and neural network to study the uncertainty and imprecise information, which improved the efficiency of classification. Zhang [8] established a relation data model about tunnel wall rock evaluation based on rough set theory and extension theory.

Although the rough set theory and unascertained measurement theory have been successfully applied in many fields, the method which combined the two theories is rarely used. The purpose of the present paper is to introduce the method combining rough set theory and unascertained measurement theory to classify the wall rock. Therefore, the remainder of this paper is organized as follows. In Section 1, the method combining rough set theory and unascertained measurement theory is presented, whose evaluation process is described in detail. In Section 2, the method is applied to the classification of the wall rock and verified with the measured samples. In Section 3, a comparison study is performed based on various evaluation methods and the classification of the wall rock for the Xihualing tunnel of Zhu-Yong Expressway in China.

2. Rough Set Theory and Unascertained Measurement Theory

2.1. Rough Set Theory. As a mathematical method, rough set theory is proposed to deal with incomplete and uncertain data. The main idea is based on indiscernibility relations that describe indistinguishability of objects. The theory focuses on the knowledge that can be approximated from the data [20–22]. The parameters most commonly found in the rough set theory are shown below.

Set the pair $K=(U,R)$ as the knowledge base, where U is called the universe, which is finite and nonempty, and R is called the nonempty set of attributes. $[x]_R$ denotes the equivalence class of R containing an element x of U. $X \subseteq U$, $\underline{R}X = \{x \in U \mid [x]_R \subseteq X\}$, and $\overline{R}X = \{x \in U \mid [x]_R \cap X \notin \varnothing\}$ are defined as the lower approximation and the upper approximation of X, respectively.

Let P and Q be subsets of condition attribute C and decision attribute D, respectively; then

$$K = \gamma_P(Q) = \frac{1}{|U|}\sum_{i=1}^{m} |\gamma_P(Q_i)| \tag{1}$$

where $|U|$ represents the number of elements in U. The dependency degree of the decision set on the condition set is the value of K. If K=1, Q is entirely dependent on P. Conversely, if K=0, Q is completely independent of P.

In the decision table, different attributes have different importance. To find out the importance of certain attributes, some attributes are removed from the decision table, and then the classification changes without these attributes are analyzed. If the corresponding classification changes greatly when an attribute is removed, then the importance of this attribute is high. Therefore, the condition attribute C_i defines the importance of the decision attribute D as

$$\begin{aligned}\gamma_{C-C_i}(D) &= \frac{1}{|U|}\sum_{i=1}^{m}|\gamma_{C-C_i}(D_i)| \\ \sigma_{CD}(C_i) &= \gamma_C(D) - \gamma_{C-Ci}(D)\end{aligned} \tag{2}$$

where $\gamma_C(D)$ is called the dependency degree.

When calculating the weights of evaluation indexes, the evaluation indexes are used as the condition attribute set, and the decision results are taken as the decision attribute set. The dependency degree of the decision set on the condition set is used to determine the weights. The main procedure is as follows.

(1) Construction of initial decision table.

(2) Samples screening: the screening principles are as follows. ① If the sample information and evaluation system are contradictory, then the sample is removed. ② If the condition attributes of multiple samples are the same but the decision attributes are different, then the samples with good compatibility can be retained. ③ If the condition attributes and decision attributes of different samples are the same, the samples can be merged to retain one.

(3) Reduction of the evaluation indexes: it is to remove irrelevant and redundant knowledge with maintaining the classification ability of the knowledge base and to obtain the optimal condition attributes.

(4) Construction for optimal attribute of decision table and the weight ω_i can be calculated by (3).

$$\omega_i = \frac{\gamma_C(D) - \gamma_{C-C_i}(D)}{\sum_{i=1}^{m}\left[\gamma_C(D) - \gamma_{C-C_i}(D)\right]} \tag{3}$$

2.2. Unascertained Measurement Theory. Since the concept of the unascertained information is proposed, much work has been done by many scholars, which makes the unascertained information a systematic theory and method [10, 14]. The theory is used to describe the incomplete information and refers to the decision-making-demanded information. The parameters most commonly found in the unascertained measurement theory are briefly discussed below.

(1) Unascertained Set. Unascertained set is a collection of all unascertained numbers and plays an important role in the study of unascertained mathematical theory, written $I_{[0,1]} = \{\{[a,b],F(x)\}|a \geq 0, a \leq b \leq 1|\}$. Suppose that N is an unascertained subset of discourse domain U, and N is used to characterize the mapping of a membership function. The

mapping relationship can be written as $\mu : U \longrightarrow I_{[0,1]}, u \longrightarrow \mu(u) \in I_{[0,1]}, u \in U$, $\mu(u)$ is called the membership degree of u to N, written $N_{\mu(u)}$.

(2) Measurable Space. Measurable space is a space composed of the property domain and its topology on the domain, written (F, E). Suppose that $\{F_1, F_2, F_3, \cdots\}$ is a division of F, and let

$$E = \left\{E_i \mid E_i = \bigcup_{j=1}^{i} F_j,\ F_j \in \{\varnothing, F_1, F_2, \cdots, F_k\},\ 1 \le i \le k\right\} \tag{4}$$

where E is the topology of F, and F is the property domain on U. F satisfies $F_i \cap F_j = \varnothing, i = j$ and $\bigcup_{i=1} F_i = F$.

(3) Unascertained Measurement. When $u \in U, A \in E$, the existence mapping u makes $\mu_A(u)$ satisfy (5)-(7).

$$0 \le \mu_A(u) \le 1, \quad \forall u \in U,\ A \in E \tag{5}$$

$$\mu_F(U) = 1 \tag{6}$$

$$\mu_{\bigcup_i A_i}(u) - \sum_i \mu_{A_i}(u), \quad A_i \in E,\ A_i \cap A_j = \varnothing\ (i \ne j) \tag{7}$$

where $\mu_A(u)$ is called unascertained measurement in the measurable space.

(4) Unascertained Measurement of Single Index. Set $x_1, x_2, \cdots, x_n$ as n evaluation objects of the pending evaluation, and the evaluation object set is $X = \{x_1, x_2, \cdots, x_n\}$. Each evaluation object x_i (i= 1, 2,..., n) has m evaluation indexes $I_1, I_2, \cdots, I_m$ and $I = \{I_1, I_2, \cdots, I_m\}$, which is the evaluation index set. Then x_{ij} is expressed as a m dimensional vector $x_i = \{x_{i1}, x_{i2}, \cdots, x_{im}\}$. Set c_k (k= 1, 2,..., p) as the k-th evaluation grade, and the grade of the k-th is superior to (k+1)-th, which is denoted as $c_k > c_{k+1}$. If $\{c_1, c_2, \cdots, c_p\}$ satisfies $c_1 > c_2 > \cdots > c_p$, then $\{c_1, c_2, \cdots, c_p\}$ is called ordinal classification of the evaluation index set U.

Let $\mu_{ijk} = \mu(x_{ij} \in c_k)$ to express the degree that x_{ij} belongs to k-th evaluation grade of c_k, and μ satisfies the flowing conditions:

$$0 \le \mu\left(x_{ij} \in c_k\right) \le 1 \quad (i = 1, 2, \cdots, n,\ j = 1, 2, \cdots, m,\ k = 1, 2, \cdots, p) \tag{8}$$

$$\mu\left(x_{ij} \in U\right) = 1 \quad (i = 1, 2, \cdots, n,\ j = 1, 2, \cdots, m) \tag{9}$$

$$\mu\left(x_{ij} \in \bigcup_{l=1}^{k} c_l\right) = \sum_{l=1}^{k} \mu\left(x_{ij} \in c_l\right) \quad (k = 1, 2, \cdots, p) \tag{10}$$

where (8)-(10) are, respectively, called "nonnegative boundedness", "normalization," and "additivity" of μ in the evaluation set.

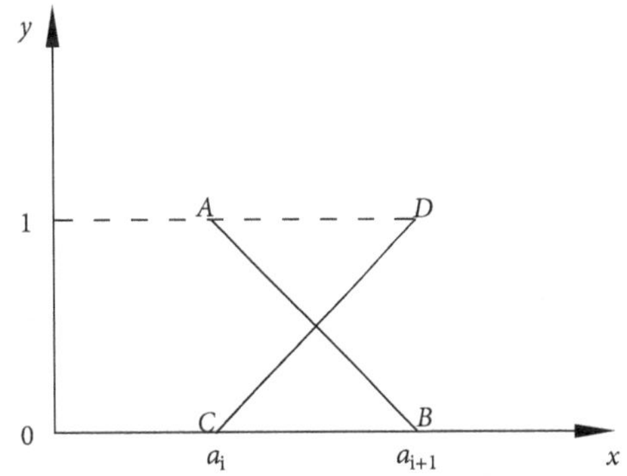

FIGURE 1: Unascertained measurement function graph of straight-line type.

Then μ is called unascertained measurement and the matrix μ_{ijk} can be written as

$$\left(\mu_{ijk}\right)_{m\times p} = \begin{bmatrix} \mu_{i11} & \mu_{i12} & \cdots & \mu_{i1p} \\ \mu_{i21} & \mu_{i22} & \cdots & \mu_{i2p} \\ \vdots & \vdots & \ddots & \vdots \\ \mu_{im1} & \mu_{im2} & \cdots & \mu_{imp} \end{bmatrix} \tag{11}$$

When "uncertainty" is described by the unascertained set, the key is to construct a rational function of the single index measurement. Normally, there are four common kinds for the construction type of single index function, such as straight-line type, para-curve type, sine curve type, and exponential curve type. Among the types, the straight-line type is the most widely used and simplest unascertained measurement function, so the function of the straight-line type is adopted to calculate the parameters in the present study. The function graph of straight-line type is shown in Figure 1.

The expression for the corresponding index function in the interval $[a_i, a_{i+1}]$ is

$$\begin{aligned} y_i(x) &= \begin{cases} \dfrac{-x}{a_{i+1} - a_i} + \dfrac{a_{i+1}}{a_{i+1} - a_i}, & a_i < x \le a_{i+1} \\ 0, & x > a_{i+1} \end{cases} \\ y_{i+1}(x) &= \begin{cases} 0, & x \le a_i \\ \dfrac{x}{a_{i+1} - a_i} - \dfrac{a_i}{a_{i+1} - a_i}, & a_i < x \le a_{i+1} \end{cases} \end{aligned} \tag{12}$$

(5) Multi-Index Unascertained Measurement. The dependence of unascertained measurement theory on each evaluation index is different. Suppose that the evaluation index weight w_{ij} ($0 \le w_{ij} \le 1, \sum_{j=1}^{m} w_{ij} = 1$) is the relative important extent of evaluation index I_j compared with other indexes, which reflects the importance of single index in the whole evaluation index set. Then w_{ij} is called the weight of I_j.

Parameter μ_{ijk} is denoted as the degree of the assessment object belonging to the k-th evaluation grade and defined as

$$\mu_{ik} = \sum_{j=1}^{m} w_{ij}\mu_{ijk}, \quad (i = 1, 2, \cdots, n;\ k = 1, 2, \cdots, p) \tag{13}$$

Table 1: Classification criteria of the structural surface.

Structural surface condition	Value
Thickness of soft rock strata > 5mm, joint width > 5mm and joint is continuous	0~2
Joint surface is smooth or thickness soft rock strata < 5mm, joint opening width is 1~5mm and joint is continuous	2~4
Joint surface is a little rough, joint width < 1mm and rock of joint surface is weak	4~7
Joint surface is a little rough, joint width < 1mm and rock of joint surface is hard	7~9
Joint surface is very rough, joint width is zero, rock of joint surface is hard, the joint surface is discontinuous	9~10

Table 2: Classification criteria of the wall rock quality.

Grade	σ_c/MPa	K_V	S_S	W/[L·(min·10m)$^{-1}$]	θ/°
I	300~250	1.0~0.75	10~9	0~5	90~80
II	250~100	0.75~0.55	9~7	5~10	80~70
III	100~50	0.55~0.35	7~4	10~25	70~30
IV	50~25	0.35~0.15	4~2	25~125	30~10
V	25~1	0.15~0	2~0	125~300	10~0

where μ_{ik} satisfies $0 \le \mu_{ik} \le 1$ and $\sum_{k=1}^{p} \mu_{ik} = 1$. $\mu_{ik} = (\mu_{i1}, \mu_{i2}, \cdots, \mu_{ip})$ is called multi-index measurement evaluation vector of the evaluation object x_i.

(6) Evaluation Criterion. It is necessary to adopt the credible degree recognition criteria to analyze the evaluation object. λ is set as credible degree, whose range of values is $0.5 \le \lambda \le 1$. If the evaluation index set U satisfies $c_1 > c_2 > \cdots > c_p$, let

$$k_0 = \min \left\{ k : \sum_{l=1}^{k} \mu_{il} \ge \lambda,\ k = 1, 2, \cdots, p \right\} \tag{14}$$

Based on the above six steps, it is judged that evaluation object x_i belongs to the k_0-th evaluation grade.

2.3. Evaluation Process of RS Theory and UM Theory. The evaluation process of rough set theory and unascertained measurement theory is depicted in Figure 2. Firstly, the evaluation system is constructed based on the rough set theory and unascertained measurement theory. Secondly, by using the measured data of the wall rock samples, the single index measurement function and measurement evaluation matrix are established. Then, according to (3), the weights of each evaluation index are obtained by significance estimating on the attributes in the rough set theory. Finally, the multi-index measurement evaluation vector is calculated, and the wall rock grade is identified according to the credible degree recognition criteria.

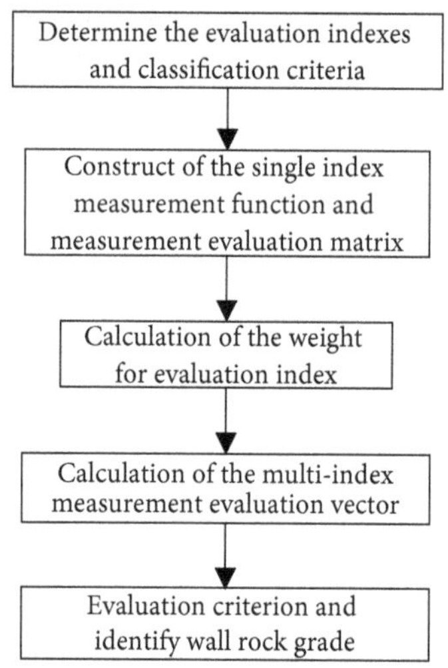

Figure 2: The evaluation process of RS theory and UM theory.

3. RS Theory and UM Theory for Quality Evaluation of Tunnel Wall Rock

3.1. Evaluation Index System of Tunnel Wall Rock Quality. The geological conditions of tunnel engineering are complex and changeable, which is influenced by many factors. The wall rock quality depends on not only the physical and mechanical properties of the wall rock, but also the external engineering geological factors such as groundwater. On the basis of relevant research results [8, 9, 25, 26], a comprehensive evaluation index system is established, which takes saturated uniaxial compressive strength σ_c, rock mass integrity index K_V, structural surface condition S_S, seepage measurement of groundwater W, and the angle between the hole axis and main structural surface θ as the evaluation indexes. The wall rock grade is divided into five grades, and its evaluation index set is $\{C_1, C_2, C_3, C_4, C_5\} = \{\text{I, II, III, IV, V}\}$. Since structural surface condition is a qualitative index, it is converted into semiquantitative index by assignment method whose corresponding assignment is listed in Table 1. The classification criteria of wall rock quality are shown in Table 2.

In this paper, 25 measured wall rock data (x_1, x_2,..., x_{25}) are chosen as the samples to test the feasibility of rock classification based on rough set theory and unascertained measurement theory. The measured wall rock samples belong to Yuanyanghui tunnel, which is located in Shanyin-Pinglu section of Rong-Wu Expressway and is a separated tunnel with four lanes. The evaluation method is applied to evaluate

TABLE 3: Raw data of the rock samples.

Sample	σ_c/MPa	K_V	S_S	$W/[L\cdot(min\cdot10m)^{-1}]$	$\theta/°$	Actual rock grade
1	62	0.35	1.5	5.3	10	IV
2	72	0.19	9.0	7.4	87	IV
3	65	0.32	9.3	11	17	III
4	61	0.34	2.5	5.2	46	IV
5	73	0.77	9.1	7.8	63	II
6	37	0.42	6.7	7.5	77	IV
7	58	0.28	5.8	2.0	42	III
8	61	0.35	9.0	7.1	25	III
9	65	0.34	2.5	4.6	67	III
10	52	0.32	9.6	6.0	86	III
11	67	0.27	8.2	0	58	III
12	63	0.33	9.6	8.7	24	III
13	28	0.20	7.5	5.8	65	IV
14	58	0.28	9.5	15.9	16	III
15	36	0.25	3.8	5.9	83	IV
16	28	0.28	3.5	6.2	82	IV
17	82	0.24	9.3	7.4	19	III
18	34	0.31	4.0	0	7	IV
19	26	0.32	3.3	0	14	IV
20	27	0.58	1.5	11.4	36	III
21	46	0.21	2.5	8.3	82	IV
22	33	0.42	5.7	5.2	82	III
23	12	0.17	1.5	5.1	67	V
24	6	0.21	1.7	6.4	53	V
25	7	0.13	3.3	5.9	46	V

the wall rock grade. The raw data of the wall rock samples are listed in Table 3.

3.2. Construction of the Single Index Measurement Function. On the basis of the straight-line type function and classification criteria of structural surface (Table 1), the single index measurement functions are constructed to get the unascertained measurement value of the evaluation index. According to the classification criteria in Table 2 and taking sample 1 as an example, the single index measurement function of saturated uniaxial compressive strength, rock mass integrity index, structural surface condition, groundwater seepage, the angle between the hole axis, and main structural surface are shown in Figure 3.

The values of evaluation indexes for sample 1 in Table 3 are substituted into the corresponding unascertained measurement functions in Figure 2. Thus, the measurement evaluation matrix of single index can be obtained in

$$\left(\mu_{1jk}\right)_{5\times5} = \begin{bmatrix} 0 & 0 & 0.65 & 0.35 & 0 \\ 0 & 0 & 0.5 & 0.5 & 0 \\ 0 & 0 & 0 & 0 & 1 \\ 0.56 & 0.44 & 0 & 0 & 0 \\ 0 & 0 & 0 & 0 & 1 \end{bmatrix} \quad (15)$$

3.3. Calculation of the Weight for Evaluation Index. (1) The initial decision table is constructed according to the evaluation criteria of the wall rock quality and the raw data of the rock samples. The evaluation indexes and actual rock grade are taken as the condition attribute set and decision attribute set, respectively. The initial decision table for evaluation indexes of wall rock quality is established in Table 4.

(2) The samples in the initial decision table are screened according to the screening principles. According to the principle ①, x_6 is removed. Because the decision attributes of x_2 and x_{10} are different, let x_{10} be retained based on the principle ②. In the initial decision table, x_3=x_{14}, x_8=x_{12}=x_{17}, x_{15}=x_{21}, x_{23}=x_{24}, which mean that the condition attributes and decision attributes of multiple samples are the same, so x_3, x_8, x_{15}, and x_{23} are retained according to the principle ③.

(3) After the attributes of the decision table are reduced, the redundancy condition attribute is not found. That is to say, saturated uniaxial compressive strength, rock mass integrity index, structural surface condition, groundwater seepage, and the angle between the hole axis and the main structural surface are core attributes.

After data processing, the optimal attribute of decision table for evaluation indexes of wall rock quality is obtained. The optimal attribute of decision table is 18 rows and 6 columns, which is listed in Table 5.

(4) Calculation of the weight for each evaluation index: firstly, according to (1), K=$\gamma_c(D)$=0.9 can be obtained by

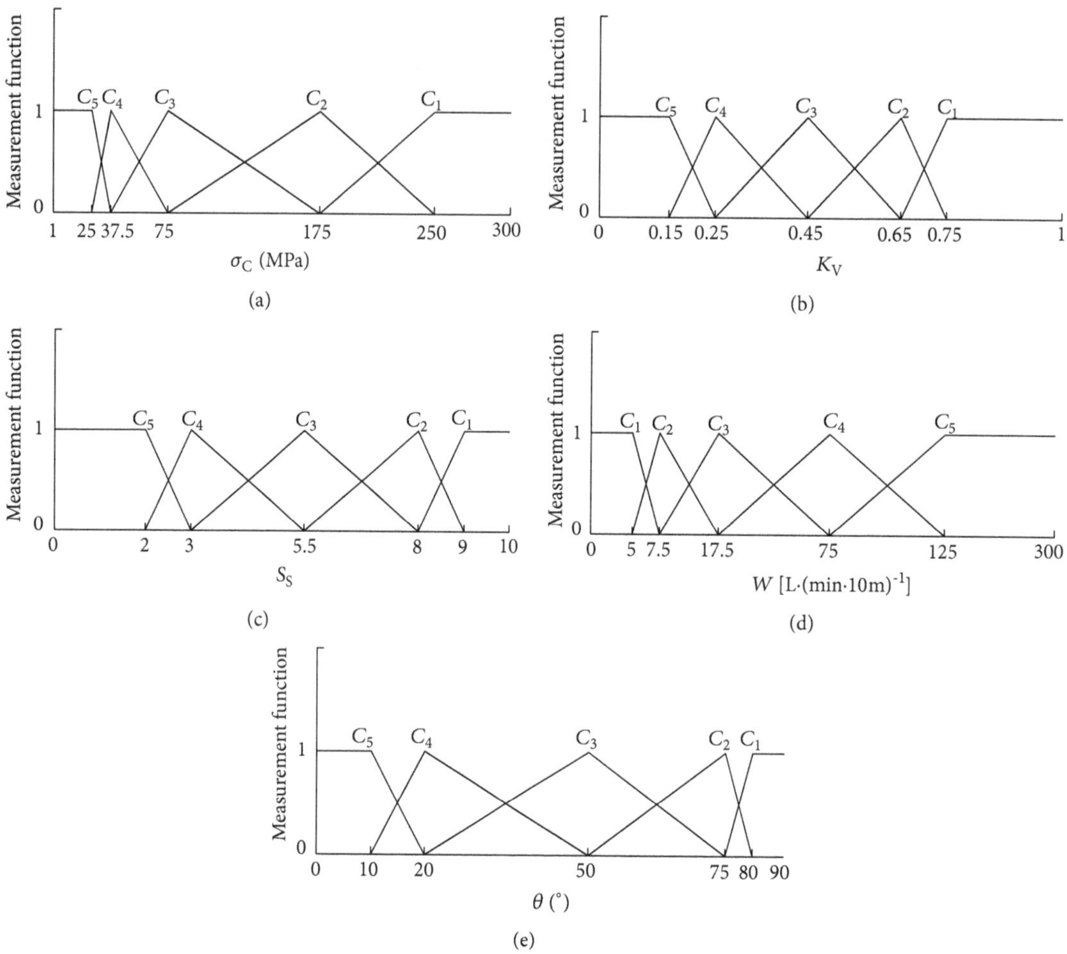

FIGURE 3: Unascertained measurement function of evaluation indexes.

calculating the optimal attribute of decision table, which indicates that the decision attribute D depends on the condition attribute C. Secondly, for each evaluation index, the dependency degree $\gamma_{C-C_i}(D)$ of the decision attribute D on the condition attribute C-C_i can be obtained by (2). $\gamma_{C-C_i}(D)$ of the evaluation indexes σ_c, K_V, S_S, W, and θ are 0.658, 0.718, 0.718, 0.823, and 0.833, respectively. Thirdly, the importance of the evaluation indexes in the indexes set can be got based on (2). Then, $\sigma_{CD}(C_i)$ of the evaluation indexes σ_c, K_V, S_S, W, and θ are 0.242, 0.182, 0.182, 0.077, and 0.067, respectively. Finally, (2) and (3) yield the weights of the evaluation index and the weight $\omega_i = (0.242/(0.242 + 0.182 + 0.182 + 0.077 + 0.067)) = 0.323$ for evaluation index σ_c. Similarly, the weights of K_V, S_S, W, and θ are 0.243, 0.243, 0.102, and 0.089.

3.4. Calculation of the Multi-Index Measurement Evaluation Vector. For sample 1, the weights are obtained on the above calculation. The multi-index measurement evaluation vector of sample 1 can be obtained according to the single index evaluation matrix (12) and (13), which is $\{0.057, 0.045, 0.331, 0.235, 0.332\}$. Similarly, the multi-index measurement evaluation vector of other samples can also be obtained.

3.5. Evaluation Criterion. Credible degree recognition is performed using (14) and multi-index measurement evaluation vector of sample 1. Here, λ is set as 0.6, k_0=0.057+0.045+0.331+0.235=0.668>λ=0.6. Therefore, the wall rock grade of the sample 1 belongs to grade IV. Similarly, the rock grade of other samples can be obtained, and the evaluation results are listed in Table 6.

From the analysis of evaluation results, the results of the wall rock quality using the evaluation method based on rough set theory and unascertained measurement theory are basically the same as these of the actual rock grade. However, the rock grade of sample 9 is classified inaccurately as grade IV. The reason may be that the actual rock grade is between grade III and grade IV. So the evaluation method determines the grade of sample 9 as a low-grade wall rock, which is allowed from the view of construction safety. Thus, the evaluation method of wall rock quality based on rough set theory and unascertained measurement theory has a high accuracy, which is feasible for determination of wall rock quality.

4. Engineering Application

Reasonable classification of the wall rock plays an important role in guiding the design and construction of the tunnel and can serve as an important basis for evaluating the stability of the wall rock. To test the accuracy of the rock quality

TABLE 4: Initial decision table for evaluation indexes of wall rock quality.

Sample	Rock grade determined by individual evaluation indexes					Actual rock grade
	σ_c	K_V	S_S	W	θ	
1	III	IV	V	II	V	IV
2	III	IV	I	II	I	IV
3	III	IV	I	III	IV	III
4	III	IV	IV	II	III	IV
5	III	I	I	II	III	II
6	IV	III	III	III	II	IV
7	III	IV	III	I	III	III
8	III	IV	I	II	IV	III
9	III	IV	IV	I	III	III
10	III	IV	I	II	I	III
11	III	IV	II	I	III	III
12	III	IV	I	II	IV	III
13	IV	IV	II	II	III	IV
14	III	IV	I	III	IV	III
15	IV	IV	IV	II	I	IV
16	IV	IV	II	II	V	IV
17	III	IV	I	II	IV	III
18	IV	IV	IV	I	V	IV
19	IV	IV	IV	I	IV	IV
20	III	II	V	II	III	III
21	IV	IV	IV	II	I	IV
22	IV	III	III	II	I	III
23	V	IV	V	II	II	V
24	V	IV	V	II	II	V
25	V	V	IV	II	III	V

evaluation method based on the rough set theory and unascertained measurement theory, the measured samples of the wall rock in the Xihualing tunnel are taken as samples for analysis. The tunnel is located in Jinhua section of Zhu-Yong Expressway, which is a separated tunnel with four lanes. Additionally, the rock quality evaluation results are based on each of the following three methods: Delphi-ideal point method [25], set pair analysis method [26], rough set theory, and unascertained measurement theory. The evaluation results of the above three methods and the actual grade can be found in Table 7.

First, the multi-index unascertained measurement of each tunnel is calculated based on rough set theory and unascertained measurement theory, the wall rock grade is determined according to the credible degree recognition. The evaluation results are consistent with those of Delphi-ideal point method, set pair analysis method, and the actual condition.

From the classification of tunnel collapse risk in Table 7, the risk grade of the mileage (K107+776~806 and K107+806~836) in the right tunnel of the Xihualing tunnel is grade III, which means that the tunnel may occasionally collapse. Hence, the problems should be taken into account during construction. The risk grade of the mileage (K107+776~806 and K107+806~836) in the left tunnel is grade IV, which means that the possibility of the tunnel collapse is greater. The evaluation results coincide with the actual wall rock conditions encountered during construction. During the tunnel construction, it was found that the wall rock exposed by the tunnel was broken and the rock quality was poor. Thus, the actions such as strengthening forepoling and shortening the excavation footage are taken to avoid the occurrence of the tunnel excavation accidents.

5. Conclusions

In this paper, a method combining the rough set theory and unascertained measurement theory is applied to the classification of tunnel wall rock for the first time. According to the rough set theory, the weights of evaluation indexes are obtained by significance estimating on the attributes. Thus, a method for classification of wall rock of tunnels is established based on the basic principles of unascertained measurement theory. The method is used to classify the measured wall rock samples, and the evaluation results are compared with the actual wall rock grade. It is proved that the method is feasible for determination of wall rock quality.

TABLE 5: Optimal attribute of decision table for evaluation indexes of wall rock quality.

Sample	Rock grade determined by individual evaluation indexes					Actual rock grade
	σ_c	K_V	S_S	W	θ	
1	III	IV	V	II	V	IV
3	III	IV	I	III	IV	III
4	III	IV	IV	II	III	IV
5	III	I	I	II	III	II
7	III	IV	III	I	III	III
8	III	IV	I	II	IV	III
9	III	IV	IV	I	III	III
10	III	IV	I	II	I	III
11	III	IV	II	I	III	III
13	IV	IV	II	II	III	IV
15	IV	IV	IV	II	I	IV
16	IV	IV	II	II	V	IV
18	IV	IV	IV	I	V	IV
19	IV	IV	IV	I	IV	IV
20	IV	II	V	III	III	III
22	IV	III	III	II	I	III
23	V	IV	V	II	II	V
25	V	V	IV	II	III	V

TABLE 6: Evaluation results of wall rock quality.

Sample	Comprehensive unascertained measurement					Evaluation results	
	C_1	C_2	C_3	C_4	C_5	RS-UM	Actual grade
1	0.057	0.045	0.331	0.235	0.332	IV	IV
3	0.243	0.066	0.357	0.307	0.027	III	III
4	0.094	0.008	0.390	0.386	0.122	IV	IV
5	0.486	0.145	0.353	0.016	0	II	II
7	0.102	0.049	0.466	0.383	0	III	III
8	0.259	0.086	0.340	0.315	0	III	III
9	0.102	0.061	0.374	0.342	0.122	IV	III
10	0.373	0.061	0.355	0.211	0	III	III
11	0.151	0.223	0.340	0.287	0	III	III
13	0.069	0.280	0.084	0.199	0.367	IV	IV
15	0.154	0.037	0.078	0.692	0.039	IV	IV
16	0.142	0.049	0.085	0.478	0.245	IV	IV
18	0.102	0	0.170	0.548	0.179	IV	IV
19	0.102	0	0.187	0.360	0.351	IV	IV
20	0.243	0.220	0.172	0.094	0.271	III	III
22	0.183	0.028	0.430	0.243	0.116	III	III
23	0.098	0.065	0.028	0.292	0.517	V	V
25	0.065	0.037	0.041	0.291	0.566	V	V

Note: RS-UM is the evaluation method based on rough set theory and unascertained measurement theory.

TABLE 7: Evaluation results of wall rock classification.

Tunnel mileage	Comprehensive unascertained measurement					Evaluation results			
	C_1	C_2	C_3	C_4	C_5	RS-UM	SPA	D-IP	Actual grade
Left tunnel, K106+760~790	0	0.056	0.355	0.589	0	IV	IV	IV	IV
Left tunnel, K106+790~820	0	0	0.333	0.546	0.121	IV	IV	IV	IV
Right tunnel, K107+776~806	0.102	0.219	0.519	0.160	0	III	III	III	III
Right tunnel, K107+806~836	0.248	0.106	0.560	0.087	0	III	III	III	III

Note: RS-UM is the evaluation method based on rough set theory and unascertained measurement theory, SPA is set pair analysis method, and D-IP is Delphi-ideal point method.

In addition, the method based on the rough set theory and unascertained measurement theory is applied to an engineering case study. The evaluation results are consistent with those of Delphi-ideal point method, set pair analysis method, and the actual condition. It provides the basis and guidance for the construction of the tunnel. Meanwhile, it is proved that the method has a good applicability and gives references for the classification of wall rock in other similar tunnel projects.

Conflicts of Interest

The authors declare that there are no conflicts of interest regarding the publication of this paper.

Acknowledgments

This work was supported by the National Natural Science Foundation of China (51374032), the Beijing Training Project for the Leading Talents in S&T (Z151100000315014), and the Fundamental Research Funds for the Central Universities (FRF-TP-16-021A3).

References

[1] D. U. Deere, "Technical description of rock cores for engineering purposes," Tech. Rep., University of Illinois, 1962.

[2] Z. Bieniawski, "Engineering classification of jointed rock masses. discussions of paper by Z.T. Bieniawski, trans. s. afr. instn. civ. engrs. v15, n12, Dec. 1973, and authors reply," *International Journal of Rock Mechanics and Mining Sciences & Geomechanics Abstracts*, vol. 11, no. 12, p. 244, 1974.

[3] A. Palmstrøm, "Characterizing rock masses by the RMi for Use in Practical Rock Engineering: Part 1: the development of the Rock Mass index (RMi)," *Tunnelling and Underground Space Technology*, vol. 11, no. 2, pp. 175–188, 1996.

[4] A. Palmstrøm, "Characterizing rock masses by the RMi for use in practical rock engineering, Part 2: some practical applications of the rock mass index (RMi)," *Tunnelling and Underground Space Technology*, vol. 11, no. 3, pp. 287–303, 1996.

[5] J. L. Rangel, U. Iturrarán-Viveros, A. G. Ayala, and F. Cervantes, "Tunnel stability analysis during construction using a neuro-fuzzy system," *International Journal for Numerical and Analytical Methods in Geomechanics*, vol. 29, no. 15, pp. 1433–1456, 2005.

[6] A. T. C. Goh and W. Zhang, "Reliability assessment of stability of underground rock caverns," *International Journal of Rock Mechanics and Mining Sciences*, vol. 55, pp. 157–163, 2012.

[7] A. Alimoradi, A. Moradzadeh, R. Naderi, M. Z. Salehi, and A. Etemadi, "Prediction of geological hazardous zones in front of a tunnel face using TSP-203 and artificial neural networks," *Tunnelling and Underground Space Technology*, vol. 23, no. 6, pp. 711–717, 2008.

[8] X. Z. Huang and C. X. She, "Research on methods of wall rock masses stability classification based on extension theory," *Rock and Soil Mechanics*, *Rock and Soil Mechanics*, vol. 27, no. 10, pp. 1800–1804, 2006.

[9] M. N. Bagde, A. K. Raina, A. K. Chakraborty, and J. L. Jethwa, "Rock mass characterization by fractal dimension," *Engineering Geology*, vol. 63, no. 1-2, pp. 141–155, 2002.

[10] G. Y. Wang, "Uncertainty information and its mathematical treatment," *Journal of Harbin Architecture and Engineering Institute*, vol. 23, no. 4, pp. 52–58, 1990.

[11] L. A. Zadeh, "Fuzzy sets," *Information and Computation*, vol. 8, pp. 338–353, 1965.

[12] G. Matheron, *Random sets and integral geometry*, John Wiley and Sons, New York, NY, USA, 1975.

[13] D. Ju-Long, "Control problems of grey systems," *Systems & Control Letters*, vol. 1, no. 5, pp. 288–294, 1982.

[14] K. Liu, L. Yi, and L. Yao, "Informational uncertainties and their mathematical expressions," *Kybernetes*, vol. 30, no. 4, pp. 378–396, 2001.

[15] K. Liu, Y. Lin, and L. Yao, "Unascertained rationals and subjective uncertain information," *Systems Analysis Modelling*

Simulation. A Journal of Mathematical Modelling and Simulation in Systems Analysis, vol. 42, no. 3, pp. 343–357, 2002.

[16] C. Wen, T. A. Parsons, and B. Al Malik, "Unascertained Measurement Evaluation on Maintenance Quality of Highway Pavement," in *Proceedings of the First International Symposium on Transportation and Development Innovative Best Practices*, pp. 228–233, Beijing, China.

[17] Z. Yao and L. Cui, "Mechanical structure design optimization by blind number theory: time-dependent reliability. World Academy of Science, Engineering and Technology," *International Journal of Mechanical, Aerospace, Industrial, Mechatronic and Manufacturing Engineering*, vol. 6, no. 2, pp. 536–539, 2012.

[18] J. Zhou and X. Li, "Integrating unascertained measurement and information entropy theory to assess blastability of rock mass," *Journal of Central South University*, vol. 19, no. 7, pp. 1953–1960, 2012.

[19] S. C. Li, J. Wu, Z. H. Xu, and L. P. Li, "Unascertained measure model of water and mud inrush risk evaluation in karst tunnels and its engineering application," *KSCE Journal of Civil Engineering*, vol. 21, no. 4, pp. 1170–1182, 2017.

[20] Z. Pawlak, "Rough sets," *International Journal of Computer & Information Science*, vol. 11, no. 5, pp. 341–356, 1982.

[21] J. W. Grzymala-Busse, *Managing Uncertainty in Expert Systems*, Springer US, Boston, MA, 1991.

[22] Z. Pawlak, *Rough Sets: Theoretical Aspects of Reasoning about Data*, Springer, Dordrecht, Netherlands, 1991.

[23] T. Beaubouef and F. E. Petry, "Uncertainty modeling for database design using intuitionistic and rough set theory," *Journal of Intelligent & Fuzzy Systems: Applications in Engineering and Technology*, vol. 20, no. 3, pp. 105–117, 2009.

[24] R. Yasdi, "Combining rough sets learning- and neural learning-method to deal with uncertain and imprecise information," *Neurocomputing*, vol. 7, no. 1, pp. 61–84, 1995.

[25] Y. C. Wang, H. Y. Sun, Y. Q. Shang, and M. Zhu, "Classification of wall rock based on Delphi method and ideal point method," *Chinese Journal of Geotechnical Engineering*, vol. 32, no. 4, pp. 651–656, 2010.

[26] Y. C. Wang, X. H. Xu, Y. L. Liu, and L. J. Zhou, "Set pair analysis model in wall rock classification of highway tunnel," *Journal of Harbin Institute of Technology*, vol. 43, no. 4, pp. 114–119, 2011.

18

Bearing Characteristics of Composite Pile Group Foundations with Long and Short Piles under Lateral Loading in Loess Areas

Tianzhong Ma[1,2] **Yanpeng Zhu,**[1,2] **Xiaohui Yang**[1,2] **and Yongqiang Ling**[1,2]

[1]*Key Laboratory of Disaster Prevention and Mitigation in Civil Engineering of Gansu Province, Lanzhou University of Technology, Lanzhou 730050, China*
[2]*Western Engineering Research Center of Disaster Mitigation in Civil Engineering of Ministry of Education, Lanzhou University of Technology, Lanzhou 730050, China*

Correspondence should be addressed to Xiaohui Yang; yxhui86@126.com

Academic Editor: Evangelos J. Sapountzakis

It is very necessary to research the bearing characteristics of composite pile group foundations with long and short piles under lateral load in loess areas, because these foundations are used widely. But few people researched this problem in loess areas up to now worldwide. In this paper, firstly, an indoor test model of a composite pile foundation with long and short piles is designed and then employed to explore the vertical load bearing characteristics and load transfer mechanisms of a single pile, a four-pile group, and a nine-pile group under different lateral loads. Secondly, ANSYS software is employed to analyze the load-bearing characteristics of the test model, and for comparison with the experimental results. The results demonstrate the following. (1) The lateral force versus pile head displacement curves of the pile foundation exhibit an obvious steep drop in section, which is a typical feature of piercing damage. A horizontal displacement limit of the pile foundation is 10 mm and 6mm for the ones sensitive to horizontal displacement. (2) The axial force along a pile and frictional resistance do not coincide, due to significant variations and discontinuities in the collapsibility of loess; a pile body exhibits multiple neutral points. Therefore, composite pile groups including both long and short piles could potentially maximize the bearing capacity and reduce pile settlement. (3) The distribution of stress and strain along the pile length is mainly concentrated from the pile head to a depth of about 1/3 of the pile length. If the lateral load is too large, short piles undergo rotation about their longitudinal axis and long piles undergo flexural deformation. Therefore, the lateral bearing capacity mainly relies on the strength of the soil at the interface with the pile or the horizontal displacement of the pile head.

1. Introduction

Pile foundation is one of the most common deep foundation forms employed for supporting superstructures in collapsible loess regions. Significant achievements have been made in the selection of pile foundations [1, 2], characterization of load bearing capacity [3, 4], and investigation of negative frictional resistance [5, 6]. However, geotechnical engineering activities conducted in loess regions often encounter multiple stiff bearing layers, which is similar to conditions encountered in coastal areas. Such discontinuities in the characteristics of collapsible loess add considerably greater complexity to the optimal design of pile foundations [7]. In response to these conditions, the lower limit value of depth collapsibility is a bit conservative [8–10], so composite pile foundations with long and short piles have gained increasing interest.

Research regarding composite pile foundation with long and short piles has typically focused on single pile deformation characteristics under vertical loading and has employed numerous research methods including theoretical analysis, field or laboratory test, and numerical simulation. However, the extent of research focused on pile groups is considerably less. S.C.Wong et al. [11] investigated the interaction of different pile lengths and suggested that the pile interaction coefficient exhibits obvious differences under these conditions. Y. F.Leung et al. [12] proposed an analysis method based on elastic theory that focused on optimizing the pile lengths of pile raft foundations to improve the overall stiffness. However, the calculation becomes very complex with an increasing number of piles. Z.J.Wang et al. [13] deduced an analytical expression for the settlement of a single pile in a homogeneous soil layer and applied it to stratified soils. J.H.Xiong et al. [14] experimentally investigated model composite pile foundations with long and short piles composed of four-pile and nine-pile groups and evaluated the load bearing properties for different pile lengths. M.S.Huang et al. [15] employed centrifugal model testing and a three-dimensional (3D) finite element method to investigate the load sharing behavior of composite pile foundation groups composed of different pile lengths under embankment loading. X.N.Ma et al. [16] conducted field testing and numerical simulations and provided a detailed analysis of the characteristic lengths of settlement for composite pile foundations. D.A.Brown et al. [17] conducted a series of 3D FEM studies on the behavior of single pile and pile group with elastic-plastic soil model. X.S.Ge et al. [18] investigated that composite pile foundations with long and short piles can ensure a more uniformly distributed foundation stress and thereby reduce the extent of foundation settlement. M.Achmus et al. [19] and H.H.Zhang et al. [20] investigated composite pile foundations under vertical and horizontal loading, evaluated the complex interaction between piles and soil, and proposed a method about common function analysis. L.Zhang et al. [21] investigated bearing capacity of single pile under horizontal load in layered soils. It was determined that the foundations of high-rise buildings under horizontal loading (e.g., loading from wind, earthquakes, and soil lateral pressure) are subject to increased horizontal shear forces and overturning moments. As such, high-rise building foundations must not only bear vertical loads, but also resist horizontal loading. Therefore, the load bearing properties of composite pile foundations with long and short piles must be more fully investigated under simultaneous vertical and horizontal loading.

This paper presents the design of an indoor composite pile foundation test model with long and short piles. The model was employed to explore the vertical load bearing characteristics and load transfer mechanism of pile groups composed of a single pile, four piles, and nine piles under different lateral loads. The ANSYS finite element software was employed to analyze the load-bearing characteristics of the test model, and comparisons verify the feasibility of applying composite pile foundations with long and short piles in loess areas. The findings of this study can provide guidance for pile foundation optimization in loess areas.

2. Test Program

2.1. Model Enclosure and Test Piles. Experimental testing employed the cubic model enclosure illustrated in Figures 1 and 2. The enclosure was 1.1m on each side. Angle iron was welded along each side of the cube, and the bottom plane was 1 cm thick steel plate in which eight drainage ports were drilled. 1 cm thick organic glass was placed all around; organic glass and angle iron were connected by bolt.

TABLE 1: Physical parameters of loess.

d_s	$w_p(\%)$	$w_L(\%)$	$\rho_{d\max}$ (g/cm^3)	$w(\%)$
2.71	16.3	29.5	1.76	12.7

Attached to the bottoms of the test piles were springs with varied stiffness coefficients to simulate soil layers of different load bearing characteristics. Collapsible loess obtained from Longxi county of China was employed as fill material in the model enclosure, and its basic physical and mechanical parameters are listed in Table 1. The foundation soil was added in 80 mm layers. The volume density and water content of the foundation soil were consistent with natural loess. All tested piles were 70 mm in diameter and were composed of C30 concrete. The length of long pile was 1,100 mm, and the length of short pile was 850 mm.

2.2. Strain Components of Pile Bodies. Strain testing of the model piles was conducted using foil type resistance strain gauges. A pair of resistance strain gauges was symmetrically placed evenly along the length of the piles, beginning with the bottom, to ensure the accuracy of the measurement results. As illustrated in Figure 3, strain gauges were placed at 8 cm, 23 cm, 38 cm, 53 cm, and 68 cm from the bottom of the short pile body, including a total of five strain gauge pairs, and strain gauges were placed at 9 cm, 23 cm, 37 cm, 51 cm, 65 cm, 79 cm, and 93 cm from the bottom of the long pile body, for a total of 7 strain gauge pairs. The strain gauges were protected from moisture contamination by coating the gauges using silica gel. Strain gauge was connected by a quarter bridge. Static strain gauge data was collected using a high speed data acquisition system.

2.3. Test Loading. Testing was conducted using a pseudo-static loading test system. A 30 t load sensor was positioned on the top surface of each pile to measure the vertical load, and a 2 cm thick steel plate was placed over the load sensors on the top surfaces of the piles. In order to make level, cushion sandbags were placed between the load sensor and pile head. Vertical settlement and horizontal displacement of the pile top were measured by displacement sensors. A hydraulic jack with a digital display was employed for applying the vertical load. To guarantee the stability of the vertical load, testing was suspended momentarily to ensure an accurate loading due to the settlement of the pile foundation.

Horizontal load testing was conducted for a single pile, four piles, and nine piles using the apparatus shown in Figure 4. Vertical load testing indicated that the vertical ultimate bearing capacities of the single pile #2 (short), single pile #7 (long), four-pile group (#4 [long], #5 [short], #9 [long], and #10 [short]), and nine-pile group (#1–#3# [long except for #2], #6–#8 [short except for #7], and #11–#13 [long, short, long, respectively]) are 3.39, 4.34, 16, and 32 kN, respectively. During horizontal load testing, a vertical load of 0.8 times the ultimate vertical bearing capacity of a pile was firstly applied, and then horizontal loading was conducted at a strain rate of 0.02 mm/s. It should be noted that, due to the level of

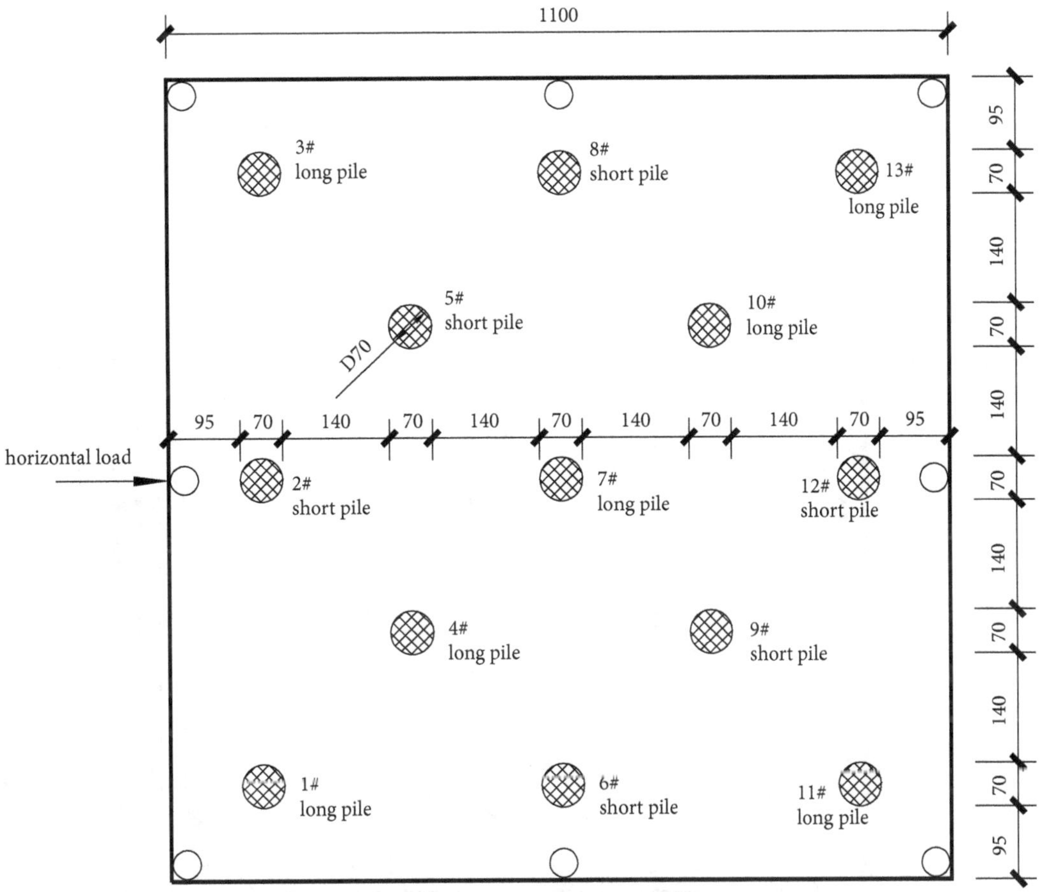

FIGURE 1: Plane arrangement of the test model.

single pile loading test, nine-pile load #2 and #7 piles and pile top force steel failed to connect, practical for seven-pile load. Because the horizontal load testing conducted was relatively limited, the horizontal displacement of a pile head was limited to a maximum of 2 cm.

3. Analysis of Test Results

3.1. Displacement of the Pile Head with respect to Horizontal Load. Applied at the top of the pile under 0.8 times vertical ultimate bearing capacity, single pile (#2 and #7 piles), four-pile (#4, #5, #9, #10 piles), and seven-pile (#1, #3, #6, #8, #11~ #13 piles) side sway displacement S of pile top change with horizontal thrust F curve is shown in Figure 5.

The displacements S of the pile head/s with respect to the applied horizontal load to the single pile, four-pile group, and seven-pile group are shown in Figure 5. We note that the F-S curves essentially present a linear relationship when the horizontal load is small. However, with increasing load, the curves gradually become nonlinear. The short piles in particular exhibit an obvious turning point at a given load, which is a typical feature of piercing damage. The turning point for the single #2 pile (short pile) occurs at a force of 2.8 kN, with a corresponding displacement of 3.97 mm. The turning point for the single #7 pile (long pile) occurs at a force of 3.5 kN, with a corresponding displacement of 4.95 mm, which is greater than that of the #2 pile. The turning point of the four-pile group occurs at a force of about 6 kN, with displacements of 3-4.5 mm. The turning point of the seven-pile group occurs at a force of about 13 kN, with displacements of 5-7.1 mm.

We observed that the short piles within a pile group are damaged with increasing horizontal loading prior to the long piles, and the short piles may be under a normal load when they are damaged. Displacement is an important factor affecting the bearing capacity of pile foundations; however, testing was discontinued after the head of the short piles exceeded the 2 cm displacement limit, even though the long piles or the overall pile structure had suffered no damage. For loess areas that lack static load test of pile group level, in conjunction with the current "technology code for building pile foundations" (JGJ 94-2008), a horizontal displacement limit of 10 mm is recommended; that for horizontal displacement sensors is 6 mm. Therefore, big thickness of loess foundation and bearing capacity of pile foundation design are not only the design standards and should be combined with regional experiences. Pile foundation design shall provide an adequate pile length-diameter ratio according to the requirements of the upper structure and the tensile strength of piles with consideration for permissible pile displacement values. To

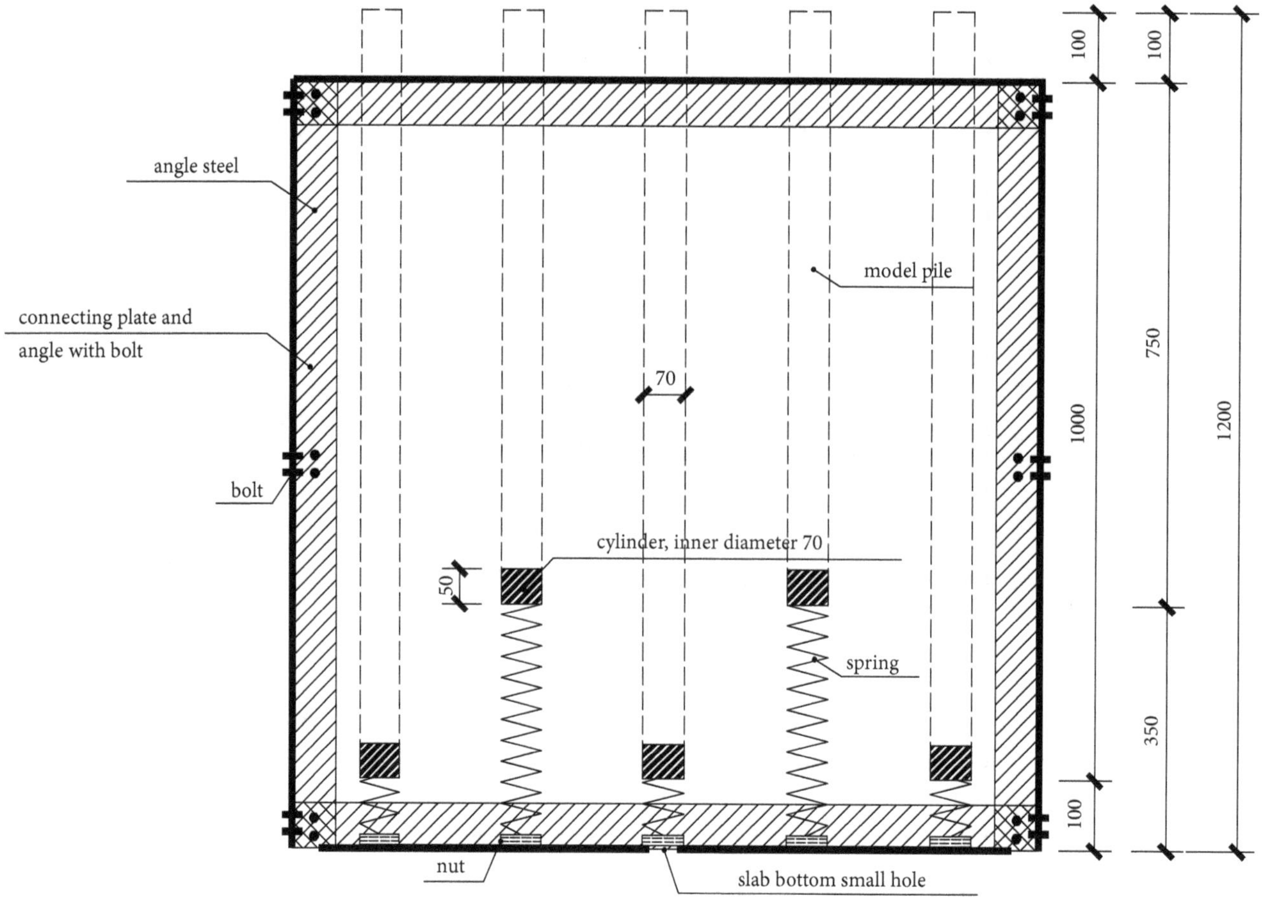

FIGURE 2: Elevation arrangement of the test model.

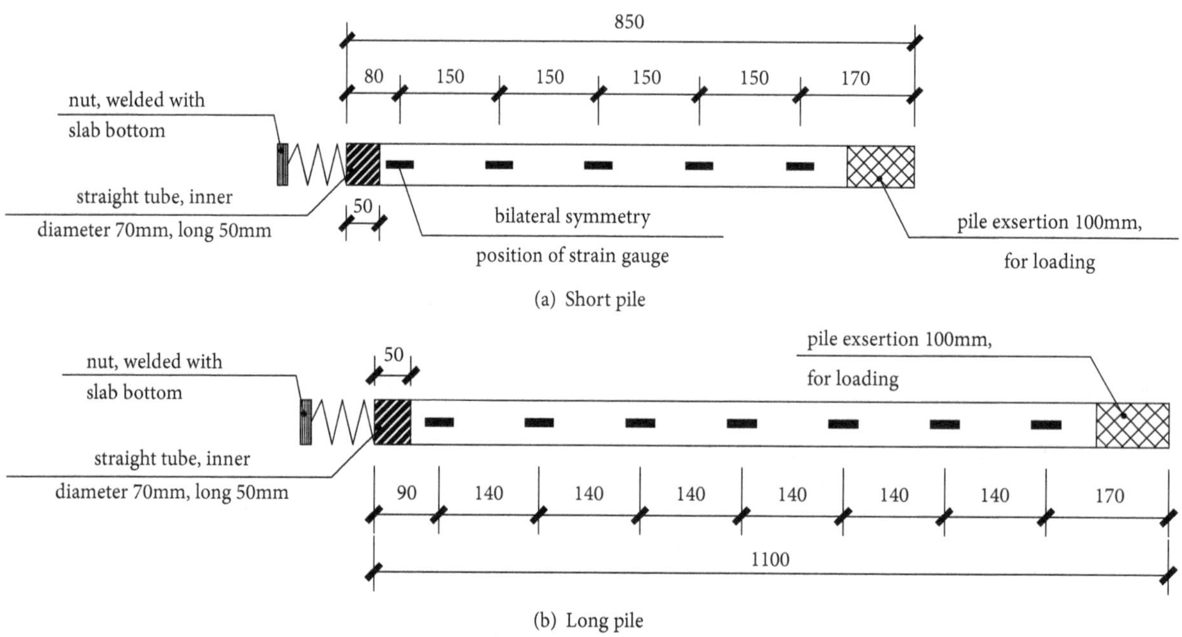

FIGURE 3: Arrangement of strain gauges on the short and long piles.

FIGURE 4: Image of testing facility.

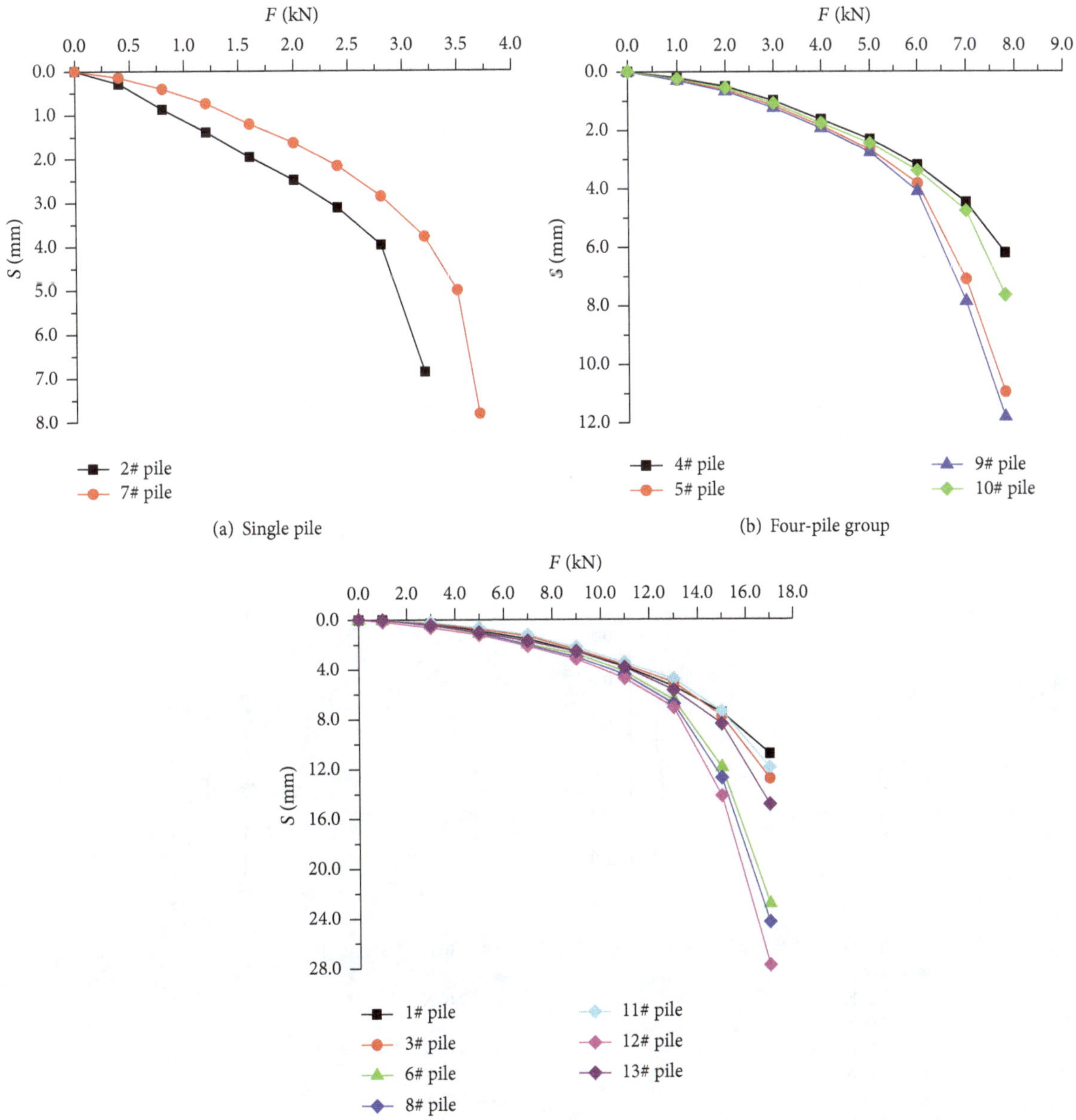

(a) Single pile

(b) Four-pile group

(c) Seven-pile group

FIGURE 5: Relations between the horizontal load F and the displacement of the pile heads S for the pile groups considered.

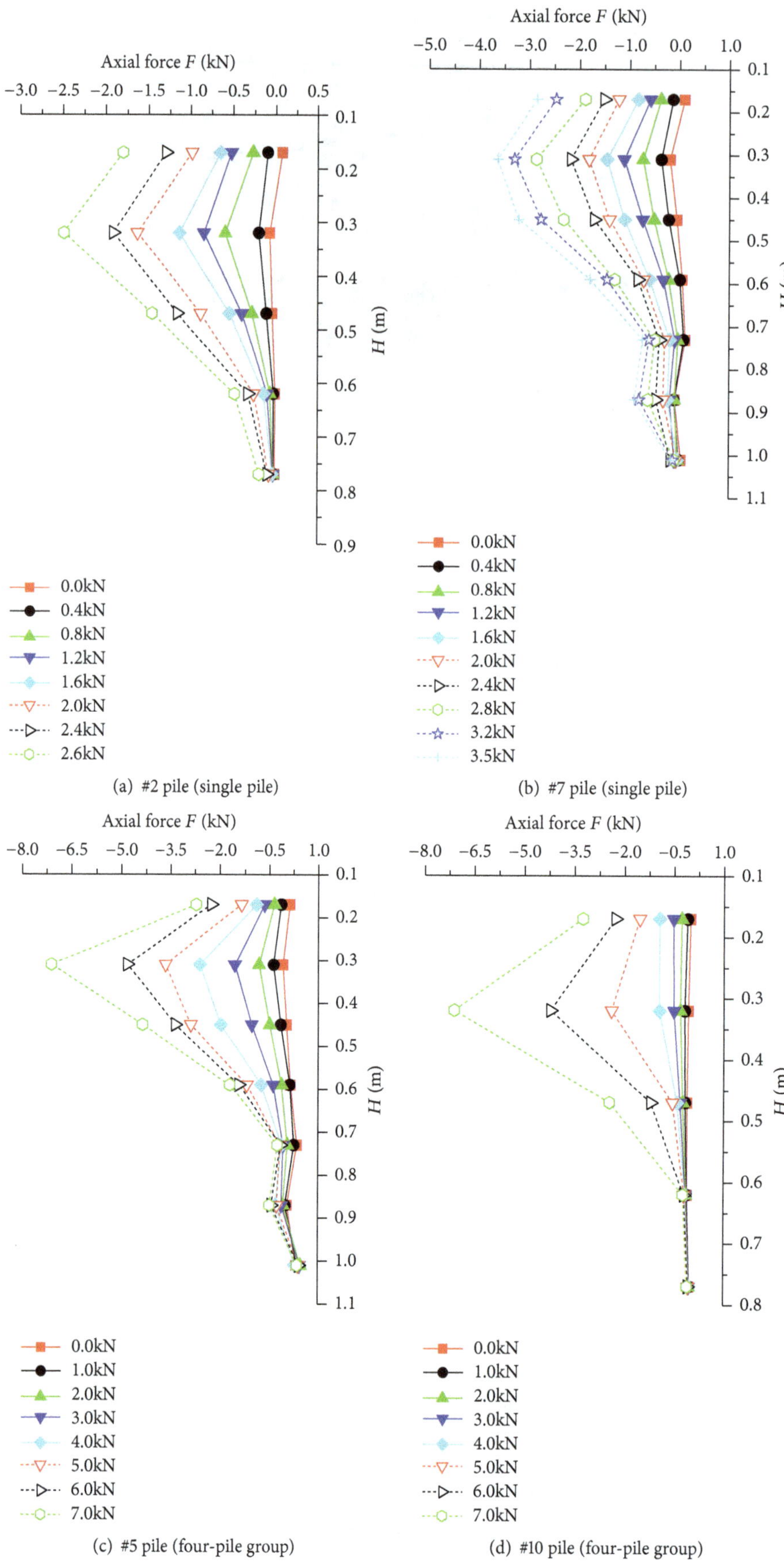

(a) #2 pile (single pile)

(b) #7 pile (single pile)

(c) #5 pile (four-pile group)

(d) #10 pile (four-pile group)

FIGURE 6: Continued.

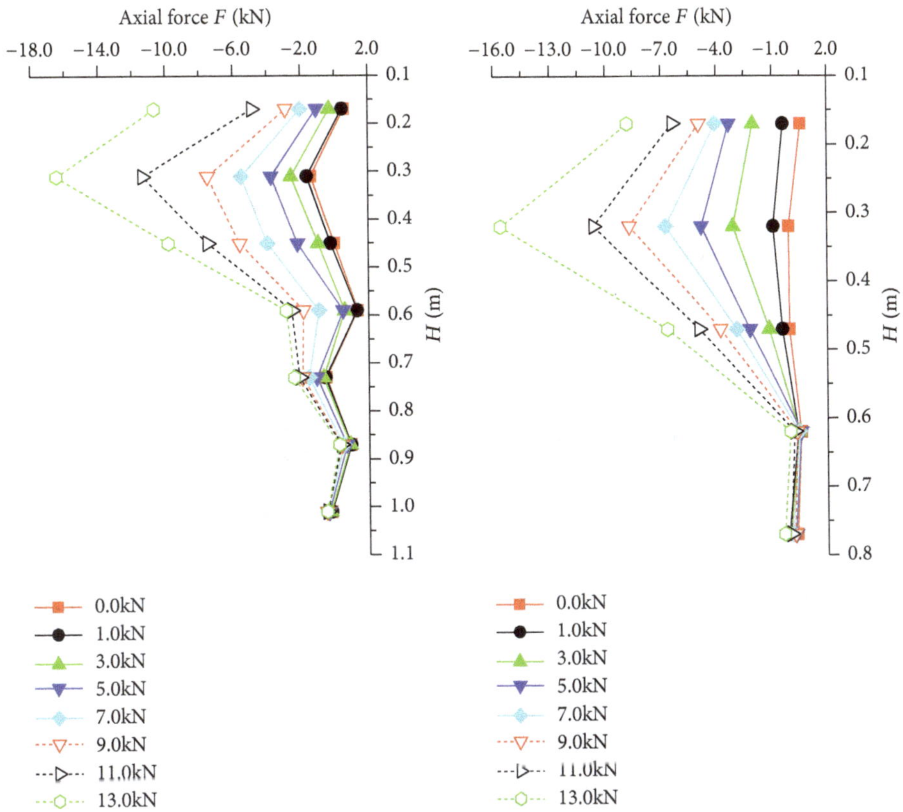

(e) #1 pile (seven-pile group)

(f) #8 pile (seven-pile group)

FIGURE 6: Relations between the axial forces acting on a pile with respect to location H along the length of the pile for various horizontal loads.

control the horizontal displacement, crack width limited, the deformation of the upper part of the pile foundation must be reduced by various measures such as increasing the pile reinforcement and improving the strength of the concrete.

3.2. Analysis of the Axial Force Acting on Test Piles with respect to Horizontal Load. Figure 6 presents the axial force distribution acting at the strain gauge locations H (m) along the lengths of the test piles for various horizontal loads. The figure shows that the axial forces increase with increasing horizontal force and that, during the early stages of loading, the horizontal loading at the pile head is mainly borne by the friction between the pile and surrounding soil near the top of the pile and type of friction pile. With the increasing load, the axial force increases, tending to end bearing pile. With increasing pile depth H, the axial stress distributions of all short piles are very similar, and a peak value is observed at about 0.3 m. Zero position of axial force constantly down shows that, the developed ability of axial force along pile length in the foundation is asynchronous.

These results differ from the axial force distribution of an in situ uniform loess foundation. Under in situ conditions, the vertical and horizontal loads are simultaneously brought to bear at the top of the pile. Moreover, the stiffness of pile end soil is small, and pile modulus is bigger. As a result, a stress concentration is easily produced, and the contact areas between the pile and surrounding soil gradually form a plastic zone. The upper portion of the pile body is firstly deformed and produced downward displacement of soil. The relative displacement between the pile and surrounding soil produces a frictional resistance at the pile-soil interface. The process of load transmission downwards along the pile body overcomes the frictional resistance, and in the diffusion process of passing it to the soil, which results in a decreasing axial force along the depth of the pile. The rate of attenuation in the axial force acting on the pile is not uniform. An increasing frictional resistance of the surrounding soil causes a more rapid decay in the axial force. This is equivalent to the change in the transverse compactness of the artificial layered fill employed in the model experiments.

Comparing the long and short piles under different horizontal loads, the long piles bear the majority of the load, but, because the length-diameter ratio is greater than that of the short piles, smaller stiffness and inflection appear, and the stress is more complicated. Compared with the single pile cases, the axial force distributions of the pile groups under the application of a horizontal force and transfer are more complicated and involve interactions between the piles and the soil surrounding the piles. The horizontal bearing capacity is related to strength of materials, the lateral resistance of the soil and foundation forms, and other factors. For example,

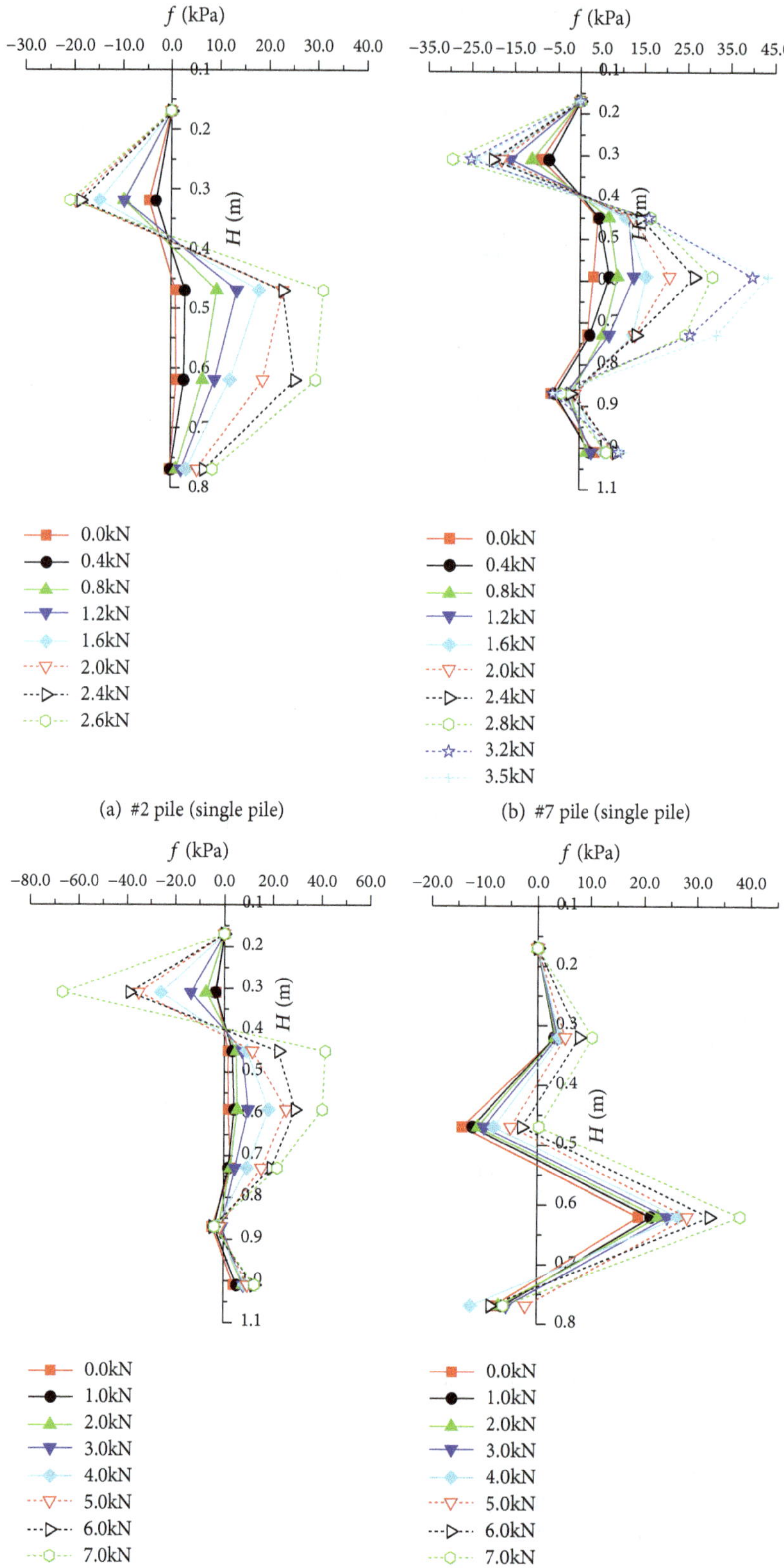

(a) #2 pile (single pile)

(b) #7 pile (single pile)

(c) #5 pile (four-pile group)

(d) #10 pile (four-pile group)

FIGURE 7: Continued.

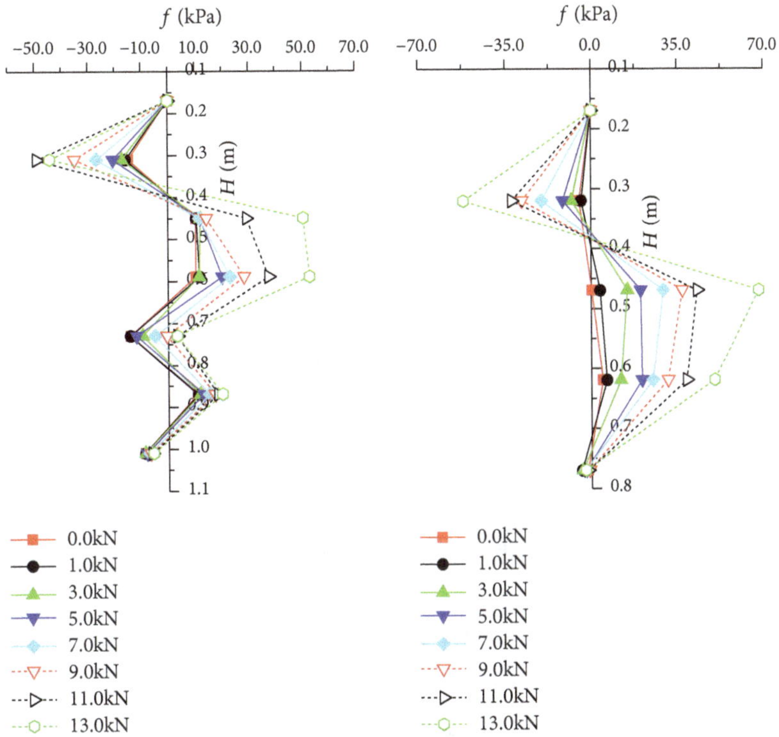

(e) #1 pile (seven-pile group) (f) #8 pile (seven-pile group)

FIGURE 7: Relations reflecting the frictional resistance f between the piles and surrounding soil with respect to H for various horizontal loads.

because the tensile strength of concrete is far lower than its compressive strength, with increasing horizontal load, the pull part of the pile side appears fracture damage caused by a lack of tensile strength. Therefore, there is the possibility of cross-sectional tensile damage under the condition of small displacements and angle displacement.

3.3. Analysis of Frictional Resistance between Test Piles and Soil with respect to Depth under Different Horizontal Loads. Figure 7 shows the frictional resistance f (kPa) between the test piles and the surrounding soil with respect to H for different horizontal loads. The figure indicates the horizontal load in the process of transfer, the upper side of soil layer prior to resistance to the lower side of soil layer. With increasing horizontal load, the lateral resistance of the lower soil is gradually motivated, and the peak value of the distribution curve has a tendency to move downward gradually. A range of steep slow extent reflects the lateral resistance of the soil at a given depth. This is closely related to the degree of filling compaction, where the slope of the curve increases with a decreasing degree of soil compaction. As such, the slope of the curve is inversely related to the lateral resistance. The average value of f for the various groups considered lies between 30 kPa and 70 kPa. The position of maximum frictional force is 0.5 m to 0.7 m. The magnitude of the negative frictional resistance and the position of neutral point are closely related to the process of water immersion, the process of consolidation settlement and pile top load, etc; for example, in the position of neutral point with the increase of immersion time down in Ningxia testing of ZH4 pile foundation, the positive and negative peak frictional resistance values increased. With increasing horizontal load in a saturated loading state, the settlement of the pile foundation is greater than that of the surrounding soil, and the neutral point moves upward sharply. However, with increasing horizontal load, the neutral point gradually moved upward, the positive frictional resistance increased, and the negative frictional resistance decreased. This indicates that the pile foundation would be costly if the neutral point position is calculated still according to the calculated thickness of collapsible soil layer. For very thick loess regions, short piles are placed on shallow supporting layers mainly to provide bearing capacity, and long piles are placed on deep supporting layers mainly to control the deformation of the pile foundation. This design conforms to the principle of a settlement-reducing pile.

In addition, although the magnitude of negative frictional resistance obtained from the test results is between 4 and 50 kPa, which is largely commensurate with the negative skin frictional resistance of collapsible loess obtained in field tests (14.2-54.0 kPa), our results present at least two neutral points of pile. If we continue to apply the experience that the largest depth of neutral point is the bottom depth of collapsible loess, obviously, it does not accord with our findings here. The filled soil in model enclosure is distributed in sandwich layers interactively, whose collapsibility is discontinuous, which is same as the actual situation. Especially there is possibility that the non-collasiblity of large thickness loess layer changes to collapsiblity, which caused by the reshape of the surrounding soil, leading to compressive deformation by pile driving or

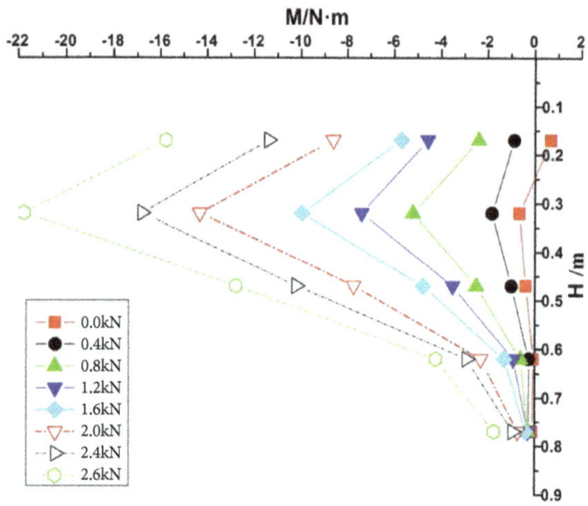

(a) #2 pile (single pile)

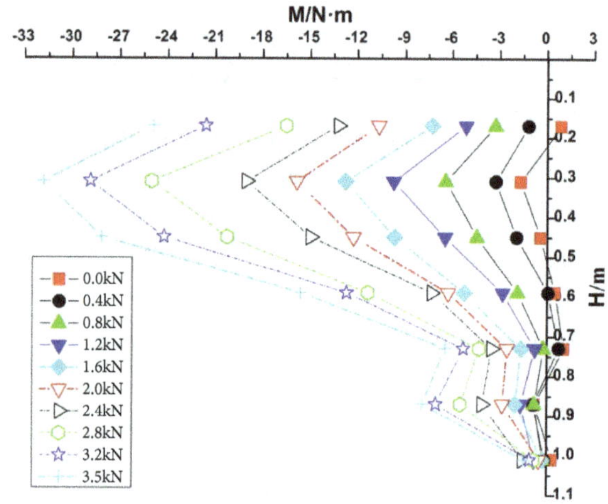

(b) #7 pile (single pile)

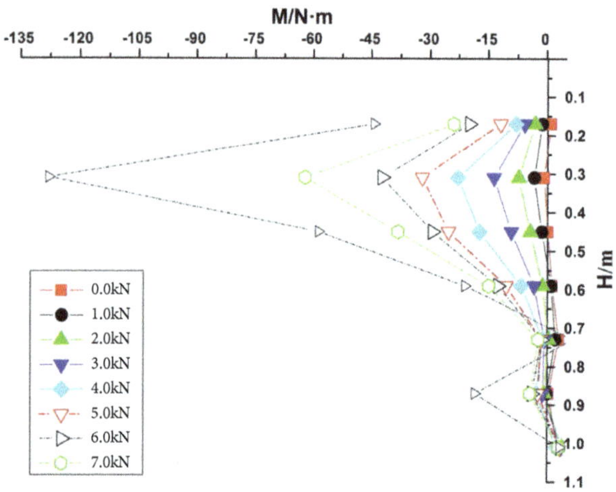

(c) #5 pile (four-pile group)

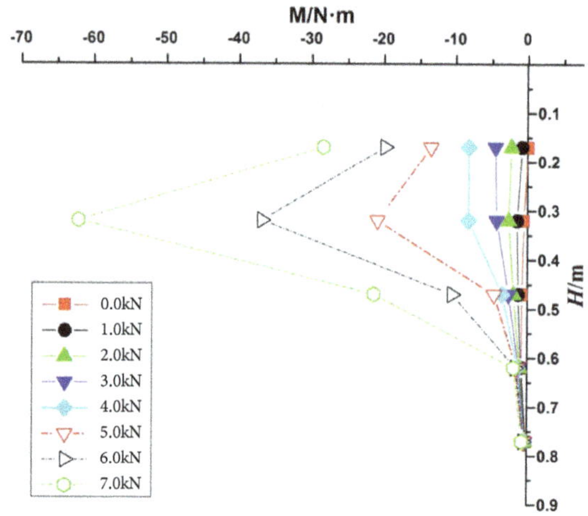

(d) #10 pile (four-pile group)

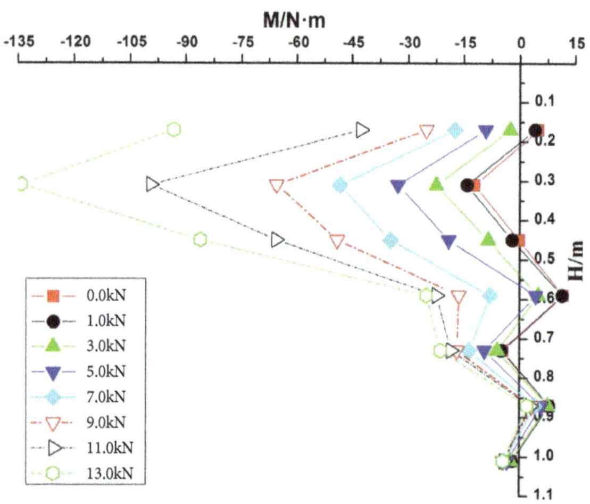

(e) #1 pile (seven-pile group)

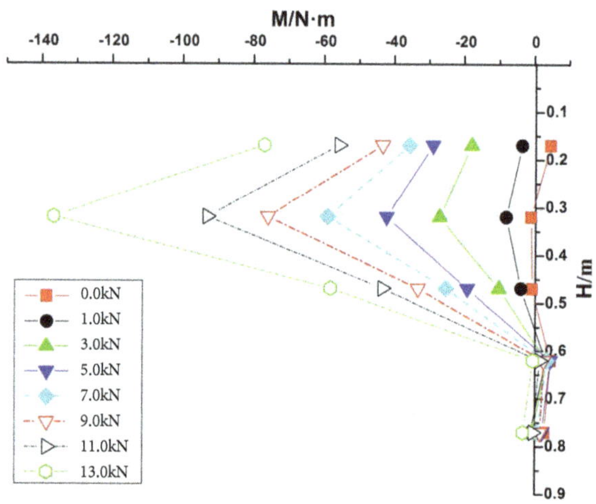

(f) #8 pile (seven-pile group)

FIGURE 8: Moments along pile depth under the different horizontal load.

transmission of upper load. Therefore, pile lengths and pile foundation design should be taken seriously.

3.4. Analysis of the Moment Acting on Test Piles with respect to Depth under Different Horizontal Loads. According to the strain measured by the test, the moments of each section of the piles can be obtained by using the bending theory of material mechanics, as shown in Figure 8.

Figure 8 shows the moment's value is basically smaller on the top of the piles and larger at the middle and upper

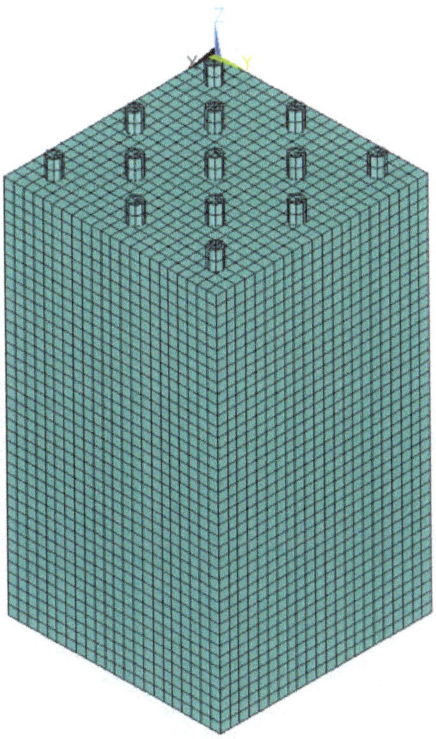

FIGURE 9: 3D model of the long-short pile composite foundation.

part of the piles. The moment of piles increases first and then decreases sharply with the pile depth, and the moment of pile approaches zero at the bottom of the piles, which shows the characteristics of flexible pile. The moment of the pile is approximately concentrated in the depth of 10d (d is the diameter of the pile) below the top of the piles, and the maximum moment is located at the depth of (3-4)d of the piles. In the process of increasing horizontal load, the maximum moment of pile does not move down obviously, but the moment on the top of piles increases continuously. In addition, Figure 8 also proved that, most of the time, there are two neutral points in the pile at least. If the design of pile foundation is carried out according to the experience that the maximum depth of the neutral point is the lower limit depth of collapsible loess layer, it is inconsistent with the actual situation.

4. Numerical Simulation Analysis

4.1. The Finite Element Model. The composite pile foundation with long and short piles shown in Figure 4 was modeled in ANSYS 14.0 for conducting 3D elastic-plastic finite element analysis using equivalent pile lengths and diameters as those employed in the indoor test model; 3D model is shown in Figure 9. Piles were modeled as elastic Beam 188 elements (3D 2-Node Beam); but the soil was modeled using Solid 185 elements (3D 8-Node Structural Solid) in accordance with the Drucker-Prager yield criterion. Because the pile and the soil material parameters were substantially different, the contact surfaces were modeled using nonlinear shear spring elements, which can overcome the errors associated with linear elastic model calculations [18]. The calculation domain was equivalent to that of the model enclosure in the horizontal direction and was 2 m high in the vertical direction. Piles were modeled using 0.05 m units and the soil was modeled using 0.05 m × 0.05 m × 0.05 m cubic units. Constraints were applied around all external sides of the model, except for the bottom, which was not constrained. The model parameters are listed in Table 2.

TABLE 2: Model parameters employed in finite element analysis.

	E (MPa)	ν	c (kPa)	φ (°)	γ (kN/m^3)
Pile	30,000	0.2	-	-	2,600
Soil layer	70	0.34	35	24	1,700

In the first step, only gravity was applied, without any other load. Then horizontal loading was separately applied to the single pile (#2 or #7), four-pile group (#4, #5, #9, and #10), and nine-pile group (#1-#3, #6-#8, and #11-#13) for different conditions, when a vertical load was forced that was 80% of the ultimate bearing capacity of the piles.

4.2. Analysis of Pile Body Strain and Displacement under Different Horizontal Loads. The strain contours of pile bodies are shown in Figure 10 under the different horizontal loads. We note that the strain increases with increasing horizontal load, while the strain gradually decreases (i.e., the minimum strain point moves continuously downward) with increasing H. Long piles exhibit regions of negative bending because of their lower stiffness and bear the majority of the horizontal load. Single piles #2 and #7 both present S values of 2.8 and 5.0 mm at horizontal loads of 2.65 and 3.49 kN, respectively, which are in good agreement with the static load test results. This indicates that the numerical analyses can accurately represent the F-S relationships obtained under experimental conditions, and represents a useful tool in engineering design. For the four-pile and nine-pile groups, the maximum values of S obtained under a horizontal load of 15 kN were 5.6 and 10 mm, respectively. For the four-pile group, the maximum strain was obtained on the short piles, and the soil between piles in the central pile (#7) strain was greater than that of the peripheral piles (#6 and #8). For the nine-pile group, the piles residing in the corners (#1, #3, #11, and #13) exhibit no obvious displacements, which is not consistent with standard pile foundation theory and the pile foundation test results. This may have been caused by the test of the pile head adjusted uneven and eccentric loading. As was observed for the four-pile group, the maximum strain was also obtained for the short piles of the nine-pile group (#2, #6, #8, and #12), which verifies the conclusion obtained from the experimental results that short piles experience greater deformation and damage before the long piles in a composite pile foundation. The strain in upper part of #4, #5, #9, and #10 piles is obvious. In conclusion, the strain of piles is mainly concentrated from the pile head to about a third of the maximum H. On a certain point in a pile, the pile body will rotate about its longitudinal axis, and the integrity of the pile foundation will be reduced to below acceptable limits when S becomes too large. For the

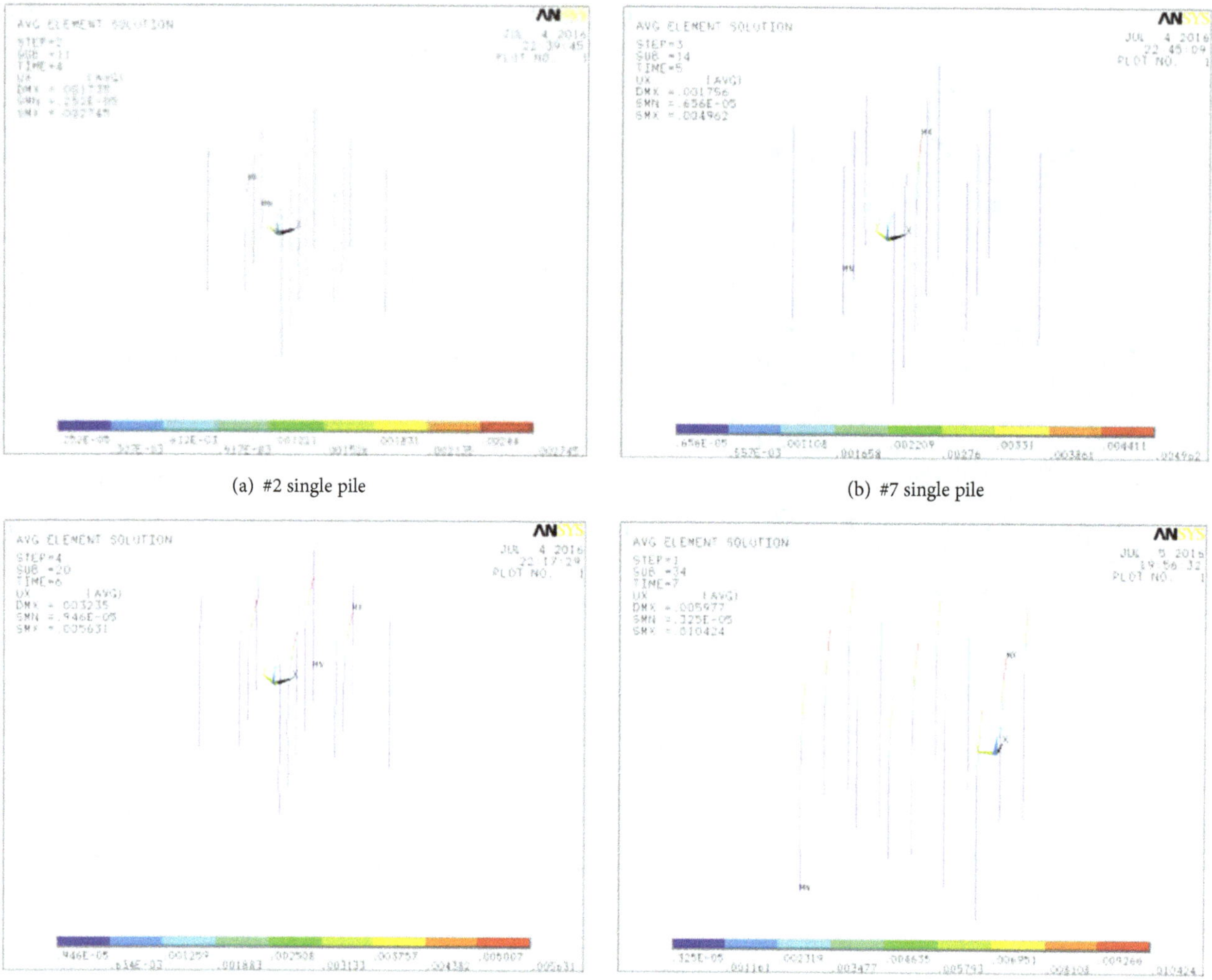

(a) #2 single pile

(b) #7 single pile

(c) Four-pile group

(d) Nine-pile group

FIGURE 10: Strain nephograms of piles under different horizontal loads.

long piles, the pile bodies undergo flexural deformation, and the pile body will suffer damage due to bending when the pile body displacement and bending moment increase beyond acceptable limits.

Owing to space limitations, we examine the displacement S with respect to H under different horizontal loads for only the nine-pile group in detail. As shown in Figures 11(a) and 11(b), the values of S for #1 and #8 pile bodies gradually increase with increasing horizontal load, and the values of S rapidly decrease with increasing H, and both exhibit a point of zero displacement. For long pile #1, the first point of zero displacement occurs at H=0.4 m, and the second point of zero displacement occurs near the bottom of the pile. The magnitude of the negative displacement obviously increases with increasing horizontal load, and the second point of zero displacement moves downward continuously. If the material strength of the pile is increased, the horizontal bearing capacity of pile foundation is controlled by the displacement. For short pile #8, the displacement is small, but, because the pile under section is not fully built-in, pile around the pile body rotates about midpoint with increasing horizontal load. With further increase in the horizontal load, the soil surrounding the pile begins to undergo local shear deformation due to a lack of shear strength and subsequently yields. The yield zone moves continuously downward until the surrounding soil suffers shear failure, which resulted in an excessively large foundation displacement, and damage to the soil at the base. Figure 11(c) shows that the relationship between the pile top displacement and horizontal load of numerical simulation and that of model tests has good consistency. At the same time, it also proved the rationality of the recommendation made in section 3.1 that a horizontal displacement limit of the pile foundation is 10 mm and 6mm for the ones sensitive to horizontal displacement.

In conclusion, the material strength of the individual piles was eventually exceeded by the applied horizontal loading and produced lateral deformation. The interplay of lateral forces between the piles and their surrounding soil represents a complex interaction. As pile body deformation gradually increases with increasing horizontal load, the stress

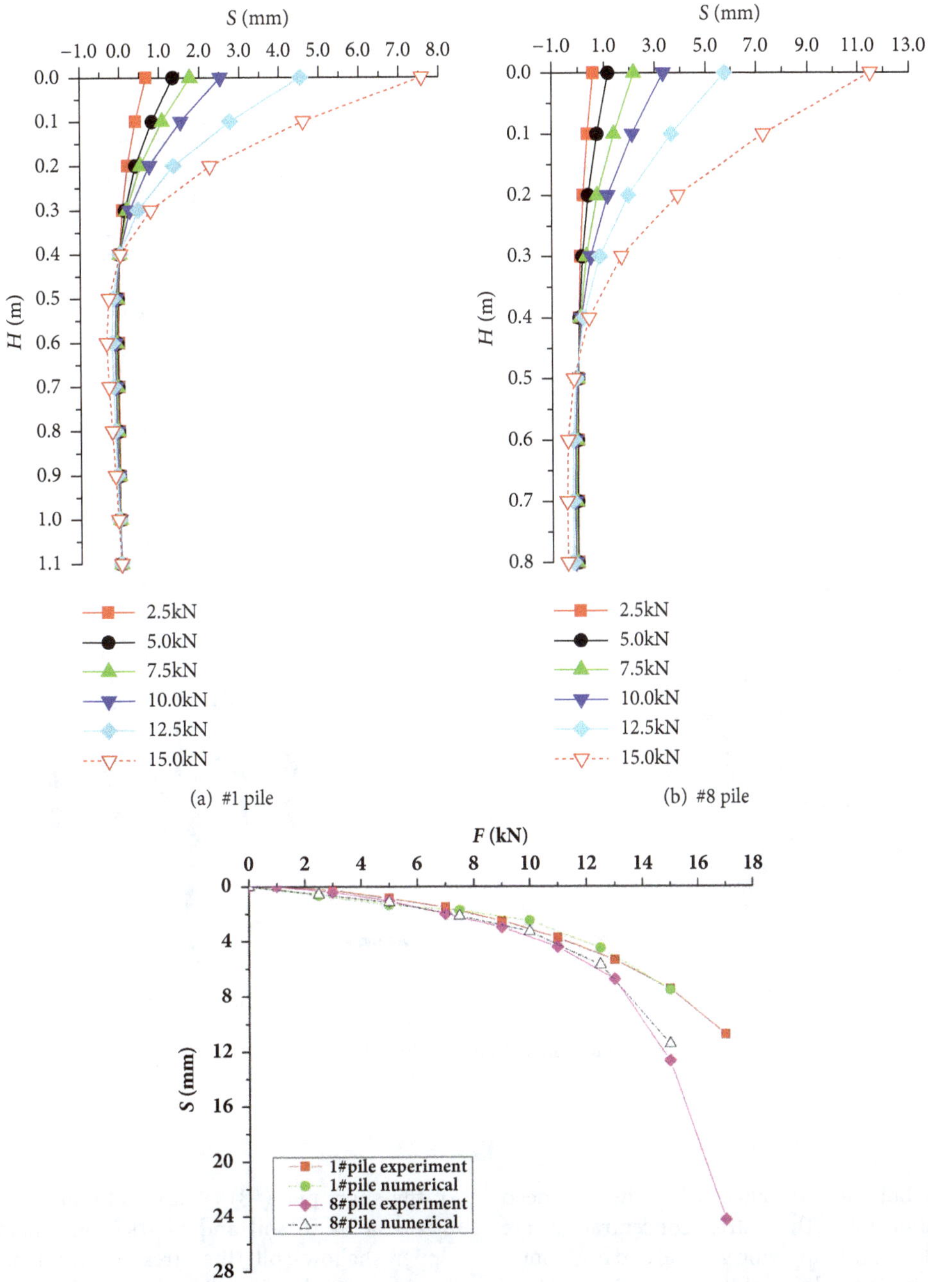

(a) #1 pile

(b) #8 pile

(c) Displacement comparison of pile between the numerical and experimental results

FIGURE 11: Displacement S with respect to H for the nine-pile group under different horizontal loads.

eventually exceeds the allowable limit or the surrounding soil loses its stability and yields. Even if the pile body strength was sufficient to accommodate the high-stress conditions, the loss of soil stability or pile displacement exceeding specifications would lead to pile foundation damage. Therefore, the horizontal bearing capacity is mainly controlled by the strength of the soil surrounding the piles or the extent of horizontal displacement of the pile head, which are more likely to occur in areas prone to collapsible loess. All these factors should be considered in the design, construction, and use of composite pile foundations with short and long piles.

4.3. Analysis of Pile Body Stress and Axial Force under Different Horizontal Loads. Figure 12 presents stress nephograms of the pile groups under different horizontal loads. Owing to space limitations, we examine the axial force with respect to *H* under different horizontal loads for only the nine-pile group in detail, as shown in Figure 13. We note that the maximum axial stress in Figure 12 was not obtained at the

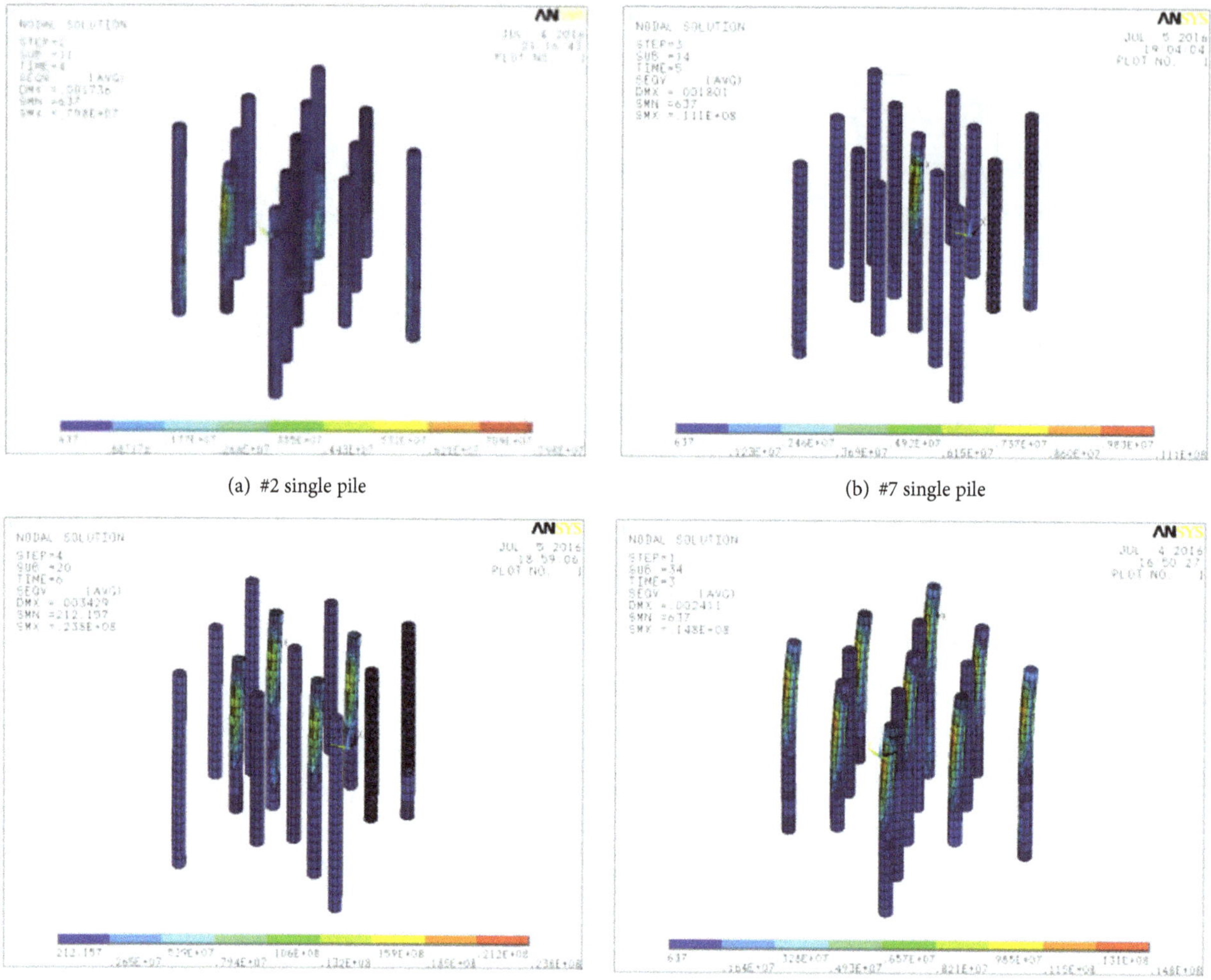

(a) #2 single pile

(b) #7 single pile

(c) Four-pile group

(d) Nine-pile group

FIGURE 12: Stress nephograms of piles under different horizontal loads.

pile head position, but some distance below the pile head for both long and short piles. These stress concentrations are formed when simultaneously applying vertical and horizontal loads because the elastic modulus of the piles is large. Here, the top soil layer surrounding the piles undergoes plastic deformation, which allows relatively easy pile displacement near the head, caused by pile on top of a certain depth range; displacement of soil mass is greater than the displacement of the pile, producing negative frictional resistance areas. As previously established, a negative skin frictional resistance above the neutral point makes the axial stress of the pile greater with increasing H, and, below the neutral point, the pile axial stress decays rapidly with increasing H owing to the frictional resistance of the surrounding soil. Near the neutral point, the pile axial force attains a maximum value, as shown in Figure 13.

The results in Figures 13(a) and 13(b) show the maximum axial force occurs near the pile head, and at least two points of zero axial force are observed. The axial force distribution is similar to an "s" type distribution, whereas that of the short pile (#8) is similar to an "s" type distribution. Stress areas in soil and plastic zone mainly range in the upper shallow soil (the stress diagram of foundation soil is abbreviated). A collapsible loess region improves the engineering properties of shallow soil layers and it not only can prevent the wet fall under the influence of water, but also can improve the horizontal bearing capacity of a pile foundation. #1 long pile deflection, #8 short pile in the overall rotation, with increasing horizontal loading, the axial force near the pile bottom can significantly increase, and, near the pile head, the amplitude of the axial force increases less rapidly with decreasing H. The load above a certain depth increment will be negligible, which can provide a reference for pile length design. It must be pointed out that, according to Figures 13(c) and 13(d), the results of the numerical simulation are basically consistent with those of the indoor model test, but because the indoor test conditions are difficult to achieve the ideal conditions of the numerical simulation, there are still some differences between the two results.

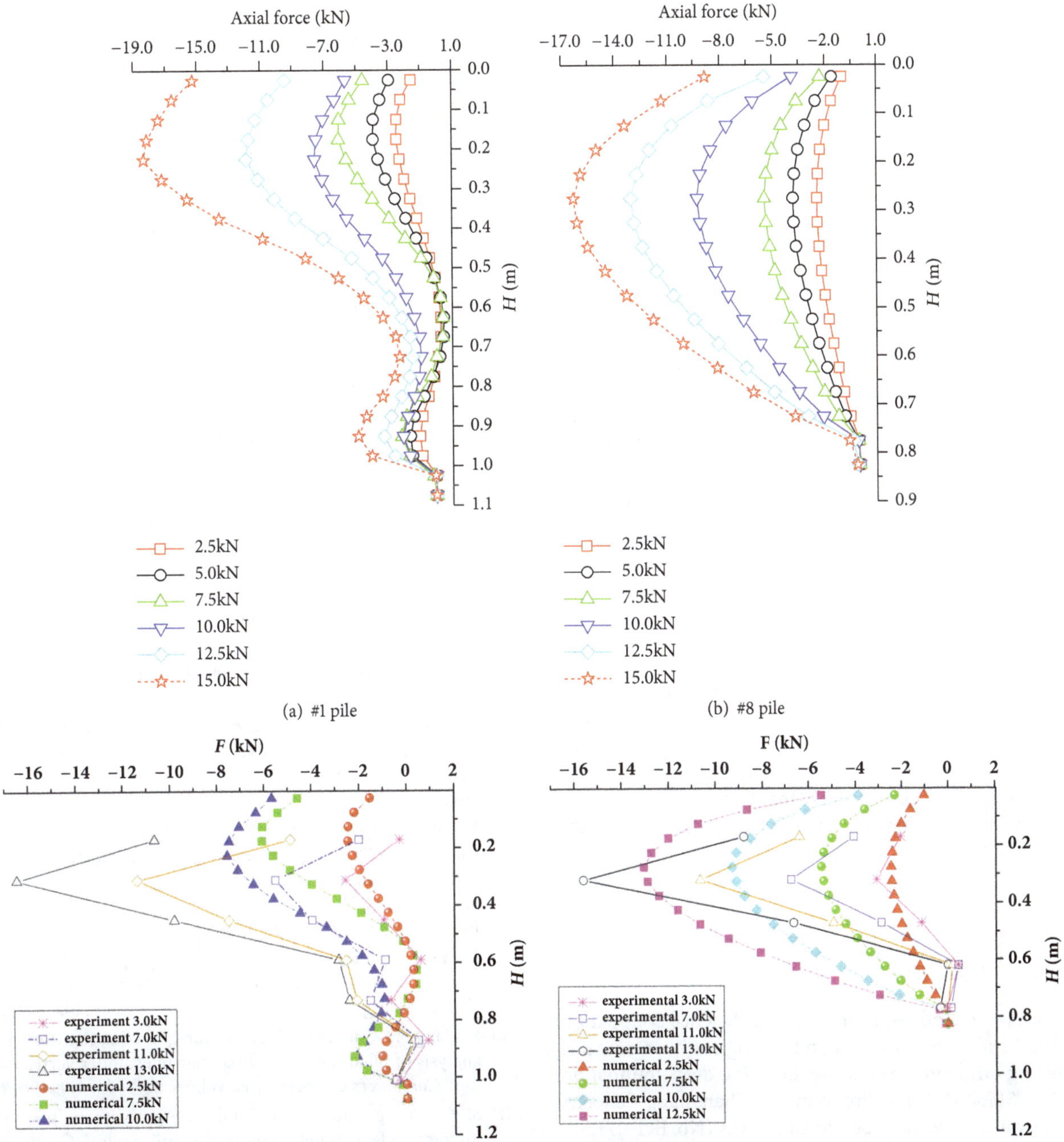

(a) #1 pile

(b) #8 pile

(c) Axial comparison of Pile 1 between the numerical and experimental results

(d) Axial comparison of Pile 8 between the numerical and experimental results

FIGURE 13: Axial force distribution with respect to H for the nine-pile group under different horizontal loads.

5. Conclusions

(1) Under the application of different horizontal loads, the F-S curves of the pile foundation exhibited an obvious steep fall in the section, which is a typical feature of piercing damage. Short piles were observed to suffer damage with increasing horizontal loading prior to long piles. A horizontal displacement limit of the pile foundation is 10 mm and 6mm for the ones sensitive to horizontal displacement.

(2) The play of pile axial force and pile-soil interfacial frictional resistance with asynchrony: the pile-soil interfacial frictional resistance under different horizontal loads ranged from 30 kPa to 70 kPa. The magnitude of the negative skin frictional resistance ranged between 4 and 50 kPa. There are multiple nuetral points along pile bodies because the soil interlayers surrounding piles distribute interactively and the distribution of collapsibility is discountinous, which leads to the determination of the lower limit depth is complex in collapsible soil layer. So the design thought applying

long-short composite pile foundation conforms to the principle of settlement reducing pile.

(3) Pile body stress and strain are mainly concentrated between the pile head and about one-third of the pile depth. Beginning at a particular depth along a short pile, the pile body will rotate about its longitudinal axis, and the integrity of the pile foundation will be reduced to below acceptable limits when the displacement becomes too large. For the long piles, the pile bodies undergo flexural deformation, and the pile body will suffer damage due to bending when the pile body displacement and bending moment increase beyond acceptable limits. The horizontal bearing capacity is mainly controlled by the strength of the soil surrounding the piles or the extent of horizontal displacement of the pile heads.

(4) The finite element analysis results of the pile foundation were in good agreement with the results of the experimental model testing. Especially the pile-soil contact surface is simulated with nonlinear shear spring elements, which can provide a useful example for future engineering design.

Conflicts of Interest

The authors declare that they have no financial and personal relationships with other people or organizations that can inappropriately influence their work; there is no professional or other personal interest of any nature or kind in any product, service, and/or company that could be construed as influencing the position presented in, or the review of, the paper entitled.

Acknowledgments

This work was supported by the National Science Foundation of China "Theoretical analysis and testing study on row pile supporting with pre-stressed anchors for deep foundation pit" (No. 51568042) and Program for Changjiang Scholars and Innovative Research Team in University (No. IRT_17R51). We thank LetPub (www.letpub.com) for its linguistic assistance during the preparation of this manuscript.

Supplementary Materials

Supplementary Materials description: 1, e.g., #5: relations between the horizontal load F and the displacement of the pile heads S for the pile groups considered. 2, e.g., #6: relations between the axial forces acting on a pile with respect to location H along the length of the pile for various horizontal loads. 3, e.g., #7: relations reflecting the frictional resistance f between the piles and surrounding soil with respect to H for various horizontal loads. 4, Fig. 10: displacement S with respect to H for the nine-pile group under different horizontal loads. 5, e.g., #12: axial force distribution with respect to H for the nine-pile group under different horizontal loads. *(Supplementary Materials)*

References

[1] Y. J. Teng, W. D. Wang, and J. W. Kang, "The new development of the technology of building foundation engineering," in *Proceedings of the set of academic paper of the national soil mechanics and geotechnical engineering in 12th*, pp. 66–95, Shanghai, China, 2015.

[2] X. Zhao, Q. Zhang, and B. Zhang, "Iterative method for calculating the settlement of pile foundation with negative friction," *Yantu Lixue/Rock and Soil Mechanics*, vol. 20, no. 2, pp. 17–21, 1999.

[3] D. Z. Li, Y. J. Teng, and Y. H. He, "Vertical bearing behaviour of large diameter belled pile in collapse loess," *Chinese Journal of Geotechnical Engineering*, vol. 16, no. 2, pp. 11–21, 1994.

[4] S. J. Feng, H. Ke, and Y. H. Chen, "Experimental study on super long bored pile in loess," *Chinese Journal of Geotechnical Engineering*, vol. 26, no. 1, pp. 110–114, 2004.

[5] X. F. Huang, Z. H. Chen, and S. Ha, "Research on bearing behaviors and negative friction force for filling piles in the site of collapsible loess with big thickness," *Chinese Journal of Geotechnical Engineering*, vol. 29, no. 3, pp. 338–346, 2007.

[6] Y. P. Zhu, T. S. Zhao, and C. L. Chen, "Field tests on changes of pile negative friction along its length," *Rock and Soil Mechanics*, vol. 34, no. S1, pp. 265–272, 2013.

[7] J. G. Zheng, G. H. Deng, and Z. H. Liu, "Influence of discontinuous distribution of collapsible loess on its deformation," *Chinese Journal of Geotechnical Engineering*, vol. 37, no. 1, pp. 165–170, 2015.

[8] X. Yang, X. Huang, Y. Zhu, Z. Yao, and S. Zhang, "Experimental study on collapsibility evaluation and treatment depths of collapsible loess upon self weight with thick depth," *Yanshilixue Yu Gongcheng Xuebao/Chinese Journal of Rock Mechanics and Engineering*, vol. 33, no. 5, pp. 1063–1074, 2014.

[9] M. Yang, H. Yang, and W. Wang, "Design philosophy and settlement analysis of the composite long-short pile foundations," *China civil engineering journal*, vol. 38, no. 12, pp. 103–108, 2005.

[10] F. Liang, L. Chen, and J. Han, "Integral equation method for analysis of piled rafts with dissimilar piles under vertical loading," *Computers & Geosciences*, vol. 36, no. 3, pp. 419–426, 2009.

[11] S. C. Wong and H. G. Poulos, "Approximate pile-to-pile interaction factors between two dissimilar piles," *Computers & Geosciences*, vol. 32, no. 8, pp. 613–618, 2005.

[12] Y. F. Leung, A. Klar, and K. Soga, "Theoretical study on pile length optimization of pile groups and piled rafts," *Journal of Geotechnical and Geoenvironmental Engineering*, vol. 136, no. 2, pp. 319–330, 2010.

[13] Z. J. Wang, X. Y. Xie, and P. F. Fang, "Analysis for calculation of nonlinear settlement of rigid long-short compound piles," *Journal of Zhejiang University (Engineering Science)*, vol. 48, no. 3, pp. 463–470, 2014.

[14] J. Xiong, F. Li, X. Zhu, and M. Yang, "Model test of composite long-short pile foundation," *Tongji Daxue Xuebao/Journal of Tongji University*, vol. 36, no. 10, pp. 1337–1343, 2008.

[15] M. Huang, B. Li, and Y. Cheng, "Centrifuge model tests and numerical simulation of load sharing law for composite embankment with long and short piles," *Yanshilixue Yu Gongcheng Xuebao/Chinese Journal of Rock Mechanics and Engineering*, vol. 29, no. 12, pp. 2543–2550, 2010.

[16] X. N. Ma, S. Z. Li, and Z. B. Tian, "Research on the Deformation and Bearing Characteristics of Composite Foundation with Long and Short Piles," *Journal of Railway Engineering Society*, vol. 3, pp. 6–11, 2016.

[17] D. A. Brown and C.-F. Shie, "Some numerical experiments with a three dimensional finite element model of a laterally loaded pile," *Computers & Geosciences*, vol. 12, no. 2, pp. 149–162, 1991.

[18] X. S. Ge, X. N. Gong, and X. M. Zhang, "FEM analysis and design of long-short-pile composite foundation," *Journal of Building Structures*, vol. 24, no. 4, pp. 91–96, 2003.

[19] M. Achmus and K. Thieken, "On the behavior of piles in non-cohesive soil under combined horizontal and vertical loading," *Acta Geotechnica*, vol. 5, no. 3, pp. 199–210, 2010.

[20] H. H. Zhang and J. C. Small, "Analysis of capped pile groups subjected to horizontal and vertical loads," *Computers & Geosciences*, vol. 26, no. 1, pp. 1–21, 2000.

[21] L. Zhang, M. Zhao, and X. Zou, "Behavior of Laterally Loaded Piles in Multilayered Soils," *International Journal of Geomechanics*, vol. 15, no. 2, pp. 1–7, 2015.

Warping Torsional and Distortional Stress of Composited Box Girder with Corrugated Steel Webs

Peng Qiao,[1] **Jin Di**,[2,3] **and Feng-jiang Qin**[2,3]

[1]*School of Civil Engineering, Chang'an University, Xi'an 710061, China*
[2]*Key Laboratory of New Technology for Construction of Cities in Mountain Area (Chongqing University), Ministry of Education, Chongqing 400045, China*
[3]*School of Civil Engineering, Chongqing University, Chongqing 400045, China*

Correspondence should be addressed to Peng Qiao; qiaopeng@chd.edu.cn

Academic Editor: Roman Wendner

The torsional behaviors of composite box girders with corrugated steel webs are more obvious than traditional concrete girders due to the lower torsional rigidity. In this paper, the torsion and distortion of this kind of composite girder are studied. The formulas for warping normal stress and shear stress were put forward according to the second theory of Umanskii, considering the accordion effect of corrugated steel webs. Then, the influences of different dimensional parameters on the torsional and distortional stress are investigated. Results show that the effect of parameters on stress is different and implicit in composite box girders with corrugated steel webs. Under eccentric loads, the warping torsional and distortional stress in this kind of girder should not be neglected. Compared with girders under corresponding symmetric loads, the total warping stress may be as big as flexural normal stress, and the total shear stress usually reaches 30 to 50 percentage of flexural shear stress. So the warping stress and additional shear stress due to warping torsion and distortion are suggested to be calculated by the proposed equations in structural analysis, which are usually not taken into account in conventional concrete box girders.

1. Introduction

The composite box girders with corrugated steel webs have been used more extensively in bridge engineering since first appearance in France. Under eccentric loads, the box girder may not only produce longitudinal flexural and shear deformation, but also the torsion and distortion. The torsional behaviors in these composite girders were found quite different from traditional concrete box girders [1]. Researchers worldwide have done much work on the torsional and distortional properties of composited box girder with corrugated steel webs. Mo et al. presented an analytical model to predict the torsional behavior of composite girders with corrugated steel webs [2] and developed an analytical model according to the test results [3]. According to experimental results of scaled models and bridge in site, Li et al. [4] and Liu et al. [5] concluded that resistance of this kind of box girder to torsion and distortion is lower than conventional box girders. Yang et al. investigated the warping and shear stress of the box girders with corrugated steel webs under torsion by the second theory of Umanskii [6]. Di et al. [7] and Wang [8] studied the torsional property of composite girders with corrugated steel webs by theoretical analysis and experiment and found its normal stress and shear stress due to torsion are 15% to 50% bigger than flexural normal stress and the distortional warping stress is larger than that due to warping torsion. Nie and Zhang [9] and Chen et al. [10] studied the influencing factors to torsional properties by finite element method, including the diaphragm number, geometry parameters of section, and corrugated web. Ding et al. investigate the behavior under pure torsion and the ultimate torsional strength considering material nonlinearity by finite element models [11] and scaled specimen experiments [12]. According to current studies, the additional normal and shear stress caused by torsion and distortion should be considered properly, which is often neglected for concrete box girders.

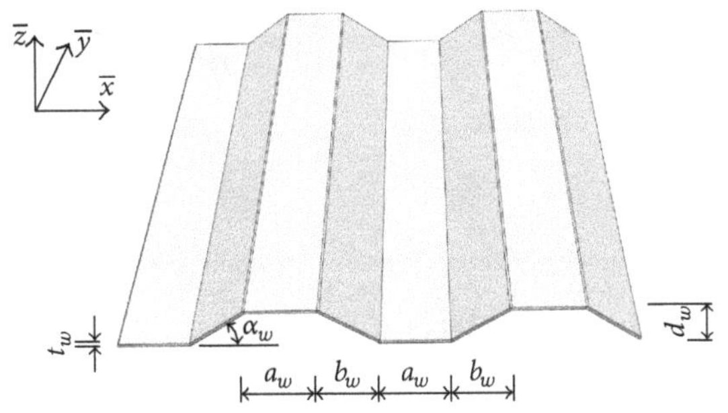

FIGURE 1: Geometry of the corrugated web.

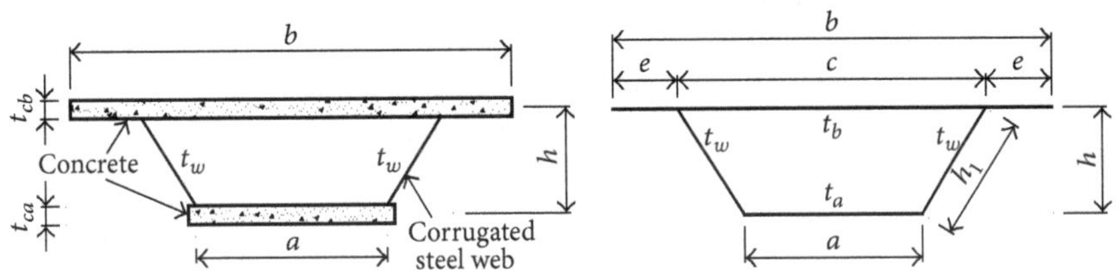

FIGURE 2: Geometry of equivalent section.

But little detailed conclusions are found on how the section dimensions influence the torsional and distortional stress.

As to composite box girders with corrugated steel webs, the 3D finite element models are rather complex to build. So, the theoretical analysis, based on the thin-walled beam theory, can be used more conveniently to study the mechanical properties of such composite structures. In the theoretical deduction process, the accordion effect of corrugated steel web should be considered correctly, as discussed by Yang et al. [6]. However, there were still some improper results from current studies. For example, the web shear stress is not influenced by the warping torsion, or the shear stress is not even in web. Additionally, the formula of distortional shear stress has not yet been proposed explicitly. The purpose of this paper is to propose formulas for warping torsional and distortional shear stress and study the influences of different factors to the warping torsional and distortional stress compared to the flexural normal stress.

2. Equivalent Section

2.1. Equivalent Orthotropic Plate of the Corrugated Steel Web. The geometry of the typical corrugated web is shown in Figure 1. The corrugated steel web is recommended to be equivalent to the orthotropic plate [13, 14], which has different elastic modulus in $\overline{x}$ and $\overline{y}$ direction:

$$E_{\overline{x}} = \frac{a_w + b_w}{4a_w}\left(\frac{t_w}{d_w}\right)^2 E_s, \tag{1}$$

$$E_{\overline{y}} = \frac{a_w + b_w \sec\alpha_w}{a_w + b_w} E_s, \tag{2}$$

where

E_s is Young's modulus of steel;

$E_{\overline{x}}$, $E_{\overline{y}}$ are the equivalent elastic modulus in $\overline{x}$ and $\overline{y}$ direction, respectively;

a_w, b_w, d_w, α_w are flat panel width, inclined panel width, wave height, angle of the corrugated web, respectively.

According to (1), the axial stiffness in longitudinal direction can be neglected, for the ratio of t_w and d_w of a corrugated web is always less than 1/100.

Additionally, the effective shear modulus of a corrugated web [15], G_e, is given by

$$G_e = \frac{a_w + b_w}{a_w + b_w \sec\alpha_w} G_s, \tag{3}$$

in which G_s is the shear modulus of steel.

2.2. Equivalent Section of Composite Box Girders with Corrugated Steel Webs. In order to directly adopt the thin-walled box girder theory to analyze the mechanical properties of the composite box girders with corrugated steel webs, the composite section is transformed to an equivalent cross section of single material, as shown in Figure 2. Considering the corrugated webs bear most of shear force in composite box girder, the cross section is assumed to be composed of steel material alone.

Assuming that the shearing force and stress remain the same, the equivalent thickness of concrete slab, t, is given by

$$t = \frac{G_c}{G_e} \cdot t_c = n' \cdot t_c, \tag{4}$$

where

t_c is the thickness of concrete slab in composite box section;

G_c is the shear modulus of concrete;

n' is the ratio of shear modulus of concrete and corrugated steel web.

3. Warping Torsional Stress

3.1. Torsional Warping Stress. According to the second theory of Umanskii, the warping displacement caused by warping torsion is given by

$$u(z) = u_0(z) - \overline{\omega}\beta'(z), \tag{5}$$

in which $\overline{\omega}$ is the principal unit warping (sectorial coordinate). Its practical formulas may be found in [16–18].

The torsional warping normal stress, σ_ω, is determined from the bimoment:

For concrete slab,

$$\sigma_\omega = \frac{B_{\overline{\omega}} \cdot \overline{\omega}}{I_{\overline{\omega}}}. \tag{6}$$

For corrugated steel web,

$$\sigma_\omega = \frac{E_{\overline{x}}}{E_{\overline{y}}} \frac{B_{\overline{\omega}} \cdot \overline{\omega}}{I_{\overline{\omega}}} \approx 0, \tag{7}$$

where

$B_{\overline{\omega}}$ is the warping bimoment caused by warping torsion;

$I_{\overline{\omega}}$ is the warping torsional inertia. For composite box girder with corrugated steel webs, it is given by

$$I_{\overline{\omega}} = \int_b \overline{\omega}^2 dA + \int_a \overline{\omega}^2 dA + 2\frac{E_{\overline{x}}}{E_{\overline{y}}} \int_w \overline{\omega}^2 dA. \tag{8}$$

Considering that the ratio $E_{\overline{x}}/E_{\overline{y}}$ of corrugated steel web is small enough to ignore, the warping torsional inertia is simplified by

$$I_{\overline{\omega}} = \int_b \overline{\omega}^2 dA + \int_a \overline{\omega}^2 dA \tag{9}$$

in which the subscripts a and b indicate the bottom and top slab, respectively. Equation (9) is used in formulas to calculate warping normal stress for composite girder with corrugated steel webs, which has already been given in [6].

3.2. Total Torsional Shear Stress. The deduction procedure of shear stress for composite box girder with corrugated steel web is similar to that of concrete box girder. However, the accordion effect of corrugated steel web should be considered correctly.

The shear flow in slab can be determined by the warping normal stress considering the equilibrium in the longitudinal direction of a plate element:

$$\frac{\partial q}{\partial s} ds = -\frac{\partial \sigma_\omega}{\partial z} t ds. \tag{10}$$

The shear flow in different plate is shown as follows.

For equivalent cantilever slab,

$$q(s) = E_{\overline{y}}\beta'''(z) \int_0^s \overline{\omega} dA = E_{\overline{y}}\beta'''(z) \cdot S_{\overline{\omega}}(s). \tag{11}$$

For equivalent top and bottom slab,

$$\begin{aligned} q(s) &= E_{\overline{y}}\beta'''(z) \int_0^s \overline{\omega} dA + q_0 \\ &= E_{\overline{y}}\beta'''(z) \cdot S_{\overline{\omega}}(s) + q_0. \end{aligned} \tag{12}$$

For corrugated steel web,

$$\begin{aligned} q(s) &= E_{\overline{x}}\beta'''(z) \int_0^s \overline{\omega} dA + q_0 \\ &= E_{\overline{x}}\beta'''(z) \cdot S_{\overline{\omega}}(s) + q_0, \end{aligned} \tag{13}$$

where

$$S_{\overline{\omega}}(s) = \int_0^s \overline{\omega} dA. \tag{14}$$

Noting that the external torque is M_k equal to the total action of q, we get

$$\begin{aligned} M_k(z) &= \oint q(s)\rho(s)\, ds \\ &= q_0\Omega + E_{\overline{y}}\beta'''(z) \cdot \oint S_{\overline{\omega}} d\omega, \end{aligned} \tag{15}$$

where

$$\Omega = \oint \rho(s)\, ds, \tag{16}$$

$$\begin{aligned} \oint S_{\overline{\omega}}(s)\, d\omega &= \int_b S_{\overline{\omega}}(s)\, d\omega + \int_a S_{\overline{\omega}}(s)\, d\omega \\ &\quad + \frac{E_{\overline{x}}}{E_{\overline{y}}} \int_w S_{\overline{\omega}}(s)\, d\omega \\ &\approx \int_b S_{\overline{\omega}}(s)\, d\omega + \int_a S_{\overline{\omega}}(s)\, d\omega. \end{aligned} \tag{17}$$

In (17), it is noted that the ratio $E_{\overline{x}}/E_{\overline{y}}$ of corrugated steel web is small enough to ignore corrugated steel webs.

From (15),

$$q_0 = \frac{M_k(z)}{\Omega} - \frac{E_{\overline{y}}\beta'''(z)}{\Omega} \oint S_{\overline{\omega}} d\omega. \tag{18}$$

The warping torsional moment derived from torsional warping bimoment is

$$M_{\overline{\omega}} = \frac{dB_{\overline{\omega}}(z)}{dz} = -E_{\overline{y}} I_{\overline{\omega}} \beta'''(z). \tag{19}$$

Substituting (18) and (19) into (11)~(13), the total shear stress considering the warping torsion can be determined from torsion and warping torsional moment:

$$\tau(s) = \frac{M_k(z)}{t\Omega} - \frac{M_{\overline{\omega}}}{I_{\overline{\omega}} t} \overline{S}_{\overline{\omega}}(s), \tag{20}$$

in which $\overline{S}_{\overline{\omega}}(s)$ is the generalized area moment of inertia.

For equivalent cantilever slab,

$$\overline{S_{\overline{\omega}}} = S_{\overline{\omega}}. \quad (21)$$

For equivalent top and bottom slab,

$$\overline{S_{\overline{\omega}}} = S_{\overline{\omega}} - \frac{\oint S_{\overline{\omega}} d\omega}{\Omega}. \quad (22)$$

For corrugated steel web,

$$\overline{S_{\overline{\omega}}} = -\frac{\oint S_{\overline{\omega}} d\omega}{\Omega}. \quad (23)$$

3.3. Differential Equation. Considering the longitudinal displacement closure condition, the differential equation is deduced:

$$\beta^{IV}(z) - k^2 \beta(z) = -\frac{m\mu}{EI_{\overline{\omega}}}, \quad (24)$$

in which m is load intensity of torque; μ is the restrained constant of section; k is the moment-torsional coefficient, defined by

$$k^2 = \mu \frac{G_e I_d}{E_{\overline{y}} I_{\overline{\omega}}}, \quad (25)$$

in which $G_e I_d$ is the torsional rigidity; $E_{\overline{y}} I_{\overline{\omega}}$ is the warping torsional rigidity. According to specific boundary conditions, the above differential equation can be solved by initial-parameter method in [19, 20].

4. Distortional Stress

4.1. Distortional Warping Stress. According to [20], the distortional warping stress of composite box girder with corrugated steel web has already been given. The most important equations are list as follows.

The distortional differential equation of distortional angle, γ, is

$$EI_{\omega_D} \gamma^{(4)} + EI_R \gamma = \frac{a^2}{a+c} P(z) \quad (26)$$

in which EI_{ω_D} is distortional warping rigidity, EI_R is distortional rigidity of transverse frame, and $P(z)$ is antisymmetric load at intersection point of top slab.

Combined with specific boundary conditions, the analytical solutions of distortional angle, γ, and distortional bimoment, B_D, can be deduced by initial-parameter method. Then the distortional warping normal stress is given as follows:

$$\sigma_{Da} = \frac{B_D}{I_{\omega_D}} \omega_{Da}, \qquad \sigma_{Db} = \frac{B_D}{I_{\omega_D}} \omega_{Db}, \quad (27)$$

in which σ_{Da}, σ_{Db} are the warping stress at the junction between bottom and top slab, respectively. $\omega_{Da} = K_4$, $\omega_{Db} = \beta K_4$. The coefficients β and K_4 can be calculated according to [20].

4.2. Distortional Shear Stress. Little has been studied on the distortional shear stress of composite box girder with corrugated steel webs. So the additional shear stress due to distortion is deduced as follows.

The distortional shear moment is given by

$$M_D(z) = \frac{dB_D(z)}{dz} = -E\hat{I}\gamma'''. \quad (28)$$

Then, the distortional shear force in different slabs can be calculated as follows.

For corrugated steel web,

$$Q_1 = \frac{M_D(z)}{a \sin\theta}. \quad (29)$$

For equivalent bottom slab,

$$Q_2 = \frac{c}{h_1} Q_1 = \frac{c M_D(z)}{h_1 a \sin\theta}. \quad (30)$$

For equivalent top slab,

$$Q_4 = \frac{a}{h_1} Q_1 = \frac{M_D(z)}{h_1 \sin\theta}. \quad (31)$$

The distortional shear stress is given by

$$\tau_{Di} = \frac{Q_i}{A_i}, \quad (32)$$

in which A_i is the cross-sectional area of equivalent top, bottom, or web slab. Because of the changing of normal stress due to distortional warping, the additional shear stress is also produced. The additional shear stress can be deduced by similar procedures like warping torsional shear stress, and it should be considered that no torsional force in section is generated under distortion. Thus, the distortional additional shear stress can be calculated by

$$\tau_{D\omega}(s) = -\frac{M_D}{I_{\omega_D} t} \overline{S}_{\omega_D}(s), \quad (33)$$

in which $\overline{S}_{\omega_D}(s)$ is the generalized area moment of inertia.

For equivalent cantilever slab,

$$\overline{S}_{\omega_D} = S_{\omega_D}. \quad (34)$$

For equivalent top and bottom slab,

$$\overline{S}_{\omega_D} = S_{\omega_D} - \frac{\oint S_{\omega_D} d\omega}{\Omega}. \quad (35)$$

For corrugated steel web,

$$\overline{S}_{\omega_D} = -\frac{\oint S_{\omega_D} d\omega}{\Omega}, \quad (36)$$

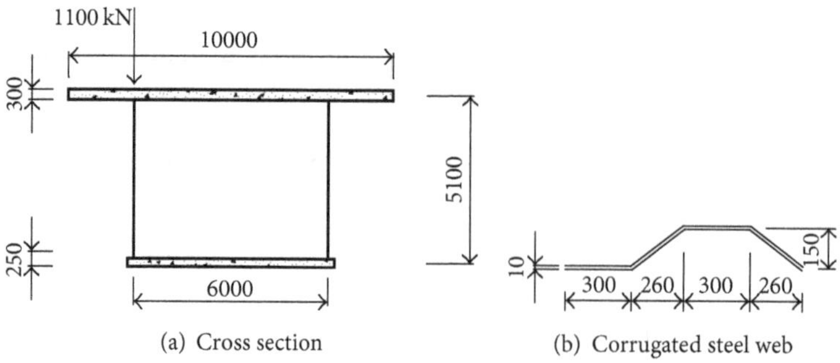

FIGURE 3: Configurations and dimension of specimens in case 1 (unit: mm).

where

$$S_{\omega_D}(s) = \int_0^s \omega_D \, dA,$$
$$\oint S_{\omega_D}(s)\, d\omega \approx \int_b S_{\omega_D}(s)\, d\omega + \int_a S_{\omega_D}(s)\, d\omega. \tag{37}$$

Finally, the total distortional shear stress is as follows:

$$\tau_{Di} = \frac{Q_i}{A_i} - \frac{M_D}{I_{\omega_D} t} \overline{S}_{\omega_D}(s). \tag{38}$$

According to the above equation, the distortional shear stress is comprised of two parts, which are the direct shear stress due to distortional shear force and the additional shear stress due to changing normal stress, just like the warping torsional shear stress.

5. Comparative Studies

5.1. Case 1. Take the simple supported composite girder with corrugated steel webs in [21] as an illustration (as shown in Figure 3). The span, L, is 43.2 m. The height, h, is 5.1 m. The top slab width, b, is 10 m and thickness, t_{cb}, is 0.30 m. The bottom slab width, a, is 6 m and thickness, t_{ca}, is 0.25 m. For the corrugated steel webs, the thickness, t_w, is 0.01 m, the panel width a_w is 0.30 m, b_w is 0.26 m, and d_w is 0.15 m. For concrete, E_c is 3.4×10^4 MPa and μ_c is 0.1667. For steel, E_s is 2.1×10^5 MPa and μ_s is 0.3. The 1100 kN eccentric load is applied at one conjunction of web on the top slab at the mid-span cross section.

Using the finite element program ANSYS, 3D models of composite girders were built to verify the above theoretical analysis expressions. The eight-node Solid element (SOLID45) was used to model the concrete slabs. The four-node Shell element (SHELL63) was chosen to simulate the corrugated steel webs. The concrete and steel materials were linear elastic and no plastic or strain hardening was taken into account. The slip and composite action between the concrete slab and steel web were neglected, so the concrete and steel nodes on their interface were merged to ensure they had the same deformation. As the girder is assumed to be simple supported at both ends, four nodes at the bottom were constrained as shown in Figure 4. Nodes no. 1 to 4 were provided constraints in different directions, but no rotating constraint was given. In finite element analysis (FEA), the torsional loads and the distortional loads were separately applied on nodes of top and bottom concrete slab and steel webs in the mid-span section of the girder.

The warping normal stress and shear stress of this composite girder are shown in Figure 5. As the normal stress in corrugated webs and shear stress in concrete slabs are rather small, only the normal stress in concrete slabs and shear stress in corrugated webs are given. In order to verify the results according to equations in this paper, the stress calculated by FEA is presented separately in bracket [].

As shown in Figure 5, the above equations of warping normal stress and shear stress for composite girder with corrugated steel webs are verified by FEA results. Compared with stress due to distortion, the warping normal stress due to warping torsion is much smaller, while the shear stress in webs is a little bigger. Noticing that the flexural normal stress in bottom slab is 1.45 MPa and the flexural shear stress in webs is 5.39 MPa, the distortional warping stress is even as big as flexural normal stress. This means the influence of both warping torsion and distortion on normal and shear stress must be considered properly. In this girder, the normal stress in bottom slab due to warping torsion has the opposite direction to stress due to distortion.

The results of warping normal stress and web shear stress were also displaced in Figure 6, comparing with the stress according to analytical equations. As shown in Figure 6, the stress from two different methods has little discrepancy, which means the analytical equations in this essay are effective.

5.2. Case 2. Take the simple supported composite girder with corrugated steel webs in [6] as another illustration (as shown in Figure 7). The span, L, is 7.5 m. The height, h, is 0.3688 m. The top slab width, b, is 1.8 m and thickness, t_{cb}, is 0.1125 m. The bottom slab width, a, is 0.4626 m and thickness, t_{ca}, is 0.11 m. For the corrugated steel webs, the thickness, t_w, is 0.003 m, the panel width a_w is 0.063 m, b_w is 0.050 m, and d_w is 0.038 m. For concrete, E_c is 3.4×10^4 MPa amd μ_c is 0.1667. For steel, E_s is 2.1×10^5 MPa and μ_s is 0.3. The 20 kN eccentric load is applied at one conjunction of web on the top slab at mid-span section.

Based on equations for warping torsional and distortional stress, the warping normal stress and shear stress of the composite girder are shown in Figure 8. To verify the results,

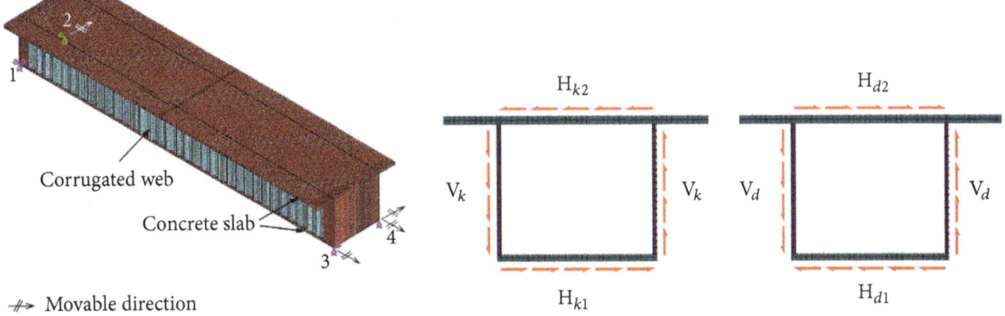

(a) Girder model

(b) Torsional loads and distortional loads

FIGURE 4: Finite element model of composite girder in [21].

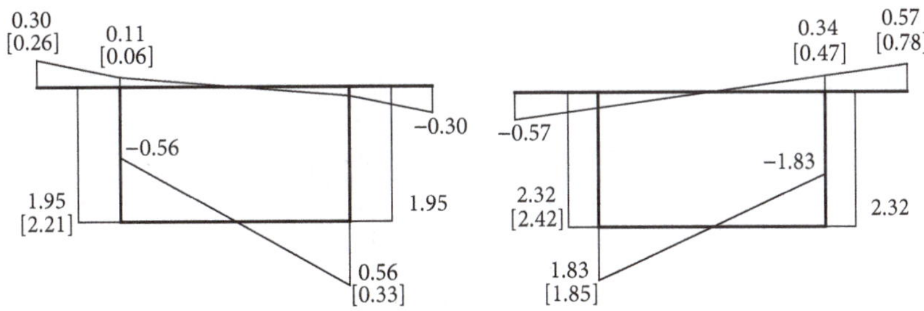

(a) Warping torsional stress

(b) Distortional stress

FIGURE 5: Warping normal stress and shear stress of composite girder in [21] (unit: MPa).

stress results from [6, 20] and stress calculated by FEA are presented separately in brackets () and [].

Figure 8 shows that the above equations of warping normal stress and shear stress are verified by FEA results again. The warping torsional shear stress calculated by equations in this paper is 80% smaller than that in [6, 20], because the accordion effect of corrugated steel web is considered to get the generalized area moment of inertia, $\overline{S}_{\overline{\omega}}(s)$. It should be also noted that the additional shear stress due to restrained torsion has the opposite direction to the free torsional stress, which may reduce the total shear stress. The shear stress due to warping torsion is smaller than that by distortion. Additionally, the warping normal stress in bottom slab is much bigger than stress in top slab for trapezoidal sections. Besides, the normal stress in bottom slab caused by warping torsion has the opposite direction to stress due to distortion.

On the corresponding symmetric load condition, the flexural normal stress in bottom slab is 1.91 MPa and the flexural shear stress in webs is 4.52 MPa. The above results show that the total warping stress in this trapezoidal composite box girder is small, but the total shear stress due to torsion and distortion is 30 percent of flexural shear stress and must be calculated properly.

6. Parametric Analysis

According to the above two examples, the influence of torsion and distortion on stress is different if dimensional parameters of composite box girder change. When calculating the warping torsional and distortional stress, the warping rigidity and transverse form rigidity are not easy to be derived. Besides the fact that the equations for normal and shear stress are complex, it is difficult to find the explicit relationship between girder dimension parameters and warping torsional and distortional rigidity or stress. So, the parametric analysis needs to be done to study how these parameters influence the warping torsion and distortion behaviors for composite girder with corrugated steel webs.

The composite girder dimensions in [21] were used to analyze the torsional and distortional stress under eccentric load at mid-span section under the simple supported boundary condition. The following analysis covers parameters including the span length and section height of girder, the width and thickness of top and bottom flange, and the dimensions of corrugated steel web. Only one parameter is changed in each analysis. The normal stress on the bottom slab at the conjunction of web and web shear stress at mid-span section were studied to compare the effect of warping torsion and distortion.

Six coefficients were defined to analyze the influence of warping torsion and distortion on the normal stress and shear stress, which are

$$\eta_1 = \left|\frac{\sigma_\omega}{\sigma_M}\right|,$$

$$\eta_2 = \left|\frac{\sigma_{D\omega}}{\sigma_M}\right|,$$

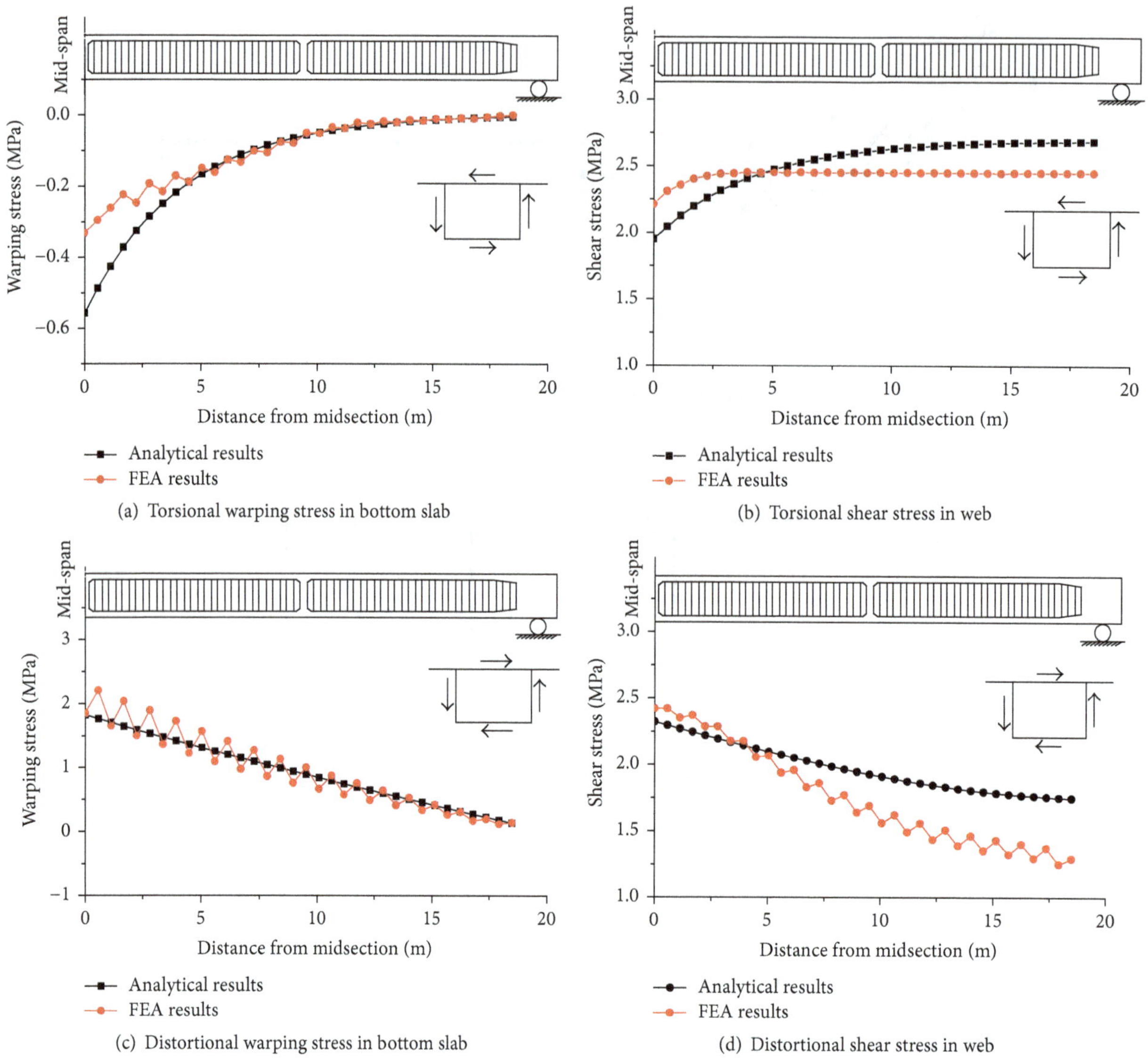

(a) Torsional warping stress in bottom slab

(b) Torsional shear stress in web

(c) Distortional warping stress in bottom slab

(d) Distortional shear stress in web

FIGURE 6: Stress by FEA and analytical equations.

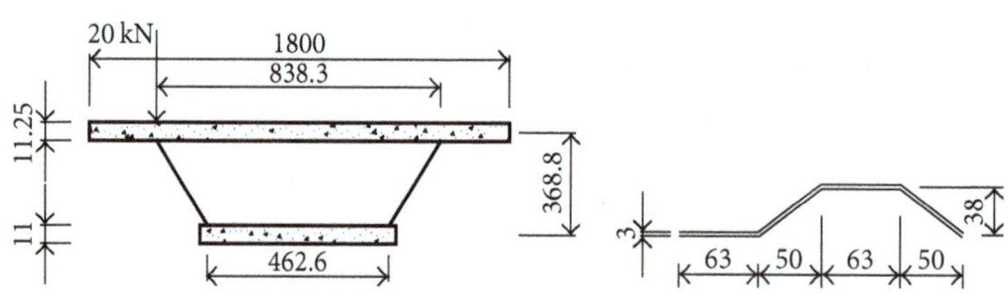

(a) Cross section

(b) Corrugated steel web

FIGURE 7: Configurations and dimension of specimens in case 2 (unit: mm).

$$\eta_3 = \frac{(\tau_k + \tau_\omega)}{\tau_M},$$

$$\eta_4 = \frac{(\tau_D + \tau_{D\omega})}{\tau_M},$$

$$\eta_5 = \frac{\tau_\omega}{\tau_k},$$

$$\eta_6 = \frac{\tau_{D\omega}}{\tau_D}, \tag{39}$$

where σ_M, σ_ω, σ_D are normal stress separately caused by bending, torsional warping, and distortional warping; τ_M, τ_k, τ_ω, τ_D, $\tau_{D\omega}$ are shear stress separately caused by bending, free

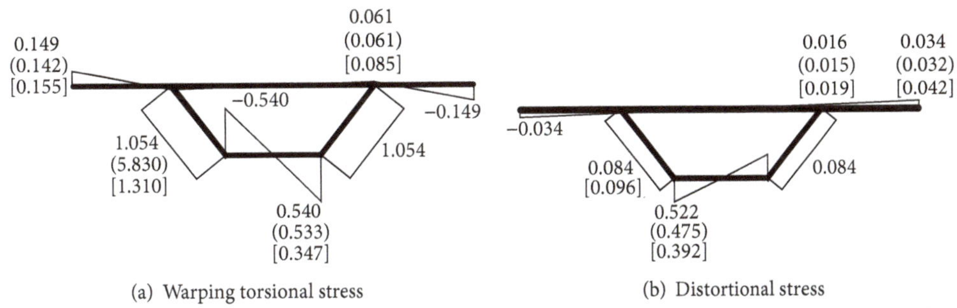

FIGURE 8: Warping normal stress and shear stress of composite girder in [6] (unit: MPa).

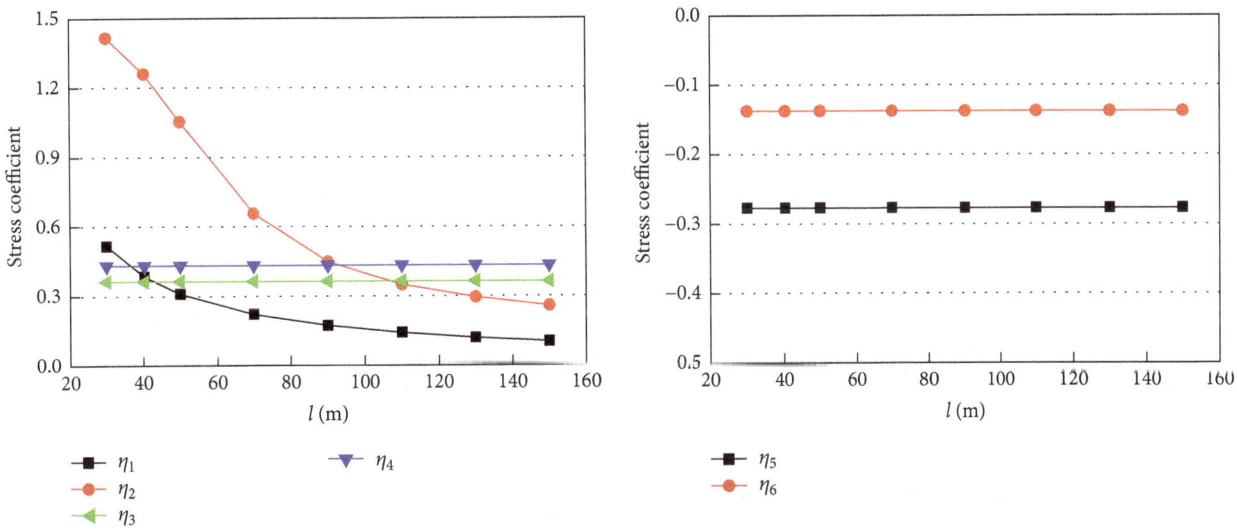

FIGURE 9: Stress coefficients versus different span length.

torsion, warping torsion, pure distortion, and distortional addition.

The coefficients η_1, η_2 imply the influence of warping torsion and distortion on normal stress, and η_3, η_4 imply the influence on shear stress. The coefficients η_5, η_6 imply the additional shear stress due to changing of normal stress caused by warping torsion and distortion.

6.1. Effect of Span Length. When the span length is separately 30, 40, 50, 70, 90, 110, 130, and 150 m, the variation of stress coefficients is given in Figure 9. The span length had obvious influence only on both warping normal stress but not shear stress. The effects of torsion and distortion were reduced if the girder span got longer; the coefficient η_1 changed from 0.52 to 0.10, and η_2 changed from 1.42 to 0.26. On most conditions of girder span within 80 m, the torsional and distortional warping stress are more than 30% of flexural normal stress. So the warping normal stress needs to be considered in normal stress check in structural design.

6.2. Effect of Section Height. The section height is generally 1/40 to 1/25 of the span length. Here, the height is set to be 2.5, 3.8, 5.1, 6.4, 7.7, 9.0, 10.3, 11.6, and 12.9 m. The variation of stress coefficients for different height is given in Figure 10. As the girder height got bigger, the coefficients of normal stress got bigger obviously. The coefficient η_1 changed from 0.19 to 0.66 and η_2 changed from 0.72 to 1.48. The shear stress coefficients η_3 and η_4 had small change by less than 0.05. It shows that the influence of section height is much more obvious to warping normal stress than shear stress.

6.3. Effect of Top Flange Width. In following analysis, the girder section is rectangular whose corrugated steel webs are vertical. When the top flange width, measured width between webs, is separately 4, 5, 6, 7, and 8 m, the variation of stress coefficients is given in Figure 11. Only the coefficient η_1 nearly remained constant as the chamber width grows, but other coefficients changed. η_2 grew from 0.95 to 1.39. The coefficients $\eta_3 \sim \eta_6$ changed separately from 0.28, 0.39, −0.44, −0.22 to 0.41, 0.46, −0.17, and −0.09. Thus, the influence of top flange width is not ignorable.

6.4. Effect of Cantilever Flange Width. The cantilever flange width (flange width between a web and a free edge) is usually designed within one-half of the top flange width between webs. As the top flange width is 6.0 m, the cantilever flange width is accordingly proposed to be 0, 0.5, 1.0, 1.5, 2.0, 2.5, and 3.0 m, respectively. The variation of stress coefficients is given in Figure 12. As the cantilever flange width got bigger, the normal stress coefficients η_1 and η_2 changed in a limit

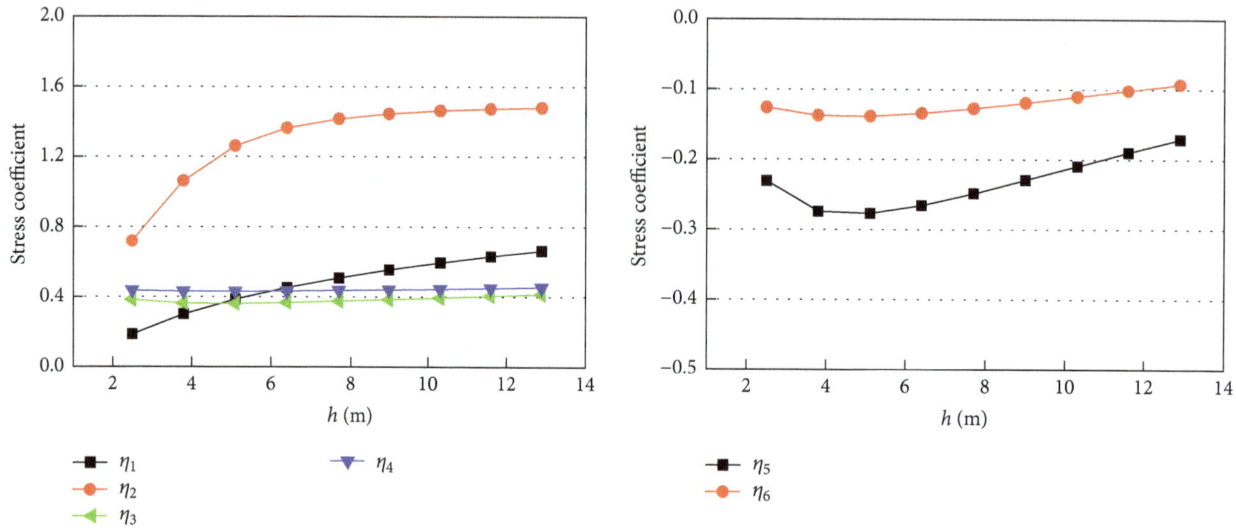

FIGURE 10: Stress coefficients versus different section height.

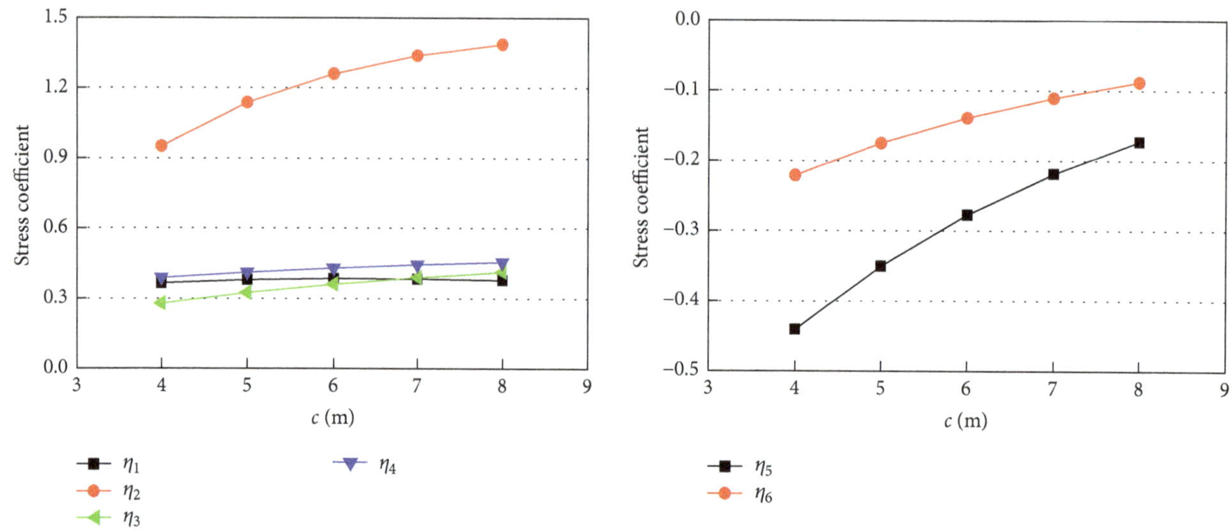

FIGURE 11: Stress coefficients versus different top flange width.

degree, while the shear stress coefficients changed obviously. The coefficients η_3 and η_4 decreased separately from 0.67, 0.59 to 0.27, and 0.39, while η_5 and η_6 changed separately from 0.34, 0.17 to −0.45, −0.22. It should be noted that the direction of shear stress due to warping torsion and distortion may change as cantilever flange width varied.

6.5. Effect of Bottom Flange Width. When the width of top flange remain unchanged, only the bottom flange width is set to be 3, 3.5, 4, 4.5, 5, 5.5, and 6 m. The variation of stress coefficients is given in Figure 13. The coefficients η_1 and η_3 nearly remained constant as the bottom flange width grew, but other coefficients changed obviously. The normal stress coefficients η_2 and η_4 grew separately from 0.69, 0.10 to 1.26, 0.43. The coefficients η_5 and η_6 decreased separately from −0.46, −0.70 to −0.28, −0.13. So, the influence of bottom flange width is not ignorable.

6.6. Effect of Flange Thickness. When the top and bottom flange thickness is separately 0.20, 0.25, 0.30, 0.35, 0.40, 0.45, 0.50 m, the variation of stress coefficients is given in Figures 14 and 15. As shown below, the coefficients η_1 and η_2 in Figure 15 and η_5 and η_6 in Figure 14 changed less than 15%. Other coefficients nearly remained constant as flange thickness varied. It is meant that the stress is almost independent of flange thickness, except that the stress of the slab whose thickness is changed.

6.7. Effect of Web Thickness. Assuming that the general construction of corrugated steel web is unchanged, the web thickness is set to be 0.006, 0.010, 0.014, 0.018, 0.022, 0.026, 0.030, 0.034, and 0.038 m. The variation of stress coefficients is given in Figure 16. The coefficients η_1 and η_2 decreased separately from 0.59, 1.34 to 0.02, 1.01, while η_3 and η_4 increase separately from 0.32, 0.41 to 0.49, 0.47. η_5 and η_6 changed

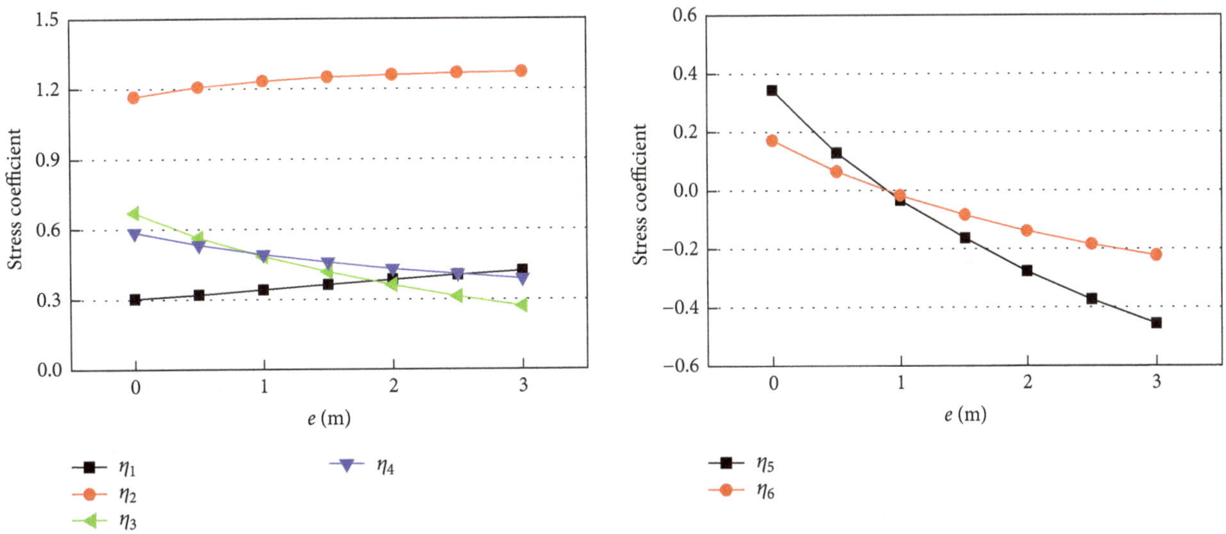

FIGURE 12: Stress coefficients versus different cantilever flange width.

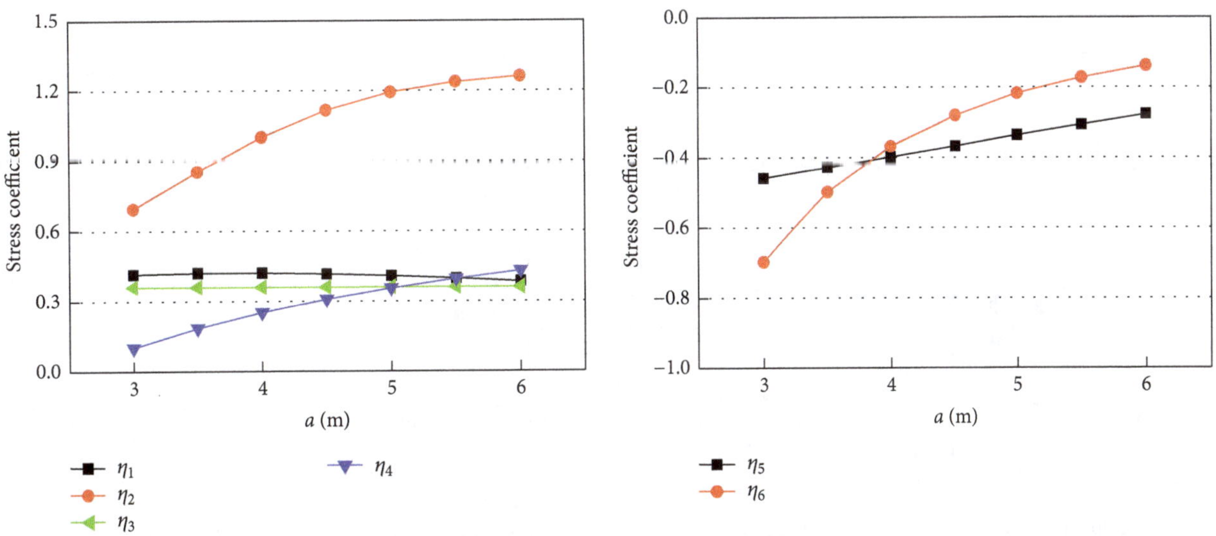

FIGURE 13: Stress coefficients versus different bottom flange width.

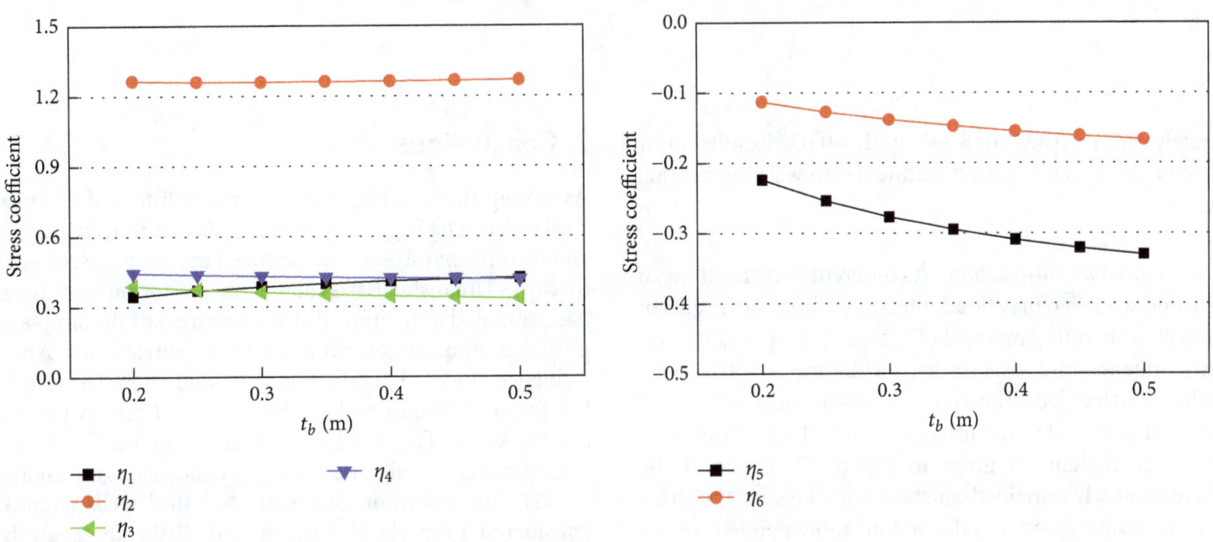

FIGURE 14: Stress coefficients versus different top flange thickness.

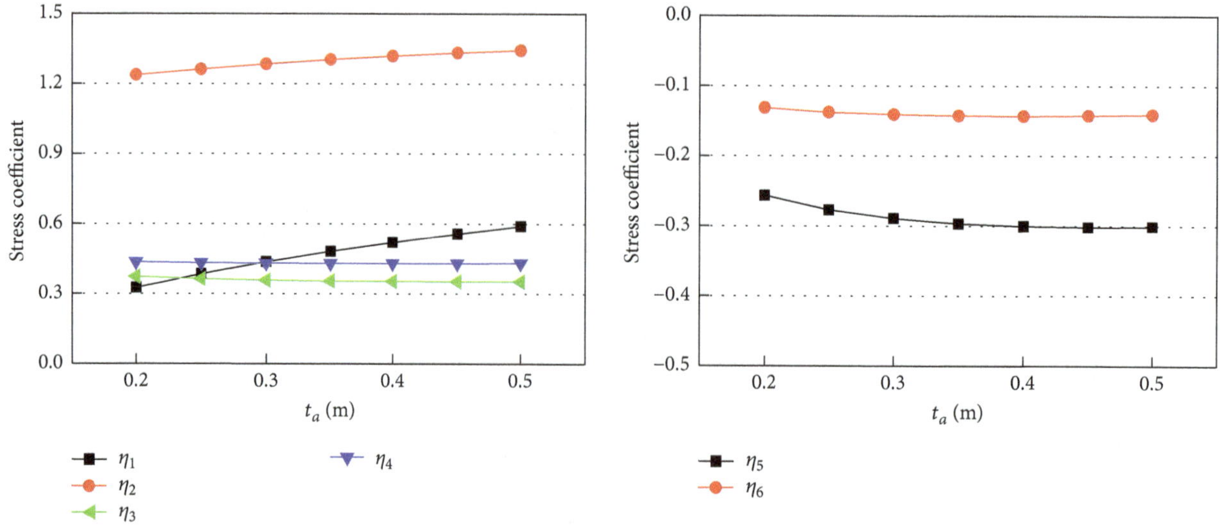

FIGURE 15: Stress coefficients versus different bottom flange thickness.

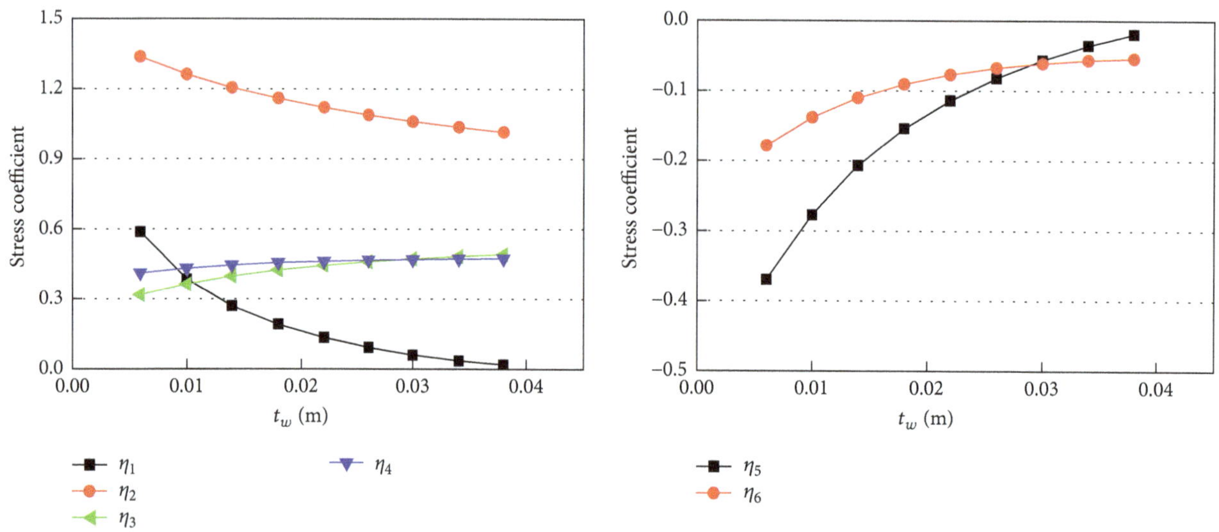

FIGURE 16: Stress coefficients versus different web thickness.

separately from −0.37, −0.18 to −0.02, −0.05. Results show that web thickness has obvious influence on warping normal stress.

6.8. Effect of Web Construction. As different constructions of corrugated steel web may be adopted in the composite girder, the out-plane bending moment of inertia, I_c, may be different. So the influence of I_c was studied, while the web thickness remained unchanged. The common used constructions of corrugated steel web are listed in Table 1. The variation of stress coefficients is given in Figure 17. It shows the coefficients nearly remained constant, which means the stress due to warping torsion and distortion is independent of the web construction.

7. Conclusions

Assuming that the longitudinal axial stiffness of corrugated steel web can be neglected, the formulas for warping torsional and distortional stress were deduced according to the second theory of Umanskii. Based on the revised formulas, this paper has presented a theoretical study on stress of the simple supported composite box girder under eccentric loads. And the influence of varied parameters were studied in detail, including the span length and section height of girder, the width and thickness of top and bottom flange, and the dimensions of corrugated steel web. The following conclusions are obtained.

(1) The accordion of corrugated steel webs should be considered properly in torsion and distortion analysis of composite box girder. The formulas of warping normal stress

TABLE 1: Common construction dimensions of corrugated steel web.

Type number	1	2	3	4	5
a_w/m	0.250	0.300	0.330	0.400	0.430
b_w/m	0.200	0.260	0.270	0.340	0.370
d_w/m	0.150	0.150	0.200	0.210	0.220
I_c/m^4	4.018×10^{-5}	4.167×10^{-5}	7.333×10^{-5}	7.946×10^{-5}	8.672×10^{-5}

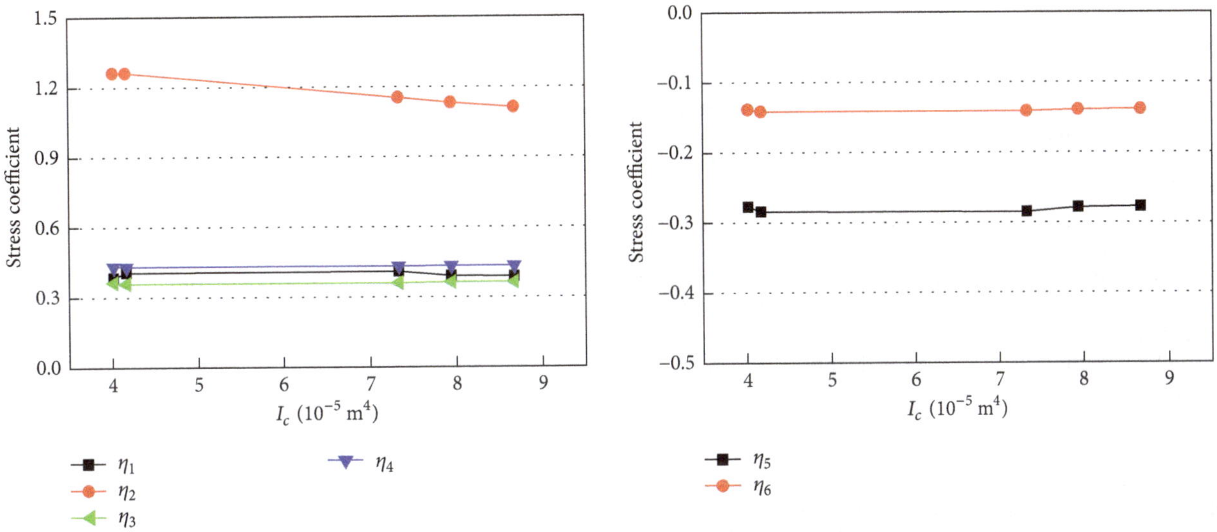

FIGURE 17: Stress coefficients versus different web construction.

and shear stress were revised for analyzing the warping torsion and distortion of the composite box girder of corrugated steel webs. Comparing with the FEA results, the revised formulas are proved to be more effective.

(2) The results show that the influences of warping torsion and distortion are more significant if the girder is shorter, which is easily acceptable. However, even though the whole girder rigidity may increase by enlarging the section dimensions, the warping normal stress and shear stress may not decrease at all. According to studies in this paper, in order to reduce the warping stress ratio, it is suggested to increase the girder span length and the thickness of corrugated steel web height or reduce the girder height and the top, bottom, and cantilever slab width. To reduce the web shear stress ratio, it is suggested to increase the cantilever flange width flange width or reduce the bottom slab width. The thickness of concrete flange and the web construction have a limited impact on both kinds of stress.

(3) As for common composite box girders with corrugated steel webs under eccentric loads, the warping torsion and distortion may cause considerable warping normal stress of shear stress. The total warping normal stress may be as big as flexural normal stress, and the total shear stress is usually 30 to 50 percent of flexural shear stress. Accordingly, it is recommended that both the warping normal stress of concrete flanges and shear stress of corrugated steel webs should be included in stress calculating, when eccentric loads are imposed on composited girders.

(4) The FEA results showed a special feature that the shear stress were different in flat panel and inclined panel of corrugated steel webs, as illustrated in Figure 6. So the detailed mechanical mechanism of corrugated steel webs under different shearing conditions should be studied in future.

Conflicts of Interest

The authors declare that there are no conflicts of interest regarding the publication of this paper.

Acknowledgments

The work described herein is part of a research project funded by the National Natural Science Foundation of China (no. 51308053 and no. 51608069) and Chinese Universities Scientific Fund of Chang'an University (no. 300102288110).

References

[1] M. Cheyrezy and J. Combault, "Composite bridges with corrugated steel webs-achievements and prospects," in *BSE Symp.: Structures including New Materials*, pp. 479–484, IABSE, Zürich, Switzerland, 1990.

[2] Y. L. Mo, C.-H. Jeng, and Y. S. Chang, "Torsional behavior of prestressed concrete box-girder bridges with corrugated steel webs," *ACI Structural Journal*, vol. 97, no. 6, pp. 849–859, 2000 (English).

[3] Y. L. Mo and Y.-L. Fan, "Torsional design of hybrid concrete box girders," *Journal of Bridge Engineering*, vol. 11, no. 3, pp. 329–339, 2006 (English).

[4] H. J. Li, J. S. Ye, S. Wan, and Y. Wang, "Mechanical performances of box girder with corrugated steel webs under eccentric loading," *Journal of Traffic and Transportation Engineering*, vol. 4, no. 2, pp. 23–26, 2004 (Chinese).

[5] B. D. Liu, W. Z. Feng, H. W. Ren, P. F. Li, and K. Mou, "Experimental study on torsion and distortion of continuous rigid frame bridge with corrugated steel webs," *China Railway Science*, vol. 36, no. 4, pp. 40–46, 2015 (Chinese).

[6] B. W. Yang, Y. L. Li, S. Wan, and J. D. Zhang, "Stress analysis of box girder with corrugate steel webs under torsion," *Journal of South China University of Technology (Natural Science Edition)*, vol. 40, no. 2, pp. 19–22, 2012 (Chinese).

[7] J. Di, X. H. Zhou, J. L. You, and X. F. Kong, "Torsional property of prestressed concrete composite beam with corrugated steel webs," *Journal of Chang'an University (Natural Science Edition)*, vol. 29, no. 3, pp. 58–63, 2009 (Chinese).

[8] S. B. Wang, "Mechanical properties of pre-stressed concrete composite box girders with corrugated steel webs under eccentric loading," *Zhongguo Gonglu Xuebao/China Journal of Highway and Transport*, vol. 25, no. 6, pp. 68–73, 2012 (Chinese).

[9] X. F. Nie and Q. H. Zhang, "Study on the torsion and distortion effect of concrete box-girder with corrugated steel webs influenced by the critical factors," *Building Science Research of Sichuan*, vol. 40, no. 5, pp. 46–49, 2014 (Chinese).

[10] Y. Y. Chen, B. C. Chen, and S. Lin, "Experiment of torsional performance and finite element analysis of PC composite box girder with corrugated steel webs," *Journal of Architecture and Civil Engineering*, vol. 28, no. 4, pp. 106–115, 2011 (Chinese).

[11] Y. Ding, K. Jiang, and Y. Liu, "Nonlinear analysis for PC box-girder with corrugated steel webs under pure torsion," *Thin-Walled Structures*, vol. 51, pp. 167–173, 2012 (English).

[12] Y. Ding, K. Jiang, F. Shao, and A. Deng, "Experimental study on ultimate torsional strength of PC composite box-girder with corrugated steel webs under pure torsion," *Structural Engineering and Mechanics*, vol. 46, no. 4, pp. 519–531, 2013 (English).

[13] C. Kondo, T. Kobayashi, Y. Shimizu, and M. Hattori, "Design and construction of a composite box girder bridge with corrugated steel webs: shinkai bridge," *Bridge and Foundation Engineering*, vol. 28, no. 9, pp. 13–20, 1994 (Japanese).

[14] L. Liu and D. S. Qian, "Behavior of corrugated steel webs under loading," *Journal of the China Railway Society*, vol. 22, supplement 1, pp. 53–56, 2000 (Chinese).

[15] R. P. Johnson and J. Cafolla, "Corrugated webs in plate girders for bridges," *Proceedings of Institute of Civil Engineering Structures and Bridge*, vol. 123, no. 2, pp. 157–164, 1997 (English).

[16] Y. H. Zhang and L. X. Lin, "Research on warping stress of restrained torsion for thin-walled box girder under eccentric axial load," *Gongcheng Lixue/Engineering Mechanics*, vol. 30, no. 3, pp. 32–36, 2013 (Chinese).

[17] Y. R. Hu, L. Liu, and Y. H. Zhang, "Torsional geometrical properties of box section with inclined webs," *Journal of Central South University (Science and Technology)*, vol. 46, no. 7, pp. 2558–2563, 2015 (Chinese).

[18] L. Li, X. Bai, and Y. H. Zhang, "Research on twist center and principal sectorial coordinate of thin-walled box section," *Journal of Lanzhou Jiaotong University*, vol. 32, no. 4, pp. 96–100, 2013 (Chinese).

[19] J. Q. Guo, Z. Z. Fang, and Z. Zheng, *Design Theory of Box Girder*, China Communication Press, Beijing, China, 2008.

[20] H. J. Li, *Experimental study and analysis on torsion and distortion of box-girder with corrugated steel webs [Ph.D. thesis]*, Southeast University, Nanjing, China, 2003 (Chinese).

[21] Q. Xu and S. Wan, *Design and Application for Prestressed Concrete Box-Girder Bridges with Corrugated Steel Webs*, China Communication Press, Beijing, China, 2009.

Permissions

All chapters in this book were first published in MPE, by Hindawi Publishing Corporation; hereby published with permission under the Creative Commons Attribution License or equivalent. Every chapter published in this book has been scrutinized by our experts. Their significance has been extensively debated. The topics covered herein carry significant findings which will fuel the growth of the discipline. They may even be implemented as practical applications or may be referred to as a beginning point for another development.

The contributors of this book come from diverse backgrounds, making this book a truly international effort. This book will bring forth new frontiers with its revolutionizing research information and detailed analysis of the nascent developments around the world.

We would like to thank all the contributing authors for lending their expertise to make the book truly unique. They have played a crucial role in the development of this book. Without their invaluable contributions this book wouldn't have been possible. They have made vital efforts to compile up to date information on the varied aspects of this subject to make this book a valuable addition to the collection of many professionals and students.

This book was conceptualized with the vision of imparting up-to-date information and advanced data in this field. To ensure the same, a matchless editorial board was set up. Every individual on the board went through rigorous rounds of assessment to prove their worth. After which they invested a large part of their time researching and compiling the most relevant data for our readers.

The editorial board has been involved in producing this book since its inception. They have spent rigorous hours researching and exploring the diverse topics which have resulted in the successful publishing of this book. They have passed on their knowledge of decades through this book. To expedite this challenging task, the publisher supported the team at every step. A small team of assistant editors was also appointed to further simplify the editing procedure and attain best results for the readers.

Apart from the editorial board, the designing team has also invested a significant amount of their time in understanding the subject and creating the most relevant covers. They scrutinized every image to scout for the most suitable representation of the subject and create an appropriate cover for the book.

The publishing team has been an ardent support to the editorial, designing and production team. Their endless efforts to recruit the best for this project, has resulted in the accomplishment of this book. They are a veteran in the field of academics and their pool of knowledge is as vast as their experience in printing. Their expertise and guidance has proved useful at every step. Their uncompromising quality standards have made this book an exceptional effort. Their encouragement from time to time has been an inspiration for everyone.

The publisher and the editorial board hope that this book will prove to be a valuable piece of knowledge for researchers, students, practitioners and scholars across the globe.

List of Contributors

Weiping Liu and Lina Hu
School of Civil Engineering and Architecture, Nanchang University, Nanchang 330031, China

Mingfu Fu
School of Civil Engineering and Architecture, Nanchang University, Nanchang 330031, China
School of Civil Engineering and Architecture, Nanchang Institute of Technology, Nanchang 330099, China

Xiaoyan Luo
Jiangxi Science and Technology Normal University, Nanchang 330013, China

Jinsong Huang
ARC Centre of Excellence for Geotechnical Science and Engineering, University of Newcastle, Callaghan, NSW 2308, Australia

Ma Xingen, Zhang Yong, Zhang Jiabin and Liu Yuxing
State Key Laboratory for Geomechanics and Deep Underground Engineering, China University of Mining and Technology (Beijing), Beijing 100083, China
School of Mechanics and Civil Engineering, China University of Mining and Technology (Beijing), Beijing 100083, China

He Manchao
State Key Laboratory for Geomechanics and Deep Underground Engineering, China University of Mining and Technology (Beijing), Beijing 100083, China
Department of Geotechnical Engineering, College of Civil Engineering, Tongji University, Shanghai 200092, China

Wang Yajun
Department of Geotechnical Engineering, College of Civil Engineering, Tongji University, Shanghai 200092, China

João P. R. Cortina, Fernando J. M. de Sousa and Luis V. S. Sagrilo
Laboratory of Analysis and Reliability of Offshore Structures, Civil Engineering Program/COPPE, Federal University of Rio de Janeiro, Rio de Janeiro, Brazil

Yu Han and Liu-Chao Qiu
College of Water Resources & Civil Engineering, China Agricultural University, Beijing 100083, China

Shu-Qing Yang and Muttucumaru Sivakumar
School of Civil, Mining & Environmental Engineering, University of Wollongong, Wollongong, NSW2522, Australia

Jian Chen
College of Engineering, China Agricultural University, Beijing 100083, China

Gan Nan and Wei Guo
Faculty of Civil Engineering and Mechanics, Kunming University of Science and Technology, Kunming, Yunnan 650224, China

Hai-ming Liu
Faculty of Civil Engineering and Mechanics, Kunming University of Science and Technology, Kunming, Yunnan 650224, China
State Key Laboratory of Geomechanics and Geotechnical Engineering, Institute of Rock and Soil Mechanics, Chinese Academy of Sciences, Wuhan 430071, China

Chun-he Yang and Chao Zhang
State Key Laboratory of Geomechanics and Geotechnical Engineering, Institute of Rock and Soil Mechanics, Chinese Academy of Sciences, Wuhan 430071, China

Liang Jia, Li Zhang, Jian Guo and Shikai He
College of Civil Engineering, Lanzhou University of Technology, Lanzhou 730050, China

Kai Yao
Department of Civil & Environmental Engineering, National University of Singapore, Singapore 117576

Ali Hosseinkhani and Davood Younesian
Center of Excellence in Railway Transportation, School of Railway Engineering, Iran University of Science and Technology, Tehran 16846-13114, Iran

Saman Farhangdoust
Department of Civil and Environmental Engineering, Florida International University, Miami, FL 33174, USA

Goo Sam Jee, Huan Ha Nguyen and Wonseok Chung
Department of Civil Engineering, Kyung Hee University, Yongin, Gyeonggi-do 17104, Republic of Korea

Heeyoung Lee
Department of Civil Engineering, Kyung Hee University, Yongin, Gyeonggi-do 17104, Republic of Korea
Department of Civil Engineering, University of Colorado Denver, 1200 Larimer Street, Denver, CO 80217, USA

Seung Yup Jang
Department of Transportation System Engineering, Graduate School of Transportation, Korea National University of Transportation, Gyeonggi-do 16106, Republic of Korea

Shunqing Liu, Xianwen Huang and Aizhao Zhou
Department of Civil and Architecture Engineering, Jiangsu University of Science and Technology, Zhenjiang, Jiangsu 212003, China

Jun Hu
College of Civil Engineering and Architecture, Hainan University, Haikou, Hainan 570228, China

Wei Wang
School of Civil Engineering, Shaoxing University, Shaoxing, Zhejiang 312000, China

Ming-yi Zhang, Zheng Kuang and Xiao-yu Bai
College of Civil Engineering, Qingdao University of Technology, Qingdao, Shandong 266033, China

Xiao-yu Chen
Faculty of Construction and Environment, The Hong Kong Polytechnic University, Hung Hom, Hong Kong

Gonzalo García-Ros and Iván Alhama
Technical University of Cartagena, Civil Engineering Department, Spain

Manuel Cánovas
Universidad Católica del Norte, Metallurgical and Mining Engineering Department, Chile

Francisco Alhama
Technical University of Cartagena, Applied Physics Department, Spain

Xu Huang, Huaifeng Peng and Yaohu Hao
Key Laboratory of Ministry of Education for Geomechanics and Embankment Engineering, Hohai University, Nanjing 210098, China

Yajun Wu
Key Laboratory of Ministry of Education for Geomechanics and Embankment Engineering, Hohai University, Nanjing 210098, China
Department of Civil Engineering, Shanghai University, Shanghai 200072, China

Chenyang Lu
Department of Civil Engineering, Shanghai University, Shanghai 200072, China

Xudong Chen
School of Civil Engineering, Suzhou University of Science and Technology, Suzhou 215011, China

Kangsheng Ye
Department of Civil Engineering, Tsinghua University, Beijing 100084, China

Lingzhi Sun, Yunyue Xie and Hongtian Xiao
College of Civil Engineering and Architecture, Shandong University of Science and Technology, China
Shandong Key Lab of Civil Engineering Disaster Prevention & Mitigation, Qingdao 266510, China

Aizhao Zhou, Chengchao Li and Pengming Jiang
Department of Civil and Architecture Engineering, Jiangsu University of Science and Technology, Zhenjiang, Jiangsu 212003, China

Kai Yao
Department of Civil & Environmental Engineering, National University of Singapore, Singapore 117576

Na Li and Wei Wang
School of Civil Engineering, Shaoxing University, Shaoxing, Zhejiang 312000, China

Yanan Tian, Yongjiang Shi and Jiayu Rong
College of Energy and Environmental Engineering, Hebei Institute of Architecture and Civil Engineering, Zhangjiakou, Hebei 075000, China

Haosen Wang, Zhenhua Ma and Cong Xie
College of Mathematics and Physics, Hebei Institute of Architecture and Civil Engineering, Zhangjiakou, Hebei 075000, China

Shuguang Tian and Jun Huang
China Railway 16th Bureau Group Co., Ltd., Beijing 100018, China

Changfu Huang
China Railway 16th Bureau Group Co., Ltd., Beijing 100018, China
School of Civil and Resources Engineering, University of Science and Technology Beijing, Beijing 100083, China

Qun Li
School of Civil and Resources Engineering, University of Science and Technology Beijing, Beijing 100083, China

Tianzhong Ma, Yanpeng Zhu, Xiaohui Yang and Yongqiang Ling
Key Laboratory of Disaster Prevention and Mitigation in Civil Engineering of Gansu Province, Lanzhou University of Technology, Lanzhou 730050, China
Western Engineering Research Center of Disaster Mitigation in Civil Engineering of Ministry of Education, Lanzhou University of Technology, Lanzhou 730050, China

Peng Qiao
School of Civil Engineering, Chang'an University, Xi'an 710061, China

Jin Di and Feng-jiang Qin
Key Laboratory of New Technology for Construction of Cities in Mountain Area (Chongqing University), Ministry of Education, Chongqing 400045, China
School of Civil Engineering, Chongqing University, Chongqing 400045, China

Index

T

U

V

www.ingramcontent.com/pod-product-compliance
Lightning Source LLC
LaVergne TN
LVHW080718210926
848485LV00003B/486